Heating, Ventilating, and Air Conditioning

ANALYSIS AND DESIGN

Heating, Ventilating, and Air Conditioning

ANALYSIS and DESIGN

SECOND EDITION

FAYE C. McQUISTON
JERALD D. PARKER
Oklahoma State University

1807 1982

JOHN WILEY & SONS
New York • Chichester • Brisbane • Toronto • Singapore

Library of Congress Cataloging in Publication Data:

McQuiston, Faye C.
  Heating, ventilating, and air conditioning.

  Includes index.
  1. Heating.   2. Ventilation.   3. Air conditioning.
I. Parker, Jerald D.   II. Title.
TH7222.M26   1982   697   81-3004
ISBN 0-471-08259-7   AACR2

Printed in the United States of America

10 9 8 7 6 5 4 3 2

Preface

The field of heating and air conditioning advances so rapidly that textbooks quickly become inadequate for classical design courses. The *ASHRAE Handbook and Product Directory* series, published by the American Society of Heating, Refrigerating and Air-Conditioning Engineers, contains the most current data needed for design purposes, and each volume is revised every four years. The normal evolution of new analysis and design procedures has been accelerated by the pressures of the current, critical energy situations and inflation. All of these reasons have made it desirable to revise the first edition. The revision has also given us an opportunity to clarify certain areas and to correct a few minor errors. The original objective in the first edition was to prepare an up-to-date text, based on ASHRAE methods, that would be a convenient classroom teaching aid. Our objective has not changed. The text is complete in itself, and the handbooks are not needed for classroom work. Even though the book is intended primarily as a teaching device, it should also be useful to practicing engineers as a ready source of background material and as a valuable aid in studying new design procedures.

Although the book is intended for use in two regular semester courses, there can be a great deal of latitude in the makeup of these courses. Following a study of the complete text, a student should be capable of participating in the design of all types of HVAC systems. The book is intended for use at the undergraduate and beginning graduate levels by students who have taken the basic engineering science courses including thermodynamics, heat transfer, and fluid mechanics.

Emphasis has been placed on the development of design procedures based on sound concepts. The chapters related to calculation of heating and cooling loads have been extensively revised to reflect procedures developed during the past four years. Chapter Two has been expanded to cover psychrometric space conditioning processes for partial load conditions. Information and data pertaining to variable-air-volume systems have been added at appropriate places in the text. New design data and procedures pertaining to infiltration have been included in Chapter Four. Chapter Five, which now pertains to solar radiation, has been revised to reflect current calculation methods and procedures for use of measured solar data. The material related to comfort, indoor air quality, and ventilation has been extensively revised to agree with the new standards in these areas. Energy conservation procedures are emphasized throughout, and the material on solar systems in Chapter Sixteen has been revised to reflect current thinking.

Additional problems have been added throughout; however, the instructor is encouraged to provide examples and problems that emphasize the requirements of the local geographical region and the instructor's design philosophies.

The computer plays an increasingly important role in HVAC calculations, especially since small personalized computers have become readily available. Most of the design methods can be adapted to a computer, and the instructor is encouraged to assign such exercises. Several problems of this type have been added.

The conversion from English to the international system of units (SI) in the United States seems to be a certainty; however, the process is proceeding slower than was anticipated five years ago. The academic community should be a leader in this process, but engineers must be able to use both systems of units when they enter practice. Therefore, this book continutes to use a dual system of units. Although emphasis is placed on the English system, all of the equations are written to be dimensionally correct in either system of units. Extensive data, examples, and problems are provided in SI units with proper data provided to solve the problem. Instructors may blend the two systems of units as they see fit.

Many years of experience in teaching, research, and applications have prepared us for this work. It is our hope that this book will contribute to the continued growth of the HVAC industry through education of both beginning and practicing engineers.

We are deeply indebted to ASHRAE for providing a great deal of material. Many other companies and individuals contributed suggestions, material, and hard work. Our thanks to everyone.

Faye C. McQuiston
Jerald D. Parker

Contents

Symbols

English Letter Symbols

A	area, ft^2 or m^2
A	constant, Eq. (2-54), Table 2-2, ft^3/lbma or m^3/kga
A	apparent solar irradiation for zero air mass, Btu/(hr-ft^2) or W/m^2
a	light fixture classification, Table 7-19
ADPI	air distribution performance index, dimensionless
B	constant, Eq. (2-55), Table 2-2, Btu/lbma or kJ/kga
B	wet bulb coefficient, Eq. (12-27), lbmv/(lbma-F) or kgv/(kga-C)
B	atmospheric extinction coefficient
b	transfer function coefficient, Btu/(hr-ft^2-F) or W/(m^2-C)
b	light fixture classification, Table 7-20
b	bypass factor, dimensionless
C	constant, Eq (2-56), Table 2-2, Btu/(lbma-F) or kJ/(kga-C)
C	concentration, lbm/ft^3 or kg/m^3
C	unit thermal conductance, Btu/(hr-ft^2-F) or W/(m^2-C)
C	discharge coefficient, dimensionless
C	constant in Eq. (10-3), dimensionless
C	loss coefficient, dimensionless
C	fluid capacity rate, Btu/(hr-F) or W/C
C	clearance factor, dimensionless
C_d	overall flow coefficient, dimensionless
C_d	draft coefficient, dimensionless
C_p	pressure coefficient, dimensionless
C_v	flow coefficient, dimensionless
CLF	cooling load factor, dimensionless
CLTD	cooling load temperature difference, F or C
COP	coefficient of performance, dimensionless
c	specific heat, Btu/(lbm-F) or J/(kg-C)
c	transfer function coefficient, Btu/(hr-ft^2-F) or W/(m^2-C)
cfm	volume flow rate, ft^3/min
clo	clothing thermal resistance, (ft^2-hr-F)/Btu or (m^2-C)/W

D	diameter, ft or m
D	diffusion coefficient, ft²/sec or m²/s
DD	degree days, F-day or C-day
DBT	dry bulb temperature, F or C
DR	daily range of temperature, F or C
d	bulb diameter, ft or m
d	sun's declination, degrees
d	transfer function coefficient, dimensionless
E	effective emittance, Eq. (4-11), dimensionless
EDT	effective draft temperature, or C
ET	effective temperature, F or C
F	configuration factor, dimensionless
F	quantity of fuel, ft³ or m³
F_c	fraction of heat gain, dimensionless
F(s)	wet surface function, Eq. (14-60), dimensionless
f	friction factor, dimensionless
f_t	Darcy friction factor with fully turbulent flow, dimensionless
FP	correlating parameter, Eq. (14-51), dimensionless
G	irradiation, Btu/(hr-ft²) or W/m²
g	local acceleration due to gravity, ft/sec² or m/s²
g	transfer function coefficient, Btu/(hr-ft) or W/C
g_c	dimensional constant, 32.17 (lbm-ft)/(lbf/sec²) or 1.0 (kg-m)/(N-s²)
H	heating value of fuel, Btu or J per unit of volume
H	head, ft or m
h	height or length, ft or m
h	heat transfer coefficient, Btu/(hr-ft²-F) or W/(m²-C) (also used for mass transfer coefficient with subscripts m, d, and i)
h	hour angle, degrees
h'	hour angle for a tilted surface, degrees
i	enthalpy, Btu/lbm or J/kg
$\bar{i}$	enthalpy correction, Eq. (2-55), Btu/lbm or J/kg
J	Joule's equivalent, 778.28 (ft-lbf)/Btu
JP	correlating parameter, Eq. (14-46), dimensionless
$J(s)$	wet surface function, Eq. (14-58), dimensionless
$J_i(s)$	wet surface function, Eq. (14-59), dimensionless
j	Colburn j factor, Eq. (12-18), dimensionless
K	color correction factor, dimensionless
K	resistance coefficient, dimensionless
K	constant in Eq. (10-2), dimensionless
K_t	unit length conductance, Btu/(ft-hr-F) or W/(m-C)
k	thermal conductivity, (Btu-ft)/(ft²-hr-F), (Btu-in)/(ft²-hr-F) or (W-m)/(m²-C)
k	isentropic exponent, c_p/c_v, dimensionless
L	fin dimension, Fig. 14-13 and 14-14, ft or m
L	total length, ft or m
Le	Lewis number, Sc/Pr, dimensionless
LM	latitude-month correction, F or C
$LMTD$	log mean temperature difference, F or C
l	latitude, degrees
l	lost head, ft or m

M	molecular weight, lbm/(lb mole) or kg/(kg mole)
M	constant in Eq. (10-1), F-min/ft or (C-s)/m
M	fin dimension, Fig. 14-13 and 14-14, ft or m
MRT	mean radiant temperature, F or C
m	mass, lbm or kg
$\dot{m}$	mass flow rate or mass transfer rate, lbm/sec or kg/s
N	number of hours or other integer
Nu	Nusselt number, hx/k, dimensionless
NC	noise criteria, dimensionless
NTU	number of transfer units, dimensionless
P	pressure, lb/ft^2 or psia or N/m^2 or Pa
P	heat exchanger parameter, Eq. (14-4), dimensionless
P	circumference, ft or m
Pr	Prandtl number, $\mu c_p/k$, dimensionless
PD	piston displacement, ft^3/min or m^3/s
p	partial pressure, lbf/ft^2 or psia or Pa
p	transfer function coefficient, dimensionless
$\dot{Q}$	volume flow rate, ft^3/sec or m^3/s
q	heat transfer, Btu/lbm or J/kg
$\dot{q}$	heat transfer rate, Btu/hr or W
R	gas constant, (ft-lbf)/(lbm-R) or J/(kg-K)
R	unit thermal resistance, (ft^2-hr-F)/Btu or (m^2-C)/W
R	heat exchanger parameter, Eq. (14-4), dimensionless
R	fin radius, ft or m
R'	thermal resistance, (hr-F)/Btu or C/W
$\overline{R}$	gas constant, (ft-lbf)/(lb mole-R) or J/(kg mole-K)
Re	Reynolds number $\rho \bar{V} D/\mu$, dimensionless
R_f	unit fouling resistance, (hr-ft^2-F)/Btu, or (m^2-C)/W
r	radius, ft or m
rpm	revolutions per minute
S	fin spacing, ft or m
S	Equipment characteristic, Btu/(hr-F) or W/C
Sc	Schmidt number, ν/D, dimensionless
Sh	Sherwood number, $h_m x/D$, dimensionless
SC	shading coefficient, dimensionless
SHF	sensible heat factor, dimensionless
$SHGF$	solar heat gain factor, Btu/(hr-ft^2) or W/m^2
s	entropy, Btu/(lbm-R) or J/(kg-K)
$\bar{s}$	entropy correction, Eq. (2-56), Btu/(lbm-R) or J/(kg-K)
T	absolute temperature, R or K
t	temperature, F or C
t^*	thermodynamic wet bulb temperature, F or C
U	overall heat transfer coefficient, Btu/(hr-ft^2-F) or W/(m^2-C)
u	velocity in x direction, ft/sec or m/s
V	volume, ft^3 or m^3
$\overline{V}$	velocity, ft/sec or m/s
v	specific volume, ft^3/lbm or m^3/kg
v	transfer function coefficient, dimensionless
v	velocity in y direction, ft/sec or m/s

$\bar{v}$	volume correction, Eq. (2-54), ft³/lbm or m³/kg
W	humidity ratio, lbmv/lbma or kgv/kga
W	equipment characteristic, Btu/hr or W
$\dot{W}$	power, Btu/hr or W
$WBGT$	wet bulb globe temperature, F or C
w	skin wettedness, dimensionless
w	work, Btu, or ft-lbf, or J
w	transfer function coefficient, dimensionless
X	normalized input, dimensionless
X	fraction of daily range
x	mole fraction
x	quality, lbmv/lbm or kgv/kg
x	throw, ft or m
x, y, z	length, ft or m
Y	normalized capacity, dimensionless

Subscripts

a	transverse dimension
a	air
a	average
a	attic
as	adiabatic saturation
as	denotes change from dry air to saturated air
avg	average
B	barometric
b	branch
b	longitudinal dimension
b	base
CD	on collector, daily
CH	on collector, hourly
c	cool or coil
c	convection
c	ceiling
c	cross section or minimum free area
c	cold
c	condenser
c	Carnot
c	collector
cl	center line
D	direct
D	diameter
d	dew point
d	total heat
d	diffuse
d	design
d	downstream
db	dry bulb
dry	dry surface

e	equivalent
e	sol-air
e	equipment
e	evaporator
f	film
f	friction
f	fin
fg	refers to change from saturated liquid to saturated vapor
fl	fluorescent light
fl	floor
fr	frontal
g	refers to saturated vapor
g	globe
H	horizontal
HD	on horizontal surface, daily
h	heat
h	hydraulic
h	head
h	heat transfer
h	hot
i	*j* factor for total heat transfer
i	inside
i	instantaneous
l	latent
l	liquid
m	mean
m	mass transfer
m	mechanical
ND	direct normal
n	integer
o	outside
o	total or stagnation
o	initial condition
oh	humid operative
P	pressure
p	constant pressure
p	pump
R	reflected
R	refrigerating
r	radiation
r	room air
SR	sunrise
SS	sunset
s	stack effect
s	sensible
s	saturated vapor or saturated air
s	supply air
s	shaft
s	static

s	surface
sc	solar constant
sh	shade
sl	sunlit
t	temperature
t	total
t	contact
t	tube
u	unheated
u	upstream
V	vertical
v	vapor
v	ventilation
v	velocity
w	wind
w	wall
w	liquid water
wb	wet bulb
wet	wet surface
x	length
x	extraction
1, 2, 3	state of substance at boundary of a control volume
1, 2, 3	a constituent in a mixture
∞	free stream condition

Greek Letter Symbols

α	angle of tilt from vertical, degrees
α	absorptivity or absorptance, dimensionless
α	total heat transfer area over total volume, ft^{-1} or m^{-1}
α	thermal diffusivity, ft^2/sec or m^2/s
β	parameter in Eq. (14-23), dimensionless
β	altitude angle, degrees
γ	wall solar azimuth angle, degrees
Δ	change in a quantity or property
δ	boundary layer thickness, ft or m
ε	heat exchanger effectiveness, dimensionless
ε	emittance or emissivity, dimensionless
η	efficiency, dimensionless
θ	angle of incidence, degrees
θ	time, sec
μ	degree of saturation, percent or fraction
μ	dynamic viscosity, lbm/(ft-sec) or (N-s)/m^2
ν	kinematic viscosity, ft^2/sec or m^2/s
ρ	mass density, lbm/ft^3 or kg/m^3
ρ	reflectivity or reflectance, dimensionless
Σ	angle of tilt from horizontal, degrees
σ	Stefan-Boltzmann constant, Btu/(hr-ft^2-R^4) or J/(s-m^2-K^4)
σ	free flow area over frontal area, dimensionless

τ	transmissivity or transmittance, dimensionless
ϕ	fin parameter, Eq. (14-22), dimensionless
ϕ	solar azimuth angle, degrees
ϕ	relative humidity, percent or fraction
ψ	wall azimuth angle, degrees
Ψ	parameter in Eq. (14-23), dimensionless
Ψ	zenith angle, degrees

List of Charts*

Chart 1   ASHRAE Psychrometric Chart No. 1 (reprinted by permission of ASHRAE).

Chart 1a   Psychrometric chart for normal temperature in SI units (reproduced by permission of Carrier Corporation).

Chart 2   Enthalpy-concentration diagram for ammonia-water solutions (from *Unit Operations* by G. G. Brown, Copyright 1951 by John Wiley and Sons, Inc. Reprinted by permission of John Wiley and Sons, Inc., New York).

Chart 3   Enthalpy-concentration diagram for lithium bromide–water solutions (courtesy of Institute of Gas Technology, Chicago, Illinois).

Chart 4   Pressure-enthalpy diagram for refrigerant 12 (reprinted by permission).

Chart 5   Pressure enthalpy diagram for refrigerant 22 (reprinted by permission).

Chart 6   Duct design worksheet (reprinted with permission of Air Research Associates).

*For the convenience of the reader all charts have been folded into a pocket located on the inside back cover.

Heating, Ventilating, and Air Conditioning

ANALYSIS AND DESIGN

CHAPTER ONE

Introduction

A comfortable and healthful environment is now considered a necessity, and many modern processes and products would not exist without precise control of environmental conditions. Therefore almost all homes, offices, and industrial structures are now designed with a means of controlling the indoor environment throughout the year. Maintenance of warm surroundings has always been considered necessary during the cold months. In our modern society, however, the maintenance of a cool environment during the warm months proves to be just as important for the utilization of human resources and productivity. The shortage of energy and escalating energy costs, however, are causing reexamination of comfort conditions once thought essential. All these considerations have placed more emphasis on the design and simulation of thermal environmental systems.

1-1 HISTORICAL NOTES

There is evidence that human beings have always struggled to make their lives more comfortable through control of the immediate environment. Because of both necessity and natural laws the first human beings were most successful in situations requiring heat. Although there is evidence of the use of evaporative effects and ice for cooling in very early times, it was not until the middle of the nineteenth century that a practical refrigerating machine was built.

By the end of the nineteenth century the concept of central heating was fairly well developed, and early in the twentieth century cooling for comfort got its start. Developments since that time have been rapid. The greatest advance has been in the change from rather artful design practices to the detailed analytical methods necessitated by complex building structures, high construction costs, and energy shortages. More detailed discussions of the history of heating, refrigeration, and air conditioning are given in articles by Willis R. Woolrich, John W. James, Walter A. Grant, and William L. McGrath (1). In these articles only the names of outstanding contributors appear. Because of the wide scope and diverse nature of the Heating, Ventilating and Air Conditioning (HVAC) field, literally thousands of engineers, their names too numerous to list, have developed the industry. The accomplishments of all these unnamed persons are summarized in the *ASHRAE* Handbook and Product Directory* consisting of four volumes entitled: *Fundamentals, Applications, Equipment,* and *Systems.* These volumes are constantly being revised by ASHRAE members. Considerable research designed to improve the handbooks is also sponsored by ASHRAE and monitored by ASHRAE members. Unquestionably the ASHRAE Handbooks are the encyclopedias of the industry throughout the world. The principles presented in this textbook follow the handbooks closely.

1-2 AIR CONDITIONING AND ENVIRONMENTAL CONTROL

Historically *air conditioning* has implied cooling or otherwise improving the indoor environment during the warm months of the year. In modern times the term has taken on a more literal meaning that can be applied to year-round environmental situations. That is, air conditioning refers to the control of temperature, moisture content, cleanliness, air quality, and air circulation, as required by occupants, a process, or product in the space. This general definition has lead to the alternate term *environmental control.* The functions required to air condition a space completely are briefly described below.

Heating

Heating is the transfer of energy to a space or to the air in a space by virtue of a difference in temperature between the source and the space or air. This process may take different forms such as direct radiation to the space, direct heating of the circulated air, or through heating of water that is circulated to the vicinity of the space and used to heat the circulated air. This type of heat transfer, which is manifested in a rise in temperature of the air, is called *sensible heat transfer.*

Humidifying

The transfer of water to atmospheric air is referred to as humidification. Heat transfer is associated with this mass transfer process; however the transfer of mass and energy are manifested in an increase in the concentration of water in the air–water vapor mixture. Here the term *latent heat transfer* is used. This process is usually accomplished by introducing water vapor or by spraying fine droplets of water into the circulating air stream. Wetted mats or plates may also be used.

Cooling

Cooling is the transfer of energy from the space or air supplied to the space by virtue of a difference in temperature between the source and the space or air. In the

---

**ASHRAE is an abbreviation for the American Society of Heating, Refrigerating and Air-Conditioning Engineers, Incorporated.

usual cooling process air is circulated over a surface maintained at a low temperature. The surface may be in the space to be cooled or at some remote location from it, the air being ducted to and from the space. Usually water or a volatile refrigerant is the cooling medium. Cooling usually denotes sensible heat transfer.

Dehumidifying

The transfer of water vapor from atmospheric air is called dehumidification. Latent heat transfer is also associated with this process. The transfer of energy is from the air; as a consequence, the concentration of water in the air–water vapor mixture is lowered. This process is most often accomplished by circulating the air over a surface maintained at a sufficiently low temperature to cause the condensation of water vapor from the mixture. It is also possible to dehumidify by spraying cold water into the air stream.

Cleaning

The cleaning of air usually implies filtering; however it is sometimes necessary to deodorize the air. Filtering is most often done by a process in which dirt particles are captured in a porous medium. Electrostatic cleaners are also used especially to remove very small particles; in some cases water sprays may be used. Odors are removed by activated charcoal or by exhaust and makeup air, for example.

Air Motion

The motion of air in the vicinity of the occupant should be sufficiently strong to remove energy generated by the body but gentle enough to be unnoticed. The desired air motion is achieved by the proper placement of air inlets to the space and by the use of various air-distributing devices. The importance of air motion especially where occupant comfort is required cannot be underestimated.

All of the functions of the air-conditioning system discussed above may not be active all of the time. Residences and many commercial establishments have inactive cooling and dehumidifying sections during the winter months and keep the heating and humidifying sections inactive during the summer. In large commercial installations, however, it is not uncommon to have all of the functions under simultaneous control for the entire year. Obviously this requires elaborate controls and sensing devices. Precise control of the moisture content of the air is difficult, and humidifying and dehumidifying are usually not done during both winter and summer unless they are absolutely necessary for process control or product preservation, even though the heating and cooling functions may be active all of the time. Because cleaning of the air and good air circulation is always necessary, these functions are used continuously.

1-3 UNITS AND DIMENSIONS

In HVAC computations as in all engineering work consistent units must be employed. A *unit* is a specific quantitative measure of a physical characteristic in reference to a standard. An example of a unit is the foot, which is used to measure the physical characteristic length. A physical characteristic, such as length, is called a *dimension*. Other dimensions of interest in HVAC computations are force, time, temperature, and mass.

In this text two systems of units will be employed. The first is called the English Engineering System and is most commonly used in HVAC work in the United States with some modification such as use of inches instead of feet. The second is the International System (or SI for Système International d'Unites) and is the system in use in engineering practice throughout most of the world and is being widely adopted in the United States.

ASHRAE has adopted the following policy and schedule (7):

1. All ASHRAE documents published after January 1, 1976 shall be prepared using only SI units, or shall be prepared using dual units, that is, SI and conventional units.
2. Except where difficulties would be encountered in achieving the stated purpose, the Handbook Series, starting with the 1981 Fundamentals Volume, shall be published using dual units.
3. Exclusive use of SI units shall be required in ASHRAE publications when it is determined to be in the best interest of the membership.

Equipment designed using U.S. conventional units will be operational for years and even decades. For the foreseeable future, then, it will be necessary for the engineer to work in either system of units and to be able to make conversions from one system to another. The base SI units important in HVAC work are given in Table 1-1 together with the comparable English Engineering unit.

The SI system is described in several documents: (2), (3), and (4). The system consists of seven base units (the four base units given in Table 1-1 plus units of electric current, amount of substance, and luminous intensity). In addition there are two supplementary units for the plane angle (rad), the solid angle (r), and a list of

Table 1-1

	Units	
Dimensions	**English Engineering System**	**International System**
Mass	pound mass (lbm)	kilogram (kg)
Length	foot (ft)	meter (m)
Time	second (sec)	second (s)
Temperature	degree Fahrenheit (F)	degree Kelvin (K)

Table 1-2 DERIVED SI UNITS FREQUENTLY USED IN HVAC

Dimension	Unit	Special Name and Symbol
Acceleration, angular	rad/s²	
Acceleration, linear	m/s²	
Area	m²	
Density	kg/m³	
Energy	N-m	joule (J)
Force	(kg-m)/s²	newton (N)
Frequency	l/s	hertz (Hz)
Power	J/s	watt (W)
Pressure	N/m²	pascal (Pa)
Specific heat capacity	J/(kg-C)	
Stress	N/m²	pascal (Pa)
Thermal conductivity	W/(m-C)	
Thermal flux density	W/m²	
Velocity, angular	rad/s	
Velocity, linear	m/s	
Viscosity, dynamic	(N-s)/m²	
Viscosity, kinematic	m/s²	
Volume	m³	

derived units. The derived units in the SI system, which are frequently used in HVAC, are given in Table 1-2.

Because there is only one unit for each dimension in the SI system, there is a carefully defined set of multiple and submultiple prefixes to avoid the use of a large number of zeroes for very large or very small quantities. These prefix names and symbols are given in Table 1-3 (4).

Conversion factors for changing to and from SI units are given in Appendix E.

Example 1-1. Describe the quantity of pressure equal to 1000 N/m² in suggested SI terminology.

Solution.

1000 can be described by the prefix kilo (k).

The N/m² has the special name of pascal.

Therefore,

$$1000 \text{ N/m}^2 = 1 \text{ kilo pascal} = 1 \text{ kPa}$$

Example 1-2. A certain corkboard has a thermal conductivity of 0.025 Btu/(hr-ft-F). Convert this quantity to the equivalent SI value.

Table 1-3   MULTIPLE AND SUBMULTIPLE SI
PREFIXES

Multiplying Factor	Prefix	Symbol
$1\ 000\ 000\ 000\ 000 = 10^{12}$	tera	T
$1\ 000\ 000\ 000 = 10^{9}$	giga	G
$1\ 000\ 000 = 10^{6}$	mega	M
$1\ 000 = 10^{3}$	kilo	k
$100 = 10^{2}$	hecto[a]	h
$10 = 10^{1}$	deka[a]	da
$0.1 = 10^{-1}$	deci	d
$0.01 = 10^{-2}$	centi[a]	c
$0.001 = 10^{-3}$	milli	m
$0.000\ 001 = 10^{-6}$	micro	μ
$0.000\ 000\ 001 = 10^{-9}$	nano	n
$0.000\ 000\ 000\ 001 = 10^{-12}$	pico	p

[a]These prefixes are to be avoided if possible.

Solution.   According to the conversion factor given in Appendix E, to convert from Btu/(hr-ft-F) to W/(m-C) you must multiply the given value by 1.7307.

$$(0.025)\ \frac{\text{Btu}}{\text{hr-ft-F}}\ (1.7307)\ \frac{\text{W}/(\text{m-C})}{\text{Btu}/(\text{hr-ft-F})} = 0.043\ \text{W}/(\text{m-C})$$

Because the prefix centi is to be avoided if possible (Table 1-3), the quantity is expressed as shown. Notice that the final quantity is not expressed to any more significant figures than the original quantity, in this case two.

In the SI system the temperature is given in degrees Kelvin (K), which is a thermodynamic absolute temperature. It is quite common to specify temperature in degrees Celsius or Centigrade even where all other units might be SI units. One degree change on the Celsius scale is identical to one degree change on the Kelvin scale. The relationship between the Kelvin scale and the Celsius scale is

$$K = C + 273.15 \tag{1-1}$$

For example, 100 C is identical to 373.15 K.

In the English Engineering system the unit of temperature is the degree Fahrenheit. When the thermodynamic absolute temperature is needed, the temperature is specified in degrees Rankine (R). The relationship between the Fahrenheit scale and the Rankine scale is

$$R = F + 459.67 \tag{1-2}$$

Since both the Rankine and Kelvin scales are absolute scales, absolute zero is identical on each scale. The relationship between the Rankine scale and the Kelvin scale is

$$K = \tfrac{5}{9}R \qquad\qquad (1\text{-}3)$$

For example 900 R is identical in temperature to 500 K. The relationship between the four temperature scales is given in Fig. 1-1.

Consistent units must always be employed in physical computations. For example, thermal and mechanical energy may be interchangeable in a given situation. Thermal energy traditionally has been specified in terms of British Thermal Units (Btu) in engineering work in English-speaking countries. Mechanical energy is frequently specified in the units of foot pounds force (ft-lbf). If the net energy (the sum of the thermal and mechanical energy) is to be computed, it is necessary to specify both types of energy in the same unit. The relationship between the Btu and the ft-lbf is

$$1 \text{ Btu} = 778.28 \text{ ft-lbf} \qquad\qquad (1\text{-}4)$$

Example 1-3.   A system is known to contain 100 Btu of thermal energy and 30,000 ft-lbf of mechanical energy. What is the total energy (mechanical plus thermal) contained by the system in Btu?

Solution.   Use Eq. (1-4) as a conversion factor.

$$100 \text{ Btu} + \frac{(30,000) \text{ ft-lbf}}{(778.28) \text{ ft-lbf/Btu}} = (100 + 38.5) \text{ Btu} = 138.5 \text{ Btu}$$

In this example notice that an equation (Eq. 1-4) was changed to a conversion factor by simple algebra.

$$1 \text{ Btu} = 778.28 \text{ ft-lbf}$$
$$1 = 778.28 \text{ ft-lbf/Btu}$$

Since 778.28 (ft-lbf)/Btu is equivalent to unity, it can be placed in the appropriate term in such a way as to cancel the undesired units, and yet not change the true value of the physical quantity represented by the term. In Example 1-3 above, 30,000 ft-lbf of energy is identical to 38.5 Btu of energy.

From Table 1-2 we see that the derived unit of energy in the SI system is the joule, which is equivalent to one newton-meter. The derived unit of power in the SI system is the watt, which is equivalent to one joule per second.

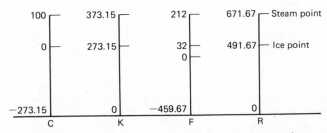

Figure 1-1.   Relationship between temperature scales.

Example 1-4. The specific heat of air at normal conditions is approximately equal to 0.241 Btu/(lbm-F). Express this value of the specific heat of air in SI units, joule per kilogram, and degree C.

Solution. From the conversion factor given in Appendix E the following relationship is true

$$(0.241) \frac{\text{Btu}}{\text{lbm-F}} (4.1868 \times 10^3) \frac{\text{J/(kg-C)}}{\text{Btu/(lbm-F)}} = 1010 \text{ J/(kg-C)}$$
$$= 1.01 \text{ kJ/(kg-C)}$$

REFERENCES

1. "Environmental Control Principles," American Society of Heating, Refrigerating, and Air-Conditioning Engineers, New York, 1978.

2. "ASHRAE SI Metric Guide for Heating, Refrigerating, Ventilating and Air-Conditioning," American Society of Heating, Refrigerating and Air-Conditioning Engineers, New York, 1976.

3. W. F. Stoecker, "Using SI Units in Heating, Air-Conditioning, and Refrigeration," Business News Publishing Co., Birmingham, Michigan, 1975.

4. "Metric Practice Guide E380-72," American Society for Testing and Materials, Philadelphia, 1972, (also ANSI Standard Z210.1).

5. "ASME Orientation and Guide for Use of Metric Units," Third Edition, The American Society of Mechanical Engineers, New York, 1973.

6. "SI Units and Recommendations for the Use of Their Multiples and of Certain Other Units," International Standard ISO 1000, International Organization for Standardization, New York, 1973.

7. ASHRAE Handbook and Product Directory-Systems, American Society of Heating, Ventilating and Air-Conditioning Engineers, New York, 1980.

PROBLEMS

1-1 Write the following quantities in the suggested SI terminology.
    (a) 11,000 newton per square meter
    (b) 12,000 watts
    (c) 1600 joule per kilogram
    (d) 101,101 pascal
    (e) 12,000 kilowatt
    (f) 0.012 meter
    (g) 0.0000012 second

1-2 Convert the following quantities from English to SI units.
    (a) 98 Btu/(hr-ft-F)
    (b) 0.24 Btu/(lbm-F)

(c) 0.04 lbm/(ft-hr)
(d) 1050 Btu/lbm
(e) 12000 Btu/hr
(f) 14.7 lbf/in.^{2}

1-3 Convert the following quantities from SI to English units.
(a) 120 kPa
(b) 100 W/(m-C)
(c) 0.8 W/(m^2-C)
(d) 10^{-6} (N-s)/m^2
(e) 1200 kW
(f) 1000 kJ/kg

1-4 The kinetic energy of a flowing fluid is proportional to the velocity squared divided by two. Compute the kinetic energy per unit mass for the following velocities in the units indicated.
(a) Velocity of 100 ft/sec; English units
(b) Velocity of 600 m/s; SI units
(c) Velocity of 400 ft/sec; SI units
(d) Velocity of 500 m/s; English units

1-5 The potential energy of a fluid is proportional to the elevation of the fluid. Compute the potential energy for the elevations given below per unit mass.
(a) Elevation of 200 ft; English units
(b) Elevation of 70 m; SI units
(c) Elevation of 120 ft; SI units
(d) Elevation of 38 m; English units

1-6 A gas is contained in a vertical cylinder with a frictionless piston. Compute the pressure in the cylinder for the following cases.
(a) Piston mass of 20 lbm and area of 7 in.^{2}
(b) Piston mass of 10 kg and diameter of 100 mm

1-7 A pump develops a total head of 50 ft of water under a given operating condition. What pressure is the pump developing in SI units and terminology?

1-8 A fan is observed to operate with a pressure difference of 4 in. of water. What is the pressure difference in SI units and terminology?

1-9 Compute the Reynolds number (Re $= \rho \overline{V} D/\mu$) for 21 C water flowing in a standard 2 in. pipe at a velocity of one m/s using SI units. Tables B-3a and D-1 will be useful. What is the Reynolds number in English units?

1-10 Compute the thermal diffusivity ($\alpha = k/\rho c_p$) of the following substances in SI units.
(a) saturated liquid water at 38 C
(b) air at 47 C and 101.3 kPa
(c) saturated liquid refrigerant 22 at 38 C
(d) saturated liquid refrigerant 12 at 38 C

1-11 Compute the Prandtl number (Pr $= \mu c_p/k$) for the following substances in SI units.
(a) saturated liquid water at 10 C
(b) saturated liquid water at 60 C
(c) air at 20 C
(d) saturated liquid refrigerant 22 at 38 C

1-12 Compute the heat transferred from water as it flows through a heat exchanger at a steady rate of one m³/s. The decrease in temperature of the water is 5 C and the mean bulk temperature is 60 C. Use SI units.

1-13 Make the following volume and mass flow rate calculations in SI units. (a) Water flowing at an average velocity of 2 m/s in nominal $2\frac{1}{2}$ in., type L copper tubing. (b) Standard air flowing at an average velocity of 4 m/s in a 30 cm diameter duct.

1-14 A room with dimensions of 3 m × 10 m × 20 m is estimated to have an infiltration rate of $\frac{1}{4}$ volume change per hour. Determine the infiltration rate in m³/s.

CHAPTER TWO

Moist Air Properties and Conditioning Processes

A thorough understanding of the properties of moist air and the ability to analyze the various processes involving air is basic to the HVAC engineer. Atmospheric air makes up the environment in almost every design situation.

In 1911 Willis H. Carrier made a significant contribution to the air conditioning field when he published relations for moist air properties together with a psychrometric chart. These formulas became fundamental to the industry.

In about 1945 Goff and Gratch (1) published thermodynamic properties for moist air that are still generally recognized as the most accurate available. However, Threlkeld (2) has shown that errors in calculation of the major properties will be less than 0.7 percent when perfect gas relations are used. This chapter emphasizes the use of the perfect gas relations; the final section is devoted to the Goff and Gratch formulations.

2-1   MOIST AIR AND THE STANDARD ATMOSPHERE

Atmospheric air is a mixture of many gases plus water vapor and countless pollutants. Aside from the pollutants, which may vary considerably from place to place, the composition of the dry air alone is relatively constant, varying slightly with time, location, and altitude. The *ASHRAE Handbook of Fundamentals* (3) gives the following approximate composition of dry air by volume fraction:

Nitrogen	0.78084
Oxygen	0.20948
Argon	0.00934
Carbon dioxide	0.00031
Neon, helium, methane, sulfur dioxide, hydrogen, and other minor gases	0.00003

11

Table 2-1   COMPOSITION OF DRY AIR

Constituent	Molecular Weight	Volume Fraction
Oxygen	32.000	0.2095
Nitrogen	28.016	0.7809
Argon	39.944	0.0093
Carbon dioxide	44.010	0.0003

In 1949 a standard composition of dry air was fixed by the International Joint Committee on Psychrometric Data (4) as shown in Table 2-1.

Based on the composition of air in Table 2-1, the molecular weight M_a of dry air is 28.965, and the gas constant R_a is

$$R_a = \frac{\overline{R}}{M_a} = \frac{1545.32}{28.965} = 53.352 \text{ (ft-lbf)/(lbm-R)} \quad \text{or} \quad 287 \text{ J/(kg-K)} \quad (2\text{-}1)$$

where $\overline{R}$ is the universal gas constant; $\overline{R} = 1545.32$ (ft-lbf)/(lb mole-R) or 8314 J/(kg mole-K).

The basic medium in air conditioning practice is a mixture of dry air and water vapor. The amount of water vapor may vary from zero to a maximum determined by the temperature and pressure of the mixture. The latter case is called saturated air, a state of neutral equilibrium between the moist air and the liquid or solid phases of water. The molecular weight of water is 18.015 and the gas constant for water vapor is

$$R_v = \frac{1545.32}{18.015} = 85.78 \text{ (ft-lbf)/(lbm-R)} \quad \text{or} \quad 462 \text{ J/(kg-K)} \quad (2\text{-}2)$$

The *ASHRAE Handbook* gives the following definition of the U.S. Standard Atmosphere:

1. Acceleration due to gravity is constant at 32.174 ft/sec² or 9.807 m/s².
2. Temperature at sea level is 59.0 F, 15 C or 288.1 K.
3. Pressure at sea level is 29.921 in. of mercury or 101.039 kPa.*
4. The atmosphere consists of dry air, which behaves as a perfect gas.

$$Pv = \frac{P}{\rho} = R_a T \quad (2\text{-}3)$$

Standard sea level density computed using Eq. (2-3) with the standard temperature and pressure is 0.0765 lbm/ft³ or 1.225 kg/m³. The *ASHRAE Handbook* summarizes standard atmospheric data for altitudes up to 60,000 ft or 18,291 m.

---

*Standard atmospheric pressure is also commonly taken to be 14.696 lbf/in.² or 101.325 kPa, which corresponds to 30.0 in. of mercury.

2-2   FUNDAMENTAL PARAMETERS

Moist air up to about three atmospheres pressure obeys the perfect gas law with sufficient accuracy for engineering calculations. The Gibbs Dalton law for a mixture of perfect gases states that the mixture pressure is equal to the sum of the partial pressures of the constituents.

$$P = p_1 + p_2 + p_3 \tag{2-4}$$

For moist air

$$P = p_{N_2} + p_{O_2} + p_{CO_2} + p_A + p_v \tag{2-5}$$

Because the various constituents of the dry air may be considered to be one gas, it follows that the total pressure of moist air is the sum of the partial pressures of the dry air and the water vapor.

$$P = p_a + p_v \tag{2-6}$$

Each constituent in a mixture of perfect gases behaves as if the others were not present. Consider a saturated mixture of dry air and water vapor at 80 F. Table A-1 shows that the partial pressure of the water vapor is 0.5073 lbf/in.^{2} and the mass density is 1/632.8 or 0.001580 lbm/ft^3. By using Eq. (2-3) we get

$$\frac{1}{v} = \rho = \frac{p_v}{R_v T} = \frac{0.5073(144)}{85.78(459.67 + 80)}$$

or

$$\rho = 0.001578 \frac{\text{lbm}}{\text{ft}^3}$$

This result agrees with the more accurate table values within about 0.1 percent. For superheated conditions the agreement is generally better. In some cases the data of Table A-1 are most convenient to use, whereas in other cases the perfect gas law is expedient.

Humidity ratio W (sometimes called the specific humidity) is the ratio of the mass of the water vapor m_v to the mass of the dry air m_a in the mixture

$$W = \frac{m_v}{m_a} \tag{2-7}$$

Relative humidity ϕ is the ratio of the mole fraction of the water vapor x_v in a mixture to the mole fraction x_s of the water vapor in a saturated mixture at the same temperature and pressure

$$\phi = \left. \left| \frac{x_v}{x_s} \right| \right|_{t,P} \tag{2-8}$$

For a mixture of perfect gases the mole fraction is equal to the partial pressure ratio of each constituent. The mole fraction of the water vapor is

$$x_v = \frac{p_v}{P} \tag{2-9}$$

Using Eq. (2-8) and letting p_s stand for the partial pressure of the water vapor in a saturated mixture, we may express the relative humidity as

$$\phi = \frac{p_v/P}{p_s/P} = \frac{p_v}{p_s} \tag{2-10}$$

Since the temperature of the dry air and the water vapor are assumed to be the same in the mixture,

$$\phi = \frac{p_v/R_v T}{p_s/R_v T} = \left[\frac{\rho_v}{\rho_s} \right]_{t,P} \tag{2-11}$$

where the densities ρ_v and ρ_s are referred to as the absolute humidities of the water vapor (mass of water per unit volume of mixture).

Degree of saturation μ is the ratio of the humidity ratio W to the humidity ratio W_s of a saturated mixture at the same temperature and pressure.

$$\mu = \left[\frac{W}{W_s} \right]_{t,P} \tag{2-12}$$

This parameter is most useful when the data of Goff and Gratch, Table A-2, are used.

Dew point temperature t_d is the temperature of saturated moist air at the same pressure and humidity ratio as the given mixture.

By using the perfect gas law we can derive a relation between the relative humidity ϕ and the humidity ratio W:

$$m_v = \frac{p_v V}{R_v T} = \frac{p_v V M_v}{\overline{R} T} \tag{2-13}$$

and

$$m_a = \frac{p_a V}{R_a T} = \frac{p_a V M_a}{\overline{R} T} \tag{2-13a}$$

then

$$W = \frac{M_v p_v}{M_a p_a} \tag{2-14}$$

For the air-water vapor mixture, Eq. (2-14) reduces to

$$W = 0.6219 \frac{p_v}{p_a} \tag{2-14a}$$

Combining Eqs. (2-10) and (2-14a) gives

$$\phi = \frac{W p_a}{0.6219 p_s} \tag{2-15}$$

The enthalpy of a mixture of perfect gases is equal to the sum of the enthalpies of each constituent and is usually referenced to a unit mass of one constituent. For the air-water vapor mixture dry air is used as the reference because the amount of water vapor may vary during some processes.

$$i = i_a + Wi_v \qquad (2\text{-}16)$$

Each term has the units of energy per unit mass of dry air. With the assumption of perfect gas behavior the enthalpy is a function of temperature only. If zero Fahrenheit or Celsius is selected as the reference state where the enthalpy of dry air is zero, and if the specific heats c_{pa} and c_{pv} are assumed to be constant, simple relations result

$$i_a = c_{pa}t \qquad (2\text{-}17)$$
$$i_v = i_g + c_{pv}t \qquad (2\text{-}18)$$

where the enthalpy of saturated water vapor i_g at 0 F is 1061.2 Btu/lbm and 2501.3 kJ/kg at 0 C.

Using Eqs. (2-16), (2-17), and (2-18) with c_{pa} and c_{pv} taken as 0.240 and 0.444 Btu/(lbm-F), respectively,

$$i = 0.240t + W(1061.2 + 0.444t) \text{ Btu/lbma} \qquad (2\text{-}19)$$

In SI units Eq. (2-19) becomes

$$i = 1.0t + W(2501.3 + 1.86t) \text{ kJ/kga} \qquad (2\text{-}19a)$$

where c_{pa} and c_{pv} are 1.0 and 1.86 kJ/(kg-C), respectively.

Example 2-1. Compute the enthalpy of saturated air at 60 F or 15.56 C and standard atmospheric pressure in both English and SI units.

Solution. Equations (2-19) will be used to compute enthalpy; however, the humidity ratio W_s must first be determined from Eq. (2-14a).

$$W_s = 0.6219 \frac{p_s}{p_a} = 0.6219 \frac{p_s}{p - p_s}$$

From Table A-1, $p_s = 0.2563$ psia or 1767.9 Pa. Then

$$W_s = 0.6219 \left| \frac{0.2563}{14.696 - 0.2563} \right| = 0.01104 \text{ lbmv/lbma}$$

or

$$W_s = 0.6219 \left| \frac{1767.9}{101325 - 1767.9} \right| = 0.01104 \text{ kgv/kga}$$
$$i_s = (0.24)60 + 0.01104[1061.2 + (0.444)60] = 26.41 \text{ Btu/lbma}$$

and Eq. (2-19a) gives

$$i_s = 15.56 + 0.01104[2501.3 + (1.86)15.56] = 43.49 \text{ kJ/kga}$$

The results of Example 2-1 may be compared with the precise data of Table A-2. The enthalpy calculated using ideal gas relations is about 0.25 percent low but quite satisfactory for engineering calculations.

2-3 ADIABATIC SATURATION

Examination of the parameters discussed in the previous section indicates that *at a given pressure and temperature of an air-water vapor mixture one additional property is required to completely specify the state,* except at saturation. Theoretically any of the parameters discussed would be acceptable; however, there is no practical way to determine any one of them in a real situation. The concept of adiabatic saturation provides a convenient solution to this problem.

Consider the device shown in Fig. 2-1. The apparatus is assumed to operate such that the moist air leaving at point 2 is saturated. The temperature, t_2 is the *adiabatic saturation temperature t_{as},* and the relative humidity ϕ_2 is 100 percent. The term *Thermodynamic Wet Bulb Temperature* is also used to describe t_{as}. If we assume that the device operates in a steady flow–steady state manner, an energy balance on the control volume yields

$$i_{a1} + W_1 i_{v1} + (W_2 - W_1)i_w = W_2 i_{v2} + i_{a2} \tag{2-20}$$

or

$$W_1(i_{v1} - i_w) = c_{pa}(t_2 - t_1) + W_2(i_{v2} - i_w) \tag{2-20a}$$

and

$$W_1(i_{v1} - i_w) = c_{pa}(t_2 - t_1) + W_2 i_{fg2} \tag{2-20b}$$

This result is significant because it shows that the adiabatic saturation temperature t_2 is a function of the relative humidity ϕ_1, pressure P_1, temperature t_1, and the pressure P_2.

$$t_2 = t_{as} = f(\phi_1, P_1, t_1, P_2) \tag{2-21}$$

It is then conceivable that the relative humidity ϕ_1 can be expressed as

$$\phi_1 = f'(P_1, t_1, P_2, t_2) \tag{2-22}$$

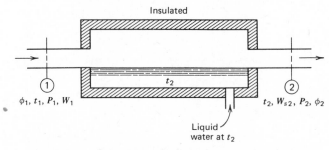

Figure 2-1. Schematic of adiabatic saturation device.

Because the measurement of pressure and temperature is relatively easy, the adiabatic saturation process provides a convenient method of determining the state of moist air. Consider the following example.

Example 2-2. The pressure entering and leaving an adiabatic saturator is 14.696 $lbf/in.^2$, the entering temperature is 80 F, and the leaving temperature is 64 F. Compute the humidity ratio W_1, and the relative humidity ϕ_1.

Solution. Because the mixture leaving the device is saturated, $p_{v2} = p_{s2}$, and W_2 can be calculated using Eq. (2-14a).

$$W_2 = 0.6219 \left| \frac{0.2952}{14.696 - 0.2952} \right| = 0.0127 \text{ lbmv/lbma}$$

Now using Eq. (2-20b) and data from Table A-1, we get

$$W_1 = \frac{c_{pa}(t_2 - t_1) + W_2 i_{fg2}}{(i_{v1} - i_w)}$$

$$W_1 = \frac{0.24(64 - 80) + (0.0127 \times 1057.3)}{(1096.4 - 32.09)} = 0.0090 \text{ lbmv/lbma}$$

Then using Eq. (2-14a)

$$W_1 = 0.6219 \frac{p_{v1}}{14.696 - p_{v1}} = 0.0090 \text{ lbmv/lbma}$$

$$p_{v1} = 0.2098 \text{ psia}$$

Finally, from Eq. (2-10)

$$\phi_1 = \frac{p_{v1}}{p_{s1}} = \frac{0.2098}{0.5073} = 0.414$$

It is therefore evident that the state of moist air can be completely determined from pressure and temperature measurements. However, the adiabatic saturator is not a practical device because it would have to be infinitely long in the flow direction. A practical method of determining the approximate adiabatic saturation temperature is discussed in the next section.

2-4   WET BULB TEMPERATURE AND THE PSYCHROMETRIC CHART

A practical device used in place of the adiabatic saturator is the *psychrometer*. This apparatus consists of two thermometers, or other temperature-sensing elements, one of which has a wetted cotton wick covering the bulb (Fig. 2-2). The temperatures indicated by the psychrometer are called the *wet bulb* and the *dry bulb* temperatures. The dry bulb temperature corresponds to t_1 in Fig. 2-1 and the wet bulb temperature is an approximation to t_2 in Fig. 2-1, whereas P_1 and P_2 are equal to barometric pressure. The combination heat-and-mass-transfer process from the wet bulb thermometer is not the same as the adiabatic saturation process; however the error is relatively small when the wet bulb thermometer is used under suitable conditions. The difference between the wet bulb and thermodynamic wet bulb temperature

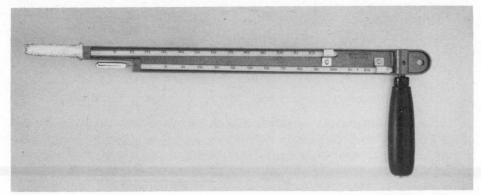

Figure 2-2.   Typical hand-held psychrometer. (Courtesy of Central Scientific Company, Chicago, Illinois.)

should be less than about 0.5 F or 0.25 C for atmospheric pressure and temperatures above 32 F or 0 C, where the difference between the dry bulb and wet bulb temperatures is less than about 20 F or 11 C, and where no unusual radiation conditions exist; an unshielded wet bulb thermometer is assumed to be in air that has a velocity greater than 100 ft/min or 0.5 m/s. If thermocouples are used, the velocity may be somewhat lower with similar accuracy. Figure 2-3 shows a psychrometer properly installed in a duct to meet the above conditions.

To facilitate engineering computations, a graphical representation of the properties of moist air has been developed and is known as a *psychrometric chart*. Richard Mollier was the first to use such a chart with enthalpy as a coordinate. Modern-day charts are somewhat different but still retain the enthalpy coordinate feature. ASHRAE has developed five of the Mollier-type charts to cover the necessary range of variables. Figure 2-4 is an abridgment of ASHRAE Chart 1 that covers the normal range of variables at standard atmospheric pressure. The charts are based on the precise data of Table A-2; within the readability of the charts, however, agreement

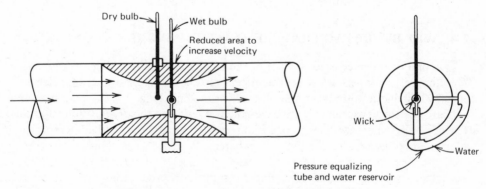

Figure 2-3.   A psychrometer installed in a duct.

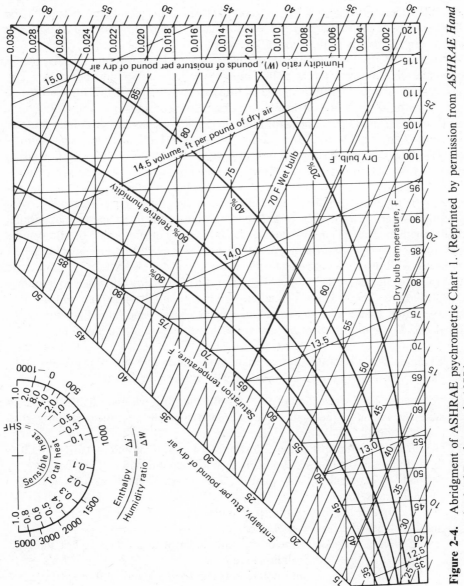

Figure 2-4. Abridgment of ASHRAE psychrometric Chart 1. (Reprinted by permission from *ASHRAE Hand book of Fundamentals*, 1977.)

with the perfect gas relations is very good. Details of the actual construction of the charts may be found in references 2 and 5. Charts for both English and SI units are provided in the packet in the back of the book.

In Fig. 2-4 dry bulb temperature is plotted along the horizontal axis in degrees Fahrenheit or Celsius. The dry bulb temperature lines are straight but not exactly parallel and incline slightly to the left. Humidity ratio is plotted along the vertical axis on the right-hand side of the chart in lbmv/lbma or kgv/kga. The scale is uniform with horizontal lines. The saturation curve with values of the wet bulb temperature curves upward from left to right. Dry bulb, wet bulb, and dew point temperatures all coincide on the saturation curve. Relative humidity lines with a shape similar to the saturation curve appear at regular intervals. The enthalpy scale is drawn obliquely on the left of the chart with parallel enthalpy lines inclined downward to the right. Although the wet bulb temperature lines appear to coincide with the enthalpy lines, they diverge gradually in the body of the chart and are not parallel to one another. The spacing of the wet bulb lines is not uniform. Specific volume lines appear inclined from the upper left to the lower right and are not parallel. A protractor with two scales appears at the upper left of ASHRAE Chart 1. One scale gives the sensible heat ratio and the other the ratio of enthalpy difference to humidity ratio difference. This feature is handled differently on Chart 1a and will be fully discussed later. Notice that the enthalpy, specific volume, and humidity ratio scales are all based on a unit mass of dry air and not a unit mass of the moist air.

Example 2-3.   Read the properties of moist air at 75 F db, 60 F wb, and standard sea level pressure from ASHRAE Psychrometric Chart 1.

Solution.   The intersection of the 75 F db and 60 F wb lines defines the given state. This point on the chart is the reference from which all the other properties are determined.

Humidity Ratio, W. Move horizontally to the right and read $W = 0.0077$ lbmv/lbma on the vertical scale. It is not uncommon for W to be given in "grains." One pound mass is equal to 7000 grains. Then $W = 7000 \times 0.0077 = 53.9$ grains/lbma.

Relative Humidity, ϕ. Interpolate between the 40 and 50 percent relative humidity lines and read $\phi = 41$ percent.

Enthalpy, i. Follow a line of constant enthalpy upward to the left and read $i = 26.4$ Btu/lbma on the oblique scale.

Specific Volume, v. Interpolate between the 13.5 and 14.0 specific volume lines and read $v = 13.65$ ft^3/lbma.

Dew Point Temperature, t_d. Move horizontally to the left from the reference point and read $t_d = 50$ F on the saturation curve.

Enthalpy, i (alternate method): The nomograph in the upper left-hand corner of Chart 1 gives the difference D between the enthalpy of unsaturated moist air and the enthalpy of saturated air at the same wet bulb temperature. Then $i = i_s + D$. For this example $i_s = 26.5$ Btu/lbma, $D = -0.1$ Btu/lbma, and $i = 26.5 - 0.1 = 26.4$ Btu/lbma. The enthalpy deviation lines are plotted directly on the body of Chart 1a.

2-5 CLASSIC MOIST AIR PROCESSES

The most powerful analytical tools of the air conditioning design engineer are the *first law of thermodynamics* or energy balance, and the *conservation of mass* or mass balance. These conservation laws are the basis for the analysis of moist air processes. It is customary to analyze these processes by using the bulk average properties at the inlet and outlet of the device being studied. In actual practice the properties may not be uniform across the flow area especially at the outlet, and a considerable length may be necessary for complete mixing.

Heating or Cooling of Moist Air

When air is heated or cooled without the loss or gain of moisture, the process yields a straight horizontal line on the psychrometric chart because the humidity ratio is constant. Such processes can occur when moist air flows through a heat exchanger. In cooling, if the surface temperature is below the dew point temperature of the moist air, dehumidification will occur. This process will be considered later. Figure 2-5 shows a schematic of a device used to heat or cool air. Under steady flow–

steady state conditions the energy balance becomes

$$\dot{m}_a i_2 + \dot{q} = \dot{m}_a i_1 \qquad (2\text{-}23)$$

The energy balance technique yields a positive number for $\dot{q}$ for both cooling and heating, and the direction of the heat transfer is implied by the terms heating and cooling. It is to be emphasized that

$$i_1 = i_{a1} + W_1 i_{v1} \qquad (2\text{-}24)$$

and

$$i_2 = i_{a2} + W_2 i_{v2} \qquad (2\text{-}25)$$

where i_1 and i_2 may be obtained from the psychrometric chart. The convenience of the chart is evident. Figure 2-6 shows heating and cooling processes. Because the moist air has been assumed to be a perfect gas, Eq. (2-23) may be rearranged and written

$$\dot{q} = \dot{m}_a c_p (t_1 - t_2) \text{ (cooling or heating)} \qquad (2\text{-}26)$$

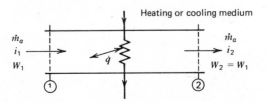

Figure 2-5. Schematic of a heating or cooling device.

Randy H. Cook

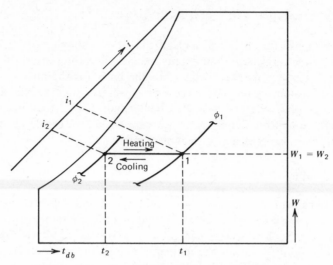

Figure 2-6.   Sensible heating and cooling process.

where

$$c_p = c_{pa} + W c_{pv} \qquad (2\text{-}27)$$

In the temperature range of interest, c_{pa} = 0.24 Btu/(lbma-F) or 1.0 kJ/(kga-C), c_{pv} = 0.45 Btu/(lbmv-F) or 1.86 kJ/(kga-C) and W is the order of 0.01. Then c_p is about 0.245 Btu/(lbma-F) or 1.02 kJ/(kga-C).

Example 2-4.   Find the heat transfer rate required to warm 1500 cfm (ft³/min) air at 60 F and 90 percent relative humidity to 120 F without the addition of moisture.

Solution.   Equations (2-23) or (2-26) may be used to find the required heat transfer rate. First it is necessary to find the mass flow rate of the dry air.

$$\dot{m}_a = \frac{\overline{V}_1 A_1}{v_1} = \frac{\dot{Q}}{v_1} \qquad (2\text{-}28)$$

The specific volume is read from Chart 1 at t_1 = 60 F and ϕ = 90 percent as 13.31 ft³/lbma:

$$\dot{m}_a = \frac{1500(60)}{13.31} = 6762 \text{ lbma/hr}$$

Also from Chart 1, i_1 = 25.3 Btu/lbma and i_2 = 40 Btu/lbma. Then by using Eq. (2-23), we get

$$\dot{q} = 6762(40.0 - 25.3) = 99{,}400 \text{ Btu/hr}$$

or from Eq. (2-26)

$$\dot{q} = (6762)(0.245)(120 - 60) = 99{,}400 \text{ Btu/hr}$$

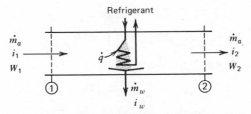

Figure 2-7. Schematic of a cooling and dehumidifying device.

Example 2-4 shows that the relative humidity decreases when the moist air is heated. The reverse process of cooling results in an increase in relative humidity.

Cooling and Dehumidifying of Moist Air

When moist air is cooled to a temperature below its dew point, some of the water vapor will condense and leave the air stream. Figure 2-7 shows a schematic of a cooling and dehumidifying device and Fig. 2-8 shows the process on the psychrometric chart. Although the actual process path will vary considerably depending on the type surface, surface temperature, and flow conditions, the heat and mass transfer can be expressed in terms of the initial and final states. By referring to Fig. 2-7, we see that the energy balance becomes

$$\dot{m}_a i_1 = \dot{q} + \dot{m}_a i_2 + \dot{m}_w i_w \tag{2-29}$$

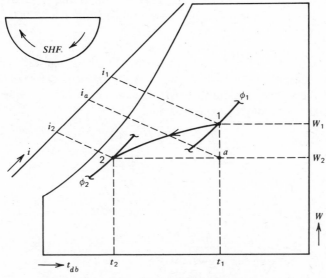

Figure 2-8. Cooling and dehumidifying process.

and the steady flow mass balance for the water is

$$\dot{m}_a W_1 = \dot{m}_w + \dot{m}_a W_2 \tag{2-30}$$

Combining Eqs. (2-29) and (2-30) we get

$$\dot{q} = \dot{m}_a(i_1 - i_2) - \dot{m}_a(W_1 - W_2)i_w \tag{2-31}$$

Equation (2-31) represents the total amount of heat transfer from the moist air. The last term on the right-hand side of Eq. (2-31) is usually small compared to the others and is often neglected. The following example illustrates this point.

Example 2-5. Moist air at 80 F db and 67 F wb is cooled to 58 F db and 80 percent relative humidity. The volume flow rate is 2000 cfm and the condensate leaves at 60 F. Find the heat transfer rate.

Solution. Equation (2-31) applies to this process, which is similar to Fig. 2-8. The following properties are read from Chart 1: $v_1 = 13.85$ ft^3/lbma, $i_1 = 31.6$ Btu/lbma, $W_1 = 0.0112$ lbmv/lbma, $i_2 = 22.9$ Btu/lbma, $W_2 = 0.0082$ lbmv/lbma. The enthalpy of the condensate is obtained from Table A-1, $i_w = 28.08$ Btu/lbmw. The mass flow rate $\dot{m}_a$ is obtained from Eq. (2-28).

$$\dot{m}_a = \frac{2000(60)}{13.85} = 8664 \text{ lbma/hr}$$

Then

$$\dot{q} = 8664[(31.6 - 22.9) - (0.0112 - 0.0082)28.08]$$
$$\dot{q} = 8664[(8.7) - (0.084)]$$

The last term, which represents the energy of the condensate, is quite insignificant in this case. For most cooling and dehumidifying processes this will be true. Finally, $\dot{q} = 74,600$ Btu/hr. A *ton of refrigeration* is 12,000 Btu/hr. Then $\dot{q} = 6.22$ tons.

The cooling and dehumidifying process involves both sensible and latent heat transfer where sensible heat transfer is associated with the decrease in dry bulb temperature and the latent heat transfer is associated with the decrease in humidity ratio. These quantities may be expressed as

$$\dot{q}_s = \dot{m}_a c_p(t_1 - t_2) \tag{2-32}$$

and

$$\dot{q}_l = \dot{m}_a(W_1 - W_2)i_{fg} \tag{2-33}$$

By referring to Fig. 2-8 we may also express the latent heat transfer as

$$\dot{q}_l = \dot{m}_a(i_1 - i_a) \tag{2-34}$$

and the sensible heat transfer is given by

$$\dot{q}_s = \dot{m}_a(i_a - i_2) \tag{2-35}$$

The energy of the condensate has been neglected. Obviously

$$\dot{q} = \dot{q}_s + \dot{q}_l \tag{2-36}$$

The sensible heat factor *SHF* is defined as $\dot{q}_s/\dot{q}$. This parameter is shown on the semicircular scale of Fig. 2-8. The use of this feature of the chart is shown later.

Heating and Humidifying Moist Air

A device to heat and humidify moist air is shown schematically in Fig. 2-9. This process is generally required during the cold months of the year. An energy balance on the device yields

$$\dot{m}_a i_1 + \dot{q} + \dot{m}_w i_w = \dot{m}_a i_2 \tag{2-37}$$

and a mass balance on the water gives

$$\dot{m}_a W_1 + \dot{m}_w = \dot{m}_a W_2 \tag{2-38}$$

Equations (2-37) and (2-38) may be combined to obtain

$$\frac{i_2 - i_1}{W_2 - W_1} = \frac{\dot{q}}{\dot{m}_a(W_2 - W_1)} + i_w \tag{2-39}$$

or

$$\frac{i_2 - i_1}{W_2 - W_1} = \frac{\dot{q}}{\dot{m}_w} + i_w \tag{2-39a}$$

Equation (2-39) or (2-39a) gives the direction of a straight line that connects the initial and final states on the psychrometric chart. Figure 2-10 shows a typical combined heating and humidifying process.

Example 2-6. Moist air at 60 F db and 20 percent relative humidity flows through a heater and humidifier at the rate of 1600 cfm. Heat is transferred to the air at the rate of 100,000 Btu/hr. Saturated water vapor at 212 F is injected at the rate of 50 lbm/hr. Determine the final state of the moist air if the process occurs at 14.696 psia.

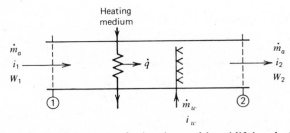

Figure 2-9. Schematic of a heating and humidifying device.

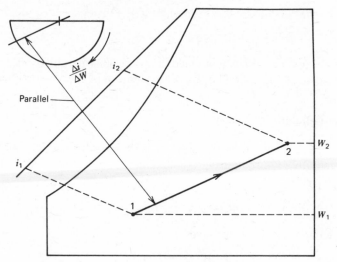

Figure 2-10.   Typical heating and humidifying process.

Solution.   Equation (2-38) may be used to determine W_2 and Eq. (2-37) will give the value of i_2. These properties will determine the state, and other properties may be determined as required. The initial state properties are read from Chart 1 as: $W_1 = 0.0022$ lbmv/lbma, $v_1 = 13.15$ ft^3/lbma, and $i_1 = 16.8$ Btu/lbma. From Table A-1, $i_w = 1151$ Btu/lbmw. Then

$$\dot{m}_a = \frac{1600 \times 60}{13.15} = 7300 \text{ lbma/hr}$$

Using Eq. (2-38), we get

$$W_2 = W_1 + \frac{\dot{m}_w}{\dot{m}_a} = 0.0022 + \frac{50}{7300} = 0.009 \text{ lbmv/lbma}$$

Then from Eq. (2-37)

$$i_2 = i_1 + \frac{\dot{q}}{\dot{m}_a} + \frac{\dot{m}_w}{\dot{m}_a} i_w$$

$$i_2 = 16.8 + \frac{100,000}{7300} + \frac{50}{7300}(1151) = 38.4 \text{ Btu/lbma}$$

The final state is fixed by i_2 and W_2 and $t_2 = 118$ F. $t_{2wb} = 75$ F from Chart 1.

A graphical procedure makes use of the circular scale on Chart 1 to solve for state 2. The ratio of enthalpy to humidity ratio $\Delta i/\Delta W$ is defined as

$$\frac{\Delta i}{\Delta W} = \frac{i_2 - i_1}{W_2 - W_1} = \frac{\dot{q}}{\dot{m}_w} + i_w \qquad (2\text{-}39a)$$

For Example 2-6,

$$\frac{\Delta i}{\Delta W} = \frac{100,000}{50} + 1151 = 3151$$

Figure 2-10 shows the procedure where a straight line is laid out parallel to the line on the protractor through state point 1. The intersection of this line with the computed value of W_2 determines the final state.

Humidifying Moist Air

Moisture is frequently added to moist air without the addition of heat. Equation (2-39a) then becomes

$$\frac{\Delta i}{\Delta W} = \frac{i_2 - i_1}{W_2 - W_1} = i_w \qquad (2\text{-}40)$$

The direction of the process on the psychrometric chart can therefore vary considerably. If the injected water is saturated vapor at the dry bulb temperature, the process will proceed at a constant dry bulb temperature. If the water enthalpy is greater than saturation, the air will be heated and humidified. If the water enthalpy is less than saturation the air will be cooled and dehumidified. Figure 2-11 shows these processes. One other situation is worthy of mention. When liquid water at the wet bulb temperature is injected, the process follows a line of constant wet bulb temperature as shown by Fig. 2-11.

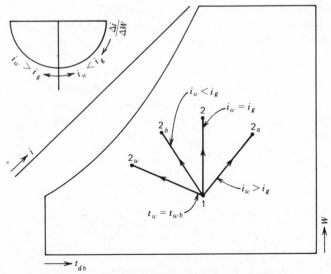

Figure 2-11. Humidification processes without heat transfer.

Adiabatic Mixing of Two Streams of Moist Air

The mixing of air streams is quite common in air-conditioning systems. The mixing process usually occurs under adiabatic conditions and with steady flow. Figure 2-12 illustrates the mixing of two air streams. An energy balance gives

$$\dot{m}_{a1}i_1 + \dot{m}_{a2}i_2 = \dot{m}_{a3}i_3 \tag{2-41}$$

The mass balance on the dry air is

$$\dot{m}_{a1} + \dot{m}_{a2} = \dot{m}_{a3} \tag{2-42}$$

and the mass balance on the water vapor is

$$\dot{m}_{a1}W_1 + \dot{m}_{a2}W_2 = \dot{m}_{a3}W_3 \tag{2-43}$$

By combining Eqs. (2-41), (2-42), and (2-43) and eliminating $\dot{m}_{a3}$, we obtain the following result.

$$\frac{i_2 - i_3}{i_3 - i_1} = \frac{W_2 - W_3}{W_3 - W_1} = \frac{\dot{m}_{a1}}{\dot{m}_{a2}} \tag{2-44}$$

The form of Eq. (2-44) shows that the state of the mixed streams must lie on a straight line between states 1 and 2. This is shown on Fig. 2-13. It may be further inferred from Eq. (2-44) that the length of the various line segments are proportional to the masses of dry air mixed.

$$\frac{\dot{m}_{a1}}{\dot{m}_{a2}} = \frac{\overline{32}}{\overline{13}}; \quad \frac{\dot{m}_{a1}}{\dot{m}_{a3}} = \frac{\overline{32}}{\overline{12}}; \quad \frac{\dot{m}_{a2}}{\dot{m}_{a3}} = \frac{\overline{13}}{\overline{12}} \tag{2-45}$$

This fact provides a very convenient graphical procedure for solving mixing problems in contrast to the use of Eqs. (2-41), (2-42), and (2-43).

It should be noted that the mass flow rate is used when the graphical procedure is employed; however the volume flow rates may be used to obtain approximate results.

Example 2-7. Two thousand cfm of air at 100 F db and 75 F wb are mixed with 1000 cfm of air at 60 F db and 50 F wb. The process is adiabatic, at a steady flow rate and at standard sea level pressure. Find the condition of the mixed streams.

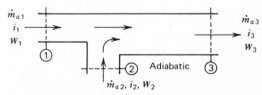

Figure 2-12. Schematic adiabatic mixing of two air streams.

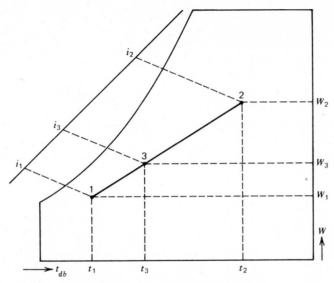

Figure 2-13.   Adiabatic mixing process.

Solution.   A combination graphical and analytical solution is first obtained. The initial states are first located on Chart 1 as illustrated on Fig. 2-13 and connected with a straight line. Equations (2-42) and (2-43) are combined to obtain

$$W_3 = W_1 + \frac{\dot{m}_{a2}}{\dot{m}_{a3}} (W_2 - W_1) \qquad (2\text{-}46)$$

By using the property values from Chart 1, we obtain

$$\dot{m}_{a1} = \frac{1000(60)}{13.21} = 4542 \text{ lbma/hr}$$

$$\dot{m}_{a2} = \frac{2000(60)}{14.4} = 8332 \text{ lbma/hr}$$

$$W_3 = 0.0053 + \frac{8332}{(4542 + 8332)} (0.013 - 0.0053)$$

$$W_3 = 0.0103 \text{ lbmv/lbma}$$

The intersection of W_3 with the line connecting states 1 and 2 gives the mixture state 3. The resulting dry bulb temperature is 86 F and the wet bulb temperature is 68 F.

The complete graphical procedure could also be used where

$$\frac{\overline{13}}{\overline{12}} = \frac{\dot{m}_{a2}}{\dot{m}_{a3}} = \frac{8332}{(8332 + 4542)} = 0.65$$

and

$$\overline{13} = 0.65(\overline{12})$$

The length of line segments $\overline{12}$ and $\overline{13}$ depends on the scale of the psychrometric chart used.

2-6 SPACE AIR CONDITIONING—DESIGN CONDITIONS

The complete air-conditioning system may involve two or more of the processes just considered. For example, in the air conditioning of a space during the summer the air supplied must have a sufficiently low temperature and moisture content to absorb the total heat gain of the space. Therefore, as the air flows through the space it is heated and humidified. If the system is a closed loop, the air is then returned to the conditioning equipment where it is cooled and dehumidified and supplied to the space again. If fresh air is required in the space, outdoor air may be mixed with the return air before it goes to the cooling and dehumidifying equipment. During the winter months the same general processes occur but in reverse. Figure 2-14 illustrates a residential or small commercial system; large commercial systems operate in basically the same way. During the summer months the heating and humidifying elements are inactive, and during the winter the cooling and dehumidifying coil is inactive. With appropriate controls, however, all of the elements may be continuously active to maintain precise conditions in the space.

Sensible Heat Factor

The *sensible heat factor (SHF)* was defined in Section 2-5 as the ratio of the sensible heat transfer to the total heat transfer for a process:

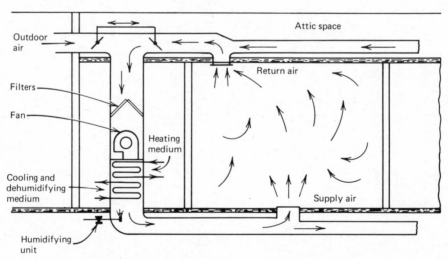

Figure 2-14.   Schematic of an air-conditioning system.

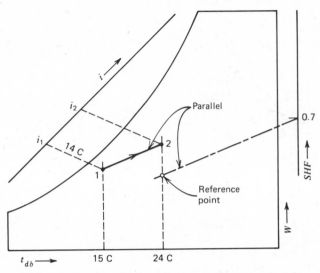

Figure 2-15. Psychrometric process for Example 2-8.

$$SHF = \frac{\dot{q}_s}{\dot{q}_s + \dot{q}_l} = \frac{\dot{q}_s}{\dot{q}} \qquad (2\text{-}47)$$

If we recall Eqs. (2-34) and (2-35) and refer to Fig. 2-8, it is evident that the *SHF* is related to the parameter $\Delta i/\Delta W$. The *SHF* is plotted on the inside scale of the protractor on Chart 1 and on the right-hand coordinate of Chart 1*a*. A reference point located on the chart at 24 C db and 17 C wb is used in conjunction with the *SHF* scale on the extreme right axis of the chart to establish the slope of the process curve. The following examples will demonstrate the usefulness of the *SHF*.

Example 2-8.   Conditioned air is supplied to a space at 15 C db and 14 C wb at the rate of 0.5 m³/s. The sensible heat factor for the space is 0.70 and the space is to be maintained at 24 C db. Determine the sensible and latent cooling loads for the space.

Solution.   Chart 1*a* can be used to solve this problem conveniently. As noted before, the scale for the *SHF* is different on this chart. For this example a line is drawn from the reference point to a value of 0.7 on the *SHF* scale. A parallel line is then drawn from the initial state (15 C db and 14 C wb) to the intersection of the 24 C db line, which defines the final state. Figure 2-15 illustrates the procedure. The total heat transfer rate for the process is given by

$$\dot{q} = \dot{m}_a(i_2 - i_1)$$

and the sensible heat transfer is given by

$$\dot{q}_s = (SHF)\dot{q}$$

and

$$m_a = \dot{Q}/v_1 = \frac{0.5}{0.827} = 0.605 \text{ kg/s}$$

where v_1 is read from Chart $1a$. Also from Chart $1a$, $i_1 = 39.3$ kJ/kg dry air and $i_2 = 52.6$ kJ/kg dry air. Then

$$\dot{q} = 0.605(52.6 - 39.3) = 8.04 \text{ kJ/s} = 8.04 \text{ kW}$$
$$\dot{q}_s = \dot{q}(SHF) = 8.04(0.7) = 5.63 \text{ kW}$$

and

$$\dot{q}_l = \dot{q} - \dot{q}_s = 2.41 \text{ kW}$$

Example 2-9. A given space is to be maintained at 78 F db and 65 F wb. The total heat gain to the space has been determined to be 60,000 Btu/hr of which 42,000 Btu/hr is sensible heat transfer. The outdoor air requirement of the occupants is 500 cfm. The outdoor air has a temperature and relative humidity of 90 F and 55 percent, respectively. Determine the quantity and state of the air supplied to the space and the required capacity of the cooling and dehumidifying equipment.

Solution. The schematic of Fig. 2-14 may be simplified for purposes of problem solution as shown in Fig. 2-16. The various quantities given are shown and stations are numbered for reference. Losses in connecting ducts will be neglected. Let us first consider the steady flow process for the conditioned space. By Eq. (2-47) the sensible heat factor is

$$SHF = \frac{42,000}{60,000} = 0.7$$

The state of the air entering the space lies on the line defined by the *SHF* on psychrometric Chart 1. Therefore, state 3 is located as shown on Fig. 2-17 and a line drawn through the point parallel to the *SHF* = 0.7 line on the protractor.

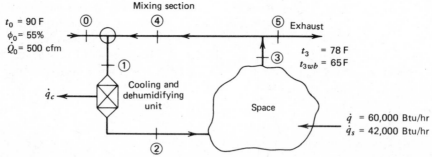

Figure 2-16. Single line sketch of cooling and dehumidifying system for Example 2-9.

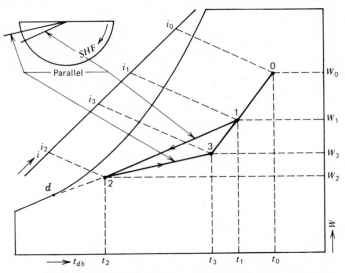

Figure 2-17. Psychrometric processes for Example 2-9.

State 2 may be any point on the line and is determined by the operating characteristics of the equipment and by what will be comfortable for the occupants. Assume that the dry bulb temperature t_2 cannot be more than 20 F less than t_3 for this case. Then $t_2 = 58$ F and state 2 is determined. The air quantity required may now be found from an energy balance on the space

$$\dot{m}_{a2} i_2 + \dot{q} = \dot{m}_{a3} i_3$$

or

$$\dot{q} = \dot{m}_{a2}(i_3 - i_2)$$

and

$$\dot{m}_{a2} = \frac{\dot{q}}{(i_3 - i_2)}$$

From Chart 1, $i_3 = 30$ Btu/lbma, $i_2 = 23$ Btu/lbma, and

$$\dot{m}_{a2} = \dot{m}_{a3} = \frac{60{,}000}{(30 - 23)} = 8570 \text{ lbma/hr}$$

Also from Chart 1, $v_2 = 13.21$ ft^3/lbma and

$$\dot{Q}_2 = \dot{m}_{a2} v_2 = \frac{8570(13.21)}{60} = 1890 \text{ cfm}$$

Attention is now directed to the cooling and dehumidifying unit. However, state 1 must be determined before continuing. A mass balance on the mixing section yields

$$\dot{m}_{a0} + \dot{m}_{a4} = \dot{m}_{a1} = \dot{m}_{a2}$$

$$\dot{m}_{a0} = \frac{\dot{Q}_0}{v_0} \,; \qquad v_0 = 14.23 \text{ ft}^3/\text{lbma}$$

$$\dot{m}_{a0} = \frac{(500 \times 60)}{14.23} = 2110 \text{ lbma/hr}$$

Then

$$\dot{m}_{a4} = \dot{m}_{a2} - \dot{m}_{a0} = 8570 - 2110 = 6460 \text{ lbma/hr}$$

By using the graphical technique discussed in Example 2-7 and referring to Fig. 2-17, we see that

$$\frac{\overline{31}}{\overline{30}} = \frac{\dot{m}_{a0}}{\dot{m}_{a1}} = \frac{2110}{8570} = 0.246$$

$$\overline{31} = 0.246(\overline{30})$$

State 1 is located at 81 F db and 68 F wb. A line constructed from state 1 to state 2 on Chart 1 then represents the process taking place in the conditioning equipment. An energy balance gives

$$\dot{m}_{a1} i_1 = \dot{q}_c + \dot{m}_{a2} i_2$$

or

$$\dot{q}_c = \dot{m}_{a1}(i_1 - i_2)$$

From Chart 1, $i_1 = 32.4$ Btu/lbma

$$\dot{q}_c = 8570(32.4 - 24) = 72,000 \text{ Btu/hr} = 6 \text{ tons}$$

The sensible heat factor (*SHF*) for the cooling unit is found to be 0.6 using the protractor of Chart 1 (Fig. 2-17). Then

$$\dot{q}_{cs} = 0.6(72,000) = 43,200 \text{ Btu/hr}$$

and

$$\dot{q}_{cl} = 72,000 - 43,200 = 28,800 \text{ Btu/hr}$$

The sum of $\dot{q}_{cs}$ and $\dot{q}_{cl}$ is known as the coil refrigeration load in contrast to the space cooling load.

An alternate approach to the analysis of the cooling coil in Example 2-9 uses the so-called coil *bypass factor*. Note that when line 1-2 of Fig. 2-17 is extended, it intersects the saturation curve at point *d*. This point represents the apparatus dew-point temperature of the cooling coil. The coil cannot cool all of the air passing through it to the coil surface temperature. This fact makes the coil perform in a manner similar to what would happen if a portion of the air was brought to the coil temperature and the remainder bypassed the coil unchanged. A dehumidifying coil thus produces unsaturated air at a higher temperature than the coil temperature.

Again referring to Fig. 2-17, notice that in terms of the length of the line d-2, the length d-1 is proportional to the air bypassed, and the length 1-2 is proportional to the air not bypassed. Because of the chart construction, it is approximately true that,

$$b = \frac{t_2 - t_d}{t_1 - t_d} \tag{2-48}$$

and

$$1 - b = \frac{t_1 - t_2}{t_1 - t_d} \tag{2-49}$$

where b is the fraction of air bypassed, or coil bypass factor, expressed as a decimal, and where the temperatures are dry bulb values. Now,

$$\dot{q}_{cs} = \dot{m}_{a1} c_p(t_1 - t_2) \tag{2-50}$$

or

$$\dot{q}_{cs} = \dot{m}_{a1} c_p(t_1 - t_d)(1 - b) \tag{2-50a}$$

Example 2-10. Find the bypass factor for the coil of Example 2-9 and compute the sensible and latent heat transfer rates.

Solution. The apparatus dewpoint temperature obtained from Chart 1 as indicated in Fig. 2-17 is 46 F. Then from Eq. (2-48) the bypass factor is

$$b = \frac{58 - 46}{81 - 46} = 0.343 \quad \text{and} \quad 1 - b = 0.657$$

Equation (2-50a) expresses the sensible heat transfer rate as

$$\dot{q}_{cs} = 8570(0.245)(81 - 46)(0.657) = 48,280 \text{ Btu/hr}$$

The coil sensible heat factor is used to compute the latent heat transfer rate. From Example 2-9, the *SHF* is 0.6, then the total heat transfer rate is

$$\dot{q}_t = \dot{q}_{cs}/SHF = 48,280/0.6 = 80,470 \text{ Btu/hr}$$

and

$$\dot{q}_{cl} = \dot{q}_t - \dot{q}_{cs} = 80,470 - 48,280 = 32,190 \text{ Btu/hr}$$

It should be noted that the bypass factor approach results in values about 12 percent high in this case.

The contrasting problem of space air conditioning during the winter months may be solved using Example 2-8 as an approximate model.

Example 2-11. A space is to be maintained at 75 F and 50 percent relative humidity. Heat losses from the space are 225,000 Btu/hr sensible and 56,250 Btu/hr latent. The latent heat transfer is due to the infiltration of cold dry air.

The outdoor air required is 1000 cfm and is at 35 F and 80 percent relative humidity. Determine the quantity of air supplied at 120 F, the state of the supply air, the size of the furnace or heating coil, and the humidifier characteristics.

Solution.   Figure 2-18 is a schematic for the problem; it contains the given information and reference points. First consider the conditioned space

$$SHF = \frac{225,000}{(225,000 + 56,250)} = 0.80$$

The state of the supply air lies on a line drawn through state point 3 parallel to the *SHF* = 0.8 line on the protractor of Chart 1. Figure 2-19 shows this construction. State 2 is located at 120 F dry bulb and the intersection of this line. An energy balance on the space gives

$$\dot{m}_{a2}i_2 = \dot{q} + \dot{m}_{a3}i_3$$

or

$$\dot{q} = \dot{m}_{a2}(i_2 - i_3)$$

From Chart 1, i_2 = 42 Btu/lbma, i_3 = 28.2 Btu/lbma, and

$$\dot{m}_{a2} = \frac{\dot{q}}{(i_2 - i_3)} = \frac{281,250}{(42 - 28.2)}$$

$$\dot{m}_{a2} = 20,400 \text{ lbma/hr}$$

From Chart 1, v_2 = 14.89 ft³/lbma and

$$\dot{Q}_2 = \frac{20,400}{60} \times 14.89 = 5060 \text{ cfm}$$

To find the conditions at state 1 the mixing process must be considered. A mass balance on the mixing section yields

$$\dot{m}_{a0} + \dot{m}_{a4} = \dot{m}_{a1} = \dot{m}_{a2}$$

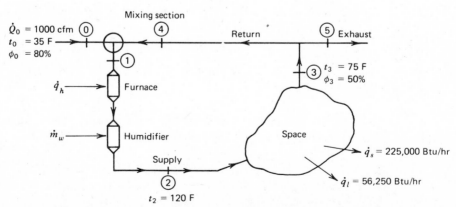

Figure 2-18.   The heating and humidifying system for Example 2-11.

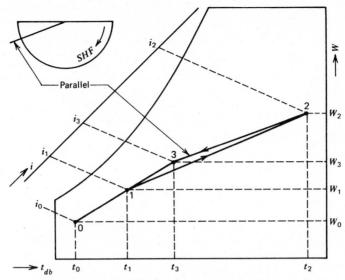

Figure 2-19.   Psychrometric processes for Example 2-11.

or

$$\dot{m}_{a4} = \dot{m}_{a2} - \dot{m}_{a0}$$

$$\dot{m}_{a0} = \frac{\dot{Q}_0}{v_0} \quad \text{and} \quad v_0 = 12.53 \text{ ft}^3/\text{lbma}$$

$$\dot{m}_{a0} = \frac{1000 \times 60}{12.53} = 4790 \text{ lbma/hr}$$

and

$$\dot{m}_{a4} = 20,400 - 4790 = 15,600 \text{ lbma/hr}$$

Using the graphical technique and referring to Fig. 2-19, we obtain

$$\frac{\overline{31}}{\overline{30}} = \frac{\dot{m}_{a0}}{\dot{m}_{a1}} = \frac{4790}{20,400} = 0.235$$

$$\overline{31} = 0.235(\overline{30})$$

State 1 is then located at 66 F db and 57 F wb. The line $\overline{12}$ constructed on Chart 1, Fig. 2-19, represents the heating and humidifying process that must take place in the heating and humidifying unit. An energy balance gives

$$\dot{m}_{a1} i_1 + \dot{q}_h + \dot{m}_w i_w = \dot{m}_{a2} i_2$$

and a mass balance on the water gives

$$\dot{m}_{a1} W_1 + \dot{m}_w = \dot{m}_{a2} W_2$$

or

$$\dot{m}_w = \dot{m}_{a1}(W_2 - W_1)$$

Then

$$q_h = \dot{m}_{a1}(i_2 - i_1) - \dot{m}_{a1}(W_2 - W_1)i_w$$

Assume that ordinary tap water at 55 F is used in the humidifier. Then $i_w = 23.08$ Btu/lbm from Table A-1. From Chart 1, $i_1 = 24.6$ Btu/lbma, $W_1 = 0.008$ lbmw/lbma, and $W_2 = 0.0119$ lbmw/lbma.

$$q_h = 20,400[(42 - 24.6) - (0.0119 - 0.008)23.08]$$
$$q_h = 353,000 \text{ Btu/hr}$$

The amount of water supplied to the humidifier is

$$\dot{m}_w = \dot{m}_{a1}(W_2 - W_1) = 20,400(0.0119 - 0.008)$$
$$\dot{m}_w = 79.6 \text{ lbm/hr} = 1.33 \text{ lbm/min}$$

2-7 SPACE AIR CONDITIONING—OFF-DESIGN CONDITIONS

The previous section treated the common space air conditioning problem assuming that the system was operating steadily at the design condition. Actually the space requires only a part of the designed capacity of the conditioning equipment most of the time. A control system functions to match the required cooling or heating of the space to the conditioning equipment by varying one or more system parameters. For example, the quantity of air circulated through the coil and to the space may be varied in proportion to the space load. This approach is known as *variable air volume* (VAV). Another approach is to circulate a constant amount of air to the space, but some of the return air is diverted around the coil and mixed with air coming off the coil to obtain a supply air temperature which is proportional to the space load. This is known as face and bypass control, because face and bypass dampers are used to divert the flow. Another possibility is to vary the coil surface temperature with respect to the required load by changing the temperature or the amount of heating or cooling fluid entering the coil. This technique is usually used in conjunction with VAV and face and bypass systems. However, control of the coolant temperature or quantity may be the only variable in some small systems.

Each of these control methods will be considered separately below. Figure 2-20a illustrates what might occur when the load on a variable air volume system decreases. The solid lines represent the full load design condition, whereas the broken lines illustrate a part load condition where the amount of air circulated to the space and across the coil has decreased. Note that the outdoor air has a lower temperature and humidity and the air is cooled to a lower temperature and humidity by the cooling coil. The thermostat maintains the space temperature but the humidity in the space will decrease. This explains why control of the water temperature and/or flow rate is desirable. Increasing the water temperature will cause point 2′ to move upward and to the right to a position where the room process curve will terminate at point 3.

The behavior of a constant air volume, face and bypass system is shown in Fig. 2-20b. The state points and process curves are quite similar to the VAV system of

Fig. 2-20*a*. The difference is related to the different amounts of air flowing at states 2, 2', 3, and 3'. The total design air flow rate is flowing at state 2, 3, and 3' but a lower flow rate occurs at state 2', leaving the coil. Air at states 2' and 1' is mixed downstream of the coil to obtain state 4. The total design flow rate and the enthalpy difference, $h_{3'} - h_4$, then match the space load. Note again that the humidity at state 4 is lower than necessary, which lowers state 3' below the design value. Increasing the coil water temperature would correct this situation as discussed above for the *VAV* system.

A constant air volume system with either water temperature or flow rate control is shown in Fig. 2-20*c*. In this case both the temperature and humidity of the air leaving the coil increase and the room process curve 2' - 3' may not terminate at state 3. In fact, there will be cases when state 3' will lie above state 3 causing an uncomfortable condition in the space. For this reason, water control alone is not usually used in commercial applications, but is used in conjunction with *VAV* and face and bypass systems as discussed above. The following example illustrates the analysis of a *VAV* system with variable water temperature.

Example 2-12.   A variable air volume system operates as shown in Fig. 2-21. The solid lines show the full load design condition of 100 tons with a room *SHF* of 0.75. At the estimated minimum load of 15 tons with *SHF* of 0.9, the air

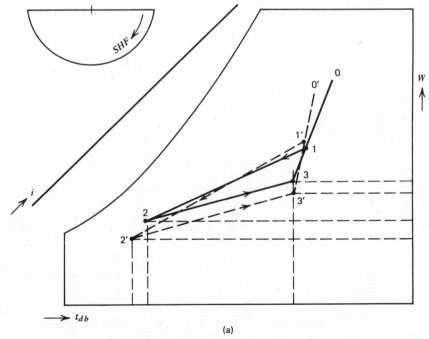

(a)

Figure 2-20.   Psychrometric processes for off-design conditions. (*a*) Variable air volume control.

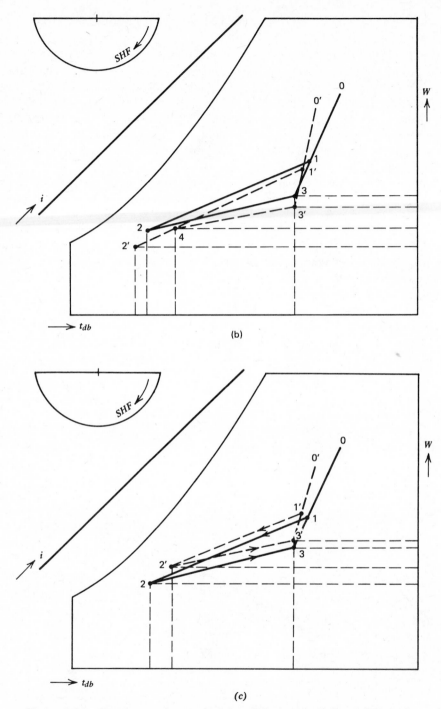

Figure 2-20. Psychrometric processes for off-design conditions (*continued*).
(*b*) Face and bypass control.
(*c*) Water temperature control.

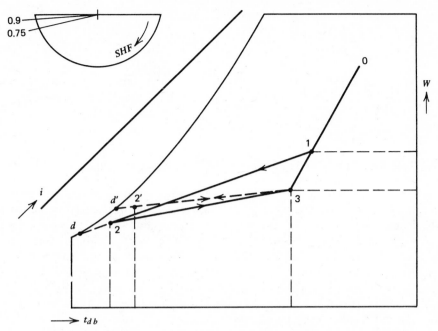

Figure 2-21.   Schematic psychrometric processes for Example 2-12.

flow rate is decreased to 20 percent of the design value and all outdoor air is shut off. Estimate the supply air and apparatus dew point temperatures of the cooling coil for minimum load assuming that state 3 does not change.

Solution.   The solution is best carried out using Chart 1 as shown in Fig. 2-21. Because the outdoor air is off during the minimum load condition, the space and coil process lines will coincide as shown by line 3-2'- d'. This line is constructed by using the protractor of Chart 1 with a *SHF* of 0.9. The apparatus dew point is seen to be 56 F, as compared with 50 F for the design condition. The air flow rate for the design condition is given by

$$\dot{m}_d = \dot{m}_2 = \dot{q}/(i_3 - i_2)$$
$$\dot{m}_d = 100(12,000)/(30 - 23.7) = 190,476 \text{ lbma/hr}$$

or

$$\dot{Q}_d = m_d v_2/60 = 190,476(13.24)/60 = 42,032 \text{ cfm}$$

Then, the minimum volume flow rate is

$$\dot{Q}_m = 0.2(42,032) = 8,406 \text{ cfm}$$

and the minimum mass flow rate may be estimated by assuming a value for $v_{2'}$.

$$\dot{m}_m = 8406(60)/14.28 = 35,320 \text{ lbma/hr}$$

State point 2' may then be determined by computing $h_{2'}$.

$$h_{2'} = h_3 - \frac{\dot{q}_m}{\dot{m}_m} = 30 - 15(12{,}000)/35{,}320 = 24.9 \text{ Btu/lbma}$$

Then from Chart 1, the air condition leaving the coil is 59.5 F db and 57.5 F wb. Calculation of the coil water temperature is beyond the scope of this analysis; however, the water temperature would be increased about 7 degrees from the design to the minimum load condition.

2-8   THE GOFF AND GRATCH TABLES FOR MOIST AIR

It was mentioned earlier that accurate thermodynamic properties of moist air were developed by Goff and Gratch (1). These properties were calculated using statistical mechanics. Table A-2 is an abridgment of these properties taken from the *ASHRAE Handbook of Fundamentals* (3). In Table A-2 the humidity ratio for saturated air, the specific volume of dry and saturated air, the enthalpy of dry and saturated air, and the entropy of dry and saturated air are given as a function of temperature. The data apply to standard atmospheric pressure. To calculate properties at states other than saturation, the following relations are used. The degree of saturation μ is useful in this connection.

$$v = v_a + \mu v_{as} + \bar{v} \tag{2-51}$$
$$i = i_a + \mu i_{as} + \bar{i} \tag{2-52}$$
$$s = s_a + \mu s_{as} + \bar{s} \tag{2-53}$$

where

$$\bar{v} = \frac{\mu(1 - \mu)A}{1 + 1.6078\mu W_s} \tag{2-54}$$

$$\bar{i} = \frac{\mu(1 - \mu)B}{1 + 1.6078\mu W_s} \tag{2-55}$$

$$\bar{s} = \frac{\mu(1 - \mu)C}{1 + 1.6078\mu W_s} \tag{2-56}$$

Below temperatures of about 150 F or 66 C, $\bar{v}$ and $\bar{i}$ may be taken as zero. Equation (2-53) for s is accurate only at temperatures considerably below 150 F or 66 C, and the correction given by Eq. (2-56) accounts for only a small part of the error. A more complex relation is required for $\bar{s}$ (3). Table 2-2 gives the values of A, B, and C for use in Eqs. (2-54), (2-55) and (2-56).

Figure 2-22 illustrates the errors in calculating specific volume, enthalpy, and specific humidity of saturated air by the perfect gas relations. It is evident that the perfect gas relations are adequate for usual work. However, laboratory experiments or precise mass transfer studies may require the more precise data of Table A-2.

Example 2-12.   Compute the specific volume, enthalpy, and humidity ratio for moist air at 44 C with a degree of saturation of 0.4.

Table 2-2   CONSTANTS A, B, AND C FOR EQS. (2-54) to (2-56) (STANDARD ATMOSPHERIC PRESSURE)[a]

Temperature		A		B		C	
F	C	ft³/lbma	m³/kga × 10⁴	Btu/lbma	kJ/kga	Btu/(lbma-F) × 10⁴	kJ/(kga-C) × 10⁴
96	36	0.0018	1.124	0.0268	0.0623	0.4	1.675
112	45	0.0042	2.622	0.0650	0.151	0.9	3.768
128	54	0.0096	5.993	0.1439	0.335	2.0	8.374
144	63	0.0215	13.42	0.3149	0.733	4.2	17.58
160	71	0.0487	30.40	0.6969	1.62	9.1	38.10
176	80	0.1169	72.98	1.636	3.81	20.7	86.67
192	89	0.3363	209.9	4.608	10.7	56.7	237.4

[a]Adapted by permission from *ASHRAE Handbook of Fundamentals*, 1977.

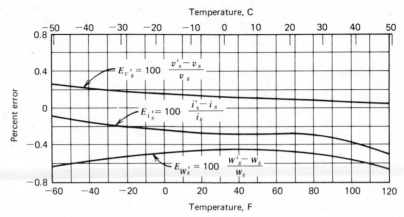

Figure 2-22. Error of perfect gas relations in calculation of humidity ratio, enthalpy, and volume of saturated air at standard atmospheric pressure. (James L. Threlkeld, *Thermal Environmental Engineering*, 2nd edition © 1970, p. 175. Reprinted by permission of Prentice-Hall, Inc. Englewood Cliffs, N.J.

Solution. The data of Table A-2a together with Eqs. (2-51) and (2-52) are used to obtain specific volume and enthalpy. The correction terms $\bar{v}$ and $\bar{i}$ are neglected because the temperature is well below 66 C. Equation (2-12) is used to find the humidity ratio.

$$W = \mu W_s = 0.4(0.06176) = 0.02470 \text{ kgv/kga}$$

where W_s is obtained from Table A-2a at 44 C. Using Eq. (2-51)

$$v = v_a + \mu v_{as} = 0.8983 + 0.4(0.0887)$$
$$v = 0.9338 \text{ m}^3/\text{kga}$$

where v_a and v_{as} are obtained from Table A-2a at 44 C. In a similar manner

$$i = i_a + \mu i_{as} = 44.278 + 0.4(159.331)$$
$$i = 108.01 \text{ kJ/kga}$$

REFERENCES

1. J. A. Goff and S. Gratch, "Thermodynamics of Moist Air," *ASHVE Transactions, 51,* 1945.

2. James L. Threlkeld, *Thermal Environmental Engineering,* Second Edition, Prentice-Hall, Englewood Cliffs, N.J., 1970.

3. *ASHRAE Handbook of Fundamentals,* American Society of Heating, Refrigerating and Air-Conditioning Engineers, New York, 1977.

4. J. A. Goff, "Standardization of Thermodynamic Properties of Moist Air," *Transactions ASHVE, 55,* 1949.

5. E. P. Palmatier, "Construction of the Normal Temperature ASHRAE Psychrometric Chart," *ASHRAE Journal,* May 1963.

PROBLEMS

2-1 Calculate values of humidity ratio, enthalpy, and specific volume for saturated air at 14.696 psia pressure using perfect gas relations and Table A-2 for temperatures of (a) 70 F and (b) 20 F.

2-2 The atmosphere within a room is at 70 F db, 50 percent relative humidity, and 14.696 psia pressure. The inside surface temperature of the windows is 40 F. Will moisture condense upon the window glass?

2-3 Assume that the dimensions of the room of Problem 2-2 are 30 by 15 by 8 ft. Calculate the mass of water vapor in the room.

2-4 Moist air exists at a dew point temperature of 18 C, a relative humidity of 60.0 percent and a pressure of 96.5 kPa. Determine (a) the humidity ratio, and (b) the volume in m^3/kga.

2-5 Calculate the enthalpy in Btu/lbma of moist air at 70 F db, 34 F dew point, and 14.696 psia pressure.

2-6 Air is supplied to a certain room from the outside where the temperature is 20 F and the relative humidity is 70 percent. It is desired to keep the room at 70 F and 60 percent relative humidity. How many pounds of water must be supplied to each pound of air entering the room if these conditions are to be met? The barometric pressure is 29.5 in. Hg.

2-7 The temperature of a certain room is 22 C and the relative humidity is 50 percent. The barometric pressure is 100 kPa. Find (a) the partial pressures of the air and water vapor, (b) the vapor density, and (c) the humidity ratio of the mixture.

2-8 Air with a dry bulb temperature of 70 F and a wet bulb temperature of 65 F is at a barometric pressure of 29.92 in. Hg. Without making use of the psychrometric chart find (a) the relative humidity of the air, (b) the vapor density, (c) the dew point temperature, (d) the humidity ratio, and (e) the volume occupied by the mixture associated with a pound of dry air.

2-9 Air is supplied to a room at 72 F db and 68 F wb from outside air at 40 F db and 37 F wb. The barometric pressure is 29.92 in. Hg. Find (a) the dew point temperatures of the inside and outside air, (b) the moisture added to each pound of dry air, (c) the enthalpy of the outside air, and (d) the enthalpy of the inside air.

2-10 Air is cooled from 75 F db and 70 F wb until it is saturated at 55 F. Find (a) the moisture removed per pound of dry air, (b) the heat removed to condense the moisture, (c) the sensible heat removed, and (d) the total amount of heat removed.

2-11 Air at 100 F db and 65 F wb is humidified adiabatically with steam. The steam supplied contains 20 percent moisture at 14.7 psia. When steam is added to humidify the air to 60 percent relative humidity, what is the dry bulb temperature of the humidified air? The barometer is at 29.92 in. Hg.

2-12 Air at 84 F db and 60 F wb and at 29.92 in. Hg is humidified with the dry bulb temperature remaining constant. Wet steam is supplied for humidification at 14.7 psia. What quality must the steam have (a) to provide saturated air, and (b) to provide air at 70 percent relative humidity?

2-13 Air at 38 C db and 20 C wb and 101.325 kPa is mixed adiabatically with water supplied at 60 C, in such proportions that the mixture has a relative humidity of 80 percent. Find the dry bulb temperature of the mixture.

2-14 Air at 40 F db and 35 F wb is mixed with warm air at 100 F db and 77 F wb in the ratio of 2 lb m cool air to 1 lb m of warm air. (a) Compute the humidity ratio and enthalpy of the mixed air and (b) find the humidity ratio and enthalpy using the psychrometric chart.

2-15 Air at 10 C db and 5 C wb is mixed with air at 25 C db and 18 C wb in a steady flow process at standard atmospheric pressure. The volume flow rates are 10 m³/s and 6 m³/s, respectively. (a) Compute the mixture conditions. (b) Find the mixture conditions using Chart 1*a*.

2-16 An auditorium is to be maintained at a temperature not to exceed 75 F and a relative humidity of 60 percent. The sensible heat load is 450,000 Btu/hr and 171 lbmv/hr of moisture must be removed. Air is supplied to the auditorium at 65 F. (a) How many lbma/hr must be supplied? (b) What is the dew point temperature of the entering air and what is its relative humidity? (c) How much latent heat is picked up in the auditorium? (d) What is the sensible heat ratio?

2-17 A meeting hall is to be maintained at 25 C db and 18 C wb. The barometric pressure is 101.3 kPa. The space has a load of 58.6 kW sensible and 58.6 kW latent. The temperature of the supply air to the space cannot be lower than 18 C db. (a) How many kg/s of air must be supplied? (b) What is the required wet bulb temperature of the supply air? (c) What is the sensible heat ratio?

2-18 A building with a heat loss of 200,000 Btu/hr is heated by warm air that is supplied at 135 F and a humidity ratio of 0.006 lbmv/lbma. Air returns to the furnaces at 65 F with no significant change in humidity ratio. Find (a) the mass of air that must be circulated per hour for heating, and (b) the cfm measured at inlet conditions.

2-19 For pipes carrying water at 50 F through a room that has an air temperature of 70 F, what is the maximum relative humidity in the room to prevent condensation on the pipes?

2-20 A space is to be maintained at 21 C dry bulb. It is estimated that the inside wall surface temperature could be as low as 7 C. What maximum relative and specific humidities can be maintained without condensation on the walls?

2-21 Outdoor air with a temperature of 40 F db and 35 F wb and with a barometric pressure of 29 in. Hg is heated and humidified under steady flow conditions to a final temperature of 70 F db and 40 percent relative humidity. (a) Find the weight of water vapor added to each pound of dry air. (b) If the water is supplied at 50 F, how much heat is added per pound of dry air?

2-22 Outdoor air at 95 F db and 79 F wb and at a barometric pressure of 29.92 in. Hg is cooled and dehumidified under steady conditions until it becomes saturated at 60 F. (a) Find the mass of water condensed per pound of dry air. (b) If the condensate is removed at 60 F, what quantity of heat is removed per pound of dry air?

2-23   Outdoor air at 38 C db and 26 C wb is cooled to 15 C db and 14 C wb. The process occurs at 101.3 kPa barometric pressure. Determine (a) the mass of water condensed per kilogram of dry air, (b) the sensible and latent heat transfer per kilogram of dry air and the sensible heat factor.

2-24   Moist air enters a refrigeration coil at 89 F db and 75 F wb temperature and at a rate of 1400 cfm. The apparatus dew point temperature of the coil is 55 F. (a) If 3.5 tons of refrigeration are available, find the dry bulb temperature of the air leaving the coil. (b) Compute the bypass factor. Assume sea level pressure.

2-25   Saturated steam at a pressure of 25 psia is sprayed into a stream of moist air. The initial condition of the air is 55 F db and 35 F dew point temperature. The mass rate of air flow is 2000 lbma/min. Barometric pressure is 14.696 psia. Determine (a) how much steam must be added in lbm/min to produce a saturated air condition, and (b) the resulting temperature of the saturated air.

2-26   Saturated water vapor at 100 C is used to humidify a stream of moist air. The air enters the humidifier at 13 C db and 2 C wb at a flow rate of 2.5 m³/s. The pressure is 101.35 kPa. Determine (a) the mass flow rate of the steam required to saturate the air, and (b) the temperature of the saturated air.

2-27   Moist air at 70 F db and 45 percent relative humidity is recirculated from a room and mixed with outdoor air at 97 F db and 83 F wb. Determine the mixture dry bulb and wet bulb temperatures if the volume of recirculated air is three times the volume of outdoor air.

2-28   A structure has a calculated cooling load of 10 tons of which 2.5 tons is latent load. The space is to be maintained at 76 F db and 50 percent relative humidity. Ten percent by volume of the air supplied to the space is outdoor air at 100 F db and 50 percent relative humidity. The air supplied to the space cannot be less than 56 F db. Find (a) the minimum amount of air supplied to the space in cfm, (b) the amounts of return air and outdoor air in cfm, (c) the conditions and volume flow rate of the air entering the cooling coil, and (d) the capacity and *SHF* for the cooling coil.

2-29   A building has a calculated cooling load of 410 kW. The latent portion of the load is 100 kW. The space is to be maintained at 25 C db and 50 percent relative humidity. Outdoor air is at 38 C and 50 percent relative humidity, and 10 percent by mass of the air supplied to the space is outdoor air. Air is to be supplied to the space at not less than 18 C. Find (a) the minimum amount of air supplied to the space in m³/s, (b) the volume flow rates of the return air, exhaust air, and outdoor air, (c) the condition and volume flow rate of the air entering the cooling coil, and (d) the capacity, apparatus dewpoint temperature, bypass factor, and *SHF* of the cooling coil.

2-30   A building has a total heating load of 200,000 Btu/hr. The sensible heat factor for the space is 0.8. The space is to be maintained at 72 F db and 40 percent relative humidity. Outdoor air at 40 F db and 20 percent relative humidity in the amount of 1000 cfm is required. Air is supplied to the space at 120 F db. Find (a) the conditions and amount of air supplied to the space, (b) the temperature rise of the air through the furnace, (c) the amount of water at 50 F required by the humidifier, and (d) the capacity of the furnace.

2-31   Calculate the humidity ratio, enthalpy, and specific volume of moist air at 78 F db with degree of saturation equal to 0.4 using Table A-2. Barometric pressure is 29.92 in. Hg.

2-32 Moist air at 90 F db has an enthalpy of 40 Btu/lbma. Compute the specific volume and humidity ratio using the data of Table A-2. Barometric pressure is 14.69 psia.

2-33 Moist air enters a cooling coil at 26 C db and is 50 percent saturated. (a) Find the enthalpy, specific volume, and entropy of the air using the data of Table A-2a, and (b) compare the v and i obtained in (a) with the values read from Chart 1a.

2-34 To save energy, the environmental conditions in a room are to be regulated so that the dry bulb temperature will be greater than or equal to 78 F and the dew point temperature will be less than or equal to 64 F. Compute the maximum relative humidity that can occur for standard barometric pressure.

2-35 A space is to be maintained at 78 F db and 68 F wb. The cooling system is a variable air volume *(VAV)* type where the quantity of air supplied and the supply air temperature are controlled. Under design conditions, the total cooling load is 150 tons with a sensible heat factor of 0.6 and the supply air temperature is 60 F db. At minimum load, about 18 tons with *SHF* of 0.8, the air quantity may be reduced no more than 80 percent by volume of the full load design value. Determine the supply air conditions for minimum load. Show all of the conditions on a psychrometric chart.

2-36 A 50 ton space air conditioning system is constant volume using face and bypass and water temperature control. At the design condition the space is to be maintained at 77 F db and 50 percent relative humidity with 55 F db supply air at 90 percent relative humidity. Outdoor air is supplied at 95 F db, 60 percent relative humidity with a ratio of one lbm to five lbm return air. A part load condition exists where total space load decreases by 50 percent and the *SHF* increases to 90 percent. The outdoor air condition changes to 85 F db and 70 percent relative humidity.
(a) At what temperature must the air be supplied to the space under the part load condition?
(b) If the air leaving the coil has a dry bulb temperature of 60 F, what is the ratio of the air bypassed to that flowing through the coil?
(c) What is the apparatus dew point temperature for both the design and part load conditions?
(d) Show all the processes on a psychrometric chart.

2-37 An unusual condition exists where it is necessary to cool and dehumidify air from 80 F db, 67 F wb to 60 F db and 54 F wb.
(a) Discuss the feasibility of doing this in one process with a cooling coil (*Hint:* Determine the apparatus dew point temperature for the process.)
(b) Describe a practical method of achieving the required process and sketch on a psychrometric chart.

2-38 A plant has a chilled water air handling unit serving one section of the factory. Five thousand cfm of outside air at 90 F db and 74 F wb are precooled to 64 F db and 90 percent relaive humidity. This air is then mixed with return air at 80 F db and 50 percent relative humidity and enters a face and bypass chilled water coil, leaving at 59.F db and 50 percent relative humidity. The flow rate of supply air is 20,000 cfm.
(a) Assume the bypass dampers are closed and compute the sensible and latent heat transfer from the conditioned space. What is the SHF for the space?
(b) With the bypass dampers still closed, what refrigeration load does the complete system impose on the water chiller?
(c) Assume that one-third of the air bypasses the coil and compute the latent and

sensible heat transfer from the conditioned space; the SHF; and the total refrigeration load on the chiller.

2-39 Investigate the feasibility of conditioning a space with 100 percent outdoor air and a cooling coil which requires a fixed volume flow rate of air per ton of cooling capacity. Assuming this is not feasible with fixed flow rate, describe a different coil arrangement to accomplish the process.

2-40 A chilled water cooling coil receives 2.5 m³/s of air at 25 C db, 20 C wb. It is necessary for the air to leave the coil at 13 C db, 12 C wb.
(a) Determine the *SHF* and the apparatus dew point temperature.
(b) Compute the bypass factor.
(c) Compute total and sensible heat transfer rates from the air using the enthalpy difference and the *SHF*.
(d) Compute the total and sensible heat transfer rate from the air using the bypass factor and the *SHF*.
(e) Compare the results of (c) and (d).

2-41 During the winter months it is possible to cool and dehumidify a space using outdoor air. Suppose an interior zone of a large building is designed to have a supply air flow rate of 5000 cfm which can be all outdoor air. The cooling load is constant at 10 tons with a *SHF* of 0.8 the year round. Indoor conditions are 78 F db and 67 F wb.
(a) What is the maximum outdoor air dry bulb temperature and humidity ratio that would satisfy the load condition?
(b) Consider a different time when the outdoor air has a temperature of 40 F db and 20 percent relative humidity. Return air and outdoor air may be mixed to cool the space, but humidification will be required. Assume that saturated water vapor at 14.7 psia is used to humidify the mixed air and compute the amounts of outdoor and return air in cfm.
(c) At another time, outdoor air is at 70 F db with a relative humidity of 90 percent. The cooling coil is estimated to have a minimum apparatus dew point of 50 F. What amount of outdoor and return air should be mixed before entering the coil to satisfy the given load condition?
(d) What is the refrigeration load in the coil of part (c) above?

2-42 An economizer mixes outdoor air with room return air to reduce the refrigeration load on the cooling coil.
(a) For a space condition of 25 C db and 20 C wb, describe the maximum wb and db temperatures that will reduce the coil load.
(b) Suppose a system is designed to supply 5 m³/s at 18 C db and 17 C wb to a space maintained at the conditions given in (a) above. What amount of outdoor air at 20 C db and 90 percent relative humidity can be mixed with the return air if the coil *SHF* is 0.6?
(c) What is the apparatus dew point and the bypass factor in (b) above?
(d) Compare the coil refrigeration load in (b) above with the outdoor air to that without outdoor air.

CHAPTER THREE

Comfort and Health

Air conditioning was defined in Chapter 1 as the simultaneous control of temperature, humidity, cleanliness, odor, and air circulation as required by the occupants of the space. In this chapter, we shall be concerned with the conditions that actually provide a comfortable and healthful environment. Experience has shown that not everyone within a given space can be made completely comfortable by one set of conditions. This is due to a number of factors, many of which cannot be completely explained. However, clothing, age, sex, and the level of activity of each person are considerations. The factors that influence comfort, in their order of importance are: temperature, radiation, humidity, air motion, and the quality of the air with regard to odor, dust, and bacteria. With a complete air conditioning system all of these factors may be controlled simultaneously. It is found that in most cases a reasonably comfortable environment can be maintained when two or three of these factors are controlled. The *ASHRAE Handbook of Fundamentals* (1) is probably the most up-to-date and complete source of information relating to the physiological aspects of thermal comfort. ASHRAE comfort standard 55 (11) defines acceptable thermal comfort as an environment that at least 80 percent of the occupants will find thermally acceptable. This definition gives insight into the complex problem of making as many people as possible comfortable with one set of conditions.

3-1 PHYSIOLOGICAL CONSIDERATIONS

A complex regulating system in the body acts to maintain the deep body temperature at approximately 98.6 F or 36.9 C. If the environment is maintained at suitable conditions so that the body can easily maintain an energy balance, a feeling of comfort will result. When the environmental conditions transfer energy away from the body too rapidly or if the environment is so warm that there is a problem in transferring energy from the body, discomfort will result. Obviously, if these conditions are carried to an extreme and the regulating system cannot maintain the normal body temperature, unhealthy conditions result.

Two basic mechanisms within the body seem to control the body temperature. The first is a decrease or increase in the internal energy production as the body temperature rises or falls. This internal process is generally called metabolism. The second is the control of the rate of heat dissipation by changing the rate of cutaneous blood circulation (the blood circulation near the surface of the skin). In this way heat transfer from the body can be increased or decreased.

All forms of heat transfer as well as mass transfer are important in maintaining body temperature. Heat transfer to or from the body is principally by convection and conduction and, therefore, the air motion in the immediate vicinity of the body is a very important factor. Radiation exchange between the body and surrounding surfaces, however, can be a deciding factor in achieving comfort if the surfaces surrounding the body are at a different temperature than the air.

Another very important regulatory function of the body is sweating. Under very warm conditions great quantities of moisture can be released by the body to help cool itself.

The metabolic rate depends on the level of activity such as rest, work, or exercise. Naturally the higher the metabolic rate, the higher the heat transfer rate from the body to the environment to maintain an energy balance. A general knowledge of the amounts of energy expended is useful to engineering. Table 3-1, which is taken from the *ASHRAE Handbook of Fundamentals*, lists approximate metabolic rates for various types of activities. Notice that a new unit is introduced called the "met." One met is equal to 18.4 Btu/(hr-ft^2) or 58.2 W/m^2 of surface area, and is based on the approximate energy generated by an average sedentary man. This corresponds to approximately 350 Btu/hr or 105 W. A normal healthy man 20 years of age has a maximum energy capacity of approximately 12 mets, and he can maintain about 50 percent of his maximum capacity on a continuous basis. The maximum levels for

Table 3-1 METABOLIC RATE AT DIFFERENT TYPICAL ACTIVITIES*

Activity	Metabolic Rate in Met Units
Reclining	0.8
Seated, quietly	1.0
Sedentary activity (office, dwelling, lab, school)	1.2
Standing, relaxed	1.2
Light activity, standing (shopping, lab, light industry)	1.6
Medium activity, standing (shop assistant, domestic work, machine work)	2.0
High activity (heavy machine work, garage work)	3.0

*Abridged by permission from the *ASHRAE Handbook of Fundamentals*, 1977.

women tend to be about 30 percent lower than those for men. Well-trained athletes may have a maximum energy capacity as high as 20 mets.

Tolerance to high temperature and humidity depends on ability to transfer heat to such an environment. This takes place to a large extent through sweating. There is a limit to the length of time that sweating can continue because this process means that water in the body will be depleted. This dehydration leads to reduced sweating, and a rise in body temperature; the result is a heat stroke. A practical cooling limit is about 110 Btu/(ft²-hr) or 350 W/m², which corresponds to approximately 0.037 lbm/min or 0.28 g/s of sweating for the average man.

Survival in extreme cold is determined by the ability to maintain a thermal energy balance. The problem is to reduce the rate of heat loss from the body, or to generate energy at a rate equal to that at which it is being lost. A person usually maintains this heat balance by wearing adequate clothing. From a practical viewpoint the lower limit for useful outdoor activity while wearing the best clothing is about −30 F or −34 C.

3-2   ENVIRONMENTAL INDICES

There are many parameters to describe the environment in terms of comfort. These indices are generally subdivided into three groups entitled "the direct," "the rationally derived," and "the empirical." Not all of these indices are directly useful by the design engineer in his or her day-to-day work and many of these are defined mainly for research purposes. However, we shall mention all of them to provide some background for the discussion to follow.

Direct Indices

The direct indices are most useful to the design engineer, with *dry bulb temperature* the single most important index of comfort. This is especially true when the relative humidity is between 40 and 60 percent. The dry bulb temperature is especially important for comfort in the colder regions. When humidity is high, the significance of the dry bulb temperature is less.

The *dew point temperature* is a good single measure of the humidity of the environment, and is directly related to the water vapor pressure in saturated air. The usefulness of the dew point temperature in specifying comfort conditions is, however, limited.

The *wet bulb temperature* is useful in describing comfort conditions in the regions of high temperature and high humidity where dry bulb temperature has less significance. For example, the upper limit for tolerance of the average individual with normal clothing is a wet bulb of about 86 F or 30 C when the air movement is in the neighborhood of 50 to 75 ft/min or 0.25 to 0.38 m/s.

Relative humidity, although considered to be a direct index, has no real meaning in terms of comfort unless the accompanying dry bulb temperature is also known. Very high or very low relative humidity is generally associated with discomfort, however.

The last direct index, *air movement,* is the most difficult of the direct indices to describe. The convective heat transfer from the body depends on the velocity of the air moving over it. Evidence shows that one is more comfortable in a warm humid environment if the air movement is high. Yet if the temperature is low, one becomes uncomfortable if the air movement is too high. Generally, when air motion in the vicinity of an individual is in the neighborhood of 50 ft/min or 0.25 m/s, the average person will be comfortable.

Rationally Derived Indices

The rationally derived indices have less direct use in design. Nevertheless they form a basis from which we can draw various conclusions about comfort conditions. The *mean radiant temperature,* for example, is the uniform surface temperature of an imaginary black enclosure with which a person, also assumed to be a black body, exchanges the same heat by radiation as in the actual environment. This parameter is important when surrounding surfaces are at a temperature different from the body.

The *operative temperature* is the uniform temperature of an imaginary enclosure with which an individual exchanges the same heat by radiation and convection as in the actual environment. This index attempts to include the effect of convection as well as radiation. For usual practical applications the operative temperature is the mean of the dry bulb and mean radiant temperatures at a given location in the space and is referred to as the *adjusted dry bulb* temperature (11).

The *humid operative temperature* is the uniform temperature of an environment at 100 percent relative humidity with which a person will exchange the same heat from the skin surface by radiation, convection, conductance through clothing, and evaporation as in the actual environment.

Note that the three temperatures just defined progressively bring in the effect of radiation, convection, and mass transfer in the cooling or heating of the human body, as the case may be. The humid operative temperature is particularly important because it takes into account all three of the heat transfer mechanisms employed by the body to maintain a constant temperature. The humid operative temperature also involves *skin wettedness,* defined below, which is related to sweating and is a physiological rather than an environmental variable.

The *heat stress index* is the ratio of the total evaporative heat loss required for thermal equilibrium of the body to the maximum possible evaporative heat lost to the environment. Closely related to this is the index of *skin wettedness,* which is the ratio of the observed skin sweating to the maximum sweating possible for the environment as defined by the skin temperature, air temperature, humidity, air movement, and clothing. Skin wettedness has been shown to be a more reliable parameter for describing a sense of discomfort or unpleasantness than the usual temperature sensation.

Figure 3-1 shows a comparison of the humid operative temperature, lines of constant wettedness, comfort, and tolerance plotted on a psychrometric chart (4). This plot is an initial step toward obtaining a more directly useful comfort chart. At the lower dry bulb temperatures in the neighborhood of the comfortable region, the

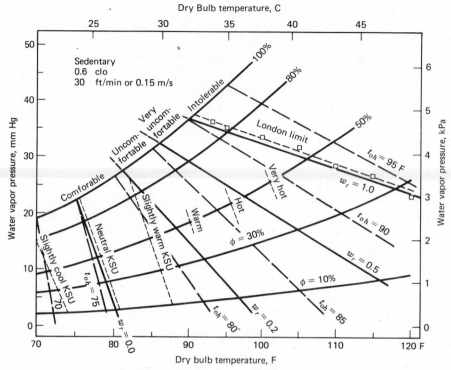

Figure 3-1. Loci of constant humid operative temperature t_{oh} and of constant wettedness w_r, compared with KSU measurement of temperature sensation, Pierce Laboratory observations of warm discomfort, and the London Limit for Heat Tolerance. (Adapted by permission from *ASHRAE Handbook of Fundamentals*, 1977.)

humid operative temperatures and the lines of constant wettedness have essentially the same trends. The divergence of these two indices may be noted in the upper right of the chart. This indicates the increased importance of humidity in the higher temperature regions.

Empirical Indices

The empirical indices are attempts to use one particular parameter to include all the effects in the environment that influence comfort. The first of these and probably the oldest is the *effective temperature, ET*. It attempts to combine the effect of dry bulb, wet bulb, and air movement to yield equal sensations of warmth or cold. In other words, as humidity becomes high, the dry bulb temperature needs to be lower and as humidty becomes lower, dry bulb temperature needs to be higher to give the same sensation of comfort. The original effective temperature concept overemphasizes the effect of humidity in the cooler regions, underemphasizes the effect of humidity under warm conditions, and does not fully account for air motion under hot humid conditions. Finally its use is limited to sedentary conditions.

The *black globe temperature* (5) is the equilibrium temperature of a 6 in. diameter black globe, and is used as a single temperature index describing the combined physical effects of dry bulb temperature, air movement, and the radiant energy received from various surrounding areas.

The *corrected effective temperature* (5) is the black globe temperature used in place of the dry bulb temperature of the original effective temperature scale to include the effects of any intense radiant heat source in the surrounding environment.

The *wet bulb globe temperature* (6) is a weighted average of the dry bulb, a naturally convected wet bulb, and the globe temperature according to the relationship

$$WBGT = 0.7 t_{wb} + 0.2 t_g + 0.1 t_{db} \qquad (3\text{-}1)$$

The index includes the combined effect of low temperature radiant heat, solar radiation, and air movement.

The last empirical index is the *wind chill index* (7). This parameter attempts to include the effects of air motion and dry bulb temperature on comfort.

3-3 COMFORT CONDITIONS

Intensive work in the field of comfort has been conducted at Kansas State University (2, 3, 8) and most of the following is based on that work.

The perception of comfort, temperature, and thermal acceptability is related to one's rate of metabolic heat production, its rate of transfer to the environment, and the resulting physiological adjustments and body temperatures. The heat transfer rate is influenced by the environmental factors of air temperature, thermal radiation, air movement, and humidity, and by the personal factors of activity and clothing. Thermal sensations can be described by feelings of hot, warm, slightly warm, neutral, slightly cool, cool, and cold. Judgment as to whether the environment is thermally acceptable is related to environmental parameters and to thermal sensation. In a uniform thermal environment the 80 percent thermal acceptability limits occur at conditions that produce thermal sensations near slightly cool and slightly warm. Clothing, through its insulation properties, is an important modifier of body heat loss and comfort. Clothing insulation can be described in terms of its clo value [1 clo = 0.88 (ft²-hr-F)/Btu = 0.155 (m²-C)/W]. A heavy two-piece business suit and accessories has an insulation value of about 1 clo, whereas a pair of shorts is about 0.05 clo. The operative or adjusted db temperatures and clo values corresponding to the optimum sensation of neutral, and the 80 percent thermal acceptability limits of Standard 55 (11) are given in Fig. 3-2.

The insulation value of clothing worn indoors is influenced by the season and outside weather conditions. During the summer months, typical clothing in commercial establishments consists of lightweight dresses, lightweight slacks, short-sleeved shirts or blouses, and accessories. These ensembles have insulation values ranging from 0.35 to 0.6 clo. In the winter heating season thicker, heavier clothing is worn.

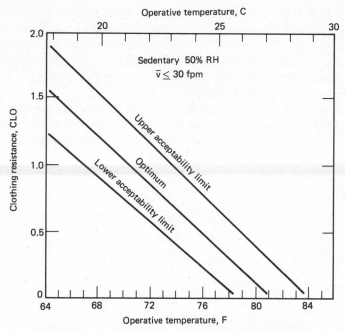

Figure 3-2. Clothing insulation necessary for various levels of comfort at a given temperature. (Adapted by permission from ASHRAE Standard 55-81, 1981.)

A typical winter ensemble may include heavy slacks or skirt, long-sleeved shirt or blouse, warm sweater or jacket, and appropriately warm accessories such as stockings, shoes, and underwear. These ensembles have insulation values ranging from 0.8 to 1.2 clo. At other times, between seasons, the clothing may have an insulation value in the range of 0.6 to 0.8 clo. Because of the seasonal clothing habits of building occupants, the temperature range for comfort in summer is higher than for winter. The acceptable range of operative temperatures and humidities for the winter and summer is defined on the psychrometric chart of Fig. 3-3. The zones overlap in the 73–75 F (23–24 C) range. In this region people in summer dress would tend to be slightly cool, whereas those in winter clothing would be near the slightly warm sensation. Due to individual, clothing, and activity differences, the boundaries of each comfort zone are not actually as sharp as shown in Fig. 3-3.

The air temperature in a room generally increases from floor to ceiling. If this increment is sufficiently large, local discomfort can occur. To prevent this, the vertical air temperature difference between head and ankles should not exceed 5 F (3 C).

To minimize foot discomfort, the surface temperature of the floor for people wearing appropriate indoor footwear should be between about 65 F (18.3 C) and 84 F (29 C).

Comfort conditions for clothing levels different from those described above can be determined approximately by lowering the temperature ranges of Fig. 3-3 by 1 F (0.6 C) for each 0.1 clo of increased clothing.

In general, children below 12 years of age require a reduction in temperature of 1 F or 0.6 C, whereas adults over 60 may require an increase of 1 F or 0.6 C.

Humidity is described in terms of dew point temperature. In the zone occupied by sedentary or near sedentary people the dew point temperature should not be less than 35 F (1.7 C) or greater than 62 F (16.7 C). The thermal effect of humidity on the comfort of sedentary persons is small, Fig. 3-3. The upper and lower humidity limits are based on considerations of comfort, respiratory health, mold growth, and other moisture related phenomena. It should be noted that humidification in winter may need to be limited to prevent condensation on windows and other building surfaces. To minimize respiratory distress in winter, relative humidity should be as high as the structure will allow without condensation, but not more than 50 percent.

Within the thermally acceptable temperature ranges of Fig. 3-3, there is no minimum air movement that is necessary for thermal comfort. The maximum average air movement allowed in the occupied zone is lower in winter than in summer.

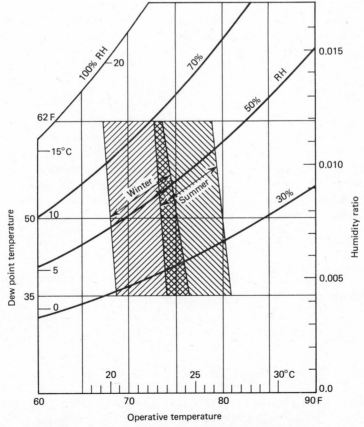

Figure 3-3.   Acceptable ranges of operative temperature and humidity for persons clothed in typical summer and winter clothing, at light, mainly sedentary, activity. (Adapted by permission from ASHRAE Standard 55-81, 1981.)

In winter, the average air movement in the occupied zone should not exceed 30 fpm (0.15 m/s). If the temperature is less than the optimum, the maintenance of low air movement is important to prevent local draft discomfort.

In summer, the average air movement in the occupied zone should not exceed 50 fpm (0.25 m/s). The comfort zone can be extended above 79 F (26 C), however, if the average air movement is increased 30 fpm (0.152 m/s) for each degree F (0.6 C) of increased temperature to a maximum temperature of 82.5 F (28 C).

The mean radiant temperature can be as important as air temperature in effecting heat loss and comfort. For an indoor environment, where air movement is low, the operative temperature is approximately the average of air temperature and mean radiant temperature. When the mean radiant temperature in an occupied zone differs from the air temperature adjustments should be made to keep the operative temperature within the appropriate comfort zone. With the increasing need to conserve energy, adjustments may be mandated in the allowable winter and summer temperatures. The extent to which comfort can be achieved by clothing adjustments is indicated in Fig. 3-2.

It is often necessary for an engineer to inspect a project and determine the acceptability of the comfort conditions by making measurements. ASHRAE Standard 55 describes procedures for making these measurements and analyzing the data.

High Activity

The comfort conditions discussed so far relate to sedentary activity, which accounts for most applications. However, in research laboratories, machine shops, assembly lines, and general manufacturing areas, for example, the occupants may be engaged in rather active work. The comfort zone temperatures of Fig. 3-3 should be decreased when the average steady state activity level of the occupants is higher than sedentary or slightly active (1.2 met). The acceptable operative temperature depends on both the time average activity level and the clothing insulation. The temperature can be calculated with the following equation, which is only appropriate between 1.1 and 3 mets:

$$t_{o_{active}} = t_{o_{sedentary}} - 5.4(1 + clo)(met - 1.1); \text{ deg F} \qquad (3\text{-}2)$$

The minimum allowable operative temperature for these activities is 59 F (15 C). Table 3-1 gives the met levels of some common activities. One might expect people to remove some of their normal clothing when working hard.

3-4   CONTROL OF INDOOR AIR QUALITY

Smell plays an important role in human survival. Through it a person is able to detect danger, food, and other things relating to safety and health. However, even pleasant odors when confined to a given space will become objectionable, when the intensity reaches an excessive level. Until recently the control of air quality came about rather naturally. Most structures had considerable infiltration, and excessive amounts of outdoor air were often used because of poor understanding of the needs for contaminant control.

The indoor air quality question has become increasingly important as a result of several factors. Modern construction has greatly reduced infiltration, and there is very little diffusion of air through the building materials themselves. Also outdoor air requirements have been reduced considerably because the quality of outdoor air is now lower and because excessive outdoor air may add considerably to the cooling and heating load, a very important consideration in these times of energy shortages.

To eliminate or reduce an odor or gas so that it is undetectable, it is necessary either to remove the offending material or to dilute the air until the concentration is too low for perception. The sources of odors are many. Outdoor air as a source of odors has increased in importance because it contains a high percentage of pollutants from automotive exhausts, furnace effluents, and other industrial sources. In recent years problems have arisen from contaminants generated by the building and its contents. Examples include asbestos dust, radon gas, and vapors of formaldehyde and mercury. In industrial spaces odors may arise from printing ink, dyes, synthetics, plastics, and rubber products. In offices, arenas, theaters, and other enclosed spaces, obnoxious odors consist mainly of body odors and tobacco smoke. Tobacco smoke contains many irritants, some of which may even impair visibility. Other sources are paint, upholstery, drapes, rugs, food, food preparation, and pets. The air-conditioning equipment may also contribute odors. Cooling coils collect dirt and lint, both of which are moistened by the condensate; this helps mildew form and leads to consequent objectionable odors.

Humidity has an effect on odor perception, and research has shown that increasing humidity at constant dry bulb temperature has the same effect as lowering the odor level of cigarette smoke as well as that of pure vapors. This effect is more pronounced for some odors than others. Dry bulb temperature has some effect on odor perception. For example, an increase in dry bulb temperature at constant specific humidity seems to lower the odor level of cigarette smoke slightly. It has been shown that if the conditioned space is held at a relative humidity of about 45 to 60 percent, odor perception and irritation will be at a minimum.

Contaminants may be removed by physical or chemical means. The principal methods involve ventilation, air washing or scrubbing, condensation, adsorption, chemical reaction, and odor modification. Other methods are chemical absorption, oxidation, vapor neutralization, and combustion. Ventilation, adsorption, chemical reaction, and odor modification are most effectively used in air-conditioning applications. If the contaminant is a gas or vapor, like ammonia, water sprays are most practical for removal. But, if the gases are insoluble in water, these methods are fairly ineffective. It should also be kept in mind that when various odorants are collected in water, they must ultimately be disposed of and cannot generally be dumped in the environment.

A method closely related to ventilation is isolation or exhaust. This method is generally used in kitchen and restroom areas where air is exhausted directly from the space to outdoors. This requires that makeup air be introduced, which takes on the same function as outdoor air. The various methods of air quality control will now be discussed.

Ventilation

The dominating function of outdoor air is to control air quality, and spaces that are more or less continuously occupied require some outdoor air. Research shows that the required outdoor air is dependent on the rate of contaminant generation and the maximum acceptable contaminant level. Misunderstanding of these factors has led to considerable confusion in the past concerning the quantity of outdoor air required. It can generally be stated that in most cases more outdoor air than necessary is supplied. However, some overzealous attempts to save energy through reduction of outdoor air have caused poor quality indoor air. ASHRAE Standard 62 (10) defines acceptable air quality as ambient air in which there are no known contaminants at harmful concentrations and with which a substantial majority of the people exposed do not express dissatisfaction. Standard 62 defines ventilation requirements and specifies allowable contaminant concentrations for ventilation air. Table 3-2, reprinted from Standard 62, prescribes the basic requirements for acceptable air quality and Table 3-3, also from Standard 62, gives the minimum outdoor air requirements. Standard 62 should be consulted for additional details and definitions. However, some critical definitions will be considered before proceeding. *Ventilation air* is the combination of outdoor air, of acceptable quality, and recirculated air from the conditioned space which after passing through the air-conditioning unit becomes *supply air*. The ventilation air may be 100 percent outdoor air. The term *makeup air* may be used synonymously with *outdoor* air and the terms *return* and *recirculated* air are often used interchangably.

The discussion here pertains to the required ventilation air to maintain indoor air quality. A situation could exist where the supply air required to match the heating

Table 3-2   AMBIENT AIR QUALITY STANDARDS[a,*]

Contaminant	Long Term		Short Term	
	Level	Time	Level	Time
Carbon Monoxide			40 mg/m^3	1 hr
			10 mg/m^3	8 hr
Hydrocarbons			160 μg/m^3	3 hr (6– 9 AM)
Lead	1.5μg/m^3	3 mo.		
Nitrogen Dioxide	100 μg/m^3	Year		
Oxidants (Ozone)			240 μg/m^3	1 hr
Particulates	75 μg/m^3	Year	260 μg/m^3	24 hr
Sulfur Dioxide	80 μg/m^3	Year	365 μg/m^3	24 hr

[a]U.S. Environmental Protection Agency, National Primary and Secondary Ambient Air Quality Standards, Code of Federal Regulations, Title 40 Part 50 (40C.F.R. 50). Pertinent local regulations should also be checked. Some regulations may be more restrictive than those given here, and additional substances may be regulated.
*Reprinted by permission from ASHRAE Standard 62-81, 1981.

Table 3-3 OUTDOOR AIR REQUIREMENTS FOR VENTILATION*

	Estimated occupancy, persons per 1000 ft² or 100 m²/floor area. Use only when design occupancy is not known.	Outdoor Air Requirements				Comments
		Smoking	Nonsmoking	Smoking	Nonsmoking	
COMMERCIAL FACILITIES		*cfm/person*		*(L/s)/person*		
Retail Stores						
Sales floors and showrooms						
Basement and street floors	30	25	5	12.5	2.5	
Upper floors	20	25	5	12.5	2.5	
Storage areas (serving sales and storerooms)	15	25	5	12.5	2.5	
Dressing rooms	—	25	5	12.5	3.5	
Malls and arcades	20	10	5	5	2.5	
Shipping and receiving areas	10	10	5	5	2.5	
Warehouses	5	10	5	5	2.5	
Elevators	—	—	15	—	7.5	
Smoking rooms	70	50	—	25	—	
		cfm/ft² floor		*(L/s)/m² floor*		
Public Spaces						
Corridors and utility rooms		0.02	0.02	0.001	0.001	
		cfm/stall or urinal		*(L/s)/stall or urinal*		
Public restrooms	100	75		37.5		
		cfm/locker		*(L/s)/locker*		
Locker and dressing rooms	50	35	15	17.5	7.5	
		cfm/person		*(L/s)/person*		
Sports and Amusement Facilities						
Ballrooms and discos	100	35	7	17.5	3.5	
Bowling alleys (seating area)	70	35	7	17.5	3.5	
Playing floors (gymnasium)	30	—	20	—	10	
Spectator areas	150	35	7	17.5	3.5	
Game rooms (e.g., card and billiard rooms)	70	35	7	17.5	3.5	

61

Table 3-3 OUTDOOR AIR REQUIREMENTS FOR VENTILATION, continued

	Estimated occupancy, persons per 1000 ft² or 100 m²/floor area. Use only when design occupancy is not known.	Outdoor Air Requirements				Comments
		Smoking	Nonsmoking	Smoking	Nonsmoking	
COMMERCIAL FACILITIES						
Swimming pools		*cfm/ft² area*		*(L/s)/m² area*		
Pool and deck areas	—	—	0.5	—	0.025	
Spectators area	70	*cfm/person* 35	7	*(L/s)/person* 17.5	3.5	
		cfm/room		*(L/s)/room*		
Hotels, Motels, Resorts, Dormitories, and Correctional Facilities						
Bedrooms (single, double)	5	30	15	15	7.5	
Living rooms (suites)	20	20	10	10	5	
Baths, toilets (attached to bedrooms)		50	50	25	25	
		cfm/person		*(L/s)/person*		
Lobbies	30	15	5	7.5	2.5	
Conference rooms (small)	50	35	7	17.5	3.5	
Large assembly rooms, and gambling casinos	120	35	7	17.5	3.5	
Offices						
Office space	7	20	5	10	2.5	
Meeting and waiting spaces	60	35	7	17.5	3.5	
INSTITUTIONAL FACILITIES *(for areas not listed, refer to Commercial Facilities)*						
Hospitals, Nursing, and Convalescent Homes		*cfm/bed*		*(L/s)/bed*		
Patient rooms	10	35	7	17.5	3.5	

		cfm/person		(L/s)/person		
Medical procedure areas	10	35	7	17.5	3.5	Operable windows or mechanical ventilation systems shall be provided for use when occupancy is greater than usual conditions or when unusual contaminant levels are generated within the space.
Operating rooms, delivery rooms	20	—	10	—	10	
Recovery and intensive care rooms	30	—	15	—	7.5	
Autopsy rooms	20	—	60	—	30	
Physical therapy areas	20	—	15	—	7.5	

Educational Facilities		cfm/person		(L/s)/person		
Classrooms	50	25	5	12.5	2.5	
Laboratories	30	—	10	—	5	
Training shops	30	35	7	17.5	3.5	
Music rooms	50	35	7	17.5	3.5	
Libraries	20	—	5	—	2.5	

RESIDENTIAL FACILITIES *(private dwelling places, single or multiple, low or high rise)*

	cfm/room	(L/s)/room	
General living areas, Bedrooms	10	5	Independent of room size.
All other rooms	10	5	
Kitchens	100	50	Independent of room size; installed capacity for intermittent use, may be a window of 2 ft² (0.2 m²) or greater opening area, with no closeable door between it and the kitchen range or bath.
Baths, toilets	50	25	

	cfm/car space	(L/s)/car space	
Garages (separate for each dwelling unit)	10	5	Independent of room size

	cfm/ft² floor	(L/s)/m² floor	
Garages (common for several units)	1.5	0.07	

*Adapted by permission from ASHRAE Standard 62-81, 1981.

or cooling load is greater than the ventilation air. In that case an increased amount of air would be recirculated to meet this condition.

Standard 62 describes two procedures by which indoor air quality may be controlled. The first, known as the *Ventilation Rate Procedure,* achieves indoor air quality indirectly by prescribing the minimum amount of ventilation air and various means to condition that air. The second approach, known as the *Indoor Air Quality Procedure,* specifies maximum permissible concentrations of certain contaminants in indoor air but does not prescribe ventilation rates or air treatment methods. The ventilation rate procedure is the most commonly used approach.

The Ventilation Rate Procedure has five elements:

1. Acceptable outdoor air quality, Table 3-2
2. Outdoor air treatment
3. Minimum ventilation rates, Table 3-3
4. Criteria for reduction of outdoor air when mixed with recirculated air.
5. Criteria for noncontinuous ventilation

Acceptable outdoor air quality is generally established in Table 3-2 although some additional contaminants are also specified in Standard 62. Several steps are outlined in Standard 62 to determine acceptability of local outdoor air.

When the outdoor air contaminant levels exceed the values given in Table 3-2, the air must be treated to control the offending contaminants. Filtering equipment necessary to do this is described in Article 3-5.

Indoor air quality is considered acceptable when the required rates of acceptable outdoor air of Table 3-3 are provided. The table lists the required volume flow rates per person for a variety of indoor spaces with and without smoking.

It is presumed that the contamination produced is in proportion to the number of persons in a space. The table lists estimated density of people for design purposes where such values may not be known. In residential applications, volume flow rate per room, regardless of size, is specified.

Higher ventilation rates are specified for spaces where smoking is permitted because tobacco smoke is one of the most difficult contaminants to control.

A minimum supply of outdoor air is necessary to dilute the carbon dioxide produced by metabolism and expired from the lungs. This value, 5 cfm or 2.5 L/s per person, allows an adequate factor of safety to account for health variations and some increased activity levels. Therefore, outdoor air requirements should never be less than 5 cfm or 2.5 L/s per person regardless of the treatment of the recirculated air. At this time no practical equipment exists to remove carbon dioxide.

In some cases exhaust air from one space can be used as supply air to another space where different contaminants are generated (corridors and office spaces exhausted through toilet rooms, or dining areas exhausted through kitchens); this air is considered equivalent to acceptable outdoor air.

The number of persons cannot always be estimated accurately or may vary considerably. In other cases, a space may require ventilation to remove contamination generated within the space, but unrelated to human occupancy (e.g., radon or formaldehyde outgassing from building materials or furniture). For these cases, Table 3-3 lists quantities in cfm/ft^2 or L/(s-m^2).

When spaces are unoccupied, ventilation is not generally required. Ventilation may be required, however, to prevent accumulation of contaminants injurious to people, contents, or structure. The requirements for ventilation quantities given in Table 3-3 are for 100 percent outdoor air when the outdoor air meets the specifications for air quality given in Table 3-2. Although these quantities are for 100 percent outdoor air, they also set the amount of air required to dilute contaminants to acceptable levels. Therefore, it is necessary that at least this amount of air be circulated at all times that the building is in use. If proper air filters and adequate temperature control are provided, part of this air may be recirculated, but the outdoor air portion must never be less than 5 cfm or 2.5 L/s per person. The filtering system for the recirculated air may be located in the recirculated air stream or in the plenum, which mixes the outdoor air and return air.

The filtering system used to reclaim recirculated air should be designed to remove both particulate and gaseous contaminants. The required air cleaner efficiency and recirculation rate for the system should be sufficient to provide an indoor air quality equivalent to that obtained using outdoor air as specified in Table 3-2 and at a rate specified in Table 3-3. The filter efficiency and recirculation rate will be discussed in Section 3-5, "The Cleaning of Air."

Ventilating systems for spaces not continuously occupied may be shut off under certain conditions. Start-up of the system should lag or lead occupancy depending on the source of contaminants. When contaminants are associated only with occupants and are dissipated by natural means during unoccupied periods, start-up may lag occupancy. When contaminants generated in the space are independent of occupants or their activities do not present health hazards during short-term exposure, start-up should lead occupancy so that acceptable conditions will exist at the start of occupancy. In some situations local codes may override the limits given above. For energy conservation, ASHRAE Standard 90, Energy Conservation in New Building Design (9), states that minimum amounts of outdoor air shall be used, and the outdoor air portion of the ventilation air should be reduced as discussed above. However, the standard also specifies that allowance for 100 percent outside air shall be provided for cooling purposes when conditions are appropriate. This is called an economizer cycle because it saves energy at off-design conditions, see Article 8-10.

Washing

The air washer is used to control temperature and humidity; it is also used to remove particulate matter. As these functions are carried out, removal of odorous vapors or particles is also accomplished.

Generally air washers work in three ways. Probably the most important method is absorption. When absorption occurs, the concentration of a vapor in the air is high in relation to the concentration of that chemical in the liquid, and the contaminant is absorbed in the liquid. The second method is by condensation of the vapor, much the same as water vapor is condensed from air. The third method referred to above is removal of particles that contain odorous materials from the air by a direct washing action. Although water, with or without additives, is a common liquid used in scrubbers and washers, other liquids may be used. Obviously the liquids used in an air washer or scrubber must be maintained in an odorfree condition or odors will be transferred to the air. Therefore, new water must be continuously added or the liquid must be regenerated. In air-conditioning applications it is generally required that large quantities of air be moved through the washer or scrubber without an excessive drop in pressure.

Adsorption

Adsorption is the adhesion of molecules to the surface of a solid called the adsorbent; this contrasts to absorption, which is basically the dissolving of molecules into a substance. Because the process of adsorption is greatly dependent on the available surface area, the materials used must contain large surface areas usually taking the form of minute pores in a solid substance. There are wide varieties of adsorbents available, including charcoal, zeolite, silica gel, activated alumina, mica, and others. Activated charcoal can adsorb a variety of gases, vapors, and a broader spectrum of organic substances than most sorbents; and is therefore the most nearly universal adsorbent in general use. The basic raw material from which the activated charcoal is manufactured has little significance to the user; however, the basic material may have some influence on the density, pore structure, and hardness. Charcoal has its greatest selectivity toward molecules of high molecular weight and is less effective for lighter gases such as ammonia or ethylene. Chemically active gases may be better accommodated if the charcoal is impregnated with some other substance.

Because there must be intimate contact between the gaseous odorants in the air and the adsorbent, the air to be cleaned is usually passed through a bed of the adsorbent material. These beds may range from one-half to four inches in depth. The charcoal or other adsorbent has a limited life and must be removed or replaced as its capacity decreases. Further information relating to the use of adsorbents may be found in the Systems volume of the *ASHRAE Handbook* series.

Chemical Reactions

A great many chemical reactions are possible to neutralize airborne odorants or other contaminants. In general these may be categorized as specific reagents for specific contaminants, or broad spectrum reagents for a broad range of contaminants. These reactions are usually accomplished in a manner similar to the air washer except that solutions of reagent chemicals are used instead of water. It is also possible to use solid reagents or activated solids containing reagents in the same way in which the activated charcoal was described above. Another method is to inject a reagent gas into the contaminated air. This is usually done only when air is being exhausted because the concentration of reagents can also cause a problem. For additional information in the use of this method, the readers should refer to the Systems volume of the *ASHRAE Handbook* series.

Combustion

Two types of combustion methods are used for odor control. The first method utilizes high temperature oxidation in a combustion chamber fired directly with a flame; the second employs oxidation at lower but still elevated temperatures on catalytic surfaces. Combustion methods have merit when the recovered odor would not have significant economic value, the concentration of odorant is sufficient to contribute appreciable heat of combustion, the effluent is already well above ambient temperature, and the wasted heat can be used economically. The combustion methods of controlling odors in air-conditioning and heating applications are not very numerous, and again the reader is referred to the Systems volume of the *ASHRAE Handbook* series.

Odor Masking and Counteraction

Odor masking is the technique of introducing a pleasant odor to cover an unpleasant odor. Obviously this can be a hazardous undertaking because what might be pleasant to one person may be unpleasant to another. However, many odorants for this purpose are available along with equipment for introducing them into the conditioned air. In some cases there is no alternative but to use this type of control.

Odor counteraction is the art of mixing two odorous vapors together such that both odors tend to be diminished. This type of control tends to be inexact, and it is not an area for amateur experimentation. Where other methods fail or cannot be used, equipment can be obtained to use this type of control. Odor masking and counteraction should be considered as methods of last resort.

3-5   THE CLEANING OF AIR

Atmospheric dust is a complex mixture of smoke, mists, fumes, dry granular particules, and fibers suspended in air. In addition air may contain living organisms such as mold spores, bacteria, and plant pollens that may cause diseases or allergic responses. Particles in the atmosphere range in size from less than 10^{-6} m up to the order of magnitude of the dimension of leaves and insects. The particles exist in all sizes and shapes and generally it is impossible to design one air cleaner that is best for all applications.

Different degrees of air cleanliness are required for various applications. The minimum standards for air quality are defined by ASHRAE Standard 62 as discussed in Section 3-4. Another consideration is the fact that dust particles in the air may damage the conditioned spaces or its furnishings. for example, the walls may be discolored if even the smallest components of the atmospheric dust are not removed. Unfortunately the smallest particles cause the most discoloring. Therefore, in this case, electronic air cleaners or high performance dry filters are required. In cleanroom applications or when radioactive particles are present, extremely high performance filters should be employed. One of the most effective ways of controlling allergic disorders is to remove the allergens from the air by using high performance filters. Of all the characteristics affecting the performance of an air filter, particle size is the most important; however, the cleaning efficiency of a filter is affected to some extent by the velocity of the air stream. The major factor influencing filter design and selection is the degree of air cleanliness required. In general the cost of the filter or filter system will increase as the size of the particles to be removed decreases. Maintenance will also increase in the same way.

There are three operating characteristics that distinguish the various types of air cleaners. These are the *efficiency,* the *air flow resistance,* and the *dust-holding capacity.* Efficiency measures the ability of the air cleaner to remove particulate matter from an air stream and the average efficiency over the life of the filter is the most meaningful for most applications. The air flow resistance is the loss in total pressure across the filter at a given air flow rate. This of course is a very important consideration. The dust-holding capacity defines the amount of a particular type of dust an air cleaner can hold while operating at a specific air flow rate and maximum air flow resistance. Sometimes the dust-holding capacity refers to the value attained before the filter efficiency is seriously reduced. Therefore, to have a complete rating of an air cleaner, data should be made available on efficiency, resistance, and dust-holding capacity. The effect of the dust loading on the efficiency and resistance is also important. Details of the standardized methods by which air cleaners are tested and rated are given in ASHRAE Standard 52 (12). In general, however, the gravimetric efficiency of a filter gives the ratio of the mass of the dust arrested by the filter to the total mass of the dust fed to the filter. The second type of efficiency is called the dust-spot efficiency. This is designed to rate filters for their ability to prevent discoloration. In this case white filter paper targets are used and the efficiency is determined through light transmission measurements and expressed in terms of efficiency. The dust-holding capacity is actually given as an absolute measure of the mass of dust held by the filter. This rating gives an indication of how often the filter may have to be renewed. Of course this will vary considerably depending on the application. Air flow resistance or pressure drop is probably the simplest of all the tests made and is usually given in terms of the clean filter. Naturally, as the filter becomes loaded, pressure drop will increase. The pressure drop or resistance depends on the square of the velocity; therefore, if a filter is rated for a certain velocity, it is a simple matter to convert to some other velocity. The efficiency of a filter depends on the particle sizes, and manufacturers make different kinds of filter media for different applications. Figure 3-4 is an example of gravimetric efficiency as a function of particle size. Here it can be seen that with the use of various filter media the the efficiency is different. With the larger particle sizes all of the media are fairly efficient, but when the particle sizes become very small, efficiency varies considerably with the media selected.

Mechanisms of Particle Collection

As air passes through a filter the airstream is made to turn. The particles, being more dense than the air, tend to keep going in a straight line. When the particles collide with the fibers they stick or are embedded in the fiber. This mechanism is

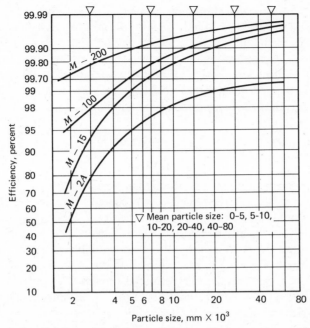

Figure 3-4. Gravimetric efficiency of high performance dry media filters.

generally called viscous impingement. Interception is a special case where the particles collide head on with the fibers and the effects of inertia are unimportant. This method is independent of velocity. Another important mechanism is referred to as "straining." Here the fibers are so close together that a particle trying to go between them is arrested. This mechanism usually occurs in the collection of large particles and lint. Diffusion can be an important effect in the case of very fine particles. The very small particles are bombarded by the random motion of the air molecules and are driven into the filter fibers where they are arrested. Turbulence in the airstream helps in this regard. Electrostatic effects are used in one type of filter. Charged particles are forced toward a plate or grid having an opposite charge. This mechanism is employed by the electronic air cleaner and is usually most effective with very small particles.

There are three broad categories of air cleaners. The first, the *fibrous media unit filter,* is characterized by a gradual increase in accumulated dust load that increases pressure drop up to some maximum permissible value. During this time the efficiency also increases. If allowed to load too heavily with dust, however, the efficiency will again fall off.

The second category, the *renewable media filter,* uses a media usually of a fibrous type. In this case new media is continually introduced into the airstream, and loaded media is removed so that efficiency, resistance, and loading remain essentially constant.

The third classification is referred to as *electronic air cleaners.* These units have essentially constant pressure drop and efficiency unless their elements become severely dustloaded. Various combinations of the above types may be used. For example, a fibrous media filter may be placed both upstream and downstream of an electronic filter, the upstream unit to remove large particles, and the downstream mechanical unit to catch any particles that may blow off the plates of the electronic precipitator. Another attractive combination is to use a lower efficiency filter upstream of a very high efficiency filter to extend the life of the more expensive high efficiency filter.

Types of Filter Media

Viscous impingement filters are usually flat panels made up of coarse fibers and have a high porosity. Usually the filter media is coated with a viscous substance that acts as an adhesive for particles impinging on the fibers. Design air velocity through the media is usually in the range of 250 to 700 ft/min or 1.3 to 3.6 m/s. These filters are characterized by low pressure drop, low cost, and high efficiency on lint, but low efficiency on normal atmospheric dust. The most common media are glass fibers, metallic wools, expanded metals, foils, crimped screens, and random matted wire. Table 3-4 gives some typical performance data for viscous impingement filters. This type of filter varies so much in material, quality, and construction that precise performance data cannot be given.

Another type is called the dry media air filter. The absence of an adhesive on the filter medium does not necessarily make it a dry-type air filter. The most distinguishing characteristic of this type of filter is the smaller and much closer spaced fibers that form a dense mat or pad. The efficiency of dry-type air filters is usually much higher than viscous impingement filters, and there is usually a much larger variety of media available. An accordionlike arrangement of the media is usually employed. Table 3-5 contains some typical performance data for dry media filters. Table 3-6 gives engineering data related to the efficiency data of Fig. 3-4. The equipment volume of the *ASHRAE Handbook* series may be consulted for additional information concerning various types of media and application.

Electronic Air Cleaners

The term *electronic air cleaner* usually refers to the types of cleaners used in air-conditioning applications. However, by using much higher potentials, the same general-type apparatus is used for cleaning of stack gases. Electronic air cleaners are of two general types. The first of these is the "ionizing plate electronic air cleaner" shown schematically in Fig. 3-5. The positive ions, generated at the high potential ionizer wire, flow across the airstream, striking and adhering to any dust particles carried by the airstream. These particles then pass into the system of charged and grounded plates where they are driven to the plates by the force exerted by the electric field on the charges they carry. As the dust particles reach the plates, they are removed from the airstream. A typical electronic air cleaner will use a dc potential of 12,000 volts for the ionizing field and will have about 6000 volts maintained

Table 3-4 PERFORMANCE OF UNIT VISCOUS-IMPINGEMENT FILTERS*

Filter Thickness, Inches	ASHRAE Weight Arrestance, Percent	ASHRAE Atmospheric Dust-Spot Efficiency, Percent	ASHRAE Dust-Holding Capacity Pounds per 1000 cfm Cell	Pressure Loss, Inches of Water at 350 fpm
1	20–50	5–10	0.16–0.32	0.08
$1\frac{3}{4}$	50–75	5–15	0.26–0.8	0.14
$2\frac{1}{2}$	60–80	5–20	0.4–1.2	0.20
4	70–85	10–25	0.53–1.7	0.32

*Adapted by permission from *ASHRAE Handbook and Product Directory—Equipment*, 1979.

Table 3-5 PERFORMANCE OF DRY MEDIA FILTERS

Filter Media Type	ASHRAE Weight Arrestance, Percent	ASHRAE Atmospheric Dust-Spot Efficiency, Percent	ASHRAE Dust-Holding Capacity, kg per m³/s Cell
Finer open cell foams and textile denier nonwovens	70–80	15–30	0.380–0.900
Thin, paperlike mats of glass fibers, cellulose	80–90	20–35	0.190–0.380
Mats of glass fiber multiply cellulose, wool felt	85–90	25–40	0.190–0.380
Mats of 5 to 10 micron fibers of 6 to 12 mm thickness	90–95	40–60	0.570–1.150
Mats of 3 to 5 micron fibers, 6 to 12 mm thickness	>95	60–80	0.380–0.950
Mats of 1 to 4 micron fibers, mixture of various fibers and asbestos	>95	80–90	0.380–0.760
Mats of $\frac{1}{2}$ to 2 micron fibers (usually glass fibers)	NA	90–98	0.190–0.570
Wet laid papers of mostly submicron glass and asbestos fibers (HEPA filters)	NA	NA	1.060–2.120
Membrane filters (membranes of cellulose acetate, nylon, etc. having micron size holes)	NA	NA	NA

NA or not applicable means that test method cannot be applied to this level of filter
Adapted by permission from *ASHRAE Handbook and Product Directory—Equipment*, 1979.

Table 3-6 ENGINEERING DATA—HIGH PERFORMANCE DRY MEDIA FILTERS (CORRESPONDS TO EFFICIENCY DATA OF FIGURE 3-4)

Standard Size	Meter Inch	0.3 × 0.6 × 0.2 12 × 24 × 8		0.3 × 0.6 × 0.3 12 × 24 × 12		0.6 × 0.6 × 0.2 24 × 24 × 8		0.6 × 0.6 × 0.3 24 × 24 × 12		Pressure Loss	
Rated Capacity[a]		ft³/min	m³/s	ft³/min	m³/s	ft³/min	m³/s	ft³/min	m³/s	Inches of Water	Pa
Media Type	M-2A[b]	900	0.42	1025	0.48	1725	0.81	2000	0.94	0.15	37.4
	M-15	900	0.42	1025	0.48	1725	0.81	2000	0.94	0.35	87.2
	M-100	650	0.30	875	0.41	1325	0.62	1700	0.80	0.40	100.0
	M-200	450	0.21	630	0.29	920	0.43	1200	0.56	0.40	100.0
Effective filtering area All media types		ft² 14.5	m² 1.35	ft² 20.8	m² 1.93	ft² 29.0	m² 2.69	ft² 41.7	m² 3.87		

[a] Filters may be operated from 50 to 120 percent of the rated capacities with corresponding changes in pressure drop.
[b] The M-2A is available in 2 in. thickness and standard sizes with a nominal rating of 0.28 in. water at 500 fpm face velocity

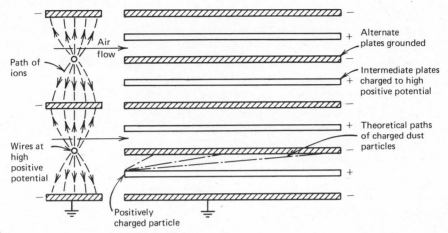

Figure 3-5.   Diagrammatic cross-section of ionizing-type electronic air cleaner. (Reprinted by permission from *ASHRAE Handbook and Product Directory-Equipment*, 1979.)

between the plates. This type of air cleaner is an efficient, low pressure drop device for removing fine dust and smoke particles. Sometimes the collector plates are coated with a special adhesive to improve the dust-holding capacity. Cleaning is generally accomplished by washing the plates in hot water with or without a detergent. Some units are supplied with servicing equipment from the factory.

The second type of electronic air cleaner is called the "charged media nonionizing electronic air cleaner." This type combines certain characteristics of both the dry filter and electronic air cleaner. It consists of a dielectric filtering medium usually arranged in pleats as in a typical dry filter. The filtering medium is in contact with a grid consisting of alternately grounded and charged members, usually at a potential of about 12,000 volts. An intense but nonuniform electrostatic field is thus created through the dielectric media. Airborne particles approaching the field are polarized and drawn to the fibers of the medium. In this type of filter the medium is replaced as with other media filters. It has a higher pressure drop than the ionized plate type of filter. The charged medium is ineffective when the relative humidity becomes greater than about 70 percent.

A drawback to the electronic-type air cleaner is that charged particles can sometimes be carried through the cleaner into the space. If this happens on a large scale, a considerable buildup of space charge will result, which drives the charged dust particles to the wall and causes blackening of the surface.

Selection and Maintenance

In selecting a filter system for a particular application, three factors should be carefully weighed: one, the degree of air cleanliness required; two, disposal of the dirt after it is removed from the air; and three, the amount and type of dust in the air to be filtered. These factors determine the initial cost, the operating cost, and the extent of maintenance required. Viscous impingement filters do not have efficiencies as high as can be expected from dry types of filters but their initial cost and upkeep are generally low. This unit type of filter requires more attention than the moving type of filter, which replenishes itself. However, the moving-type filter has a higher initial cost. Developments have improved the efficiencies of dry-type filters so that they approximate the efficiency of electronic types and cost less. However, they are more expensive than the viscous-impingement type and pressure loss is the greatest for this type of filter. Electronic air cleaners have a high initial cost. But they have higher efficiencies, especially when the particles are very small. They also have very low pressure drop.

Filtering and Recirculation

It was mentioned in Section 3-4 that the amount of outdoor air can be reduced and mixed with recirculated air when proper filtering is done. Under normal conditions when the concentrations of contaminants in the outdoor air are much less than those specified in Table 3-2 and the concentrations indoors are greater than those outdoors, the recirculation rate is given by

$$\dot{Q}_r = (\dot{Q}_o - \dot{Q}_m)/\eta \tag{3-3}$$

where

$\dot{Q}_o$ = Outdoor air rate from Table 3-3, ft³/min-person or L/s-person
$\dot{Q}_m$ = Minimum outdoor air rate to provide acceptable indoor air quality
        but not less than 5 ft³/min-person or 2.5 L/s-person
η = Efficiency of filter or gas removal device

the total air supply rate then becomes

$$\dot{Q}_s = \dot{Q}_r + \dot{Q}_m \tag{3-4}$$

The following examples demonstrate the calculation procedure.

Example 3-1.   Consider an office space where smoking is permitted. A ventilation rate of 35 cfm per person would be required. However, only 10 cfm per person would be required if smoking were not allowed. Assume the M-15 filter media of Fig. 3-4 is 90 percent efficient in removing tobacco smoke and compute the recirculation and total air supply rates if the outdoor air rate is reduced to 10 cfm per person. The outdoor air concentration of total suspended particulates is negligible.

Solution.   Equations (3-3) and (3-4) apply to this problem where $\dot{Q}_o$ is 35 cfm per person, $\dot{Q}_m$ is 10 cfm per person, and η is 90 percent. Using Eq. (3-3):

$$\dot{Q}_r = (35 - 10)/0.9 = 28 \text{ cfm/person}$$

then from Eq. (3-4):

$$\dot{Q}_s = 28 + 10 = 38 \text{ cfm/person}$$

Note that the total supply rate has been increased from 35 to 38 cfm/person; however, the outdoor air rate is only 10 cfm/person.

Example 3-2.   According to Table 3-3, dormitory bedrooms require 15 L/s-room when smoking is allowed. To save energy it is desirable to reduce the outdoor air rate to a minimum of 5 L/s-room. If the outdoor air has negligible contaminants, compute the recirculation and total air supply rates, assuming the M-2A media of Fig. 3-4 is used and the efficiency is for the 0.5×10^{-3} mm particle size range.

Solution.   The required air flow rates may be computed using Equations (3-3) and (3-4). The filter efficiency is 79 percent as read from Fig. 3-4 and the flow rates $\dot{Q}_o$ and $\dot{Q}_m$ are 15 and 5 L/s-room, respectively. Then from Equation (3-3)

$$\dot{Q}_r = (15 - 5)/0.79 = 25 \text{ L/s-room}$$

and

$$\dot{Q}_s = 25 + 5 = 30 \text{ L/s-room}$$

This approach would result in significant energy savings because the outdoor air rate has been reduced by 67 percent.

Filter System Design

After the filter type and media have been selected as discussed above, the filter system must be sized to accommodate the required air flow rate and flow resistance. Filters may be placed at different locations in the air-conditioning system. For example, outdoor air and return air may be filtered separately and then mixed, or they may be mixed and then filtered. The air is usually filtered before entering the air-conditioning equipment rather than afterward. The following examples are typical.

Example 3-3.   Suppose that the office space of Example 3-1 involves 95 people and the outdoor air and recirculated air are mixed before filtering. Design a filter system which has a pressure loss of not more than 0.30 in. water in the clean condition.

Solution.   Table 3-6 gives application data for the M-15 filter media cited in Example 3-1. Several sizes are shown, none of which match the requirements of this problem. Therefore, two or more modules must be combined and the capacity adjusted to obtain the required pressure loss performance. Let us try the 12 × 24 × 12 size, which is rated at 1025 cfm with 0.35 in. water pressure loss. To a close approximation the pressure loss is proportional to the square of the air velocity or flow rate. Then the capacity with a pressure loss of 0.3 in. water is

$$Q = Q_c \left[\frac{\Delta P}{\Delta P_c} \right]^{1/2} = 1025(0.3/0.35)^{1/2} = 949 \text{ cfm}$$

where subscript c refers to the rated conditions. The total air flow rate is

$$Q = Q_s \times 95 = 38 \times 95 = 3610 \text{ cfm}$$

and the number of filter modules is

$$N = 3610/949 = 3.8$$

Because N must be an integer, four modules must be used. A somewhat lower pressure loss will result.

Example 3-4. An air handler has been selected which is designed for 5400 cfm with provisions for six 16 in. $\times$ 20 in. $\times$ 2 in. M-2A filters (see Table 3-6). Assuming that the unobstructed face area of the filters is 90 percent of the actual area, estimate the pressure loss for the clean filters.

Solution. Footnote b of Table 3-6 indicates that these filters have a pressure loss of 0.28 in. water when the face velocity is 500 ft/min. The pressure loss is proportional to the face velocity squared. The face velocity is given by

$$\overline{V}_f = Q/A_f = 5400 \times 144/(6 \times 16 \times 20 \times 0.9) = 450 \text{ ft/min.}$$

Then

$$\Delta P = 0.28(450/500)^2 = 0.23 \text{ in. water}$$

REFERENCES

1. *ASHRAE Handbook of Fundamentals,* American Society of Heating, Refrigerating and Air-Conditioning Engineers, New York, 1977.

2. R. G. Nevins, "Psychrometrics and Modern Comfort," presented at the joint ASHRAE-ASME meeting, 1961.

3. Rohles, F. H., J. E. Woods, and R. G. Nevins, "The Effect of Air Movement and Temperature on the Thermal Sensations of Sedentary Man," ASHRAE Trans., *80*, Part I, 1974.

4. R. G. Nevins et al., "Temperature-Humidity Chart for Thermal Comfort of Seated Persons," *ASHRAE Trans., 72*, Part I, 1966.

5. T. Bedford and C. G. Warner, "The Globe Thermometer in Studies of Heating and Ventilating," *J. Hygiene, 35,* 1935.

6. C. P. Yalou and D. Minard, "Control of Heat Casualties at Military Training Centers, AMA Archives of Industrial Health, *16,* 1957.

7. P. A. Siple and C. F. Passel, "Measurements of Dry Atmospheric Cooling in Subfreezing Temperatures," *Proc. American Philosophical Society, 89,* 1945.

8. F. H. Rohles and R. G. Nevins, "The Nature of Thermal Comfort for Sedentary Man," *ASHRAE Trans., 77,* Part I, 1971.

9. ASHRAE Standard 90A-80, "Energy Conservation in New Building Design," American Society of Heating, Refrigerating and Air-Conditioning Engineers. New York, 1979.

10. ASHRAE Standard 62-81, "Standards for Ventilation Required for Minimum Acceptable Indoor Air Quality," American Society of Heating, Refrigerating and Air-Conditioning Engineers, New York, 1981.

11. ASHRAE Standard 55-81, "Thermal Environmental Condition for Human Occupancy," American Society of Heating, Refrigerating and Air-Conditioning Engineers, New York, 1981.

12. ASHRAE Standard 52-76, "Air Cleaning Devices Used in General Ventilation for Removing Particulate Matter, Methods of Testing," American Society of Heating, Refrigerating and Air-Conditioning Engineers, New York, 1976.

13. A. P. Gagge, J. A. J. Stolwijk, and Y. Nishi, "An Effective Temperature Scale Based on a Simple Model of Human Physiological Regulatory Response," *ASHRAE Trans., 77,* Part I, 1971.

14. Gagge, A. P., G. M. Rapp, and J. D. Hardy, "The Effective Radiant Field and Operative Temperature Necessary for Comfort," ASHRAE Trans., *73,* Part I, 1967.

15. *ASHRAE Handbook and Product Directory-Equipment,* American Society of Heating, Refrigerating and Air Conditioning Engineers, New York, 1979.

PROBLEMS

3-1 A mixed group of men and women occupy a space maintained at 76 F db and 62 F wb. All are lightly clothed, with sedentary activity. (a) By using Fig. 3-3, draw a conclusion about the general comfort of the group. The *MRT* is 78 F. (b) Suppose the group is moving around (light activity) instead of sitting. What is your conclusion about their comfort? (c) Suppose the group is composed of retired people, 65 years of age or more, playing cards. How would you change the room conditions to insure their comfort?

3-2 An air-conditioning system is to be designed for a combination shop and laboratory facility. The space will be occupied by men engaged in active work such as hammering, sawing, walking, climbing, and so on. The men will be dressed in short sleeves. (a) Select comfort conditions for the space, assuming the occupants are to be comfortable. (b) It is expected that the toolroom keeper, possibly a woman, will work in a fenced-in area in the space with light activity. What is your recommendation about how he or she should dress to be comfortable?

3-3 A class room in a school is designed for 100 people. (a) What is the minimum amount of clean outdoor air required? (b) The floor area is 1500 ft². What is the outdoor air ventilation requirement on the basis of Table 3-3?

3-4 A specification calls for viscous impingement filters with ASHRAE weight arrestance efficiency of 75 percent. Pressure loss must not exceed 0.10 in. of water. Select a filter (thickness and face area) to handle 1000 cfm using Table 3-4.

3-5 Select a filter system that will have a gravimetric efficiency of at least 95 percent in the particle size range of 0 to 5×10^{-3} mm. The system must handle 2200 cfm and the maximum pressure drop across the filters must be less than or equal to 0.25 in. of water. Space limits the depth of the filters to 8 in.

3-6 A laboratory involving animal research must use either 100 percent outdoor air or 25 percent outdoor air with high performance filters for the return air. Gravimetric efficiency must be at least 99 percent in the size 0 to 5×10^{-6} m. The computed heat gain of the laboratory is 3 tons with a sensible heat factor (SHF) of 0.7. The cooling and dehumidifying unit requires a fixed air flow rate of 350 cfm per ton. Inside conditions are 78 F db and 40 percent relative humidity, whereas outdoor conditions are 95 F db and 50 percent relative humidity. (a) Investigate the feasibility, and find the required amount of air and the size of the cooling unit when 100 percent outdoor air is used. (b) Find the required amount of air and the size of the cooling unit when 25 percent outdoor air is used and the remainder is recirculated through the high performance filters. (c) Design the filter system using the data of Fig. 3-4 and Table 3-6 so that the maximum pressure loss is 0.125 in. of water with clean filters.

3-7 Suppose that a space is used for both athletic events and classroom space at different times. The relative humidity can be maintained at 40 percent. At what temperature should the thermostat be set when the space is used for: (a) Classroom? (b) Basketball practice?

3-8 A space used for sedentary activity by occupants in light clothing has a dry bulb temperature of 25 C and the air motion is about 0.1 m/s. The mean radiant temperature is about 29 C. (a) Can the occupants be expected to be comfortable? Why? (b) If not, what changes to the room air can be made to improve the comfort?

3-9 For sedentary activity, light clothing, and air temperature equal to mean radiant temperature, what is the approximate relationship between air motion (m/s) and the air temperature to maintain comfort? Consider a velocity range of 0.2 to 0.8 m/s.

3-10 An M-100, $24 \times 24 \times 12$ filter described in Table 3-6 is suggested for use with 1200 cfm of air. What pressure loss can be expected?

3-11 A filter system to handle 5600 cfm of air is to be designed using the M-2A media (Table 3-6) and the $12 \times 24 \times 8$ modules. The lost pressure must be 0.10 in. of water or less when the filters are clean. How many modules are required?

3-12 A space is to be maintained at 24 C db and 18 C wb by cooling and dehumidifying 100 percent outside air at 35 C db and 28 C wb and supplying it to the space. The space has a total cooling load of 17.6 kW and sensible heat factor of 0.6. Specify the cooling unit capacity and the air inlet condition. Assume that the cooling coil must have at least 0.04 m³/s per kW of capacity flowing through it. Explain.

3-13   The M-200, 0.6 × 0.6 × 0.2 filters of Table 3-6 are to be used with a volume flow rate of 0.38 m³/s. What minimum pressure loss can be expected?

3-14   Design a filter system to handle 20 m³/s of air using the M-15, 0.6 × 0.6 × 0.3 filters of Table 3-6. The pressure loss in the clean condition must be 70 Pa or less.

3-15   To save energy, an office building cooling system is controlled so that the minimum dry bulb temperature in the space is 78 F with a dew point temperature of 63 F. (a) Comment on the comfort expectations for the space. (b) What general type of clothing would you recommend? (c) Assuming the dry bulb temperature is fixed, what could be done to improve comfort conditions?

3-16   During an emergency energy shortage thermostats were ordered set at 65 F to conserve heating fuel. Using Figures 3-2 and 3-3 discuss the comfort expectations of the following: (a) an executive dressed in a vested business suit, (b) a secretary sitting and typing, (c) a factory worker, (d) a customer in a grocery store, and (e) a patient in doctor's examining or waiting room.

3-17   Air motion is a parameter that may be adjusted to improve comfort when space temperatures are fixed. Discuss the general effect of increased or decreased air motion when the space temperature is low in winter and high in the summer.

3-18   It is necessary to design a relatively high efficiency filter system for a residence to handle 1200 cfm of air with a maximum pressure loss of 0.085 in. water when the filters are clean. (a) Consider the M-2A media of Table 3-6 in the 2 in. thickness and determine the face area required. (b) The physical layout of the equipment would permit the use of either 12 in. × 24 in. × 2 in. or 16 in. × 20 in. × 2 in. sizes. Select the size and number of modules to best match the requirements.

3-19   A method of saving energy in large, chilled water cooling systems is to increase the temperature of the water circulating in the system. (a) How will this affect the space comfort conditions? (b) Would there be certain periods of time when this would be feasible? Discuss.

3-20   The fan power required to circulate air to and from the conditioned space is often minimized by reducing air flow to conserve energy. (a) Is this consistent with high space temperatures in the summer months? Explain. (b) Would the same be true for low space temperatures in the winter? Explain.

3-21   The productivity of people is related to their comfort. Give an example where an overzealous attempt to save energy by raising or lowering the space temperature might result in an increase in overall energy use and greater overall cost.

3-22   A physical therapy area is designed for 75 people. Design a filter and air circulation system where the outdoor air rate can be reduced to 5 cfm/person. Outdoor and recirculated air should be mixed before filtering. Outdoor air contaminants are negligible. The filter media must have a gravimetric efficiency of about 80 percent in the 0–5 × 10⁻³ mm particle range and the pressure loss must not exceed 0.10 in. water.

3-23   Consider a game room where smoking is permitted. Suppose a filter system is available with an efficiency of 80 percent in removing tobacco smoke. Compute the air supply rates assuming the outdoor air rate is set at the nonsmoking level and contaminants in the outdoor air are negligible.

3-24 Suppose the game room of problem 3-23 is designed for 40 people. The cited filter media is rated at 0.2 in. water pressure loss with a face velocity of 600 ft/min. What net filter face area is required to obtain a pressure loss of 0.15 in. water when clean?

3-25 Consider an office building where smoking is prohibited. (a) Discuss the possibility of reducing the outdoor air rate to save energy. (b) Would there be any need for a return air system? Explain.

3-26 It is desirable to reduce the outdoor air rate for a 3000 ft^2 combination gymnasium, recreation and exercise operation to a minimum by use of filtering and air recirculation. (a) Design a system using filters selected from Table 3-4. Use the minimum values of the given weight arrestance efficiency. Pressure loss should not exceed about 0.20 in. water. Outdoor air contaminants are negligible. (b) Discuss the effect of the filter efficiency on the design of the system. (c) Suppose the cooling load dictates a larger supply air rate than the ventilation rate. Discuss how this would influence your choice of a filter.

CHAPTER FOUR

Heat Transmission in Building Structures

The design of an acceptable air-conditioning system is dependent on a good estimate of the heat gain or loss in the space to be conditioned. In the usual structure the walls and roofs are rather complex assemblies of materials. Windows are often made of two or more layers of glass with air spaces between them and usually have drapes or curtains. In basements, floors and walls are in contact with the ground. Because of these conditions precise calculation of heat transfer rates is difficult, but experience and experimental data make reliable estimates possible. Because most of the calculations require a great deal of repetitive work, tables that list coefficients and other data for typical situations are used. Thermal resistance is a very useful concept and will be used extensively in the following sections. Because thermal capacitance is an important concept in all transient analysis computations, discussion of specific heat and density will also be included in this section.

Generally all three modes of heat transfer—conduction, convection, and radiation, are important in building heat gain and loss. Solar radiation will be treated in a separate section, Chapter five, because it creates some unusual complexities in geometry and consequently in computation. Long wavelength radiation, such as occurs in air gaps, will be treated in this chapter.

4-1 BASIC HEAT TRANSFER MODES

In the usual situation all three modes of heat transfer, occur simultaneously. In this section, however, they will be considered separately for clarity and ease of presentation.

Thermal conduction is the mechanism of heat transfer between parts of a continuum because of the transfer of energy between particles or groups of particles at the atomic level. The Fourier equation expresses steady state conduction in one dimension as follows:

$$\dot{q} = -kA\frac{dt}{dx} \qquad (4\text{-}1)$$

81

where

$\dot{q}$ = heat transfer rate, Btu/hr or W
k = thermal conductivity, Btu/(hr-ft-F) or W/(m-C)
A = area normal to heat flow, ft^2 or m^2
$\dfrac{dt}{dx}$ = temperature gradient, F/ft or C/m

Equation (4-1) incorporates a negative sign because $\dot{q}$ flows in the positive direction of x when dt/dx is negative.

Consider the flat wall of Fig. 4-1a where uniform temperatures t_1 and t_2 are assumed to exist on each surface. If the thermal conductivity, the heat transfer rate, and the area are constant, Eq. (4-1) may be integrated to obtain

$$\dot{q} = \frac{-kA(t_2 - t_1)}{(x_2 - x_1)} \tag{4-2}$$

Another very useful form of Eq. (4-2) is

$$\dot{q} = \frac{-(t_2 - t_1)}{R'} \tag{4-2a}$$

where R' is the thermal resistance defined by

$$R' = \frac{x_2 - x_1}{kA} = \frac{\Delta x}{kA} \tag{4-3}$$

The thermal resistance for a unit area of material is very commonly used in handbooks and in the HVAC literature. In this book this quantity, sometimes called the "R-factor," is referred to as the *unit thermal resistance,* or simply the *unit resistance, R.* For a plane wall the unit resistance is

$$R = \frac{\Delta x}{k} \tag{4-3a}$$

Notice that thermal resistance R' is analogous to electrical resistance and $\dot{q}$ and $(t_2 - t_1)$ are analogous to current and potential difference in Ohm's law. This analogy

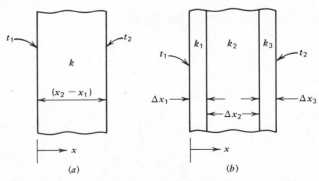

Figure 4-1.   Nomenclature for conduction in plane walls.

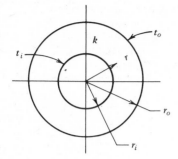

Figure 4-2.   Radial heat flow in a hollow cylinder.

provides a very convenient method of analyzing a wall or slab made up of two or more layers of dissimilar material. Figure 4-1 b shows a wall constructed of three different materials. The heat transferred by conduction is given by Eq. (4-2a) where

$$R' = R'_1 + R'_2 + R'_3 = \frac{\Delta x_1}{k_1 A} + \frac{\Delta x_2}{k_2 A} + \frac{\Delta x_3}{k_3 A} \tag{4-4}$$

Although the foregoing discussion is limited to a plane wall where the cross-sectional area is a constant, a similar procedure applies to a curved wall.

Consider the long, hollow cylinder shown in cross section in Fig. 4-2. The surface temperatures t_i and t_o are assumed to be uniform and steady over each surface. The material is assumed to be homogeneous with a constant value of thermal conductivity. Integration of Eq. (4-1) with k and $\dot{q}$ constant but A a function of r yields

$$\dot{q} = \frac{2\pi k L}{\ln\left(\dfrac{r_o}{r_i}\right)} (t_i - t_o) \tag{4-5}$$

where L is the length of the cylinder. Here the thermal resistance is

$$R' = \frac{\ln\left(\dfrac{r_o}{r_i}\right)}{2\pi k L} \tag{4-6}$$

Cylinders made up of several layers may be analyzed in a manner similar to the plane wall where resistances in series are summed as shown in Eq. (4-4) except that the individual resistances are given by Eq. (4-6).

Table 4-1 summarizes the thermal resistances for several common situations, and Tables 4-2 and 4-2a gives the thermal conductivity k for a wide variety of building and insulating materials. Other useful data are also given in Tables 4-2 and 4-2a; for example, the reciprocal of the unit thermal resistance and the *unit thermal conductance, C.* Note that k has the units of Btu-in./(ft²-hr-F) or W/(m-C). With Δx given in inches or meters, respectively, the unit thermal conductance C is given by

$$C = \frac{1}{R} = \frac{k}{\Delta x} \ \text{Btu}/(\text{hr-ft}^2\text{-F}) \ \text{or} \ \text{W}/(\text{m}^2\text{-C}) \tag{4-7}$$

Table 4-1 THERMAL RESISTANCES FOR SOME STEADY-STATE CONDUCTION PROBLEMS

Number	System	Expressions for the Resistance R' $\dot{q} = \Delta t/R'$ (Btu/hr) or W
1.	Cylinder buried in a semi-infinite medium having a temperature at a great distance t_∞. The ground surface is assumed adiabatic.	

$$R' = \frac{\left(\ln \dfrac{2L}{D} \right)\left(1 + \dfrac{\ln(L/2z)}{\ln(2L/D)} \right)}{2\pi kL}$$

| 2. | A vertical cylinder placed in a semi-infinite medium having an adiabatic surface and temperature t_∞ at a great distance. | |

$$R' = \frac{\ln(4L/D)}{2\pi kL}$$

| 3. | Conduction between inside and outside surfaces of a rectangular box having uniform inside and outside temperatures. Wall thickness Δx is less than any inside dimension. | |

$$R' = \frac{1}{k\left[\dfrac{A}{\Delta x} + 0.54\Sigma L + 1.2\Delta x \right]}$$

ΣL = sum of all 12 inside lengths
Δx = thickness of wall
A = inside surface area

Table 4-2   THERMAL PROPERTIES OF BUILDING AND INSULATING MATERIALS AT A MEAN TEMPERATURE OF 75 F (ENGLISH UNITS)*

Material	Description	Density ρ $\dfrac{\text{lbm}}{\text{ft}^3}$	Thermal Conductivity k $\dfrac{\text{Btu-in.}}{\text{ft}^2\text{-hr-F}}$	Unit Conductance C $\dfrac{\text{Btu}}{\text{hr-ft}^2\text{-F}}$	Unit Resistance Per Inch Thickness $1/k$ $\dfrac{\text{ft}^2\text{-hr-F}}{\text{Btu-in.}}$	Unit Resistance For Thickness Listed $1/C$ $\dfrac{\text{hr-ft}^2\text{-F}}{\text{Btu}}$	Specific Heat $\dfrac{\text{Btu}}{\text{lbm-F}}$
BUILDING BOARD Boards, panels, subflooring, sheathing, woodbased panel products	Asbestos-cement board						
	$\frac{1}{4}$ in. or 6 mm	120	—	16.5	—	0.06	0.24
	Gypsum or plasterboard						
	$\frac{3}{8}$ in. or 10 mm	50	—	3.10	—	0.32	0.26
	$\frac{1}{2}$ in. or 13 mm	50	—	2.22	—	0.45	—
	Plywood	34	0.80	—	1.25	—	0.29
	$\frac{1}{4}$ in. or 6 mm	34	—	3.20	—	0.31	0.29
	$\frac{3}{8}$ in. or 10 mm	34	—	2.13	—	0.47	0.29
	$\frac{1}{2}$ in. or 13 mm	34	—	1.60	—	0.62	0.29
	$\frac{3}{4}$ in. or 20 mm	34	—	1.07	—	0.93	0.29
	Insulating board, and sheathing, regular density						
	$\frac{1}{2}$ in. or 13 mm	18	—	0.76	—	1.32	0.31
	$\frac{25}{32}$ in. or 20 mm	18	—	0.49	—	2.06	0.31
	Hardboard, high density, standard tempered	63	1.00	—	1.00	—	0.32
	Particle board						
	Medium density	50	0.94	—	1.06	—	0.31
	Underlayment						
	$\frac{5}{8}$ in. or 16 mm	40	—	1.22	—	0.82	0.29
	Wood subfloor						
	$\frac{3}{4}$ in. or 20 mm	—	—	1.06	—	0.94	0.33

Table 4-2 THERMAL PROPERTIES OF BUILDING AND INSULATING MATERIALS AT A MEAN TEMPERATURE OF 75 (ENGLISH UNITS)*, continued

Material	Description	Density ρ $\dfrac{lbm}{ft^3}$	Thermal Conductivity k $\dfrac{Btu\text{-}in.}{ft^2\text{-}hr\text{-}F}$	Unit Conductance C $\dfrac{Btu}{hr\text{-}ft^2\text{-}F}$	Unit Resistance Per Inch Thickness $1/k$ $\dfrac{ft^2\text{-}hr\text{-}F}{Btu\text{-}in.}$	Unit Resistance For Thickness Listed $1/C$ $\dfrac{hr\text{-}ft^2\text{-}F}{Btu}$	Specific Heat $\dfrac{Btu}{lbm\text{-}F}$
BUILDING PAPER	Vapor—permeable felt	—	—	16.7	—	0.06	—
	Vapor—seal, two layers of mopped 15 lb felt	—	—	8.35	—	0.12	—
FINISH FLOORING MATERIALS	Carpet and fibrous pad	—	—	0.48	—	2.08	0.34
	Carpet and rubber pad	—	—	0.81	—	1.23	0.33
	Tile—asphalt, linoleum, vinyl, or rubber	—	—	20.0	—	0.05	0.30
INSULATING MATERIALS Blanket and batt	Mineral fiber—fibrous form processed from rock, slag, or glass						
	Approximately 2–2¾ in. or 50–70 mm	0.3–2.0	—	0.143	—	7	0.17–0.23
	Approximately 3–3½ in. or 75–90 mm	0.3–2.0	—	0.091	—	11	0.17–0.23
	Approximately 5¼–6½ in. or 135–165 mm	0.3–2.0	—	0.053	—	19	0.17–0.23
Board and slabs	Cellular glass	8.5	0.38	—	2.63	—	0.24
	Glass fiber, organic bonded	4–9	0.25	—	4.00	—	0.23

Category	Material						
	Expanded polystyrene—molded beads	1.0	0.28	—	3.57	—	0.29
	Expanded polyurethane—R-11 expanded	1.5	0.16	—	6.25	—	0.38
	Mineral fiber with resin binder	15	0.29	—	3.45	—	0.17
LOOSE FILL	Mineral fiber—rock, slag, or glass						
	Approximately 3.75–5 in. or 75–125 mm	0.6–2.0	—	—	—	11	0.17
	Approximately 6.5–8.75 in. or 165–222 mm	0.6–2.0	—	—	—	19	0.17
	Approximately 7.5–10 in. or 191–254 mm	—	—	—	—	22	0.17
	Approximately 7¼ in. or 185 mm	—	—	—	—	30	0.17
	Silica aerogel	7.6	0.17	—	5.88	—	—
	Vermiculite (expanded)	7–8	0.47	—	2.13	—	—
ROOF INSULATION	Preformed, for use above deck						
	Approximately ½ in. or 13 mm	—	—	0.72	—	1.39	—
	Approximately 1 in. or 25 mm	—	—	0.36	—	2.78	—
	Approximately 2 in. or 50 mm	—	—	0.19	—	5.56	—
	Cellular glass	9	0.4	—	2.5	—	0.24
MASONRY MATERIALS	Lightweight aggregates including	200	5.2	—	0.19	—	—
Concretes	expanded shale, clay, or slate;	100	3.6	—	0.28	—	—
	expanded slags; cinders;	80	2.5	—	0.40	—	—
	pumice; vermiculite; also	40	1.15	—	0.86	—	—
	cellular concretes	20	0.70	—	1.43	—	—
	Sand and gravel or stone aggregate (not dried)	140	12.0	—	0.08	—	—

Table 4-2   THERMAL PROPERTIES OF BUILDING AND INSULATING MATERIALS AT A MEAN TEMPERATURE OF 75 F (ENGLISH UNITS)*, continued

Material	Description	Density ρ $\dfrac{lbm}{ft^3}$	Thermal Conductivity k $\dfrac{Btu\text{-}in.}{ft^2\text{-}hr\text{-}F}$	Unit Conductance C $\dfrac{Btu}{hr\text{-}ft^2\text{-}F}$	Unit Resistance		Specific Heat $\dfrac{Btu}{lbm\text{-}F}$
					Per Inch Thickness $1/k$ $\dfrac{ft^2\text{-}hr\text{-}F}{Btu\text{-}in.}$	For Thickness Listed $1/C$ $\dfrac{hr\text{-}ft^2\text{-}F}{Btu}$	
MASONRY UNITS	Brick, common	120	5.0	—	0.20	—	—
	Brick, face	130	9.0	—	0.11	—	—
	Concrete blocks, three-oval core—sand and gravel aggregate						
	4. in or 100 mm	—	—	1.4		0.71	—
	8 in. or 200 mm	—	—	0.9		1.11	—
	12 in. or 300 mm	—	—	0.78		1.28	—
	lightweight aggregate (expanded shale, clay slate or slag; pumice)						
	3 in. or 75 mm	—	—	0.79		1.27	—
	4 in. or 100 mm	—	—	0.67		1.50	—
	8 in. or 200 mm	—	—	0.50		2.00	—
	12 in. or 300 mm	—	—	0.44		2.27	—
PLASTERING MATERIALS	Cement plaster, sand, aggregate	116	5.0	—	0.20	—	—
	Gypsum plaster:						
	Lightweight aggregate						
	$\frac{1}{2}$ in. or 13 mm	45	—	3.12		0.32	—
	$\frac{5}{8}$ in. or 16 mm	45	—	2.67		0.39	—

Lightweight aggregate on metal lath $\frac{3}{4}$ in. or 20 mm	—	—	2.13	—	0.47	—
ROOFING						
Asbestos-cement shingles	120	—	4.76	—	0.21	0.21
Asphalt roll roofing	70	—	6.50	—	0.15	—
Asphalt shingles	70	—	2.27	—	0.44	—
Built-up roofing						
$\frac{3}{8}$ in. or 10 mm	70	—	3.00	—	0.33	0.35
Slate, $\frac{1}{2}$ in. or 13 mm	—	—	20.00	—	0.05	—
Wood shingles—plain or plastic film faced	—	—	1.06	—	0.94	0.31
SIDING MATERIALS (on Flat Surface)						
Shingles						
Asbestos-cement	120	—	4.76	—	0.21	—
Siding						
Wood, drop, 1 in. or 25 mm	—	—	1.27	—	0.79	0.31
Wood, plywood, $\frac{3}{8}$ in. or 10 mm, lapped	—	—	1.59	—	0.59	0.29
Aluminum or steel, over sheathing, hollowbacked	—	—	1.61	—	0.61	—
Insulating board—backed nominal $\frac{3}{8}$ in. or 10 mm	—	—	0.55	—	1.82	—
Insulating board—backed nominal $\frac{3}{8}$ in. or 10 mm, foil-backed	—	—	0.34	—	2.96	—
Architectural glass	—	—	10.00	—	0.10	—
WOODS						
Maple, oak, and similar hardwoods	45	1.10	—	0.91	—	0.30
Fir, pine, and similar softwoods	32	0.80	—	1.25	—	0.33
METALS						
Aluminum (1100)	171	1536	—	0.00065	—	0.214
Steel, mild	489	314	—	0.00318	—	0.120
Steel, stainless	494	108	—	0.00926	—	0.109

*Abstracted by permission from *ASHRAE Handbook of Fundamentals*, 1977.

Table 4-2a   THERMAL PROPERTIES OF BUILDING AND INSULATING MATERIALS AT A MEAN
TEMPERATURE OF 24 C (SI UNITS)*

| | | | | | Unit Resistance | | |
| | | | | | Per Metre Thickness | For Thickness Listed | |
Material	Description	Density ρ $\dfrac{kg}{m^3}$	Thermal Conductivity k $\dfrac{W}{m\text{-}C}$	Unit Conductance C $\dfrac{W}{m^2\text{-}C}$	$1/k$ $\dfrac{m\text{-}C}{W}$	$1/C$ $\dfrac{m^2\text{-}C}{W}$	Specific Heat $\dfrac{kJ}{kg\text{-}C}$
BUILDING BOARD							
Boards, panels, subflooring, sheathing, woodbased panel products	Asbestos-cement board $\frac{1}{4}$ in. or 6 mm	1922	—	93.7	—	0.011	1.00
	Gypsum or plasterboard						
	$\frac{3}{8}$ in. or 10 mm	800	—	17.6	—	0.057	1.09
	$\frac{1}{2}$ in. or 13 mm	800	—	12.6	—	0.078	—
	Plywood	545	0.12	—	8.70	—	1.21
	$\frac{1}{4}$ in. or 6 mm	545	—	18.2	—	0.055	1.21
	$\frac{3}{8}$ in. or 10 mm	545	—	12.1	—	0.083	1.21
	$\frac{1}{2}$ in. or 13 mm	545	—	9.09	—	0.110	1.21
	$\frac{3}{4}$ in. or 20 mm	545	—	6.08	—	0.165	1.21
	Insulating board and sheathing, regular density						
	$\frac{1}{2}$ in. or 13 mm	288	—	4.32	—	0.232	1.30
	$\frac{25}{32}$ in. or 20 mm	288	—	2.78	—	0.359	1.30
	Hardboard, high density, standard tempered	1010	0.14	—	6.94	—	1.34
	Particleboard						
	Medium density	800	0.14	—	7.35	—	1.30
	Underlayment $\frac{5}{8}$ in. or 16 mm	640	—	6.93	—	0.144	1.21
	Wood subfloor $\frac{3}{4}$ in. or 20 mm	—	—	6.02	—	0.166	1.38

BUILDING PAPER						
Vapor—permeable felt	—		94.8	—	0.011	—
Vapor—seal, two layers of mopped 15 lb felt	—		47.4	—	0.021	—
FINISH FLOORING MATERIALS						
Carpet and fibrous pad	—		2.73	—	0.367	1.42
Carpet and rubber pad	—		4.60	—	0.217	1.38
Tile—asphalt, linoleum, vinyl, or rubber	—		113.0	—	0.009	1.26
INSULATING MATERIALS						
Blanket and batt						
Mineral fiber—fibrous form processed from rock, slag, or glass. Approximately 2-2¾ in. or 50-70 mm	4.8-32		0.812	—	1.23	0.71-0.96
Approximately 3-3½ in. or 75-90 mm	4.8-32		0.517	—	1.94	0.71-0.96
Approximately 5¼-6½ in. or 135-165 mm	4.8-32		0.301	—	3.32	0.71-0.96
Board and slabs						
Cellular glass	136	0.0548	—	18.2	—	1.0
Glass fiber, organic bonded	64-144	0.036	—	27.8	—	0.96
Expanded polystyrene—molded beads	16	0.040	—	25.0	—	1.2
Expanded polyurethane—R-11 expanded.	24	0.023	—	43.5	—	1.6
Mineral fiber with resin binder	240	0.042	—	23.9	—	0.71
LOOSE FILL						
Mineral fiber—rock, slag, or glass						
Approximately 3.75-5 in. or 75-125 mm	9.6-32		0.45	—	1.94	0.71
Approximately 6.5-8.75 in. or 165-222 mm	9.6-32		0.28	—	3.35	0.71
Approximately 7.5-10 in. or 191-254 mm	—		—	—	3.87	0.71

Table 4-2a THERMAL PROPERTIES OF BUILDING AND INSULATING MATERIALS AT A MEAN
TEMPERATURE OF 24 C (SI UNITS*, continued

Material	Description	Density ρ $\frac{kg}{m^3}$	Thermal Conductivity k $\frac{W}{m\text{-}C}$	Unit Conductance C $\frac{W}{m^2\text{-}C}$	Unit Resistance		Specific Heat $\frac{kJ}{kg\text{-}C}$
					Per Metre Thickness $1/k$ $\frac{m\text{-}C}{W}$	For Thickness Listed $1/C$ $\frac{m^2\text{-}C}{W}$	
LOOSE FILL (*cont.*)	Approximately 7¼ in. or 185 mm	—	—	0.23	—	5.28	0.71
	Silica aerogel	122	0.025	—	40.8	—	—
	Vermiculite (expanded)	122	0.068	—	14.8	—	—
ROOF INSULATION	Preformed, for use above deck						
	Approximately ½ in. or 13 mm	—	—	4.1	—	0.24	1.0
	Approximately 1 in. or 25 mm	—	—	2.0	—	0.49	2.1
	Approximately 2 in. or 50 mm	—	—	1.1	—	0.93	3.9
	Cellular glass	144	0.058	—	17.3	—	1.0
MASONRY MATERIALS Concretes	Lightweight aggregates including expanded shale, clay, or slate; expanded slags; cinders; pumice; vermiculite; also cellular concretes	3200	0.75	—	1.32	—	—
		1600	0.52	—	1.94	—	—
		1280	0.36	—	2.77	—	—
		640	0.17	—	6.03	—	—
		320	0.10	—	10.0	—	—
	Sand and gravel or stone aggregate (not dried)	2242	1.73	—	0.58	—	—

MASONRY UNITS						
Brick, common	1922	0.72	—	1.39	—	—
Brick, face	2082	1.30	—	0.77	—	—
Concrete blocks, three-oval core—sand and gravel aggregate						
4 in. or 100 mm	—	—	8.0	—	0.13	—
8 in. or 200 mm	—	—	5.1	—	0.20	—
12 in. or 300 mm	—	—	4.4	—	0.23	—
lightweight aggregate (expanded shale, clay slate or slag; pumice)						
3 in. or 75 mm	—	—	4.5	—	0.22	—
4 in. or 100 mm	—	—	3.8	—	0.26	—
8 in. or 200 mm	—	—	2.8	—	0.35	—
12 in. or 300 mm	—	—	2.5	—	0.40	—
PLASTERING MATERIALS						
Cement plaster, sand aggregate	1858	0.72	—	1.39	—	—
Gypsum plaster:						
Lightweight aggregate						
$\frac{1}{2}$ in. or 13 mm	721	—	17.7	—	0.056	—
$\frac{5}{8}$ in. or 16 mm	721	—	15.2	—	0.066	—
Lightweight aggregate on metal lath						
$\frac{3}{4}$ in. or 20 mm	—	—	12.1	—	0.83	—
ROOFING						
Asbestos-cement shingles	1922	—	27.0	—	0.037	—
Asphalt roll roofing	1121	—	36.9	—	0.027	—
Asphalt shingles	1121	—	12.9	—	0.078	—
Built-up roofing						
$\frac{3}{8}$ in. or 10 mm	1121	—	17.0	—	0.059	—
Slate, $\frac{1}{4}$ in. or 13 mm	—	—	113.6	—	0.009	—
Wood shingles—plain or plastic film faced	—	—	6.02	—	0.166	—

Table 4-2a THERMAL PROPERTIES OF BUILDING AND INSULATING MATERIALS AT A MEAN TEMPERATURE OF 24 C (SI UNITS*), continued

Material	Description	Density ρ $\dfrac{kg}{m^3}$	Thermal Conductivity k $\dfrac{W}{m\text{-}C}$	Unit Conductance C $\dfrac{W}{m^2\text{-}C}$	Unit Resistance Per Metre Thickness $1/k$ $\dfrac{m\text{-}C}{W}$	Unit Resistance For Thickness Listed $1/C$ $\dfrac{m^2\text{-}C}{W}$	Specific Heat $\dfrac{kJ}{kg\text{-}C}$
SIDING MATERIALS (On Flat Surface)	Shingles						
	Asbestos-cement	1922	—	27.0	—	3.70	—
	Siding						
	Wood, drop, 1 in. or 25 mm	—	—	7.21	—	0.139	1.30
	Wood, plywood, $\frac{3}{8}$ in. or 10 mm, lapped	—	—	9.03	—	0.111	1.21
	Aluminum or steel, over sheathing, hollowbacked	—	—	9.14	—	0.109	—
	Insulating board—backed nominal, $\frac{3}{8}$ in. or 10 mm	—	—	3.12	—	0.320	—
	Insulating board backed nominal, $\frac{3}{8}$ in. or 10 mm, foil-backed	—	—	1.93	—	0.518	—
	Architectural glass	—	—	56.8	—	0.018	—
WOODS	Maple, oak, and similar hardwoods	721	0.159	—	6.3	—	1.26
	Fir, pine, and similar softwoods	513	0.115	—	8.67	—	1.38
METALS	Aluminum (1100)	2739	221.5	—	0.0045	—	0.896
	Steel, mild	7833	45.3	—	0.022	—	0.502
	Steel, stainless	7913	15.6	—	0.064	—	0.456

*Adapted by permission from *ASHRAE Handbook of Fundamentals*, 1977.

Thermal convection is the transport of energy by mixing in addition to conduction. Convection is associated with fluids in motion generally through a pipe or duct or along a surface. In the very thin layer of fluid next to the surface the transfer of energy is by conduction. In the main body of the fluid mixing is the dominant energy transfer mechanism. A combination of conduction and mixing exists between these two regions. The transfer mechanism is complex and highly dependent on whether the flow is laminar or turbulent.

The usual, simplified approach in convection is to express the heat transfer rate as

$$\dot{q} = hA(t - t_w) \tag{4-8}$$

where

$\dot{q}$ = heat transfer rate from fluid to wall, Btu/hr or W
h = film coefficient, Btu/(hr-ft²-F) or W/(m²-s)
t = bulk temperature of the fluid, F or C
t_w = wall temperature, F or C

The film coefficient h is sometimes called the unit surface conductance or alternatively the convective heat transfer coefficient. Equation (4-8) may also be expressed in terms of thermal resistance:

$$\dot{q} = \frac{t - t_w}{R'} \tag{4-8a}$$

where

$$R' = \frac{1}{hA} \qquad (hr\text{-}F)/Btu \ \ or \ \ C/W \tag{4-9}$$

or

$$R = \frac{1}{h} = \frac{1}{C}^* \qquad (hr\text{-}ft^2\text{-}F)/Btu \ \ or \ \ (m^2\text{-}C)/W \tag{4-9a}$$

The thermal resistance given by Eq. (4-9) may be summed with the thermal resistances arising from pure conduction given by Eqs. (4-3) or (4-6).

The film coefficient h appearing in Eqs. (4-8) and (4-9) depends on the fluid, the fluid velocity, the flow channel, and the degree of development of the flow field. (That is, whether the fluid has just entered the channel or is relatively far from the entrance.) Many correlations exist for predicting the film coefficient under various conditions. Correlations for forced convection are given in Chapter 2 of the *ASHRAE Handbook* (1).

In convection it is important to recognize the mechanism that is causing the fluid motion to occur. When the bulk of the fluid is moving relative to the heat transfer surface, the mechanism is called *forced convection,* because such motion is usually caused by a blower, fan, or pump which is forcing the flow. In forced convection

---

*Note that the symbol for conductance is C, in contrast to the symbol for the temperature in Celsius degrees, C.

buoyancy forces are negligible. In *free convection,* on the other hand, the motion of the fluid is due entirely to buoyancy forces, usually confined to a layer near the heated or cooled surface. The surrounding bulk of the fluid is stationary and exerts a viscous drag on the layer of moving fluid. As a result inertia forces in free convection are usually small. Free convection is often referred to as natural convection.

Natural or free convection is an important part of HVAC applications. However, it is one of the most difficult physical situations to describe, and the predicted film coefficients have a greater uncertainty than those of forced convection. Various empirical relations for natural convection film coefficients can be found in reference 1.

Most building structures have forced convection along outer walls or roofs, and natural convection occurs inside narrow air spaces and on the inner walls. There is considerable variation in surface conditions, and both the direction and magnitude of the air motion on outdoor surfaces are very unpredictable. The film coefficient for these situations usually ranges from about 1.0 Btu/(hr-ft^2-F) or 6 W/(m^2-C) for free convection up to about 6 Btu/(hr-ft^2-F) or 35 W/(m^2-C) for forced convection with an air velocity of about 15 miles per hour, 20 ft/sec or 6 m/s. Because of the low film coefficients, especially with free convection, the amount of heat transferred by thermal radiation may be equal or larger than that transferred by convection.

Thermal radiation is the transfer of thermal energy by electromagnetic waves and is an entirely different phenomenon from conduction and convection. In fact thermal radiation can occur in a perfect vacuum and is actually impeded by an intervening medium. The direct net transfer of energy by radiation between two surfaces which see only each other and which are separated by a nonabsorbing medium is given by

$$\dot{q}_{12} = \frac{\sigma(T_1^4 - T_2^4)}{\dfrac{1 - \epsilon_1}{A_1\epsilon_1} + \dfrac{1}{A_1 F_{12}} + \dfrac{1 - \epsilon_2}{A_2\epsilon_2}} \tag{4-10}$$

where

σ = Boltzmann constant, 0.1713 $\times$ 10^{-8} Btu/(hr-ft^2-R^4) or 5.673 $\times$ 10^{-8} W/(m^2-K^4)

T = absolute temperature, R or K

ϵ = emittance

A = surface area, ft^2 or m^2

F = configuration factor, a function of geometry only (Chapter 5)

In Eq. (4-10) we assume that both surfaces are "gray" (where the emittance ϵ equals the absorptance α). This assumption can often be justified. The student is referred to Parker et al. (2) or other textbooks on heat transfer for a more complete discussion of thermal radiation. Figure 4-3 shows situations where radiation may be a significant factor. For the wall

$$\dot{q}_i = \dot{q}_w = \dot{q}_r + \dot{q}_o$$

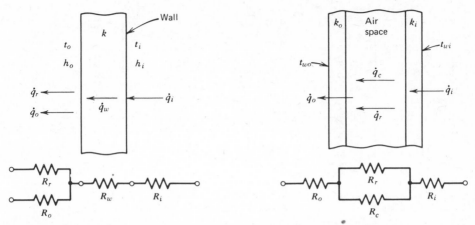

Figure 4-3. Wall and air space illustrating thermal radiation effects.

and for the air space

$$\dot{q}_i = \dot{q}_r + \dot{q}_c = \dot{q}_o$$

The resistances can be combined to obtain an equivalent overall resistance R' with which the heat transfer rate can be computed using Eq. (4-2a):

$$\dot{q} = \frac{-(t_o - t_i)}{R'} \tag{4-2a}$$

The thermal resistance for radiation is not easily computed, however, because of the fourth power temperature relationship of Eq. (4-10). For this reason and because of the inherent uncertainty in describing the physical situation, theory and experiment have been combined to develop combined or effective unit thermal resistances and unit thermal conductances for many typical surfaces and air spaces. Table 4-3 gives effective film coefficients and unit thermal resistances as a function of wall position, direction of heat flow, air velocity, and surface emittance for exposed surfaces such as outside walls. Table 4-3a gives representative values of emittance ϵ for some building and insulating materials. For example a typical vertical brick wall exposed to the outdoors has an effective emittance E of about 0.8 to 0.9. In still air the average film coefficient, from Table 4-3, is about 1.46 Btu/(hr-ft²-F) or 8.29 W/(m²-C) and the unit thermal resistance is 0.68 (hr-ft²-F)/Btu or 0.12 (m²-C)/W. If the wind velocity were to increase to 15 mph or about 7 m/s, the average film coefficient would increase to about 6 Btu/(hr-ft²-F) or 34 W/(m²-C). For still air the film coefficients are low and the effect of surface emittance is great. With higher air velocities the radiation effect diminishes.

Tables 4-4, 4-5, and 4-6 give conductances and resistances for air spaces as a function of position, direction of heat flow, air temperature, and the effective emittance of the space. The effective emittance E is given by

Table 4-3   SURFACE UNIT CONDUCTANCES AND UNIT RESISTANCES FOR AIR*,ᵃ

Position of Surface	Direction of Heat Flow	Surface Emittances ε = 0.9 h Btu hr-ft²-F	ε = 0.9 h W m²-C	ε = 0.9 R hr-ft²-F Btu	ε = 0.9 R m²-C W	ε = 0.2 h Btu hr-ft²-F	ε = 0.2 h W m²-C	ε = 0.2 R hr-ft²-F Btu	ε = 0.2 R m²-C W	ε = 0.05 h Btu hr-ft²-F	ε = 0.05 h W m²-C	ε = 0.05 R hr-ft²-F Btu	ε = 0.05 R m²-C W
STILL AIR													
Horizontal	Upward	1.63	9.26	0.61	0.11	0.91	5.2	1.10	0.194	0.76	4.3	1.32	0.232
Sloping—45 degrees	Upward	1.60	9.09	0.62	0.11	0.88	5.0	1.14	0.200	0.73	4.1	1.37	0.241
Vertical	Horizontal	1.46	8.29	0.68	0.12	0.74	4.2	1.35	0.238	0.59	3.4	1.70	0.298
Sloping—45 degrees	Downward	1.32	7.50	0.76	0.13	0.60	3.4	1.67	0.294	0.45	2.6	2.22	0.391
Horizontal	Downward	1.08	6.13	0.92	0.16	0.37	2.1	2.70	0.476	0.22	1.3	4.55	0.800
MOVING AIR (Any Position)													
wind is 15 mph or 6.7 m/s (for winter)	Any	6.0	34.0	0.17	0.029								
Wind is 7½ mph or 3.4 m/s (for summer)	Any	4.0	22.7	0.25	0.044								

*Adapted by permission from *ASHRAE Handbook of Fundamentals*, 1977.

ᵃConductances are for surfaces of the stated emittance facing virtual blackbody surroundings at the same temperature as the ambient air. Values are based on a surface-air temperature difference of 10 deg F and for surface temperature of 70 F.

Table 4-3a   REFLECTANCE AND EMITTANCE OF VARIOUS SURFACES AND EFFECTIVE
EMITTANCES OF AIR SPACE*

Surface	Reflectance in Percent	Average Emittance ϵ	Effective Emittance E of Air Space	
			With One Surface Having Emittance ϵ and Other 0.90	With Both Surfaces of Emittance ϵ
Aluminum foil, bright	92–97	0.05	0.05	0.03
Aluminum sheet	80–95	0.12	0.12	0.06
Aluminum coated paper, polished	75-84	0.20	0.20	0.11
Steel, galvanized, bright	70–80	0.25	0.24	0.15
Aluminum paint	30–70	0.50	0.47	0.35
Building materials—wood, paper, glass, masonry, nonmetallic paints	5–15	0.90	0.82	0.82

*Adapted by permission from *ASHRAE Handbook of Fundamentals*, 1977.

Table 4-4 UNIT THERMAL RESISTANCE OF A PLANE $\frac{3}{4}$ INCH (20 mm) AIR SPACE*,a

Position of Air Space	Direction of Heat Flow	Mean Air Temperature F	C	Temperature Difference F	C	E = 0.05 ft²-hr-F/Btu	m²-C/W	E = 0.2 ft²-hr-F/Btu	m²-C/W	E = 0.82 ft²-hr-F/Btu	m²-C/W
Horizontal	Up	90	32	10	6	2.22	0.391	1.61	0.284	0.75	0.132
		50	10	30	17	1.66	0.292	1.35	0.238	0.77	0.136
		50	10	10	6	2.21	0.389	1.70	0.299	0.87	0.153
		0	−18	20	11	1.79	0.315	1.52	0.268	0.93	0.164
		0	−18	10	6	2.16	0.380	1.78	0.313	1.02	0.180
45° Slope	Up	90	32	10	6	2.78	0.490	1.88	0.331	0.81	0.143
		50	10	30	17	1.92	0.338	1.52	0.268	0.82	0.144
		50	10	10	6	2.75	0.484	2.00	0.352	0.94	0.166
		0	−18	20	11	2.07	0.364	1.72	0.303	1.00	0.176
		0	−18	10	6	2.62	0.461	2.08	0.366	1.12	0.197
Vertical	Horizontal	90	32	10	6	3.24	0.571	2.08	0.366	0.84	0.148
		50	10	30	17	2.77	0.488	2.01	0.354	0.94	0.166
		50	10	10	6	3.46	0.609	2.35	0.414	1.01	0.178
		0	−18	20	11	3.02	0.532	2.32	0.408	1.18	0.210
		0	−18	10	6	3.59	0.632	2.64	0.465	1.26	0.222
45° Slope	Down	90	32	10	6	3.27	0.576	2.10	0.370	0.84	0.148
		50	10	30	17	3.23	0.569	2.24	0.394	0.99	0.174
		50	10	10	6	3.57	0.629	2.40	0.423	1.02	0.180
		0	−18	20	11	3.57	0.629	2.63	0.463	1.26	0.222
		0	−18	10	6	3.91	0.689	2.81	0.495	1.30	0.229

*Adapted by permission from *ASHRAE Handbook of Fundamentals*, 1977.

[a] Effective emittance of the space E is given by Eq. (4-11). Credit for an air space resistance value cannot be taken more than once and only for the boundary conditions established. Resistances of horizontal spaces with heat flow downward are substantially independent of temperature difference.

Table 4-5 UNIT THERMAL RESISTANCES OF A PLANE 3.5 INCH (89 mm) AIR SPACE*ᵃ

Position of Air Space	Direction of Heat Flow	Mean Air Temperature F	Mean Air Temperature C	Temperature Difference F	Temperature Difference C	E = 0.05 ft²-hr-F/Btu	E = 0.05 m²-C/W	E = 0.2 ft²-hr-F/Btu	E = 0.2 m²-C/W	E = 0.82 ft²-hr-F/Btu	E = 0.82 m²-C/W
Horizontal	Up	90	32	10	6	2.66	0.468	1.83	0.322	0.80	0.141
		50	10	30	17	2.01	0.354	1.58	0.278	0.84	0.148
		50	10	10	6	2.66	0.468	1.95	0.343	0.93	0.164
		0	−18	20	11	2.18	0.384	1.79	0.315	1.03	0.181
		0	−18	10	6	2.62	0.461	2.07	0.365	1.12	0.197
45° Slope	Up	90	32	10	6	2.96	0.521	1.97	0.347	0.82	0.144
		50	10	30	17	2.17	0.382	1.67	0.294	0.86	0.151
		50	10	10	6	2.95	0.520	2.10	0.370	0.96	0.169
		0	−18	20	11	2.35	0.414	1.90	0.335	1.06	0.187
		0	−18	10	6	2.87	0.505	2.23	0.393	1.16	0.204
Vertical	Horizontal	90	32	10	6	3.40	0.598	2.15	0.479	0.85	0.150
		50	10	30	17	2.55	0.449	1.89	0.333	0.91	0.160
		50	10	10	6	3.40	0.598	2.32	0.409	1.01	0.178
		0	−18	20	11	2.78	0.490	2.17	0.382	1.14	0.201
		0	−18	10	6	3.33	0.586	2.50	0.440	1.23	0.217
45° Slope	Down	90	32	10	6	4.33	0.763	2.49	0.438	0.90	0.158
		50	10	30	17	3.30	0.581	2.28	0.402	1.00	0.176
		50	10	10	6	4.36	0.768	2.73	0.481	1.08	0.190
		0	−18	20	11	3.63	0.639	2.66	0.468	1.27	0.224
		0	−18	10	6	4.32	0.761	3.02	0.532	1.34	0.236

*Adapted by permission from *ASHRAE Handbook of Fundamentals*, 1977.
ᵃEffective emittance of the space E is given by Eq. (4-11). Credit for an air space resistance value cannot be taken more than once and only for the boundary conditions established. Resistances of horizontal spaces with heat flow downward are substantially independent of temperature difference.

Table 4-6   UNIT THERMAL RESISTANCES OF PLANE HORIZONTAL AIR SPACES WITH HEAT FLOW DOWNWARD, TEMPERATURE DIFFERENCE 10 DEG F (6 DEG C)*

Air Space Thickness		Mean Temperature		E = 0.05		E = 0.2		E = 0.82	
Inch	mm	F	C	$\dfrac{\text{ft}^2\text{-hr-F}}{\text{Btu}}$	$\dfrac{\text{m}^2\text{-C}}{\text{W}}$	$\dfrac{\text{ft}^2\text{-hr-F}}{\text{Btu}}$	$\dfrac{\text{m}^2\text{-C}}{\text{W}}$	$\dfrac{\text{ft}^2\text{-hr-F}}{\text{Btu}}$	$\dfrac{\text{m}^2\text{-C}}{\text{W}}$
$\frac{3}{4}$	20	90	32	3.29	0.579	2.10	0.370	0.85	0.150
		50	10	3.59	0.632	2.41	0.424	1.02	0.180
		0	−18	4.02	0.708	2.87	0.505	1.31	0.231
$1\frac{1}{2}$	38	90	32	5.35	0.942	2.79	0.491	0.94	0.166
		50	10	5.90	1.04	3.27	0.576	1.15	0.203
		0	−18	6.66	1.17	4.00	0.707	1.51	0.266
$3\frac{1}{2}$	89	90	32	8.19	1.44	3.41	0.601	1.00	0.176
		50	10	9.27	1.63	4.09	0.720	1.24	0.218
		0	−18	10.32	1.82	5.08	0.895	1.64	0.289

*Adapted by permission from *ASHRAE Handbook of Fundamentals*, 1977.

$$\frac{1}{E} = \frac{1}{\epsilon_1} + \frac{1}{\epsilon_2} - 1 \tag{4-11}$$

where ϵ_1 and ϵ_2 are for each surface of the air space. The effect of radiation is quite apparent in Tables 4-4, 4-5, and 4-6. Interpolation may be used in Table 4-6 with less than 5 percent error. Straight line interpolation between Tables 4-4 and 4-5 generally results in errors of less than plus or minus 10 percent except in the 45 degree case with heat flow down where the error may be as high as plus or minus 30 percent.

The preceding paragraphs cover thermal resistances arising from conduction, convection, and radiation. Because thermal resistance is similar to electrical resistance, thermal resistances may be combined using methods analogous to electrical theory. Resistors connected in series may be replaced by an equivalent resistor equal to the sum of the series resistors; it will have a equivalent effect on the circuit.

$$R'_e = R'_1 + R'_2 + R'_3 + \cdots + R'_n \tag{4-12}$$

Figure 4-4 is an example of a wall being heated or cooled by a combination of convection and radiation on each surface and having five different resistances through which the heat must be conducted. The equivalent thermal resistance R'_e for the wall is given by Eq. (4-12) as

$$R'_e = R'_i + R'_1 + R'_2 + R'_3 + R'_o \tag{4-13}$$

Each of the resistances may be expressed in terms of fundamental variables using Eqs. (4-3) and (4-9).

$$R'_e = \frac{1}{h_i A_i} + \frac{\Delta x_1}{k_1 A_1} + \frac{\Delta x_2}{k_2 A_2} + \frac{\Delta x_3}{k_3 A_3} + \frac{1}{h_o A_o} \tag{4-14}$$

The film coefficients may be read from Table 4-3 and the thermal conductivities may be obtained from Table 4-2. For this case, a plane wall, the areas in Eq. (4-14) are all equal and the areas cancel.

In the general case the area that is properly a part of the resistance may vary and unit thermal resistances may have to be adjusted. Consider the insulated pipe shown in Fig. 4-5. Convection occurs on the inside and outside surfaces while heat

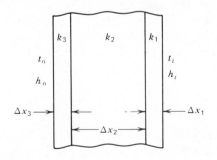

Figure 4-4.   Wall with thermal resistances in series.

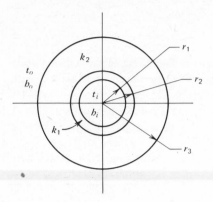

Figure 4-5.   Insulated pipe in convective environ
ment.

is conducted through the pipe wall and insulation. The overall thermal resistance for
the pipe of Fig. 4-5 is

$$R'_e = R'_o + R'_2 + R'_1 + R'_i \tag{4-15}$$

or using Eqs. (4-6) and (4-9)

$$R'_e = \frac{1}{h_o A_o} + \frac{\ln\left(\dfrac{r_3}{r_2}\right)}{2\pi k_2 L} + \frac{\ln\left(\dfrac{r_2}{r_1}\right)}{2\pi k_1 L} + \frac{1}{h_i A_i} \tag{4-16}$$

Equation (4-16) may also be written as

$$R'_e = \frac{1}{h_o A_o} + \frac{(r_3 - r_2)}{k_2 A_{m2}} + \frac{(r_2 - r_1)}{k_1 A_{m1}} + \frac{1}{h_i A_i} \tag{4-16a}$$

where

$$A_m = \frac{2\pi L(r_o - r_i)}{\ln\left(\dfrac{r_o}{r_i}\right)} \tag{4-17}$$

Equation (4-16a) has a form quite similar to Eq. (4-14), however the areas are all
unequal. The thermal resistance on the outside surface is reduced considerably by
the relatively large area. Where area changes occur in the direction of heat flow, unit
resistances or conductances can only be used with appropriate area weighting factors.

Thermal resistances may also occur in parallel. In theory the parallel resistances
can be combined into an equivalent thermal resistance in the same way as electrical
resistances.

$$\frac{1}{R'_e} = \frac{1}{R'_1} + \frac{1}{R'_2} + \frac{1}{R'_3} + \cdots \frac{1}{R'_n} \tag{4-18}$$

In most heat transfer situations with parallel heat flow paths, however, lateral flow
occurs, which may invalidate Eq. (4-18). The effect of lateral heat transfer between

two thermal conductors is to lower the equivalent resistance of the system. However, when the ratio of the larger to the smaller of the thermal resistances is less than about 5, Eq. (4-18) gives a reasonable approximation of the equivalent thermal resistance. For large variations in the thermal resistance of parallel conduction paths, numerical techniques using the computer are suggested.

The concept of thermal resistance is very useful and convenient in the analysis of complex arrangements of building materials. After the equivalent thermal resistance has been determined for a configuration, however, the overall unit thermal conductance, usually called the *overall heat transfer coefficient U,* is frequently used.

$$U = \frac{1}{R'A} = \frac{1}{R} \text{ Btu/(hr-ft}^2\text{-F) or W/(m}^2\text{-C)} \qquad (4\text{-}19)$$

The heat transfer rate is then given by

$$\dot{q} = UA\Delta t \qquad (4\text{-}20)$$

where

UA = conductance, Btu/(hr-F) or W/C
A = surface area, ft^2 or m^2
Δt = overall temperature difference, F or C

For a plane wall the area A is the same throughout the wall. In dealing with a curved wall, we select the area for convenience of calculation. For example, in the problem of heat transfer through the ceiling-attic-roof combination, it is usually most convenient to use the ceiling area. The area selected is then used to determine the appropriate value of U from Eq. (4-19).

4-2 TABULATED OVERALL HEAT TRANSFER COEFFICIENTS

For convenience of the designer, tables have been constructed that give overall coefficients for many common building sections including walls and floors, doors, windows, and skylights. The tables used in the *ASHRAE Handbook* have a great deal of flexibility and are summarized in the following pages.

Walls and Roofs

Tables 4-7 to 4-7e give data for typical wall and roof constructions. In each case the unit thermal resistance and the overall heat transfer coefficient have been computed for one set of conditions. It is then a simple matter to add or delete unit resistances to account for differences in construction.

Example 4-1. The frame wall shown in Table 4-7 is modified to have $3\frac{1}{2}$ in. of mineral fiber insulation between the studs. Compute the overall heat transfer coefficient, U.

Table 4-7　COEFFICIENTS OF TRANSMISSION U OF FRAME WALLS, Btu/(hr-ft²-F) [a],*

Replace Air Space with 3.5-in. R-11 Blanket Insulation (New Item 4)

| | Resistance (R) | | | |
| | 1 | | 2 | |
Construction	Between Framing	At Framing	Between Framing	At Framing
1. Outside surface (15 mph wind)	0.17	0.17	0.17	0.17
2. Siding, wood, 0.5 in. × 8 in. lapped (average)	0.81	0.81	0.81	0.81
3. Sheathing, 0.5-in. asphalt impregnated	1.32	1.32	1.32	1.32
4. Nonreflective air space, 3.5 in. (50 F mean; 10 deg F temperature difference)	1.01	—	11.00	—
5. Nominal 2-in. × 4-in. wood stud	—	4.38	—	4.38
6. Gypsum wallboard, 0.5 in.	0.45	0.45	0.45	0.45
7. Inside surface (still air)	0.68	0.68	0.68	0.68
Total Thermal Resistance (R)	$R_i = 4.44$	$R_s = 7.81$	$R_i = 14.43$	$R_s = 7.81$

Construction No. 1: $U_i = 1/4.44 = 0.225$; $U_s = 1/7.81 = 0.128$. With 20% framing (typical of 2-in. × 4-in. studs @ 16-in. o.c.), $U_{av} = 0.8 (0.225) + 0.2 (0.128) = 0.206$

Construction No. 2: $U_i = 1/14.43 = 0.069$; $U_s = 0.128$. With framing unchanged, $U_{av} = 0.8(0.069) + 0.2(0.128) = 0.081$

*Adapted by permission from *ASHRAE Handbook of Fundamentals*, 1977.
[a]U factor may be converted to W/(m²·C) by multiplying by 5.68.

Table 4-7a   COEFFICIENTS OF TRANSMISSION U OF SOLID MASONRY WALLS, Btu/(hr-ft²-F) [a]*

Replace Furring Strips and Air Space with 1-in. Extruded Polystyrene (New Item 4)

Construction	Resistance (R) 1		2
	Between Furring	At Furring	
1. Outside surface (15 mph wind)	0.17	0.17	0.17
2. Common brick, 8 in.	1.60	1.60	1.60
3. Nominal 1-in. × 3-in. vertical furring	—	0.94	—
4. Nonreflective air space, 0.75 in. (50 F mean; 10 deg F temperature difference)	1.01	—	5.00
5. Gypsum wallboard, 0.5 in.	0.45	0.45	0.45
6. Inside surface (still air)	0.68	0.68	0.68
Total Thermal Resistance (R)	$R_i = 3.91$	$R_s = 3.84$	$R_i = 7.90 = R_s$

Construction No. 1: $U_i = 1/3.91 = 0.256$; $U_s = 1/3.84 = 0.260$. With 20% framing (typical of 1-in. × 3-in. vertical furring on masonry @ 16-in. o.c.), $U_{av} = 0.8 (0.256) + 0.2 (0.260) = 0.257$

Construction No. 2: $U_i = U_s = U_{av} = 1/7.90 = 0.127$

[a]U factor may be converted to $W/(m^2\text{-}C)$ by multiplying by 5.68.
*Adapted by permission from *ASHRAE Handbook of Fundamentals,* 1977.

Table 4-7b   COEFFICIENTS OF TRANSMISSION U OF MASONRY CAVITY WALLS, Btu/(hr·ft²·F) [a],*

Replace Cinder Aggregate Block with 6-in. Light-weight Aggregate Block with Cores Filled (New Item 4)

| | Resistance (R) | | | |
| | 1 | | 2 | |
Construction	Between Furring	At Furring	Between Furring	At Furring
1. Outside surface (15 mph wind)	0.17	0.17	0.17	0.17
2. Face brick, 4 in.	0.44	0.44	0.44	0.44
3. Cement mortar, 0.5 in.	0.10	0.10	0.10	0.10
4. Concrete block, cinder aggregate, 8 in.	1.72	1.72	2.99	2.99
5. Reflective air space, 0.75 in. (50 F mean; 30 deg F temperature difference)	2.77	—	2.77	—
6. Nominal 1-in. × 3-in. vertical furring	—	0.94	—	0.94
7. Gypsum wallboard, 0.5 in, foil backed	0.45	0.45	0.45	0.45
8. Inside surface (still air)	0.68	0.68	0.68	0.68
Total Thermal Resistance (R)	R_i = 6.33	R_s = 4.50	R_i = 7.60	R_s = 5.77

Construction No. 1: U_i = 1/6.33 = 0.158; U_s = 1/4.50 = 0.222. With 20% framing (typical of 1-in. × 3-in. vertical furring on masonry @ 16-in. o.c.), U_{av} = 0.8 (0.158) + 0.2 (0.222) = 0.171

Construction No. 2: U_i = 1/7.60 = 0.132, U_s = 1/5.77 = 0.173. With framing unchanged, U_{av} = 0.8(0.132) + 0.2(0.173) = 0.140

[a] U factor may be converted to W/(m²·C) by multiplying by 5.68.

*Adapted by permission from *ASHRAE Handbook of Fundamentals*, 1977.

Table 4-7c   COEFFICIENTS OF TRANSMISSION U OF CEILINGS AND FLOORS, ^{a} Btu/(hr-ft²-F) ^{b,*}

Assume Unheated Attic Space above Heated Room with Heat Flow Up—Remove Tile, Felt, Plywood, Subfloor and Air Space—Replace with R-19 Blanket Insulation (New Item 4)

| Heated Room Below Unheated Space | Resistance (R) | | | |
| | 1 | | 2 | |
Construction (Heat Flow Up)	Between Floor Joists	At Floor Joist	Between Floor Joists	At Floor Joists
1. Bottom surface (still air)	0.61	0.61	0.61	0.61
2. Metal lath and lightweight aggregate, plaster, 0.75 in.	0.47	0.47	0.47	0.47
3. Nominal 2-in. × 8-in. floor joist	—	9.06	—	9.06
4. Nonreflective airspace, 7.25-in.	0.93^a	—	19.00	—
5. Wood subfloor, 0.75 in.	0.94	0.94	—	—
6. Plywood, 0.625 in.	0.78	0.78	—	—
7. Felt building membrane	0.06	0.06	—	—
8. Resilient tile	0.05	0.05	—	—
9. Top surface (still air)	0.61	0.61	0.61	0.61
Total Thermal Resistance (R)	$R_i = 4.45$	$R_s = 12.58$	$R_i = 20.69$	$R_s = 10.75$

Construction No. 1 $U_i = 1/4.45 = 0.225$; $U_s = 1/12.58 = 0.079$. With 10% framing (typical of 2-in. joists 16-in. o.c.), $U_{av} = 0.9$ (0.225) + 0.1 (0.079) = 0.210

Construction No. 2 $U_i = 1/20.69 = 0.048$; $U_s = 1/10.75 = 0.093$. With framing unchanged, $U_{av} = 0.9$ (0.048) + 0.1 (0.093) = 0.053

^{a}Use largest air space (3.5 in.) value shown in Table 4-5.

$^b U$ factor may be converted to W/(m²·C) by multiplying by 5.68.

*Adapted by permission from *ASHRAE Handbook of Fundamentals*, 1977.

Table 4-7d   COEFFICIENTS OF TRANSMISSION U OF FLAT BUILT UP ROOFS,[a] Btu/(hr-ft²-F)[c,*]

Add Rigid Roof Deck Insulation, $C = 0.24$ $(R = 1/C)$ (New Item 7)

Construction (Heat Flow Up)	1	2
	Resistance (R)	
1. Inside surface (still air)	0.61	0.61
2. Metal lath and lightweight aggregate plaster, 0.75 in.	0.47	0.47
3. Nonreflective air space, greater than 3.5 in. (50 F mean; 10 deg F temperature difference)	0.93[a]	0.93[a]
4. Metal ceiling suspension system with metal hanger rods	0[b]	0[b]
5. Corrugated metal deck	0	0
6. Concrete slab, lightweight aggregate, 2 in.	2.22	2.22
7. Rigid roof deck insulation (none)	—	4.17
8. Built-up roofing, 0.375 in.	0.33	0.33
9. Outside surface (15 mph wind)	0.17	0.17
Total Thermal Resistance (R)	4.73	8.90

Construction No. 1: $U_{av} = 1/4.73 = 0.211$
Construction No. 2: $U_{av} = 1/8.90 = 0.112$

[a]Use largest air space (3.5 in.) value shown in Table 4-5.
[b]Area of hanger rods is negligible in relation to ceiling area.
[c]U factor may be converted to $W/(m^2\text{-}C)$ by multiplying by 5.68.
*Adapted by permission from *ASHRAE Handbook of Fundamentals*, 1977.

Table 4-7e   COEFFICIENTS OF TRANSMISSION U OF PITCHED ROOFS, [a] Btu/(hr-ft^2-F) [b,*]

Find U_{av} for same Construction 2 with Heat Flow Down (Summer Conditions)

Construction 1 (Heat Flow Up) (Reflective Air Space)	Resistance (R) 1		Resistance (R) 2	
	Between Rafters	At Rafters	Between Rafters	At Rafters
1. Inside surface (still air)	0.62	0.62	0.76	0.76
2. Gypsum wallboard 0.5 in., foil backed	0.45	0.45	0.45	0.45
3. Nominal 2-in. × 4-in. ceiling rafter	—	4.38	—	4.38
4. 45 deg slope reflective air space, 3.5 in. (50 F mean, 30 deg F temperature difference)	2.17	—	4.33	—
5. Plywood sheathing, 0.625 in.	0.78	0.78	0.78	0.78
6. Felt building membrane	0.06	0.06	0.06	0.06
7. Asphalt shingle roofing	0.44	0.44	0.44	0.44
8. Outside surface (15 mph wind)	0.17	0.17	0.25	0.25
Total Thermal Resistance (R)	$R_i = 4.69$	$R_s = 6.90$	$R_i = 7.07$	$R_s = 7.12$

Construction No. 1: $U_i = 1/4.69 = 0.213$; $U_s = 1/6.90 = 0.145$. With 10% framing (typical of 2-in rafters @ 16-in. o.c.), $U_{av} = 0.9$ $(0.213) + 0.1(0.145) = 0.206$

Construction No. 2: $U_i = 1/7.07 = 0.141$; $U_s = 1/7.12 = 0.140$. With framing unchanged, $U_{av} = 0.9(0.141) + 0.1(0.140) = 0.141$.

[a]Heat flow upward; Roof pitch is 45 degrees.
[b]U factor may be converted to $W/(m^2\cdot C)$ by multiplying 5.68.
*Adapted by permission from ASHRAE Handbook of Fundamentals, 1977.

Solution.

Total unit resistance given in Table 4-7	4.44
Deduct the air space unit resistance, item 4	−1.01
Add insulation unit resistance given in Table 4-2	11.00
Total R	14.43

Then based on one square foot, we see that

$$U = \frac{1}{R} = \frac{1}{14.43} = 0.07 \text{ Btu/(hr-ft}^2\text{-F)}$$

Equation (4-18) may be used to correct for framing (2 × 4 studs on 16 in. centers)

$$\frac{1}{R'_c} = \frac{1}{R'} + \frac{1}{R'_f} \qquad \text{or} \qquad U_c A_t = U A_b + U_f A_f$$

where

A_t = total area
A_b = area between studs
A_f = area occupied by the studs

The unit thermal resistance of a section through the 2 × 4 stud is equal to the total resistance from Table 4-7 less the resistance of the air gap plus the resistance of the stud from Table 4-2. A 2 × 4 stud is only $3\frac{1}{2}$ in. deep.

$$R_f = \frac{1}{U_f} = 4.4 - 0.97 + (1.25 \times 3.5) = 7.81$$

or

$$U_f = 0.128 \text{ Btu/(hr-ft}^2\text{-F)}$$

Then using Eq. (4-18) we get

$$U_c = \frac{(0.07)(14.5) + (0.128)(1.5)}{16} = 0.075 \text{ Btu/(hr-ft}^2\text{-F)}$$

Example 4-2.   Compute the overall average coefficient for a ceiling floor combination shown in Table 4-7c but with 3.5 in. of mineral wool insulation in the space and with the floor joists at 24 in. on centers.

Solution.   Total unit resistance of ceiling floor combination in Table 4-7c with no insulation:

	Between Floor Joists	At Floor Joists
	4.45	12.58
Add mineral fiber insulation, 3.5 in.	11.00	
total R	15.45	12.58
total U	0.0647	0.0795

Using Eq. (4-18) to correct for framing (2 × 8 in. on 24 in. centers), 7.3 percent framing

$$U_{avg} = (0.927)(0.0647) + (0.073)(0.0795) = 0.0658 \text{ Btu/(hr - ft}^2\text{F)}$$

It should be noted that the data given in Tables 4-7 are based on:

1. Steady state heat transfer
2. Air spaces not insulated
3. Surrounding surfaces at ambient air temperature
4. Variation of thermal conductivity with temperature negligible.

Some caution should be exercised in applying calculated overall heat transfer coefficients such as those of Table 4-7 because the effects of poor workmanship and materials are not included. Although a safety factor is not usually applied, a moderate increase in U may be justified in some cases.

The overall heat transfer coefficients obtained from Tables 4-7 should always be adjusted for framing, as shown in Examples 4-1 and 2, using Eq. (4-18). This adjustment will normally be 5 to 12 percent of the unadjusted coefficient.

The coefficients of Tables 4-7 have all been computed for a 15 mph wind velocity on outside surfaces and should be adjusted for other velocities. The data of Table 4-3 may be used for this purpose. The *ASHRAE Handbook* (1) has conversion tables for this purpose also.

The following example illustrates the calculation of an overall heat transfer coefficient for an unvented attic.

Example 4-3.   Compute the overall heat transfer coefficient for a roof-attic-ceiling combination such as the one shown in Fig. 4-6. The ceiling is made of $\frac{1}{2}$ in. gypsum board with 2 × 6 in. joists on 16 in. centers with 6 in. mineral fiber insulation batts between the joists. The 2 × 6 in. rafters are on 24 in. centers. There is a $\frac{1}{2}$ in. plywood deck, a single layer of building paper, and asphalt shingles over the rafters. The attic is unvented. The slope of the roof is approximately 1:1.

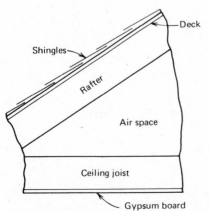

Shingles

Deck

Rafter

Air space

Ceiling joist

Gypsum board

Figure 4-6.   Section of a pitched roof-attic-ceiling combination.

Solution.   The solution is best accomplished using the thermal resistance concept. The overall thermal resistance may be expressed as

$$R' = \frac{1}{U_o A} = \frac{1}{U_c A_c} + \frac{1}{U_r A_r}$$

It is most convenient to base the overall coefficient U_o on the area of the ceiling. Therefore, let $A = A_c = 1$ ft²; the overall thermal resistance per square foot of ceiling area becomes

$$R_0 = \frac{1}{U_o} = \frac{1}{U_c} + \frac{1}{U_r\left(\dfrac{A_r}{A_c}\right)}$$

The ceiling resistance for a unit area is computed by using data from Tables 4-2 and 4-3.

	Unit Resistance	
Construction	**Between Framing**	**At Framing**
Still air, heat flow upward, $\epsilon =$		
0.9	0.61	0.61
Gypsum Board, $\frac{1}{2}$ in.	0.45	0.45
Insulation, mineral fiber, 6 in.	19.00	
Nominal 2 × 6 ceiling joist		6.88
Still air, heat flow upward, $\epsilon =$		
0.2	1.10	
Still air, heat flow upward, $\epsilon =$		
0.9		0.61
Total R	21.16	8.55
Total U	0.0473	0.117

The roof resistance per unit area is obtained from Table 4-7e.

	Between Rafters	At Rafters
Total unit resistance given	4.69	6.90
Deduct items 2 and 4	−2.62	−0.45
Total R	2.07	6.45
Total U	0.483	0.155

For 2 × 6 joists @ 16 in. centers (9.4% framing) $U_{avg} = (0.906)(0.0473) + (0.094)(0.117) = 0.0539$ Btu/(hr-ft^2-F) Using the roof unit resistance previously calculated from Table 4-7e, the average overall conductance may be determined.

For 2 × 6 rafters @ 24 in. centers (7.3 percent framing) $U_{avg} = (0.927)(0.483) + (0.073)(0.155) = 0.459$ Btu/(hr-ft^2-F) Because the slope is 1:1, the ratio of roof area to ceiling area, A_r/A_c is given by

$$A_r^2 = 2A_c^2$$

or

$$\frac{A_r}{A_c} = \sqrt{2} = 1.414$$

The equivalent unit thermal resistance for the roof-attic-ceiling combination is then

$$R_0 = \frac{1}{U_0} = \frac{1}{0.0539} + \frac{1}{(0.459)(1.414)} = 20.1$$

and

$$U_0 = \frac{1}{R_0} = 0.050 \text{ Btu/(hr-ft}^2\text{-F)}$$

Attics are usually vented to remove moisture from the insulation, but only moderate ventilation rates are required. Ventilating grilles located in the gables or under the eaves are adequate for this purpose. The effect of attic ventilation on the transfer of heat through the attic is not significant provided the ceiling is insulated with a unit thermal resistance of about 11 or more. This is true for both winter and summer conditions. It once was thought that increased ventilation of attics during the summer would dramatically reduce the heat gain to the inside space; however, this is apparently incorrect (4). It is generally not economically feasible to use power ventilation. The main reason for the ineffectiveness of ventilation is the fact that most of the heat transfer through the attic is by thermal radiation between the roof and the ceiling insulation. The use of reflective surfaces is therefore much more useful in reducing heat transfer. It is recommended that calculation of the overall transmission coefficient for attics under summer conditions be computed using the approach of Example 4-3 with appropriate unit resistances and assuming no ventilation.

Windows and Light Transmitting Panels

Table 4-8 contains overall heat transfer coefficients for windows, skylights, and other light-transmitting panels. Notice that there are values for summer and winter coefficients. This difference arises because coefficients for use in the winter assume a different velocity than in the summer. Table 4-8 applies only for air-to-air heat transfer and does not account for solar radiation, which will be discussed in Chapter 5.

Table 4-8 OVERALL COEFFICIENTS OF HEAT TRANSMISSION (*U*-FACTOR) OF WINDOWS AND SKYLIGHTS, Btu/(hr-ft²-F) [a],*

Description	Exterior Vertical Panels				Exterior Horizontal Panels (Skylights)	
	Summer[j]		Winter[i]		Summer[l]	Winter[m]
	No Indoor Shade	Indoor Shade[k]	No Indoor Shade	Indoor Shade[k]		
Flat Glass[b]						
Single Glass	1.04	0.81	1.10	0.83	0.83	1.23
Insulating Glass, Double[c]						
3/16 in. air space[d]	0.65	0.58	0.62	0.52	0.57	0.70
1/4 in. air space[d]	0.61	0.55	0.58	0.48	0.54	0.65
1/2 in. air space[e]	0.56	0.52	0.49	0.42	0.49	0.58
1/2 in. air space, low emittance coating[f]						
e = 0.20	0.38	0.37	0.32	0.30	0.36	0.48
e = 0.40	0.45	0.44	0.38	0.35	0.42	0.52
e = 0.60	0.51	0.48	0.43	0.38	0.46	0.56

Insulating Glass, Triple[c]					
1/4 in. air space[d]	0.44	0.40	0.39		
1/2 in. air space[g]	0.39	0.36	0.31	0.26	
Storm Windows					
1 in. to 4 in. air spaces[d]	0.50	0.48	0.50	0.42	
Plastic bubbles[n]					
Single Walled				0.80	1.15
Double Walled				0.46	0.70

---

*Reprinted by permission from *ASHRAE Handbook of Fundamentals*, 1977.

[a]See Table 4-8a for adjustments for various windows and sliding patio doors.

[b]Emittance of uncoated glass surface = 0.84.

[c]Double and triple refer to number of lights of glass.

[d]0.125-in. glass.

[e]0.25-in. glass.

[f]Coating on either glass surface facing air space; all other glass surfaces uncoated.

[g]Window design: 0.25-in. glass, 0.125-in. glass, 0.25-in. glass.

[m]For heat flow up.

[l]For heat flow down.

[n]Based on area of opening, not total surface area.

[i]15 mph outdoor air velocity; 0 F outdoor air; 70 F inside air temp natural convection.

[j]7.5 mph outdoor air velocity; 89 F outdoor air; 75 F inside air natural convection; solar radiation 248.3 Btu/hr·ft².

[k]Values apply to tightly closed venetian and vertical blinds, draperies, and roller shades.

Table 4-8a　ADJUSTMENT FACTORS FOR
　　　　　　COEFFICIENTS U OF TABLE 4-8*

Description	Single Glass	Double or Triple Glass	Storm Windows
Windows			
All glass	1.0	1.0	1.0
Wood sash—80% glass	0.9	0.95	0.90
Metal sash—80% glass	1.0	1.20	1.20
Sliding glass doors			
Metal frame	1.0	1.10	
Wood frame	0.95	1.0	

*Reprinted by permission from *ASHRAE Handbook of Fundamentals*, 1977.

Doors

Table 4-9 gives overall heat transfer coefficients for common slab doors. Again note the difference between summer and winter. Solar radiation has not been included.

Concrete Floors and Walls Below Grade

The heat transfer through basement walls and floors depends on the temperature difference between the inside air and the earth, the wall or floor material, and the conductivity of the ground. All of these factors involve considerable uncertainty. Experience has shown that overall coefficients, as given in Tables 4-10 and 4-11 give satisfactory results for basement walls and floors 3 ft or more below grade in winter. The proper temperature difference for use with the coefficients is the inside temperature minus the deep ground temperature. This ranges from 40 to 60 F or 4 to 16 C for most of the United States. When below grade spaces are conditioned as living space, the walls should be furred and finished with a vapor barrier, insulating board, and some type of finish layer such as paneling. This will add thermal resistance to the wall and reduce the effect of the uncertainty of the data given in Table 4-10. The basement floor should also be finished by laying down an insulating barrier and floor tile or carpet. The overall coefficients for the finished wall or floor may be computed as

$$R'_a = R' + R'_f = \frac{1}{UA} + R'_f = \frac{1}{U_a A} \qquad (4\text{-}21)$$

Floor Slabs at Grade Level

When considering the heat losses for floor slabs at or near grade level, we must take into account two situations. The first is the unheated slab where heat is supplied

Table 4-9 COEFFICIENTS OF TRANSMISSION U FOR SLAB DOORS*

Nominal Thickness	Solid Wood No Storm Door		Storm Door[a]				Summer No Storm Door	
			Wood		Metal			
	Btu / hr-ft²-F	W / m²-C	Btu / hr-ft²-F	W / m²-C	Btu / hr-ft²-F	W / m²-C	Btu / hr-ft²-F	W / m²-C
1 in. or 25 mm	0.64	3.63	0.30	1.70	0.39	2.21	0.61	3.46
1¼ in. or 30 mm	0.55	3.12	0.28	1.59	0.34	1.93	0.53	3.01
1½ in. or 38 mm	0.49	2.78	0.27	1.53	0.33	1.87	0.47	2.67
2 in. or 50 mm	0.43	2.44	0.24	1.36	0.29	1.65	0.42	2.38
Steel Door Heavy Type								
1¾ in. or 44 mm								
Mineral fiber core	0.59	3.35					0.58	3.29
Solid urethane foam core	0.40	2.27					0.39	2.21
Solid polystyrene core with thermal break	0.47	2.67					0.46	2.61

[a] Approximately 50 percent glass for wood doors; values for metal doors are independent of glass percentage.

*Adapted by permission from *ASHRAE Handbook of Fundamentals*, 1977.

Table 4-10 HEAT LOSS THROUGH BASEMENT FLOORS (FOR FLOORS MORE THAN 3 FT (0.91 m) BELOW GRADE)[a],*

Depth of Floor Below Grade		Narrowest Width of House							
		20 ft (6.1 m)		24 ft (7.3 m)		28 ft (8.5 m)		32 ft (9.8 m)	
ft	m	Btu/(hr-ft²-F)	W/(m²-C)	Btu/(hr-ft²-F)	W/(m²-C)	Btu/(hr-ft²-F)	W/(m²-C)	Btu/(hr-ft²-F)	W/(m²-C)
4	1.22	0.035	0.198	0.032	0.182	0.027	0.153	0.024	0.136
5	1.52	0.032	0.182	0.029	0.165	0.026	0.148	0.023	0.131
6	1.83	0.030	0.170	0.027	0.153	0.025	0.142	0.022	0.125
7	2.13	0.029	0.165	0.026	0.148	0.023	0.131	0.021	0.119

*Adapted by permission from *ASHRAE GRP 158 Cooling and Heating Load Calculation Manual*, 1979.
[a]For a depth below grade of 3 ft or less, treat as a slab on grade.

Table 4-11   HEAT LOSS RATE FOR BELOW-GRADE WALLS WITH INSULATION ON INSIDE SURFACE[a,*]

Insulation over Full Surface

Depth Wall Extends Below Grade*		Resistance					
		R-4 $(hr\text{-}ft^2\text{-}F)/Btu$	R-0.7 $(m^2\text{-}C)/W$	R-8	R-1.4	R-13	R-2.3
ft	m	$Btu/(hr\text{-}ft^2\text{-}F)$	$W/(m^2\text{-}C)$	$Btu/(hr\text{-}ft^2\text{-}F)$	$W/(m^2\text{-}C)$	$Btu/(hr\text{-}ft^2\text{-}F)$	$W/(m^2\text{-}C)$
4	1.52	0.110	0.625	0.075	0.426	0.057	0.324
5	1.83	0.102	0.579	0.071	0.403	0.054	0.307
6	2.13	0.095	0.539	0.067	0.380	0.052	0.295
7	1.22	0.089	0.505	0.064	0.363	0.050	0.284
Wall Insulated to a Depth of 2 ft Below Grade							
4	1.52	0.136	0.772	0.102	0.579	0.090	0.511
5	1.83	0.128	0.727	0.100	0.568	0.091	0.517
6	2.13	0.120	0.681	0.097	0.550	0.089	0.505
7	1.22	0.112	0.636	0.093	0.528	0.086	0.488

*Adapted by permission from *ASHRAE GRP 158 Cooling and Heating Load Calculation Manual*, 1979.
[a]For a depth below grade of 3 ft or less, treat as a slab on grade.

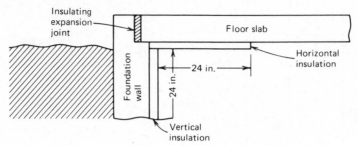

Figure 4-7.   Ideal edge insulation for a floor slab.

to the space from above. The second situation results when the air duct system is installed beneath the slab with air discharged around the perimeter of the structure. In both cases most of the heat loss is from the edge of the slab. When compared with the total heat losses of the structure this loss may not be significant; however, from the viewpoint of comfort the loss is important. Proper insulation around the perimeter of the slab is essential in severe climates to insure a warm floor.

Figure 4-7 shows ideal placement of edge insulation for a floor slab. Table 4-12 contains heat loss factors for the unheated slab and Table 4-13 contains factors for the case where the duct system is installed below the slab. Note that the heat loss factors are expressed as heat transfer rate per unit length of perimeter. For summer conditions the heat transfer to the floor slab is negligible.

Horizontal Pipes and Flat Surfaces

It is often necessary to compute the heat transfer from pipes, ducts, or flat surfaces. Typically these surfaces are in a free convection environment where radiation may also be important. This problem is complicated because the free convection heat transfer coefficient and the thermal radiation both depend on the surface temperature, which may not be accurately known. Other variables are the surface finish, ambient air temperature, and the air motion. Therefore it is customary to present overall heat transfer coefficients as shown in Table 4-14, which is for bare steel pipe and flat surfaces. The values given are based on experimental data and analytical solutions for combined free convection and thermal radiation. They are reliable when the surface temperature is known and the surrounding air is still. If the average bulk temperature of the fluid flowing in the pipe is known, a good estimate of the surface temperature can be made because of the low thermal resistance of the pipe wall.

Buried Pipe

To make calculations of the heat transfer to or from buried pipes it is necessary to know the thermal properties of the earth. The thermal conductivity of soil varies considerably with the analysis and moisture content. Typically the range is 0.33 to 1.33 Btu/(ft-hr-F) or 0.58 to 2.3 W/(m-C). A reasonable estimate can be made

Table 4-12   HEAT LOSS OF CONCRETE FLOORS LESS THAN 3 FT (0.91 m) BELOW GRADE*

Heat Loss per Unit Length of Exposed Edge

Outdoor Design Temperature		Edge Insulation^a						$(hr\text{-}ft^2\text{-}F)/Btu$ $(m^2\text{-}C)/W$
		R = 5		R = 2.5		None		
		R = 0.88		R = 0.44				
F	C	Btu/(hr-ft)	W/m	Btu/(hr-ft)	W/m	Btu/(hr-ft)	W/m	
−20 to −30	−29 to −34	50	48	60	58	75	72	
−10 to −20	−23 to −29	45	43	55	53	65	62	
0 to −10	−18 to −23	40	38	50	48	60	58	
+10 to 0	−12 to −18	35	34	45	43	55	53	
+20 to +10	−7 to −12	30	29	40	38	50	48	

*Adapted by permission from *ASHRAE GPP 158 Heating and Cooling Load Calculation Manual*, 1979.
^aInsulation is assumed to extend 2 ft (0.61 m) either horizontally under slab or vertically along foundation wall.

Table 4-13 FLOOR HEAT LOSS FOR CONCRETE SLABS WITH EMBEDDED WARM AIR PERIMETER HEATING DUCTS (PER UNIT LENGTH OF HEATED EDGE)[a],*

Outdoor Design Temperature		Edge Insulation					
		$R = 2.5$ (hr-ft²-F)/Btu $R = 0.44$ (m²-C)/W Vertical Extending Down 18 in. or 0.46 m Below Floor Surface		$R = 2.5$ (hr-ft²-F)/Btu $R = 0.44$ (m²-C)/W L-Type Extending at Least 12 in. or 0.3 m Deep and 12 in. or 0.3 m Under		$R = 5$ (hr-ft²-F)/Btu $R = 0.88$ (m²-C)/W L-Type Extending at Least 12 in. or 0.3 m Deep and 12 in. or 0.3 m Under	
F	C	Btu/(hr-ft)	W/m	Btu/(hr-ft)	W/m	Btu/(hr-ft)	W/m
−20	−29	105	101	100	96	85	82
−10	−23	95	91	90	86	75	72
0	−18	85	82	80	77	65	62
10	−12	75	72	70	67	55	53
20	−7	62	60	57	55	45	43

*Adapted by permission from ASHRAE GRP-158 Cooling and Heating Load Calculation Manual, 1979.
[a]Includes loss downward through inner area of slab.

Table 4-14 TRANSMISSION COEFFICIENTS U FOR HORIZONTAL BARE STEEL PIPES AND FLAT SURFACES,[a] Btu/(hr-ft²-F)[b],*

Pipe Size Inches	Temperature Difference F Between Pipe Surface and Surrounding Air, Air at 80 F[c]									
	50	100	150	200	250	300	350	400	450	500
$\frac{1}{2}$	2.12	2.48	2.80	3.10	3.42	3.74	4.07	4.47	4.86	5.28
$\frac{3}{4}$	2.08	2.43	2.74	3.04	3.35	3.67	4.00	4.40	4.79	5.21
1	2.04	2.38	2.69	2.99	3.30	3.61	3.94	4.33	4.72	5.14
$1\frac{1}{4}$	2.00	2.34	2.64	2.93	3.24	3.55	3.88	4.27	4.66	5.07
$1\frac{1}{2}$	1.98	2.31	2.61	2.90	3.20	3.52	3.84	4.23	4.62	5.03
2	1.95	2.27	2.56	2.85	3.15	3.46	3.78	4.17	4.56	4.97
$2\frac{1}{2}$	1.92	2.23	2.52	2.81	3.11	3.42	3.74	4.12	4.51	4.92
3	1.89	2.20	2.49	2.77	3.07	3.37	3.69	4.08	4.46	4.87
$3\frac{1}{2}$	1.87	2.18	2.46	2.74	3.04	3.34	3.66	4.05	4.43	4.84
4	1.85	2.16	2.44	2.72	3.01	3.32	3.64	4.02	4.40	4.81
$4\frac{1}{2}$	1.84	2.14	2.42	2.70	2.99	3.30	3.61	4.00	4.38	4.79
5	1.83	2.13	2.40	2.68	2.97	3.28	3.59	3.97	4.35	4.76
Vertical surface	1.84	2.14	2.42	2.70	3.00	3.30	3.62	4.00	4.38	4.79
Horizontal surface Facing upward	2.03	2.37	2.67	2.97	3.28	3.59	3.92	4.31	4.70	5.12
Horizontal surface Facing downward	1.61	1.86	2.11	2.36	2.64	2.93	3.23	3.60	3.97	4.37

[a]Values are for flat surfaces greater than 4 ft² or 0.4 m².
[b]U in W/(m²-C) equals Btu/(hr-ft²-F) times 5.678.
[c]The temperature difference in C equals F divided by 1.8. An air temperature of 27 C corresponds to 80 F.
*Adapted by permission from *ASHRAE Handbook of Fundamentals*, 1977.

using the equation for thermal resistance given in Table 4-1 with a known value of k. Thermal conductivity data for various soils and moisture contents are given in reference 1.

4-3 MOISTURE TRANSMISSION

The transfer of moisture through building materials and between the building surfaces and moist air follows theory directly analogous to conductive and convective heat transfer. Fick's law, which has the same form as Eq. (4-1),

$$\dot{m}_w = -DA\frac{dC}{dx} \tag{4-22}$$

governs the diffusion of moisture in a substance. Convective transport of moisture may be expressed as

$$\dot{m}_w = h_m A(C - C_w) \tag{4-23}$$

which is similar to Eq. (4-4). This subject is discussed in detail in Chapter 12. The important point here is that moisture moves from a location where the concentration is high to one where it is low. This movement and accumulation of moisture can cause severe damage to the structure if not controlled.

During the coldest months, the moisture concentration tends to be greatest in the interior space. It is transferred to the walls and ceilings, and if not retarded, diffuses outward into the insulation. The moisture reduces the thermal resistance of the insulation and in some cases it may freeze. Ceilings have been known to fail structurally due to an accumulation of ice.

During the summer months, the moisture transfer process is reversed. This case is not as severe as that for the winter; however, the moisture is still harmful to the insulation and condensation may occur on some inside surfaces.

The transfer of moisture and the resulting damage is controlled through the use of barriers or retardants such as aluminum foil, thin plastic film, or other such material, and through the use of ventilation. Analysis of the problem shows that the moisture retarder should be near the warmest surface to prevent moisture from entering the insulation. Because the winter months are the most critical time, the barrier is usually installed between the inside finish layer and the insulation. During the summer months, the problem can usually be controlled by natural ventilation. This is the most important reason for ventilating an attic in both summer and winter. About 0.5 cfm/ft² or (0.15 m³/(m²-min)) is required to remove the moisture from a typical attic. This can usually be accomplished through natural effects. Walls sometimes have provisions for a small amount of ventilation, but usually the outer wall covering is adequate to retard moisture.

REFERENCES

1. *ASHRAE Handbook of Fundamentals,* American Society of Heating, Refrigerating and Air-Conditioning Engineers, New York, 1977.

2. J. D. Parker, J. H. Boggs, and E. F. Blick, *Introduction to Fluid Mechanics and Heat Transfer,* Addison-Westley, Reading, Massachusetts, 1969.

3. *ASHRAE GRP 158 Heating and Cooling Load Calculation Manual,* American Society of Heating, Refrigerating, and Air-Conditioning Engineers, New York, 1979.

4. NBS Special Publication 548 "Summer Attics and Whole-House Ventilation," U.S. Department of Commerce/ National Bureau of Standards, Washington, D.C., 1978.

PROBLEMS

4-1 Compute the overall thermal resistance for a 2 in. steel pipe with 1 in. of insulation. The inside and outside film coefficients are 500 and 2 Btu/(hr-ft²-F), respectively, and the insulation has a thermal conductivity of 0.2 Btu-in./(ft²-hr-F).

4-2 What is the total unit thermal resistance for an inside partition made up of $\frac{3}{8}$ in. gypsum board on each side of 2 × 4 in. studs? (Neglect the studing.) Assume still air on each side of the partition.

4-3 Compute the overall thermal resistance of a wall made up of 100 mm face brick and 100 mm common brick with a 20 mm air gap between. There is 13 mm of gypsum plaster on the inside. Assume a 7 m/s wind velocity on the outside and still air inside.

4-4 Assuming that the 2 × 4 studs form a parallel heat transfer path, compute the unit thermal resistance for the partition of Problem 4-2. The studs are on 16 in. centers.

4-5 Compute the overall heat transfer coefficient for a frame construction wall similar to the wall shown in Table 4-7, except that this one is made of brick veneer with 3 in. insulation bats between the studs; the wind velocity is $7\frac{1}{2}$ mph instead of 15 mph.

4-6 Compute the overall heat transfer coefficient for a roof-ceiling combination similar to the one shown in Table 4-7e, except that insulation is to be installed between the rafters. Assume that the rafters are 2 × 6's, 50 × 140 mm, and that 90 mm insulation batts are used with a reflective surface ($\epsilon = 0.05$) facing upward toward the air gap. Do not correct for the wood rafters.

4-7 Adjust the overall heat transfer coefficient of Problem 4-6 for framing. Assume the 2 × 6's are on 0.5 m centers.

4-8 A wall is 12 ft wide and 8 ft high and has an overall heat transfer coefficient of 0.1 Btu/(hr-ft²-F). It contains a solid wood door, 80 × 32 × $1\frac{1}{2}$ in., and a single glass window, 60 × 30 in. Assuming parallel heat flow paths for the wall, door, and window, find the overall thermal resistance and overall heat transfer coefficient for the combination. Assume winter conditions. The window is metal sash, 80 percent glass.

4-9 A wall exactly like the one described in Table 4-7b has dimensions of 10 × 3 m. The wall has a total window area of 5 m² made of double-insulating glass with a 6 mm air space. There is a steel door, 2 × 1 m, that has a mineral fiber core. Assuming winter conditions, compute the effective overall heat transfer coefficient for the combination.

4-10   A basement wall extends 6 ft below grade and 2 ft above grade and is 6 in. thick. The inside is finished with $\frac{1}{2}$ in. insulating board, plastic vapor seal, and $\frac{1}{4}$ in. plywood paneling. Assuming parallel heat flow paths for the above and below grade parts of the wall, compute the overall heat transfer coefficient for the complete wall.

4-11   Compute the heat transfer rate per square foot through a flat, built-up roof-ceiling combination similar to that shown in Fig. 4-7*d*. Instead of the suspended acoustical tile the ceiling is metal lath and plaster with 4 in. fibrous glass batts above. Indoor and outdoor temperatures are 70 and 5 F, respectively.

4-12   Estimate the overall heat transfer coefficient for a basement floor that has been covered with carpet and fibrous pad. Assume a 4 ft depth and 32 ft width.

4-13   A house is maintained at an average temperature of 70 F while the average outdoor temperature is 40 F during the heating season (October 1 through March 31). About how much natural gas could be saved by using 1 in. of insulation around the perimeter of the slab instead of none at all? The air distribution system is under the slab, and the house plan dimensions are 40 by 50 ft. About 750 Btu of heat energy are obtained from each cubic foot of natural gas.

4-14   In an effort to save energy it is proposed to change the standard frame wall construction from 2 × 4 studs on 16 in. centers to 2 × 6 studs on 24 in. centers. Make a table similar to Table 4-7 using the same material and the 2 × 6 construction. Also add $5\frac{1}{2}$ in. of fibrous glass insulation. Compare the two different constructions with and without insulation.

4-15   Determine the overall heat transfer coefficient for a single glass window with an aluminum sash. Compare this with double insulating glass ($\frac{1}{4}$ in. air space) in a wood sash (80 percent glass).

4-16   Estimate the heat loss from the floor slab of a house with a perimeter heating system. The house has dimensions of 15 × 30 m. The insulation is made of 25 mm of expanded polystyrene installed as shown in Fig. 4-7. It is designed for an outdoor temperature of minus 12 C.

4-17   Two 3-in. steel pipes cross a basement room near the ceiling. One pipe supplies hot water to a heating unit at 150 F, and the other pipe returns water from the unit at 130 F. If the room temperature is about 75 F, estimate the heat transfer from each pipe per linear foot.

4-18   Estimate the heat transfer rate to a 100 foot length of buried steel pipe carrying chilled water at 40 F. The pipe is 30 in. deep and has a 4 in. diameter. Assume the temperature of the deep ground is 50 F. The thermal conductivity of the earth is about 8 Btu-in./(hr-ft²-F).

4-19   Estimate the heat loss from 100 m of buried hot water pipe. The mean water temperature is 49 C. The copper pipe with 10 mm of insulation, $k = 0.05$ W/(m-c), is buried 1 m below the surface and is 100 mm in diameter. Assume a thermal conductivity of the earth of 1.4 W/(m-C) and a deep ground temperature of 10 C.

4-20   Refer to Example 4-3 and compute the overall transmission coefficient U for summer conditions assuming (a) there are no reflective surfaces and (b) a reflective surface on top of the ceiling insulation.

4-21 A 6 in. diameter well 100 ft deep is to be used as a heat source and sink by circulating water through it. Assume an average heat transfer coefficient between the water and the wall of 100 Btu/hr-ft²-F, an average water temperature of 80 F, and deep ground temperature of 50 F. Estimate the heat transfer rate from the water.

4-22 Consider an attic space formed by an infinite flat roof and horizontal ceiling. The inside surface of the roof has a temperature of 135 F and the top side of the ceiling insulation has a temperature of 115 F. Compute the heat transferred by radiation and convection separately and compare them. Assume both surfaces have an emittance of 0.9.

4-23 Refer to Table 4-7 and compute the overall transmission coefficient for the same construction with 1 in. sheathing of expanded polystyrene molded beads and 3½ in. insulation batts between the studs. Compare with Table 4-7 construction with 2 × 6 studs and 5½ in. insulation batts.

4-24 Consider a long rectangular duct with inside dimensions of 12 in. × 24 in. with 1 in. of fibrous glass insulation. Estimate the approximate heat gain per ft of length if the inside air temperature is 60 F and the outside air temperature is 110 F. The average film coefficient is 7 Btu/(hr-ft²-F). Estimate an approximate outside film coefficient from Table 4-3.

4-25 Consider a frame wall construction similar to that shown in Table 4-7, case 2. The wood siding is to be replaced by brick veneer with a ¾ in. (19 mm) air gap between the brick and sheathing. (a) Compute the overall heat transfer coefficient. (b) Suppose the sheathing has a reflective foil surface facing the brick and compute the overall heat transfer coefficient. (c) Compare (a) and (b) above.

4-26 Repeat Problem 4-22, assuming that the upper surface has an emittance of 0.05 and compare the results.

4-27 Compute the overall heat transfer coefficient for (a) an ordinary vertical single glass window. (b) Assume the window has a roller shade. Compute the overall heat transfer coefficient assuming a 3½ in. (89 mm) air space between the shade and the glass. (c) Compare the results of (a) and (b) with values from Table 4-8.

4-28 A 24 ft × 40 ft (7.3 m × 12.2 m) building has a full basement with walls extending 8 ft. (2.4 m) below grade. The inside of the walls is finished with R-5 (R-0.7) insulation, a thin vapor barrier, and ½ in. (12.7 mm) gypsum board. Estimate an overall heat transfer coefficient for the walls.

4-29 Assume that the deep ground temperature is 50 F (10 C) and that the inside temperature is 68 F (20 C) in Problem 4-28 and estimate the temperatures between the wall and insulation and between the gypsum board and insulation.

4-30 The floor of the basement described in Problem 4-28 is finished with a thin vapor barrier, ⅝ in. (16 mm) particle board underlayment, and carpet with rubber pad. Estimate an overall heat transfer coefficient for the floor.

4-31 Use the temperatures given in Problem 4-29 and compute the temperature between the underlayment and the carpet pad in Problem 4-30.

4-32 A floor slab with heating ducts below has R-2.5 (R-0.44), L-type insulation. Estimate the heat loss per unit length of perimeter if the outdoor design temperature is −12 F (−24 C).

4-33 A small single story office building is constructed with a concrete slab floor with an overhead type heating system. Estimate the heat loss per unit length of perimeter if the outdoor design temperature is 5 F (−15 C). Assume (a) R-5 (R-0.88) edge insulation; (b) no edge insulation. (c) Give a good reason to use the edge insulation other than to save energy.

4-34 A large beverage cooler resembles a small building and is to be maintained at about 35 F (2 C). The walls and ceiling are well insulated and are finished on the inside with plywood. Assume that the outdoor temperature drops below 35 F (2 C) for only short periods of time during the winter. Where should the vapor barrier be located? Explain what might happen if the barrier is improperly located.

CHAPTER FIVE

Solar Radiation

The sun is the source of most of the energy used by humans. It drives the winds and ocean currents, it furnishes the energy required for plants to grow, it created our oil and coal resources, and it furnishes warmth to us both directly and indirectly.

Solar radiation has important effects on both the heat gain and heat loss of a building. This effect depends to a great extent on both the location of the sun in the sky and the clearness of the atmosphere as well as on the nature and orientation of the building. It is useful at this point to discuss ways of predicting the variation of the sun's location in the sky during the day and with the seasons for various locations on the earth's surface. It is also useful to know how to predict, for specified weather conditions, the solar irradiation of a surface at any given time and location on the earth, as well as the total radiation striking a surface over a specified period of time.

5-1 THE EARTH'S MOTION ABOUT THE SUN

The earth moves in a slightly elliptical orbit about the sun. The plane in which the earth rotates around the sun (approximately once every $365\frac{1}{4}$ days) is called the *ecliptic plane* or *orbital plane*. The mean distance from the center of the earth to the center of the sun is approximately 92.9×10^6 miles or 1.5×10^8 km. The *perihelion distance,* when the earth is closest to the sun, is 98.3 percent of the mean distance and occurs on January 4. The *aphelion distance,* when the earth is farthest from the sun, is 101.7 percent of the mean distance and occurs on July 5. Because of this the earth receives about 7 percent more total radiation in January than in July.

Figure 5-1 shows the movement of the earth around the sun. As the earth moves it also spins about its own axis at the rate of one revolution each 24 hours. There is an additional motion because of a slow wobble or gyroscopic precession of the earth. The earth's axis of rotation is tilted 23.5 degrees with respect to the orbital plane. As a result of this dual motion and tilt the position of the sun in the sky, as seen by an observer on earth, varies with the observer's location on the earth's surface and

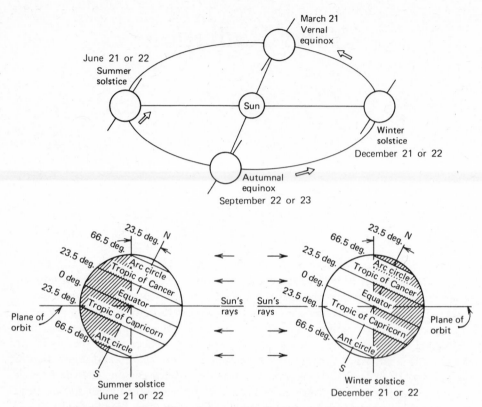

Figure 5-1. The effect of the earth's tilt and rotation about the sun.

with the time of day and the time of year. For practical purposes the sun is so small as seen by an observer on earth that it may be treated as a point source of radiation.

Figure 5-1 can be used to explain the effect of the earth's tilt and rotation about the sun. At the time of the vernal equinox (March 21) and of the autumnal equinox (September 22 or 23) the sun appears to be directly overhead at the equator and the earth's poles are equidistant from the sun. Equinox means "equal nights" and during the time of the two equinoxes all points on the earth (except the poles) have exactly 12 hours of darkness and 12 hours of daylight.

During the summer solstice (June 21 or 22) the north pole is inclined 23.5 degrees toward the sun. All points on the earth's surface north of 66.5 degrees north latitude (the Arctic Circle) are in continuous daylight, whereas all points south of 66.5 degrees south latitude (the Antarctic Circle) are in continuous darkness. Relatively warm weather occurs in the northern hemisphere and relatively cold weather occurs in the southern hemisphere. The word solstice means *sun standing still*.

During the summer solstice the sun appears to be directly overhead along the Tropic of Cancer, whereas during the winter solstice it is overhead along the Tropic of Capricorn. The *torrid zone* is the region between, where the sun is at the zenith, directly overhead, at least once during the year. In the *temperate zones* (between 23.5 and 66.5 degrees latitude in each hemisphere) the sun is never directly overhead but always appears above the horizon each day. The *frigid zones* are those zones with latitude greater than 66.5 degrees where the sun is below the horizon for at least one full day (24 hours) each year. In these two zones the sun is also above the horizon for at least one full day each year.

5-2   TIME

Because of the earth's rotation about its own axis, a fixed location on the earth's surface goes through a 24 hour cycle in relation to the sun. The earth is divided by longitudinal lines passing through the poles into 360 degrees of circular arc. Thus 15 degrees of longitude corresponds to $\frac{1}{24}$ of a day or 1 hour of time. A point on the earth's surface exactly 15 degrees west of another point will see the sun in exactly the same position as the first point after one hour of time has passed. *Universal Time* or *Greenwich Civil Time* (GCT) is the time along the zero longitude line passing through Greenwich, England. *Local Civil Time* (LCT) is determined by the longitude of the observer, the difference being four minutes of time for each degree of longitude, the more advanced time being on meridians further east. Thus when it is 12:00 noon GCT, it would be 7:00 A.M. LCT along the 75th degree west longitude meridian.

Clocks are usually set for the same reading throughout a zone covering approximately 15 degrees of longitude, although the borders of the time zone may be irregular to accommodate local geographical features. The Local Civil Time for a selected meridian near the center of the zone is called the *Standard Time*. The four standard times zones in the lower 48 states and their standard meridians are

Eastern Standard Time, EST 75 degrees
Central Standard Time, CST 90 degrees
Mountain Standard Time, MST 105 degrees
Pacific Standard Time, PST 120 degrees

In much of the United States clocks are advanced one hour during the late spring, summer, and early fall season, leading to *Daylight Savings Time.*

Whereas civil time accounts for days that are precisely 24 hours in length, solar time has slightly variable days because of nonsymmetry of the earth's orbit, irregularities of the earth's rotational speed, and other factors. Time measured by the apparent daily motion of the sun is called *solar time.*

The Local Solar Time (LST) can be calculated from the Local Civil Time (LCT) by a quantity called the *equation of time*, LST = LCT + (equation of time). Values of the equation of time are given in Table 5-1 for each week of the year (2). Values for any day may be obtained by linear interpolation.

Example 5-1.   Determine the local solar time (LST) corresponding to 11:00 A.M. CDST on February 8 in the United States at 95 degrees west longitude.

Table 5-1   THE EQUATION OF TIME*

Month	Day			
	1	8	15	22
January	−3:16	−6:26	−9:12	−11:27
February	−13:34	−14:14	−14:15	−13:41
March	−12:36	−11:04	− 9:14	− 7:12
April	− 4:11	2:07	− 0:15	1:19
May	2:50	3:31	3:44	3:30
June	2:25	1:15	0:09	− 1:40
July	− 3:33	− 4:48	− 5:45	− 6:19
August	− 6:17	5:40	− 4:35	− 3:04
September	− 0:15	2:03	4:29	6:58
October	10:02	12:11	13:59	15:20
November	16:20	16:16	15:29	14:02
December	11:14	8:26	5:13	1:47

*Reprinted from "The American Ephemeris and Nautical Almanac," U.S. Naval Observatory, Washington, D.C.

Solution.   It is first necessary to convert central daylight savings time to central standard time:

$$CST = CDST - 1 \text{ hour} = 11:00 - 1 = 10:00 \text{ A.M.}$$

Then CST is local civil time at 90 degrees west longitude. Now local civil time (LCT) at 95 degrees west is $5 \times 4 = 20$ minutes less advanced than LCT at 90 degrees west. Then

$$LCT = CST - 20 \text{ min} = 9:40 \text{ A.M.}$$

From Table 5-1 the equation of time is −14 minutes, 14 seconds. Then

$$LST = LCT + \text{equation of time}$$
$$LST = 9:40 - 0:14 = 9:26 \text{ A.M.}$$

5-3   SOLAR ANGLES

Any point on the surface of the earth can be described in relation to the sun's rays at any instant if three fundamental quantities are known. Figure 5-2 shows a point P located on the surface of the earth in the northern hemisphere. The *latitude* l is the angle between the line OP and the projection of OP on the equatorial plane. This is the same latitude that is commonly used on globes and maps to describe the location of a point with respect to the equator.

The *hour angle h* is the angle between the projection of OP on the equatorial plane and the projection on that plane of a line from the center of the sun to the center of the earth. Fifteen degrees of hour angle corresponds to one hour of time.

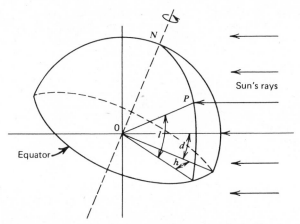

Figure 5-2.   Latitude, hour angle, and sun's declination.

The hour angle varies from zero at local solar noon to a maximum at sunrise or sunset. Solar noon occurs when the sun is at the highest point in the sky, and hour angles are symmetrical with respect to solar noon. Thus the hour angles of sunrise and sunset on a given day are identical.

The sun's *declination d* is the angle between a line connecting the center of the sun and earth and the projection of that line on the equatorial plane. Figure 5-3 shows how the sun's declination varies throughout a typical year. On a given day in the year the declination varies slightly from year to year, but for typical HVAC calculations the values from any year are sufficiently accurate. Table 5-2 shows typical values of the sun's declination for each week. More precise values may be found in reference 2.

It is convenient in HVAC computations to define the sun's position in terms of the solar altitude β, and the solar azimuth ϕ, which depend on l, h, and d.

The *solar altitude* β (sun's altitude angle) is the angle between the sun's ray and the projection of that ray on a horizontal surface, Fig. 5-4. It is the angle of the sun above the horizon. It can be shown by analytic geometry that the following relationship is true (1).

$$\sin \beta = \cos l \cos h \cos d + \sin l \sin d \qquad (5-1)$$

The *sun's zenith* angle Ψ is the angle between the sun's rays and a perpendicular to the horizontal plane at point P, Fig. 5-4. Obviously

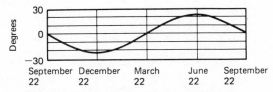

Figure 5-3.   Variation of sun's declination.

Table 5-2   THE SUN'S DECLINATION FOR A
TYPICAL YEAR (IN DEGREES AND
MINUTES)*

Month	Day			
	1	8	15	22
January	−23:08	−22:20	−21:15	−19:50
February	−17:18	−15:13	−12:55	−10:27
March	− 7:51	− 5:10	− 2:25	0:21
April	4:16	6:56	9:30	11:57
May	14:51	16:53	18:41	20:14
June	21:57	22:47	23:17	23:27
July	23:10	22:34	21:39	20:25
August	18:12	16:21	14:17	12:02
September	8:33	5:58	3:19	0:36
October	− 2:54	− 5:36	− 8:15	−10:48
November	−14:12	−16:22	−18:18	−19:59
December	−21:41	−22:38	−23:14	−23:27

*Reprinted from "The American Ephemeris and Nautical Almanac," U.S.
Naval Observatory, Washington, D.C.

$$\beta + \Psi = 90 \text{ degrees} \qquad (5\text{-}2)$$

The daily maximum altitude (solar noon) of the sun at a given location can be
shown to be

$$\beta_{noon} = 90 - |l - d| \qquad (5\text{-}3)$$

where $|(l - d)|$ is the absolute value of $(l - d)$.

The *solar azimuth* angle ϕ is the angle in the horizontal plane measured
between south and the projection of the sun's rays on that plane, Fig. 5-4. Again by
analytic geometry (1) it can be shown that

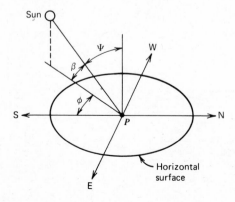

Figure 5-4.   The solar altitude, zenith angle,
and azimuth angle.

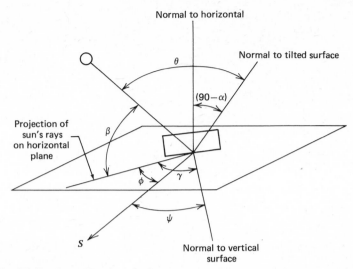

Figure 5-5. Wall solar azimuth γ, wall azimuth ψ, and angle of tilt α for an arbitrary tilted surface.

$$\cos \phi = (\sin \beta \ \sin l - \sin d) / (\cos \beta \cos l) \qquad (5\text{-}4)$$

Values of β and ϕ may be found in the *ASHRAE Handbook* or reference 3. In reference 3 the solar azimuth is measured from the north.

For a vertical surface the angle measured in the horizontal plane between the projection of the sun's rays on that plane and a normal to the vertical surface is called the *wall solar azimuth* γ. Figure 5-5 illustrates this quantity.

If ψ is the wall azimuth measured east or west from south, then obviously

$$\gamma = \phi \pm \psi \qquad (5\text{-}5)$$

The *angle of incidence* θ is the angle between the sun's rays and the normal to the surface, as shown in Fig. 5-5. The *angle of tilt* α is the angle between the surface and the normal to the horizontal surface. It may be shown that

$$\cos \theta = \cos \beta \cos \gamma \cos \alpha + \sin \beta \sin \alpha \qquad (5\text{-}5a)$$

Then for a vertical surface

$$\cos \theta = \cos \beta \cos \gamma \qquad (5\text{-}5b)$$

and for a horizontal surface

$$\cos \theta = \sin \beta \qquad (5\text{-}5c)$$

Example 5-2. Find the solar altitude and azimuth at 10:00 A.M. Central Daylight Savings Time on July 21 at 40 degrees north latitude and 85 degrees west longitude.

Solution.   The Local Civil Time (LCT) is

$$10:00 - 1:00 + 4(90 - 85) = 9:20 \text{ A.M.}$$

The equation of time is -6 min 14 sec; therefore the Local Solar Time (LST) is

$$9:20 - 0:06 = 9:14 \text{ A.M.}$$

β is calculated from Eq. (5-1).

$$h = 2 \text{ hr } 46 \text{ min} = 41.5 \text{ degrees}$$
$$\beta = \sin^{-1}(\cos 40 \cos 41.5 \cos 20.6 + \sin 40 \sin 20.6)$$
$$\beta = 49.7 \text{ degrees}$$

ϕ is calculated from Eq. (5-4)

$$\phi = \cos^{-1}\left[\frac{\sin 40 \sin 49.7 - \sin 20.6}{\cos 49.7 \cos 40}\right] = 73.8 \text{ degrees}$$

External shading of a window is effective in reducing solar heat gain to a space and may produce reductions of up to 80 percent. A window may be shaded by trees, shrubbery, another building, or by some feature of the structure itself. Figure 5-6 illustrates a window that is set back into the structure where shading may occur on the sides and top depending on the time of day and the direction the window faces. It can be shown that the dimensions x and y in Fig. 5-6 are given by

$$x = b \tan \gamma \tag{5-6}$$
$$y = b \tan \delta \tag{5-7}$$

where

$$\tan \delta = \frac{\tan \beta}{\cos \gamma}$$

and

β = sun's altitude angle from Eq. (5-1)
γ = wall solar azimuth angle $(\phi \pm \psi)$
ϕ = solar azimuth from Eq. (5-4)
ψ = wall azimuth measured east or west from the south

The following rules aid in the computation of the wall solar azimuth angle γ.
    For morning hours with walls facing east of south and afternoon hours with walls facing west of south:

$$\gamma = |(\phi - \psi)| \tag{5-8}$$

For afternoon hours with walls facing east of south and morning hours with walls facing west of south:

$$\gamma = |(\phi + \psi)| \tag{5-8a}$$

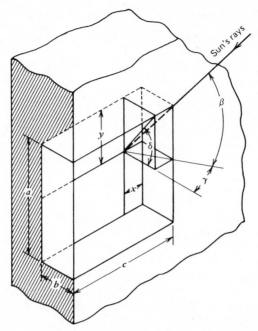

Figure 5-6. Shading of window set back from the plane of a building.

If γ is greater than 90 degrees, the surface is in the shade. Equation (5-7) can be used for an overhang at the top and perpendicular to the window provided that the overhang is wide enough for the shadow to extend completely across the window.

Example 5-2. A 4×5 ft, $\frac{1}{4}$ in. regular plate glass window faces southwest. The top of the window has a 2 ft overhang that extends a great distance on each side of the window. Compute the shaded area of the window on July 21 at 3:00 pm solar time at 40 deg N. latitude.

Solution. To find the area the dimension y from Eq. (5-7) must be computed. From Eqs. (5-1 and 5-4), β and ϕ are 47.2 and 76.7 degrees, respectively. The wall azimuth for a window facing southwest is 45 degrees. Then for a wall facing west of south and for afternoon hours:

$$\gamma = |(\phi - \psi)| = |(76.7 - 45)| = 31.7 \text{ degrees}$$

Then

$$y = b \tan \delta = \frac{b \tan \beta}{\cos \gamma}$$

$$y = \frac{2 \tan (47.2)}{\cos (31.7)} = 2.53 \text{ ft}$$

The shaded area is then

$$A_{sh} = 2.53 \times 4 = 10.2 \text{ ft}^2$$

and the sunlit portion has an area of

$$A_{sl} = A - A_{sh} = 20 - 10.2 = 9.8 \text{ ft}^2$$

The shaded portion of a window receives only diffuse radiation.

5-4 SOLAR IRRADIATION

The *mean solar constant* G_{sc} is the rate of irradiation on a surface normal to the sun's rays beyond the earth's atmosphere and at the mean earth-sun distance. According to a fairly recent study (4) the mean solar constant is

$$G_{sc} = 428 \text{ Btu}/(\text{hr-ft}^2)$$
$$= 1353 \text{ W}/\text{m}^2$$
$$= 4871 \text{ kJ}/(\text{m}^2\text{-hr})$$

Because of the inverse square law the irradiation from the sun varies about ± 3.5 percent because of the variation in distance between the sun and earth. On January 4 the irradiation is the greatest and on July 5 the least. Because of the large amount of atmospheric absorption of this radiation, and because this absorption is so variable and difficult to predict, precise values of the solar constant do not need to be used in most HVAC calculations.

The radiant energy emitted by the sun closely resembles the energy that would be emitted by a black body (an ideal radiator) at about 10,800 F or 5982 C. Figure 5-7 shows the spectral distribution of the radiation from the sun as it arrives at the outer edge of the earth's atmosphere (the upper curve). The peak radiation occurs at a wavelength of about 0.48×10^{-6} m in the green portion of the visible spectrum. Forty percent of the total energy emitted by the sun occurs in the visible portion of the spectrum, between 0.4 and 0.7×10^{-6} m. Fifty-one percent is in the near infrared region between 0.7 and 3.5×10^{-6}. About 9 percent is in the ultraviolet below 0.4×10^{-6} m.

A part of the solar radiation entering the earth's atmosphere is scattered by gas and water vapor molecules and by cloud and dust particles. The blue color of the sky is a result of the scattering of some of the shorter wavelengths from the visible portion of the spectrum. The familiar red at sunset results from the scattering of longer wavelengths by dust or cloud particles near the earth. Some radiation (particularly ultraviolet) may be absorbed by ozone in the upper atmosphere and other radiation is absorbed by water vapor near the earth's surface. That part of the radiation that is not scattered or absorbed and reaches the earth's surface is called *direct radiation*. It is accompanied by radiation that has been scattered or reemitted, called *diffuse radiation*. Radiation may also be reflected onto a surface from nearby surfaces. The total irradiation G_t on a surface normal to the sun's rays is thus made up of normal direct irradiation G_{ND}, diffuse irradiation G_d, and reflected irradiation G_R.

$$G_t = G_{ND} + G_d + G_R \tag{5-9}$$

The depletion of the sun's rays by the earth's atmosphere depends on the composition of the atmosphere (cloudiness, dust and pollutants present, atmospheric

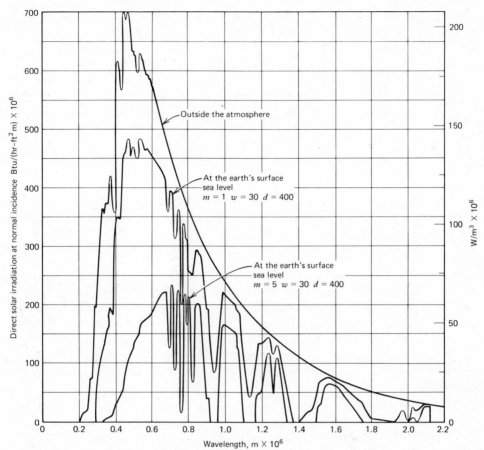

Figure 5-7.   Spectral distribution of direct solar irradiation at normal incidence during clear days. (Adapted by permission from ASHRAE Trans., 64, p. 50.)

pressure, and humidity). With a given composition on a clear day the depletion is also strongly dependent on the length of the path of the rays through the atmosphere. In the morning or evening, for example, the sun's rays must travel along a much longer path through the atmosphere than they would at noontime. Likewise the sun's rays that hit the polar regions at midday have passed through a longer atmospheric path than those that hit the tropical regions at midday. This length is described in terms of the air mass m, the ratio of the mass of atmosphere in the actual sun-earth path to the mass that would exist if the sun were directly overhead at sea level. The air mass is for practical purposes equal to the cosecant of the solar altitude β multiplied by the ratio of actual atmospheric pressure to standard atmospheric pressure.

Figure 5-7 shows the spectral distribution of direct solar radiation normally incident on a surface at sea level, with air masses equal to 1 (β = 90 degrees) and to 5 (β = 11.5 degrees), for a specified value of water vapor and dust in the air denoted by w and d. The area under each of the curves is proportional to the total irradiation that would strike a surface under that particular condition. It can be easily seen that the total radiation is significantly depleted and the spectral distribution is altered by the atmosphere.

It is important to recognize that the value of the solar constant is for a surface outside the earth's atmosphere and does not take into account the absorption and scattering of the earth's atmosphere which can be significant even for clear days. The value of the solar irradiation at the surface of the earth on a clear day is given by

$$G_{ND} = \frac{A}{\exp(B/\sin \beta)} \tag{5-10}$$

where

G_{ND} = the normal direct irradiation, Btu/(hr-ft^2) or W/m^2

A = apparent solar irradiation at air mass equal zero, Btu/(hr-ft^2) or W/m^2

B = atmospheric extinction coefficient

β = solar altitude

Values of A and B are given in Table 5-3 from reference 10 for the twenty-first day of each month. The values are representative values for average cloudless days and for an atmospheric clearness number of unity. Estimated values of atmospheric clearness numbers for various locations in the United States and values of normal direct irradiation at various times of day and for each month of the year are given in the *ASHRAE Handbook of Fundamentals*. The angle of incidence θ is used to relate the normal direct irradiation G_{ND} to the direct irradiation G_D of other surface orientations.

$$G_D = G_{ND} \cos \theta \tag{5-11}$$

Also given in Table 5-3 are values of the extraterrestrial solar radiation G_0, equation of time, solar declination, and C, the ratio of diffuse to direct normal irradiation on a horizontal surface. The parameter C does not account for the actual directional variation of the diffuse component nor its value on cloudy days. The value of C is an average for the day. The ratio actually changes with the hour of the day in a fairly predictable way, see Fig. 5-11.

To estimate the rate at which diffuse radiation $G_{d\theta}$ strikes a nonhorizontal surface on a clear day, the following equation is used

$$G_{d\theta} = CG_{ND}F_{ws} \tag{5-12}$$

In which F_{ws} is the configuration factor or angle factor between the wall and the sky. The *configuration factor* is the fraction of the diffuse radiation leaving one surface which would fall directly on another surface. This factor is sometimes referred to in the literature as the *angle factor*—or the *view, shape, interception,* or *geometrical factor*. For diffuse radiation this factor is a function only of the geometry of the surface or surfaces to which it is related. It is important to note that the configuration factor is useful for any type of diffuse radiation. For this reason information obtained in illumination, radio, or nuclear engineering studies is often useful to engineers interested in thermal radiation.

Table 5-3 EXTRATERRESTRIAL SOLAR RADIATION AND RELATED
DATA FOR TWENTY-FIRST DAY OF EACH MONTH, BASE
YEAR 1964*,[a]

	G_o Btu/(hr-ft^2)	Equation of Time, min.	Declination, deg	A Btu/(hr-ft^2)	B (Dimensionless Ratios)	C
Jan	442.7	−11.2	−20.0	390	0.142	0.058
Feb	439.1	−13.9	−10.8	385	0.144	0.060
Mar	432.5	− 7.5	0.0	376	0.156	0.071
Apr	425.3	+ 1.1	+11.6	360	0.180	0.097
May	418.9	+ 3.3	+20.0	350	0.196	0.121
June	415.5	− 1.4	+23.45	345	0.205	0.134
July	415.9	− 6.2	+20.6	344	0.207	0.136
Aug	420.0	− 2.4	+12.3	351	0.201	0.122
Sep	426.5	+ 7.5	0.0	365	0.177	0.092
Oct	433.6	+15.4	−10.5	378	0.160	0.073
Nov	440.2	+13.8	−19.8	387	0.149	0.063
Dec	443.6	+ 1.6	−23.45	391	0.142	0.057

*Reprinted by permission from *ASHRAE Handbook of Fundamentals*, 1977.
To convert Btu/(hr-ft^2) to W/m^2, multiply by 3.1525.

The symbol for configuration factor always has two subscripts describing the surface or surfaces which it describes. For example, the configuration factor, F_{12}, applies to the two surfaces numbered 1 and 2. F_{12} would be the fraction of the diffuse radiation leaving surface 1 which would fall directly on surface 2. F_{11} would be the fraction of the diffuse radiation leaving surface 1 which would fall on itself and obviously applies only to nonplane surfaces.

A very important and useful characteristic of configuration factors is the *reciprocity relationship*.

$$A_1 F_{12} = A_2 F_{21} \qquad (5\text{-}13)$$

Its usefulness is in determining configuration factors when the reciprocal factor is known or when the reciprocal factor is more easily obtained than the desired factor.

For example, the fraction of the diffuse radiation in the sky which strikes a given surface would be difficult to determine directly. The fraction of the energy that leaves the surface and "strikes" the sky directly, however, can be easily determined from the geometry

$$F_{ws} = (1 + \cos \Sigma)/2 \qquad (5\text{-}14)$$

where Σ is the tilt angle of the surface from horizontal, $\Sigma = (90 - \alpha)$.

The rate at which diffuse radiation from the sky strikes a given surface of area A_w is, per unit area of surface

$$\frac{\dot{q}}{A_w} = \frac{A_s G_d F_{sw}}{A_w}$$

by reciprocity

$$A_s F_{sw} = A_w F_{ws}$$

therefore

$$\frac{\dot{q}}{A_w} = G_d F_{ws}$$

Thus, although the computation involves the irradiation of the sky on the surface or wall, the configuration factor most convenient to use is F_{ws}, the one describing the fraction of the surface radiation that strikes the sky.

The use of the configuration factor assumes that diffuse radiation comes uniformly from the sky in all directions. This is not true for either cloudy or clear skies. In both clear and cloudy skies the location of the sun can distort the uniformity of the diffuse radiation. In the case of cloudiness, the nonuniformity of the clouds in the sky may also cause the diffuse radiation to be nonuniform. Figure 5-8 shows the variation of diffuse solar radiation with sun position on a clear day.

In determining the total rate at which radiation strikes an arbitrarily oriented surface at any time one must also consider the energy reflected onto the surface. The most common case is reflection of solar energy from the ground to a tilted surface or vertical wall. For such a case the rate at which energy is reflected to the wall is

$$G_R = G_{tH} \rho_g F_{wg} \tag{5-15}$$

where

G_R = rate at which energy is reflected onto the wall, Btu/(hr-ft²) or W/m²
G_{tH} = rate at which the total radiation (direct plus diffuse) strikes the horizontal surface or ground in front of the wall, Btu/(hr-ft²) or W/m²
ρ_g = reflectance of ground or horizontal surface
F_{wg} = configuration or angle factor from wall to ground, defined as the fraction of the radiation leaving the wall of interest that strikes the horizontal surface or ground directly. For a surface or wall at a tilt angle Σ to the horizontal

$$F_{wg} = (1 - \cos \Sigma)/2 \tag{5-16}$$

Equations (5-10, (5-11), and (5-12) are useful for design purposes where cooling loads are to be estimated because these equations are for clear days. In some cases it is desirable to estimate the actual solar radiation that might strike an arbitrarily oriented surface on a typical day or a typical series of days. An example would be the computation of the insolation on a solar collector panel. Another would be in the determination of energy requirements of a building for a simulation study. To do this accurately the direct, diffuse and reflected radiation for a variety of types of days must be determined. In many situations good insolation data may not be available

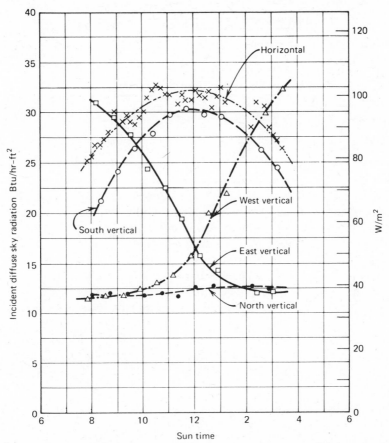

Figure 5-8. Variation of diffuse solar radiation from a clear sky. (Adapted by permission from ASHRAE Trans., 69, p. 29.)

for a given location. The best source of weather data appears to be the National Climatic Center in Ashville, North Carolina.

In the most common situation the typical weather data that are available give the total (or global) solar insolation on a horizontal surface. To use this data for making predictions of insolation on nonhorizontal surfaces, there must be some procedure for determining the relative proportion of the total horizontal radiation that is direct and that proportion that is diffuse. Each part can then be used to determine the rate at which direct and diffuse radiation strike the surface of interest. In addition, the energy reflected onto the surface must be determined.

Figure 5-9 illustrates the logic involved. The total radiation on a horizontal surface is first divided into the direct and diffuse components, step a. One way that this can be accomplished is by using the method of reference (9), shown in Fig. 5-10.

The monthly average daily total radiation is compared with the average extraterrestrial radiation (outside the earth's atmosphere). From this ratio, $\overline{K}_T$, one can

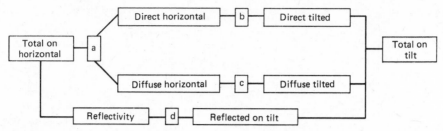

Figure 5-9. Conversion of horizontal insolation to insolation on tilted surface.

determine the ratio of the monthly average daily diffuse radiation to the monthly average daily total radiation. If hourly calculations are being made, information relating hourly diffuse to daily diffuse radiation and hourly total to daily total radiation must be used. These curves are given in Fig. 5-11.

With the total radiation thus divided one can use Eq.(5-11) to determine the direct radiation on the "tilted" surface, step b, and Eq. (5-12) to determine the diffuse radiation on the "tilted" surface, step c. Equation (5-15) is then used with the total horizontal insolation data to determine the radiation reflected, step d. The three components are then added to give the total insolation on the surface of interest.

Example 5-3. Calculate the clear day direct, diffuse and total solar radiation rate on a horizontal surface at 36 degrees north latitude and 84 degrees west longitude on June 1 at 12:00 noon CST.

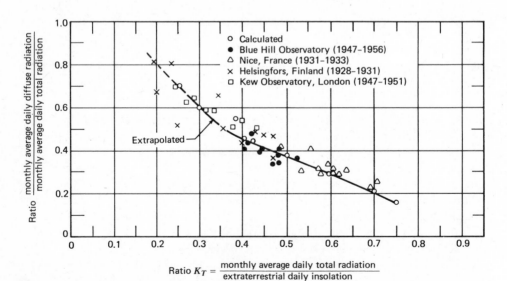

Figure 5-10. Relationship between the monthly average daily diffuse radiation and the monthly average daily total radiation for horizontal surfaces. (Reprinted by permission from *Application of Solar Energy for Heating and Cooling of Buildings,* ASHRAE GRP 170, 1977.)

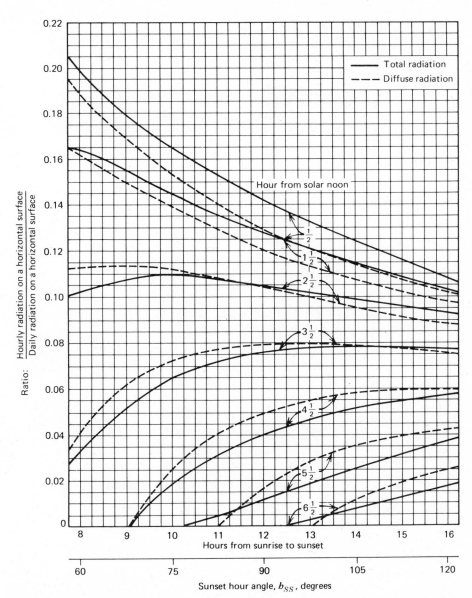

Figure 5-11.   Relationships between daily radiation and hourly radiation on a horizontal surface. (Reprinted by permission from *ASHRAE Low Temperature Engineering Application of Solar Energy*, 1967.)

Solution.   Equations (5-7) and (5-8) are used together with the parameter *A*, *B*, and *C* from Table 5-3. In order to calculate the angles β and θ used in these equations the local solar time must be calculated. The declination is obtained from Table 5-2.

$$\text{LST} = \text{LCT} + \text{Equation of time}$$

$$= 12{:}00 + \left(\frac{90 - 84}{15} \right) (60) + 0{:}02{:}25 + 12{:}26$$

The hour angle *h* is given by

$$h = \frac{(26)(15)}{60} = 6.5 \text{ deg. and } d = 21 \text{ deg. 57 min.}$$

$$\sin \beta = \cos l \cos d \cos h + \sin l \sin d$$
$$\sin \beta = (0.809)(0.994)(0.928) + (0.588)(0.374)$$
$$\sin \beta = 0.965$$

For a horizontal surface $\cos \theta = \sin \beta$

$$G_{ND} = \frac{A}{\exp (B/\sin \beta)} = \frac{345}{\exp \left(\dfrac{0.205}{0.965} \right)} = 279 \text{ Btu/ (hr-ft}^2)$$

The direct radiation is

$$G_D = G_{ND} \cos \theta = (279)(0.965) = 269 \text{ Btu/(hr-ft}^2)$$

The diffuse radiation is

$$G_d = C G_{ND} = (0.134)(279) = 37.4 \text{ Btu/(hr-ft}^2)$$

The total radiation is

$$G_t = G_D + G_d = 269 + 37.4 = 306 \text{ Btu/(hr-ft}^2)$$

A particularly useful curve, Fig. 5-12, gives the ratio of diffuse sky radiation on a vertical surface to that incident on a horizontal surface on a clear day.

Example 5-4.   Calculate the total incidence of solar radiation on a window facing south located 6 ft above the ground. In front of the window is a concrete parking area that extends 50 ft south and 50 ft to each side of the window. The window has no setback. The following parameters have been previously computed: $\beta = 69$ degrees 13 minutes, $\phi = 17$ degrees 18 minutes, $G_{ND} = 278$ Btu/(hr-ft²), $G_{tH} = 293$ Btu/(hr-ft²), $G_{dH} = 33$ Btu/(hr-ft²), $\rho_1 = 0.33$, and $F_{wg} = 0.433$.

Solution.   The angle of incidence for the window is first computed.

$$\gamma = \phi \pm \psi; \psi = 0$$
$$\gamma = \phi = 17 \text{ degrees 18 minutes}$$
$$\cos \theta = \cos \beta \cos \gamma = 0.339$$
$$G_{DV} = G_{ND} \cos \theta = 287(0.339) = 94 \text{ Btu/(hr-ft}^2)$$

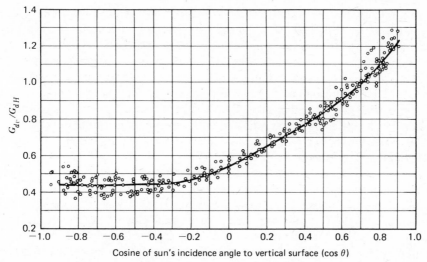

Figure 5-12. Ratio of diffuse sky radiation incident upon a vertical surface to that incident upon a horizontal surface during clear days. (Reprinted by permission from ASHRAE Trans., 69, p. 29.)

From Fig. (5-12)

$$\frac{G_{dV}}{G_{dH}} = 0.75$$

$$G_{dV} = 0.75(33) = 25 \text{ Btu}/(\text{hr-ft}^2)$$

The reflected component is given by Eq. (5-15) where

$$G_R = 0.33(293)(0.433) = 42 \text{ Btu}/(\text{hr-ft}^2)$$

Then

$$G_{tV} = G_{DV} + G_{dV} + G_R = 94 + 25 + 42 = 162 \text{ Btu}/(\text{hr-ft}^2)$$

5-5 MEASUREMENT OF DIRECT AND DIFFUSE SOLAR RADIATION

The measurement of direct, diffuse, and total radiation of the sun is essential to research in solar energy and in solar effects on HVAC systems. An instrument that measures the direct solar radiation is called a *pyrheliometer*. The term *pyranometer* describes an instrument that measures the total radiation from both sun and sky. A pyranometer can be used to measure the diffuse (sky) radiation by shielding it from the direct rays of the sun but leaving its view of the sky essentially unobstructed.

The first reliable pyrheliometer was developed by K. Angstrom of Sweden near the end of the nineteenth century (5). This instrument, still in wide use today, measures the electrical power required to heat a shaded metal strip to the same temperature as a blackened metal strip heated by the sun's rays. The response is given in the *angstrom scale of solar radiation*.

In the United States the Smithsonian Institution has done most of the developmental work on pyrheliometers. An early model used a mercury-type instrument and defined the *Smithsonian original scale of solar radiation.* In 1913 an instrument using water to cool a well-insulated, cylindrical black-body chamber was developed by C. G. Abbot and others (6). This led to a new scale called the *Smithsonian Scale of 1913.* A later improvement in which two chambers, one heated by solar rays and the other shaded and heated electrically, were both cooled by equally divided flows of water. The electric power was adjusted until the exit temperature of both streams was the same. The power required was proportional to the radiation from the sun. This led to the *Smithsonian Scale of 1932.*

In 1956 the International Radiation Conference met in Switzerland and recommended a new scale called the *International Scale of 1956.* Because it was defined in terms of two of the existing scales, no new instrument was required. The International Scale of 1956 is defined as the angstrom scale plus 1.5 percent or the Smithsonian scale of 1913 minus 2.0 percent. This new scale is thought to be within ± 1.0 percent of a true absolute scale. A comparison of the five scales described above is given in Fig. 5-13. Care should be taken to use the appropriate scale for old data or old calibrations.

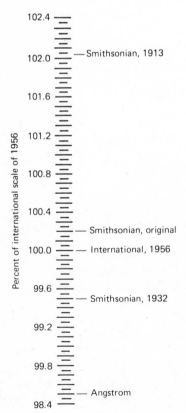

Figure 5-13.   Comparison of scales of solar radiation.

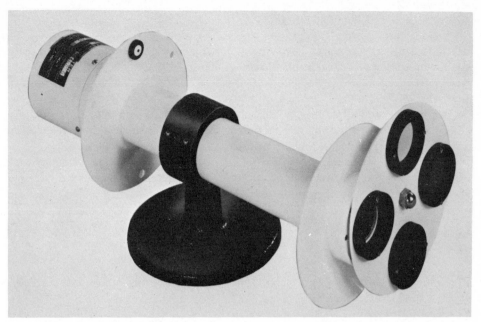

Figure 5-14. An Eppley normal incidence pyrheliometer. (Courtesy of Eppley Laboratory, Inc., Newport, Rhode Island.)

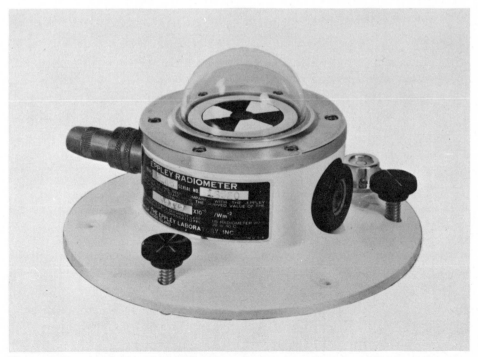

Figure 5-15. An Eppley black and white pyranometer. (Courtesy of Eppley Laboratory, Inc., Newport, Rhode Island.)

The Smithsonian water flow type of instruments require too much auxiliary equipment for convenient use in the field. A convenient, secondary-type instrument developed by Abbot at the Smithsonian is widely used and is called the *Smithsonian silver-disk pyrheliometer*. The rate of increase of temperature of a pool of mercury behind a blackened silver disk is measured by a thermopile as the sun's rays are admitted and allowed to fall on the disk. Fig. 5-14 shows this type of device used to measure normally incident direct solar radiation.

A practical device for measuring the total sun and sky radiation (pyranometer) consists of a differential thermopile with the hot junction blackened and the cold junction whitened. The thermopile is covered by a hermetically sealed, precision ground optical glass envelope (Fig. 5-15).

REFERENCES

1. James L. Threlkeld, *Thermal Environmental Engineering,* Second Edition, Prentice-Hall, Englewood Cliffs, N. J., 1970.

2. U.S. Nautical Almanac Office, "The American Ephemeris and Nautical Almanac," U.S. Naval Observatory, Washington, D.C. (published annually)

3. "Tables of Computed Altitude and Azimuth," Hydrographic Office Bulletin No. 214, U.S. Government Printing Office.

4. M. P. Thekaekara and A. J. Drummond, "Standard Values for the Solar Constant and its Spectral Components," *Nat. Phys. Sci., 6,* (1971).

5. K. Angstrom, "The Absolute Determination of the Radiation of Heat with the Electric Compensation Pyrheliometer," *Astrophysical Journal, 6,* (1899).

6. C. G. Abbot, F. E. Fowle, and L. B. Aldrich, "Improvements and Tests of Solar Constant Methods and Apparatus," *Annals of the Astrophysical Observatory, 3,* (1913).

7. John A. Duffie and William A. Beckman, *Solar Energy Thermal Processes,* Wiley-Interscience, New York, 1974.

8. *Cooling and Heating Load Calculation Manual,* ASHRAE GRP 158, American Society of Heating, Ventilating, and Air Conditioning Engineers, New York, 1979.

9. *Applications of Solar Energy for Heating and Cooling of Buildings,* ASHRAE GRP 170, American Society of Heating, Ventilating, and Air Conditioning Engineers, New York, 1977.

10. ASHRAE Handbook and Product Directory-Applications, American Society of Heating, Ventilating, and Air Conditioning Engineers, New York, 1978.

11. ASHRAE Handbook of Fundamentals, American Society of Heating, Ventilating, and Air Conditioning Engineers, New York, 1977.

PROBLEMS

5-1   Find the local solar time (LST) on August 22 for the following local times and locations:
(a) 9:00 A.M. EST and 70 degrees west longitude
(b) 1:00 P.M. CST and 95 degrees west longitude
(c) 10:00 A.M. MST and 100 degrees west longitude
(d) 3:00 P.M. PST and 125 degrees west longitude

5-2   What are the hour angles corresponding to the following local solar times:
(a) 9:32 A.M., (b) 11:38 A.M., (c) 3:20 P.M, and (d) 12:38 P.M.

5-3   Calculate the sun's altitude and azimuth angles at 9:00 A.M. solar time on September 21 at 40 degrees north latitude. Compare your results with values shown in Table 7, Chapter 26, (11).

5-4   Determine the solar time and azimuth angle for sunrise at 50 degrees north latitude on (a) June 15 and (b) December 15.

5-5   On what month, day, and time does the maximum solar altitude angle β occur in (a) Stillwater, Oklahoma, 36 degrees north latitude, (b) Chicago, Illinois, 42 degrees north latitude, and (c) central Florida, 29 degrees north latitude.

5-6   Compute the wall solar azimuth γ for a surface facing 12 degrees west of south located at 40 degrees north latitude and 90 degrees west longitude on October 1 at 3:30 P.M. central daylight saving time (CDST).

5-7   Calculate the angle of incidence for the surface of Problem 5-6 for (a) a vertical orientation, and (b) a 20 degree tilt from the vertical.

5-8   For a location at 85 degrees west longitude and 43 degrees north latitude on July 8, determine (a) the incidence angle of the sun for a horizontal surface at 4:00 P.M. central daylight saving time and (b) the time of sunset in central daylight saving time.

5-9   Calculate the angle of incidence at 10:30 A.M. EST on July 22 for 36 degrees north latitude and 78 degrees west longitude for (a) a horizontal surface, (b) a surface facing southeast, and (c) a surface inclined 40 degrees from the vertical and facing south.

5-10   Calculate the total clear day irradiation of a surface tilted at an angle of 30 degrees from the vertical located at 36 degrees north latitude and 78 degrees west longitude on July 21 at 2:00 P.M. central daylight saving time. The surface faces the southwest. Neglect reflected radiation.

5-11   Compute the reflected irradiation of a window facing southwest over a large lake on a clear day. The location is 36 degrees north latitude and 92 degrees west longitude. The time is June 15 at 3:00 P.M. CST. Assume the water has a diffuse reflectance of 0.06.

5-12   Determine the magnitudes of direct, diffuse, and reflected clear day solar radiation incident upon a small vertical surface facing south on March 1 at solar noon for a location at 36 degrees north latitude having a clearness number of 0.95. The reflecting surface is snow-covered ground of infinite extent with a diffuse reflectance of 0.85.

5-13 Estimate the total clear day irradiation of a roof with a one-to-one slope that faces southwest at 36 degrees north latitude. The date is August 22 and the time is 1:00 P.M. LST. Include reflected radiation from the ground with a reflectance of 0.3. The view factor from the roof to the ground is given approximately by $F_{rg} = (\frac{1}{2}) (1 - \cos \Sigma)$, where Σ is the slope angle of the roof.

5-14 Compute the time for sunrise and sunset on July 22 in (a) Stillwater, Oklahoma, (b) Chicago, Illinois, (c) Tallahassee, Florida, and (d) San Francisco, California.

5-15 Write a computer program to predict the altitude and zenith angles for the sun and the normal direct irradiation. Make a table of these quantities for the twenty-first day of each month for each hour between sunrise and sunset.

5-16 A south-facing window is 4 ft wide by 6 ft tall, and is set back into the wall a distance of 8 in. For Atlanta, Georgia, estimate the percent of the window that is shaded for:
(a) April 22 @ 9:00 A.M. solar time
(b) July 22 @ 12:00 noon solar time
(c) Sept. 22 @ 5:00 P.M. solar time

5-17 Work Problem 5-16 assuming a long 2 ft overhang located 2 ft above the top of the window.

5-18 Determine the amount of diffuse, direct, and total radiation that would strike a south facing surface tilted at 45 degrees on a clear December 22 in St. Louis, Missouri:
(a) at 12 noon solar time
(b) at 3 P.M. solar time
Express your answer in Btu/(hr-ft²). Neglect ground reflectance.

5-19 For the hours of 1 P.M., 2 P.M., 3 P.M., 4 P.M., 5 P.M., 6 P.M., and 7 p.m. CDST, estimate the rate at which solar energy will strike a west facing window, 3 ft wide by 5 ft high, with no setback. Assume a clear June 21 day at 40 deg N latitude and 84 deg W longitude.

5-20 Work Problem 5-19 assuming a 6 in. setback for the window.

5-21 Work Problem 5-19 for a clear day on December 21.

5-22 Work Problem 5-19 assuming a long overhang of 2 ft which is 3 ft above the top of the window.

5-23 Work Problem 5-18 assuming a clear day on July 21.

CHAPTER SIX

Space Heat Load

Prior to the design of the heating system an estimate must be made of the maximum probable heat loss of each room or space to be heated. There are two kinds of heat losses: (1) the heat transmitted through the walls, ceiling, floor, glass, or other surfaces, and (2) the heat required to warm outdoor air entering the space.

The actual heat loss problem is transient because the outdoor temperature, wind velocity, and sunlight are constantly changing. The transfer function method, discussed in Chapter 7 in connection with the cooling load, may be used under winter conditions to account for changing solar radiation, outdoor temperature, and the energy storage capacity of the structure. During the coldest months, however, sustained periods of very cold, cloudy, and stormy weather with relatively small variation in outdoor temperature may occur. In this situation heat loss from the space will be relatively constant and in the absence of internal heat gains will peak during the early morning hours. Therefore, for design purposes the heat loss is usually estimated for steady state heat transfer for some reasonable design temperature. Transient analyses are often used to study the actual energy requirements of a structure in simulation studies. In such cases solar effects and internal heat gains are taken into account.

Here is the general procedure for calculation of design heat losses of a structure:

1. Select the outdoor design conditions: temperature, humidity, and wind direction and speed.
2. Select the indoor design conditions to be maintained.
3. Estimate the temperature in any adjacent unheated spaces.
4. Select the transmission coefficients and compute the heat losses for walls, floors, ceilings, windows, doors, and floor slabs.
5. Compute the heat load due to infiltration.
6. Compute the heat load due to outdoor ventilation air. This may be done as part of the air quantity calculation.
7. Sum the losses due to transmission and infiltration.

155

6-1   OUTDOOR DESIGN CONDITIONS

The ideal heating system would provide enough heat to match the heat loss from the structure. However, weather conditions vary considerably from year to year, and heating systems designed for the worst weather conditions on record would have a great excess of capacity most of the time. The failure of a system to maintain design conditions during brief periods of severe weather is usually not critical. However, close regulation of indoor temperature may be critical for some industrial processes.

Table C-1 contains outdoor temperatures that have been recorded for selected locations in the United States and Canada. The data are based on official weather station records of the U.S. Weather Bureau, U.S. Air Force, U.S. Navy, and the Canadian Department of Transport. At this point we shall consider only data pertaining to the winter months. The $97\frac{1}{2}$ percent value is the temperature equaled or exceeded $97\frac{1}{2}$ percent of the total hours (2160) in December, January, and February. During a normal winter there would be about 54 hours at or below the $97\frac{1}{2}$ percent value. For the Canadian stations the $97\frac{1}{2}$ percent value pertains only to hours in January.

The *ASHRAE Handbook* has a more extensive tabulation of weather data; it also includes temperatures that are equaled or exceeded 99 percent of the total hours in December, January, and February.

The outdoor design temperature should generally be the $97\frac{1}{2}$ percent value as specified by ASHRAE Standard 90 A Energy Conservation in New Building Design (5). If the structure is of lightweight construction (low heat capacity), poorly insulated, has considerable glass, and space temperature contol is critical, however, the 99 percent values should be considered. The designer must remember that should the outdoor temperature fall below the design value for some extended period, the indoor temperature may do likewise. The performance expected by the owner is a very important factor, and the designer should make clear to the owner the various factors considered in the design.

Unusual wind conditions may result from the location of the structure on a hill or between two other large structures. Abnormally high wind can cause a peak heat load at outdoor temperatures above the design temperature.

6-2   INDOOR DESIGN CONDITIONS

The main purpose of Chapter Three was to define indoor conditions that make most of the occupants comfortable. Therefore the theories and data presented there should serve as a guide to the selection of the indoor temperature and humidity for heat loss calculation. It should be kept in mind, however, that the purpose of heat load calculations is to obtain data on which the heating system components are sized. In most cases, the system will never be set to operate at the design conditions. Indeed, it is practically impossible to control the environment to operate at the design point. Therefore, the use and occupancy of the space is a general consideration from the design temperature point of view. Later, when the energy requirements of the building are computed, the actual conditions in the space and outdoor environment, including internal heat gains, must be considered.

The indoor design temperature should be kept relatively low so that the heating equipment will not be oversized. ASHRAE Standard 90 A specifies 72 F or 22 C. Even properly sized equipment operates under partial load, at reduced efficiency, most of the time; therefore, any oversizing aggravates this condition and lowers the overall system efficiency. The indoor design value of relative humidity should be compatible with a healthful environment and the thermal and moisture integrity of the building envelope.

Frequently, unheated rooms or spaces exist in a strucure. These spaces will be at temperatures between the indoor and outdoor design temperatures discussed above. The temperature in an unheated space may be estimated as follows assuming steady state heat transfer:

$$\dot{q} = \frac{t_i - t_u}{R_i'} = \frac{t_u - t_o}{R_o'} = \frac{t_i - t_o}{R'} \tag{6-1}$$

where

$$R' = R_i' + R_o' = \frac{1}{U_i A_i} + \frac{1}{U_o A_o} \tag{6-2}$$

and the subscripts refer to inside, unheated, and outside. From Eq. (6-1)

$$t_u = \frac{R_o'}{R'}(t_i - t_o) + t_o \tag{6-3}$$

Note that R_i' takes into account all the surfaces through which heat is lost to the unheated space, and R_o' includes all surfaces through which heat is lost from the unheated space.

Example 6-1. Estimate the temperature in the attic of Example 4-3 if the indoor and outdoor design temperatures are 75 F and 0 F, respectively.

Solution. Referring to Example 4-3, the ceiling unit thermal resistance is

$$R_c = \frac{1}{0.051} = 19.61$$

whereas the unit thermal resistance for the roof relative to one square foot of ceiling area is

$$R_r = \frac{1}{0.464(1.414)} = 1.52$$

The total unit resistance is

$$R = R_c + R_r = 21.13$$

By using Eq. (6-3) with $R_o = R_r$, we get

$$t_u = \frac{1.52}{21.13}(75 - 0) + 0 \approx 5 \text{ F}$$

Example 6-2.   Estimate the temperature in the unheated room shown in Fig. 6-1. The structure is built on a slab. The exterior walls of the unheated room have an overall heat transfer coefficient of 0.2 Btu/(hr-ft²-F), the ceiling-attic-roof combination has an overall coefficient of 0.07 Btu/(hr-ft²-F), and the interior walls have an overall coefficient of 0.06 Btu/(hr-ft²-F). Inside and outdoor design temperatures are 72 F and −5 F, respectively.

Solution.   The heat transferred from the heated to the unheated room has a single path, neglecting the door, and may be represented as

$$\dot{q} = U_i A_i (t_i - t_u) \tag{6-4}$$

The heat transferred from the unheated room to the outdoor air has parallel paths through the ceiling and walls (neglecting the door and floor):

$$\dot{q} = U_c A_c (t_u - t_o) + U_o A_o (t_u - t_o) \tag{6-5}$$

The heat loss to the floor can be neglected because of the anticipated low temperature in the room. Equation (6-5) may be written

$$\dot{q} = \frac{(t_u - t_o)}{R'_c} + \frac{(t_u - t_o)}{R'_o} = \frac{(t_u - t_o)}{R'} \tag{6-6}$$

where

$$\frac{1}{R'} = \frac{1}{R'_c} + \frac{1}{R'_o} = U_c A_c + U_o A_o \tag{6-7}$$

By combining Eqs. (6-4) and (6-6), we obtain

$$\frac{(t_i - t_u)}{R'_i} = \frac{(t_u - t_o)}{R'} \tag{6-8}$$

and solving for t_u,

$$t_u = \frac{t_i + (R'_i/R')t_o}{1 + (R'_i/R')} \tag{6-9}$$

$$R'_i = \frac{1}{U_i A_i} = \frac{1}{0.06(8)(22)} = 0.095$$

$$\frac{1}{R'} = U_c A_c + U_o A_o = 0.07(120) + 0.2(22)8 = 43.6$$

$$R' = 0.023 \text{ (hr-F)/Btu}$$

Then from Eq. (6-9)

$$t_u = \frac{72 + (0.095/0.023)(-5)}{1 + (0.095/0.023)} = 10 \text{ F}$$

The assumption of negligible heat transfer through the slab is justified by the result.

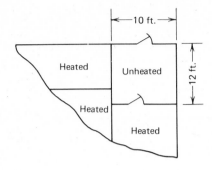

Figure 6-1. A structure with an attached un heated room.

The temperature of unheated basements is generally between the ground temperature (about 50 F or 10 C) and the inside design temperature unless there are many windows. Therefore a reasonable estimate of the basement temperature is not difficult.

The temperature in unheated crawl spaces may be estimated using Eq. (6-3) and neglecting the heat loss to the ground.

6-3 CALCULATION OF HEAT LOSSES

The heat transferred through walls, ceiling, roof, window glass, floors, and doors is all sensible heat transfer, referred to as *transmission heat loss* and computed from

$$\dot{q} = UA(t_i - t_o) \tag{6-10}$$

A separate calculation is made for each different surface in each room of the structure. To insure a thorough job in estimating the heat losses, a worksheet such as that shown in Fig. 6-2 should be used. The worksheet provides a convenient and orderly way of recording all the coefficients and areas. Summations are conveniently made by room and for the complete structure. The overall heat transfer coefficient U of Eq. (6-10) is determined as discussed in Chapter 4, whereas the area A is the net area for the given component for which U was calculated.

All structures have some *air leakage or infiltration.* This means a heat loss because the cold dry outdoor air must be heated to the inside design temperature and moisture must be added to increase the humidity to the design value. The heat required to increase the temperature is given by

$$\dot{q}_s = \dot{m}_o c_p(t_i - t_o) \tag{6-11}$$

where

$\dot{m}_o$ = mass flow rate of the infiltrating air, lbm/hr or kg/s
c_p = specific heat capacity of the moist air, Btu/(lbm-F) or J/(kg-C)

Infiltration is usually estimated on the basis of volume flow rate at outdoor conditions. Equation (6-11) then becomes

HEAT LOSS WORKSHEET

		Total House																		
1	Room																			
2	Exposed Wall, feet																			
3	Room Dimensions, feet																			
4	Ceiling Height, feet / Direction Room Faces																			

TYPE EXPOSURE	Constr.	Coeff.	Δt	Area or Length	Btuh		Area or Length	Btuh		Area or Length	Btuh		Area or Length	Btuh	
					Sens.	Lat.		Sens.	Lat.		Sens.	Lat.		Sens.	Lat.
5 Gross Exposed Walls and Partitions	a														
	b														
	c														
	d														
6 Windows and Doors	a														
	b														
	c														
7 Net Exposed Walls and Partitions	a														
	b														
	c														
	d														
8 Ceilings	a														
	b														
9 Floors	a														
	b														
10 Infiltration Windows and Doors	North														
	West														
	South														
	East														
11 Ventilation	Resid														
12 Sub Total Btuh Loss															
13 Duct Loss, Btuh															
14 Total Loss, Btuh															
15 Heat Sources, Btuh															
16 Air Quantity, cfm															
17															
18															

Figure 6-2. Heat loss worksheet.

$$\dot{q}_s = \frac{\dot{Q}c_p(t_i - t_o)}{v_o} \tag{6-11a}$$

where

$\dot{Q}$ = volume flow rate, ft³/hr or m³/s
v_o = specific volume, ft³/lbm or m³/kg

The latent heat required to humidify the air is given by

$$\dot{q}_l = \dot{m}_o(W_i - W_o)i_{fg} \tag{6-12}$$

where

$(W_i - W_o)$ = difference in design humidity ratio, lbmv/lbma or kgv/kga
i_{fg} = latent heat of vaporization at indoor conditions, Btu/lbmv or J/kgv

In terms of volume flow rate, Eq. (6-12) becomes

$$\dot{q}_l = \frac{\dot{Q}}{v_o}(W_i - W_o)i_{fg} \tag{6-12a}$$

It is easy to show, using Eqs. (6-11a) and (6-12a), that infiltration can account for a large portion of the heating load.

Two methods are used in estimating air infiltration in building structures. In one method the estimate is based on the characteristics of the windows and doors and the pressure difference between inside and outside. This is known as the *crack method* because of the cracks around window sash and doors. The other approach is the *air change method,* which is based on an assumed number of air changes per hour for each room depending on the number of windows and doors. The crack method is generally considered to be the most accurate when the window and pressure characteristics can be properly evaluated. However, the accuracy of predicting air infiltration is restricted by the limited information on the air leakage characteristics of the many components that make up a structure (8). The pressure differences are also difficult to predict because of variable wind conditions and stack effect in tall buildings.

Air Change Method

Experience and judgment are required to obtain satisfactory results with this method. The data of Table 6-1 (1) may be used with reasonable precision for residential and light commercial applications. The values given are for each room. A total allowance of one half the sum of the individual rooms should be taken since air entering on the windward side is assumed to leave the building on the leeward side. The heating load due to infiltration should generally be assigned to those rooms on the windard side of the building where the air enters.

Table 6-1   AIR CHANGES TAKING PLACE UNDER AVERAGE
CONDITIONS IN RESIDENCES, EXCLUSIVE OF AIR
PROVIDED FOR VENTILATION*,^{a}

Kind of Room or Building	Number of Air Changes Taking Place per Hour
Rooms with no windows or exterior doors	$\frac{1}{2}$
Rooms with windows or exterior doors on one side	1
Rooms with windows or exterior doors on two sides	$1\frac{1}{2}$
Rooms with windows or exterior doors on three sides	2
Entrance halls	2

*Reprinted by permission for *ASHRAE Handbook of Fundamentals,* 1977.
^{a}For rooms with weather-stripped windows or with a storm sash, use $\frac{2}{3}$ these values.

Crack Method

Outdoor air infiltrates the indoor space through cracks around doors, windows, lighting fixtures, joints between walls and floor, and even through the building material itself. The amount depends on the total area of the cracks, the type of crack, and the pressure difference across the crack. The volume flow rate of infiltration may be represented by

$$\dot{Q} = A \, C \, \Delta P^n \qquad (6\text{-}13)$$

where

A = cross-sectional area of the cracks
C = flow coefficient, which depends on the type of crack and the nature of the flow in the crack
ΔP = outside-inside pressure difference, $P_o - P_i$
n = exponent that depends on the nature of the flow in the crack, $0.4 < n < 1.0$.

Experimental data are required to use Eq. (6-13) directly; however, the relation is useful in understanding the problem. For example, Fig. 6-3 shows the leakage rate for residential-type windows and doors as a function of the pressure difference and the type of crack. The curves clearly exhibit the behavior of Eq. (6-13).

The pressure difference of Eq. (6-13) results from three different effects:

$$\Delta P = \Delta P_w + \Delta P_s + \Delta P_p \qquad (6\text{-}14)$$

where

ΔP_w = pressure difference due to the wind
ΔP_s = pressure difference due to the stack effect
ΔP_p = pressure difference due to building pressurization

All of the pressure differences are positive when each causes flow of air to the inside of the building.

The pressure difference due to the wind results from an increase or decrease in air velocity and is described by

$$\Delta P_w = \frac{\rho}{2g_c}(\overline{V}_w^2 - \overline{V}_f^2) \tag{6-15}$$

where ΔP_w has the unit of lbf/ft^2 or Pascals when consistent English or SI units are used. The velocity $\overline{V}_f$ is the final velocity of the wind at the building boundary. Note that ΔP_w is positive when $\overline{V}_w > \overline{V}_f$ which gives an increase in pressure. The velocity $\overline{V}_f$ is not known or easily predictable, therefore, it is assumed equal to zero in this application and a pressure coefficient, defined by

$$C_p = \Delta P_w / \Delta P_{wt}, \tag{6-16}$$

is used to account for the fact that $\overline{V}_f$ is not zero. The pressure difference ΔP_{wt} is the computed pressure difference when $\overline{V}_f$ is zero. The pressure coefficient may be positive or negative. Finally, Eq. (6-15) may be written

$$\frac{\Delta P_w}{C_p} = \frac{\rho}{2g_c}\overline{V}_w^2 \tag{6-17}$$

The pressure coefficient depends on the shape and orientation of the building with respect to the wind. To satisfy conditions of flow continuity, the air velocity must increase as it flows around or over a building; therefore, the pressure coefficient will change from a positive to a negative value in going from the windward to the leeward side. The pressure coefficients will also depend on whether the wind approaches normal to the side of the building or at an angle. Table 6-2 gives some approximate values for a rectangular building with normal and quartering wind.

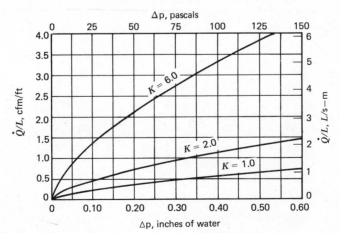

Figure 6-3. Window and residential type door infiltration characteristics. (Reprinted by permission from ASHRAE GRP 158, *Cooling and Heating Load Calculation Manual*, 1979.)

Table 6-2   PRESSURE
COEFFICIENTS FOR A
RECTANGULAR
BUILDING

Building Wall	Normal Wind	Quartering Wind
Windward	0.95	0.70
Sides	−0.4	—
Leeward	−0.15	−0.50

Notice that the pressure coefficients are consistent with the fact that air will generally flow into the building on the windward side and out on the leeward sides.

The stack effect occurs when the air density differs between the inside and outside of a building. On winter days, the lower outdoor temperature causes a higher pressure at ground level on the outside and consequent infiltration. Buoyancy of the warm inside air leads to upward flow, a higher inside pressure at the top of the building, and exfiltration of air. In the summer, the process reverses with infiltration in the upper portion of the building and exfiltration in the lower part.

Considering only the stack effect, there is a level in the building where no pressure difference exists. This is defined as the neutral pressure level. Theoretically, the neutral pressure level will be at the mid-height of the building if the cracks and other openings are distributed uniformly in the vertical direction. When larger openings predominate in the lower portion of the building, the neutral pressure level will be lowered. Similarly, the neutral pressure level will be raised by larger openings in the upper portion of the building. Normally the larger openings will occur in the lower part of the building due to doors. The theoretical pressure difference with no internal separations is given by (7)

$$\Delta P_{st} = \frac{P_o h}{R_a} \frac{g}{g_c} \left(\frac{1}{T_o} - \frac{1}{T_i} \right)$$ (6-18)

where

P_o = outside pressure, psia or Pascals
h = vertical distance from neutral pressure level, ft or m
T_o = outside temperature, R or K
T_i = inside temperature, R or K
R_a = gas constant for air, (ft-lbf)/(lbm-R) or J/(kg-K)

The floors in a conventional building offer resistance to vertical air flow. Furthermore, this resistance varies depending on how stairwells and elevator shafts are sealed. When the resistance can be assumed equal for each floor, a single correction, called the *draft coefficient,* can be used to relate the actual pressure difference ΔP_s to the theoretical value, ΔP_{st}.

$$C_d = \Delta P_s / \Delta P_{st} \tag{6-19}$$

The flow of air from floor to floor causes a decrease in pressure at each floor; therefore, ΔP_s will be less than ΔP_{st} and C_d is less than one. Using the draft coefficient, Eq. (6-18) becomes

$$\Delta P_s = \frac{C_d P_o hg}{R_a g_c} \left(\frac{1}{T_o} - \frac{1}{T_i} \right) \tag{6-20}$$

Figure 6-4 is a plot of Eq. (6-20) for an inside temperature of 75 F or 24 C, sea level outside pressure, and winter temperatures; however, Fig. 6-4 can be used for summer stack effect with little loss in accuracy.

The draft coefficient depends on the tightness of the doors in the stairwells and elevator shafts. Values of C_d range from 1.0 for buildings with no doors in the stairwells to about 0.65 for a modern office building.

Pressurization of the indoor space is accomplished by introducing more makeup

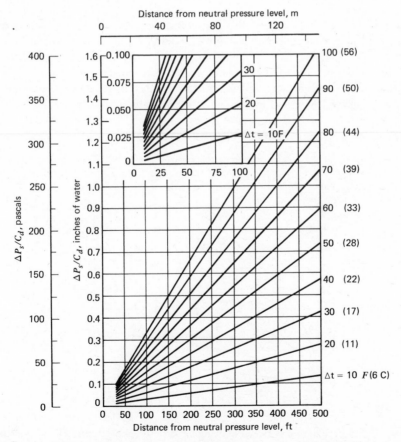

Figure 6-4. Pressure difference due to stack effect. (Reprinted by permission from ASH-RAE GRP 158, *Cooling and Heating Load Calculation Manual,* 1979.)

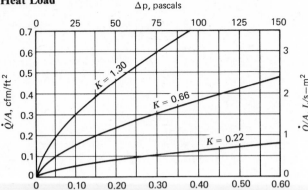

Figure 6-5. Curtain wall infiltration for one room or one floor. (Reprinted by permission from ASHRAE GRP 158, *Cooling and Heating Load Calculation Manual,* 1979.)

air than exhaust air. Prediction of this pressure difference, ΔP_p, is difficult because it depends on the pressure differences due to wind and stack effect. When building pressurization is used, it will vary from zero to about one third of the pressure difference due to wind on the windward side, except ΔP_p is negative.

Figures 6-3, 6-5, and 6-6 give the infiltration rates, based on experimental evidence, for residential windows and doors, curtain walls, and commercial swinging doors. Note that the general procedure is the same in all cases, except that curtain wall infiltration is given per unit of wall area rather than crack length. The pressure

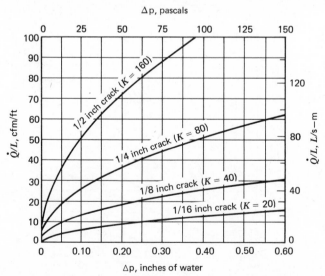

Figure 6-6. Infiltration through closed swinging door cracks. (Reprinted by permission from ASHRAE GRP 158, *Cooling and Heating Load Calculation Manual,* 1979.)

differences are estimated by the methods discussed above and the values for the coefficient K are given in Tables 6-3, 6-4, and 6-5. In the case of residences, the use of storm sash and storm doors is common. Figure 6-7 gives data for this case. Note that the addition of a storm sash with crack length and a K value equal to the prime window does not reduce the infiltration by one half but by about 35 percent.

Commercial buildings often have a rather large number of people going and coming which can increase infiltration significantly. Figures 6-8 and 6-9 have been developed to estimate this kind of infiltration for swinging doors. The infiltration rate per door is given in Fig. 6-8 as a function of the pressure difference and a traffic coefficient which depends on the traffic rate and the door arrangement. Figure 6-9 gives the traffic coefficients as a function of the traffic rate and two door types. Single bank doors open directly into the space; however, they may be two or more doors at one location. Vestibule-type doors are best characterized as two doors in series so as to form an air lock between them. These doors often appear as two pairs of doors in

Table 6-3   WINDOW CLASSIFICATION*

	Wood Double-Hung (Locked)	**Other Types**
Tight Fitting Window $K = 1.0$	Weatherstripped Average Gap (1/64 in crack)	Wood Casement and Awning Windows; Weatherstripped
		Metal Casement Windows; Weatherstripped
Average Fitting Window $K = 2.0$	Non-Weatherstripped Average Gap (1/64 in. crack)	All Types of Vertical and Horizontal Sliding Windows; Weatherstripped. Note: if average gap (1/64 in. crack) this could be tight fitting window
	or Weatherstripped Large Gap (3/32 in. crack)	Metal Casement Windows; Non-Weatherstripped Note: if large gap (3/32 in. crack) this could be a loose fitting window
Loose Fitting Window $K = 6.0$	Non-Weatherstripped Large Gap (3/32 in. crack)	Vertical and Horizontal Sliding Windows; Non-Weatherstripped

*Reprinted by permission from *ASHRAE GRP 158 Cooling and Heating Load Calculation Manual,* 1979.

Table 6-4   RESIDENTIAL-TYPE DOOR CLASSIFICATION*

Tight Fitting Door $K = 1.0$	Very small perimeter gap and perfect fit weatherstripping—often characteristic of new doors
Average Fitting Door $K = 2.0$	Small perimeter gap having stop trim fitting properly around door and Weatherstripped
Loose Fitting Door $K = 6.0$	Large perimeter gap having poor fitting stop trim and weatherstripped <div align="center">or</div>Small perimeter gap with no weatherstripping

*Reprinted by permission from *ASHRAE GRP 158 Cooling and Heating Load Calculation Manual,* 1979.

series, which amounts to two vestibule-type doors. Reference 1 gives data for revolving doors, commonly used in large high-rise buildings with high traffic rates.

Buildings may be put into two categories for calculation of infiltration. These are low rise, with less than about five stories, and high rise, with more than about five stories. The stack effect tends to be negligible in low-rise buildings and wall infiltration is usually very small; therefore, only wind effects and crackage need be considered. In high-rise buildings, the stack effect may be dominant, with a relatively large amount of leakage through the walls and around fixed window panels. All pressure effects as well as window, door, and wall leakage should be considered for high-rise buildings.

Table 6-5   CURTAIN WALL CLASSIFICATION*

Leakage Coefficient	Description	Curtain Wall Construction
$K = 0.22$	Tight Fitting Wall	Constructed under close supervision of workmanship on wall joints. When joints seals appear inadequate they must be re-done
$K = 0.66$	Average Fitting Wall	Conventional construction procedures are used
$K = 1.30$	Loose Fitting Wall	Poor construction quality control or an older building having separated wall joints

*Reprinted by permission from *ASHRAE GRP 158 Cooling and Heating Load Calculation Manual,* 1979.

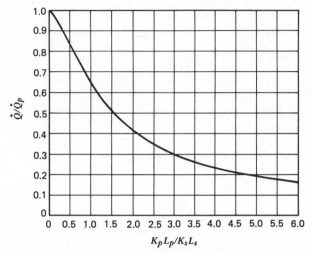

Figure 6-7. Infiltration for storm-prime combination windows. (Reprinted by permission from ASHRAE GRP 158, *Cooling and Heating Load Calculation Manual,* 1979.)

Theoretically, it is possible to predict which sides of a building will experience infiltration and those with exfiltration by use of the pressure coefficient. However, buildings usually do not have uniformly distributed openings on all sides. This will be particularly true for low-rise buildings. Because flow continuity must be satisfied, it is recommended that the infiltration for low-rise buildings be based on the crack length for the windward side of the structure, but never less than one half the total crack length for the building. The pressure coefficient approach is more feasible for high-rise buildings because the stack effect tends to cause infiltration at the lower

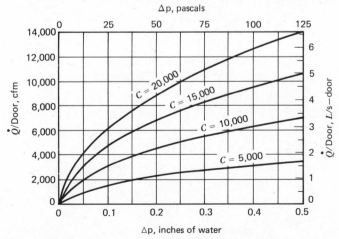

Figure 6-8. Swinging door infiltration characteristics with traffic. (Reprinted by permission from ASHRAE GRP 158, *Cooling and Heating Load Calculation Manual,* 1979.)

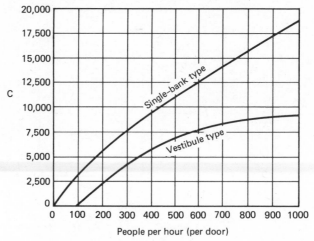

Figure 6-9.   Flow coefficient dependence on traffic rate. (Reprinted by permission from ASHRAE GRP 158, *Cooling and Heating Load Calculation Manual*, 1979.)

levels and exfiltration at the higher levels and the reverse in summer. Nonuniformity of the cracks and openings tends to be less important for flow continuity here. The following examples demonstrate the use of the data and methods described above.

Example 6-3.   A 12-story office building is 120 ft tall with plan dimensions of 120 ft by 80 ft. The structure is of conventional curtain wall construction with all windows fixed in place. There are double vestibule-type doors on all four sides. Under winter design conditions, a wind of 15 mph blows normal to one of the long dimensions. Estimate the pressure differences for all walls for the first and twelfth floors. Consider only wind and stack effects. The indoor-outdoor temperature difference is 60 F.

Solution.   The pressure difference for each effect must first be computed and then combined to find the total. First consider the wind: Equation (6-17) expresses the wind pressure difference where the pressure coefficients may be obtained from Table 6-2 for a normal wind. Then using standard sea level density:

Windward side: $C_p = 0.95$

$$\Delta P_w = \frac{0.95(0.765)(15 \times 1.47)^2 (12)}{2(32.17)62.4} = 0.106 \text{ in. wat.}$$

Leeward: $C_p = -0.15$

$$\Delta P_w = \frac{0.105}{0.95}(-0.15) = -0.017 \text{ in. wat.}$$

Sides: $C_p = -0.4$

$$\Delta P_w = \frac{0.105(-0.4)}{0.95} = -0.045 \text{ in. wat.}$$

The wind effect will be assumed constant with respect to height.

The pressure difference due to stack effect can be computed from Eq. (6-20) or more easily determined from Fig. 6-4. Because there are more openings in the lower part of the building, assume that the neutral pressure level is at the fifth floor instead of the sixth. Also assume that the draft coefficient is 0.8. Then for the first floor, $h = 50$ ft and from Fig. 6-4

$$\Delta P_s / C_d = 0.10$$

and

$$\Delta P_s = 0.10(0.8) = 0.08 \text{ in. wat.}$$

For the twelfth floor, $h = 70$ ft and

$$\Delta P_s / C_d = -0.12$$

and

$$\Delta P_s = -0.12(0.8) = -0.096 \text{ in. wat.}$$

The negative sign indicates that the pressure is greater inside the building than on the outside.

The pressure differences may now be summarized for each side where $\Delta P = (\Delta P_w + \Delta P_s)$ in. wat.

Orientation	1st Floor	12th Floor
Windward	0.186	0.010
Sides	0.035	−0.141
Leeward	0.063	−0.113

These results show that air will tend to infiltrate on all floors on the windward wall. Infiltration will occur on about the lower four floors on the leeward wall and on about the lower two floors on the side walls. All other surfaces will have exfiltration.

Example 6-4.   Estimate the infiltration rate for the leeward doors of Example 6-3. The doors have $\frac{1}{8}$ in. cracks and the traffic rate is low except at 5:00 P.M., when the traffic rate is 350 people per hour for a short time.

Solution.   This problem is solved in two steps to account for crack leakage and infiltration due to traffic. For the design condition, the effect of traffic is negligible; however, it is of interest to compute this component for 5:00 P.M. Figure

6-6 pertains to crack leakage for commercial swinging doors. For a pressure difference of 0.063 in. wat. and $\frac{1}{8}$ in. cracks, the leakage rate is 10 cfm/ft. The crack length for standard double swinging doors is

$$L = 3(6.75) + 2(6) = 32 \text{ ft}$$

Then

$$\dot{Q} = (\dot{Q}/L)L = 10(32) = 320 \text{ cfm}$$

Vestibule-type doors will tend to decrease the infiltration rate somewhat like storm sash or a storm door. Assume a 30 percent reduction, then

$$\dot{Q} = (1 - 0.3)\,320 = 224 \text{ cfm}$$

Figures 6-8 and 6-9 are used to estimate the infiltration due to traffic. The traffic coefficient C is read from Fig. 6-9 for 350 people per hour and for vestibule-type doors as 5000. Then, from Fig. 6-8 at a pressure difference of 0.063 in. wat.,

$$\dot{Q}/\text{door} = 800 \text{ cfm/door}$$

and for two doors

$$\dot{Q} = 1600 \text{ cfm}$$

A part of the crack leakage should be added to this; however, this is somewhat academic. Care should be exercised in including the traffic infiltration in the design heat load. It will usually be a short-term effect.

Example 6-5.   Estimate the net leakage rate for the twelfth floor of the building in Example 6-3.

Solution.   Referring to the pressure differences computed in Example 6-3, it is obvious that the net leakage will be from the inside out on the twelfth floor. Therefore, a great deal of air must be entering the space from the stairwells and elevator shafts. Because the twelfth floor has no movable openings, except to the roof, all leakage is assumed to be through the walls. Figure 6-5 gives data for this case where $K = 0.66$ for conventional construction.

Windward wall: $\Delta P = 0.010$ in. wat.; $\dot{Q}/A = 0.02 \text{ cfm/ft}^2$
$$\dot{Q} = 0.02(120)10 = 24 \text{ cfm}$$
Side walls: $\Delta P = 0.141$ in. wat., $\dot{Q}/A = -0.19 \text{ cfm/ft}^2$
$$\dot{Q} = -0.19(80)(10)2 = -304 \text{ cfm}$$
Leeward wall: $\Delta P = -0.113$ in. wat., $\dot{Q}/A = -0.15 \text{ cfm/ft}^2$
$$\cdot Q = -0.15(120)10 = -180 \text{ cfm}$$
The net leakage rate is then:

$$\dot{Q}_{net} = 24 - 304 - 180 = -460 \text{ cfm}$$

where the negative sign indicates that the flow is from the inside out. To compute the heat gain due to infiltration for this floor, only the 24 cfm on the windward wall should be considered. The net leakage flow of 460 cfm entered the building at other locations where the heat loss should be assigned.

Example 6-6. A single-story building of residential-type construction is oriented so that a 15 mph wind approaches at 45 degrees to the windward sides. There are 120 ft of crack for the windows and 20 ft of crack for a door on the windward sides. The leeward sides have 130 ft of window cracks and 18 ft of door crack. All windows and doors are average fitting with storm sash and doors which are classified as loose fitting.

Solution. The major portion of the infiltration for this kind of building will be through the cracks and wall leakage will be neglected. It is approximately true that air will enter on the windward sides and flow out on the leeward sides with most of the heat gain imposed on the rooms where the air enters. To be conservative, we will use at least half of the total crack length. The infiltration will therefore be based on 130 ft of window and 20 ft of door crack with a pressure difference computed for a quartering wind on the windward side. Using Eq. (6-17) and Table 6-2

$$\Delta P_w = \frac{0.70(0.0765)(15 \times 1.47)^2 (12)}{2(32.17)62.4} = 0.078 \text{ in. wat.}$$

where standard sea level air density has been used. From Tables 6-3 and 6-4, the K factor for the windows and doors is read as 2.0. Then from Fig. 6-3, the leakage per ft of crack is

$$\dot{Q}/L = 0.375 \text{ cfm/ft}$$

and the total infiltration for the prime windows and doors is

$$\dot{Q} = 0.375(130 + 20) = 56 \text{ cfm}$$

The effect of the storm sash and storm door is expressed in Fig. 6-7 where K for the storm sash is 6.0 from Table 6-3 and the crack lengths will be assumed equal for prime and storm sash. Then

$$k_p L_p / k_s L_s = 2.0 \times 150/6.0 \times 150 = 0.33$$

and

$$\dot{Q}/\dot{Q}_p = 0.88$$
$$\dot{Q} = 0.88(56) = 49 \text{ cfm}$$

This amounts to about one-quarter air change per hour for a 1600 ft^2 house. Notice that the loose fitting storm sash reduced the infiltration by only about 12 percent.

Fireplaces, chimneys, and flues can increase infiltration dramatically or necessitate the introduction of outdoor air. In either case the heat loss of the structure is increased. When the fireplace is not in use, large quantities of air will rise through the chimney if the damper is left open; this causes an increased amount of cold air to enter the structure through cracks. If the chimney damper is closed, the effect should be eliminated. However, fireplace dampers usually do not fit tightly and will allow some flow when closed. Therefore, it is recommended that at least 50 cfm or 0.024 m³/s infiltration be allowed for residential fireplaces.

When the fireplace is in use, the loss of air through the chimney is increased because of greater draft. This increases the infiltration of outdoor air for the complete structure, although the space in the immediate vicinity of the fireplace is warmed and probably will not require heat from the central system. The net effect is to increase the total heat loss from the structure. In the ideal situation, the room containing the fireplace should be provided with independent control of the warm air heating system. The additional outdoor air required for the chimney should be introduced through the central heating unit and not be allowed to infiltrate. In this way the complete space may be maintained at a comfortable temperature. The outdoor air required for the burning fireplace should be approximately 200 cfm or 0.094 m³/s, some of which may be infiltration air.

Direct fired warm air furnaces are sometimes installed within the confines of the conditioned space. If combustion air is not brought in from outdoors, conditioned air from the space will be drawn in and exhausted through the flue. Infiltration or outdoor air must then enter the structure to make up the loss and contributes to a higher heat loss. Many codes require that combustion air be introduced directly to the furnace from outdoors. Indeed this should always be the rule. For natural gas (methane) the ratio of air to gas on a volume basis is about 10. This is equivalent to 10 ft³ or 0.28 m³ of air per 1000 Btu or 1.06×10^6 J input to the furnace.

6-4   HEAT LOSSES FROM AIR DUCTS

The losses of a duct system can be considerable when the ducts are not in the conditioned space. Proper insulation will reduce these losses but cannot completely eliminate them. The loss may be estimated using the following relation.

$$\dot{q} = UA_s \Delta t_m \tag{6-21}$$

where

U = overall heat transfer coefficient, Btu/(hr-ft²-F) or W/(m²/C)
A_s = outside surface area of the duct, ft² or m²
Δt_m = mean temperature difference between the air in the duct and the environment, F or C.

When the duct is covered with 1 or 2 in. of fibrous glass insulation with a reflective covering, the heat loss will usually be reduced sufficiently to assume that the mean temperature difference is equal to the difference in temperature between the supply air temperature and the environment temperature. Unusually long ducts should not be treated in this manner and a mean air temperature should be used instead.

Example 6-7   Estimate the heat loss from 1000 cfm of air at 120 F flowing in a 16 in. round duct 25 ft in length. The duct has 1 in. of fibrous glass insulation and the overall heat transfer coefficient is 0.2 Btu/(hr-ft²-F). The environment temperature is 12 F.

Solution. Equation (6-21) will be used to estimate the heat loss assuming that the mean temperature difference is given approximately by

$$\Delta t_m = t_3 - t_a = 12 - 120 = -108 \text{ F}$$

The surface area of the duct is

$$A_s = \frac{\pi(16 + 2)(25)}{12} = 117.8 \text{ ft}^2$$

Then

$$\dot{q} = 0.2(117.8)(-108) = -2540 \text{ Btu/hr}$$

The temperature of the air leaving the duct may be computed from

$$\dot{q} = \dot{m}c_p(t_2 - t_1) = \dot{Q}\rho c_p(t_2 - t_1)$$

or

$$t_2 = t_1 + \frac{\dot{q}}{\rho c_p}$$

$$t_2 = 120 + \frac{-2540}{1000(60)(0.067)(0.24)}$$

$$t_2 = 117.4 \text{ F}$$

Although insulation drastically reduces the heat loss, the magnitude of the temperature difference and surface area must be considered in each case.

Minimum insulation of supply and return ducts is presently specified by ASHRAE Standard 90 A as follows:

All duct systems shall be insulated to provide a thermal resistance, excluding film resistance, of

$$R = \frac{\Delta t}{15} \text{ (hr-ft}^2\text{-F)/Btu} \qquad \text{or} \qquad R = \frac{\Delta t}{47.3} \text{ (m}^2\text{-C)/W}$$

where Δt is the design temperature differential between the air in the duct and the surrounding air in F or C. Heat losses from the supply ducts become part of the space heat load and should be summed with transmission and infiltration heat losses. Heat losses from the return air ducts are not a part of the space heat loss but should be added to the heating equipment load.

6-5 AUXILIARY HEAT SOURCES

The heat energy supplied by people, lights, motors, and machinery should always be estimated, but any actual allowance for these heat sources requires careful consideration. People may not occupy certain spaces in the evenings, on weekends, and during other periods and these spaces must generally be heated to a reasonably comfortable temperature prior to occupancy. In industrial plants heat sources, if available during occupancy, should be substituted for part of the heating requirement. In fact there are situations where so much heat energy is available that outdoor air must be used to cool the space. However sufficient heating equipment must still be provided to prevent freezing of water pipes during periods when a facility is shut down.

6-6   INTERMITTENTLY HEATED STRUCTURES

When a structure is not heated on a continuous basis, the heating equipment capacity may have to be enlarged to assure that the temperature can be raised to a comfortable level within a reasonable period of time. The heat capacity of the building and occupant comfort are important factors when considering the use of intermittent heating. Occupants may feel discomfort if the mean radiant temperature falls below the air temperature. To conserve energy it is becoming a common practice to set back thermostats or to completely shut down equipment during the late evening, early morning, and weekend hours. This is effective and is accompanied by only small sacrifices in comfort when the periods of shutdown are adjusted to suit outdoor conditions and the mass of the structure.

6-7   AIR REQUIRED FOR SPACE HEATING

Computing the air required for heating was discussed in Chapter 2, Example 2-8, and took into account sensible and latent effects as well as outdoor air. That procedure is always recommended when sufficient latent heat loss exists or outdoor air is used. There are many cases, especially in residential and light commercial applications, when the latent heat loss is quite small and may be neglected. The air quantity is then computed from

$$\dot{q} = \dot{m}c_p(t_s - t_r) \tag{6-22}$$

or

$$\dot{q} = \frac{\dot{Q}c_p}{v_s}(t_s - t_r) \tag{6-22a}$$

where

v_s = specific volume of supplied air, ft³/lbm or m³/kg
t_s = temperature of supplied air, F or C
t_r = room temperature, F or C

The temperature difference $(t_s - t_r)$ is normally less than 100 F or 55 C. Residential and light commercial equipment operates with a temperature rise of 60 to 80 F or 33 to 44 C, whereas commercial applications will allow higher temperatures. The temperature of the air to be supplied must not be high enough to cause discomfort to occupants before it becomes mixed with room air.

In the unit-type equipment typically used for residences and small commercial buildings each size is able to circulate a relatively fixed quantity of air. Therefore the air quantity is fixed within a narrow range when the heating equipment is selected. These units have different capacities that change in increments of 10 to 20,000 Btu/hr or about 5 kW according to the model. A slightly oversize unit is usually selected with the capacity to circulate a larger quantity of air than theoretically needed. Another condition that leads to greater quantities of circulated air for heating than needed is the greater air quantity sometimes required for cooling and dehumidifying. The same fan is used throughout the year and must therefore be large enough for the maximum air quantity required. Some units have different fan speeds for heating and for cooling.

After the total air flow rate required for the complete structure has been determined, the next step is to allocate the correct portion of the air to each room or space. This is necessary for design of the duct system. Obviously the air quantity for each room should be apportioned according to the heating load for that space; therefore,

$$\dot{Q}_{rn} = \dot{Q}(\dot{q}_{rn}/\dot{q}) \tag{6-23}$$

where

$\dot{Q}_{rn}$ = volume flow rate of air supplied to room n, ft^3/min or m^3/s
$\dot{q}_{rn}$ = total heat loss rate of room n, Btu/hr or W

The worksheet of Fig. 6-2 has provisions for recording the total air quantity for the structure and the air quantity for each room.

6-8   ESTIMATING FUEL REQUIREMENTS

Following the calculation of the design heat load and selection of the heating system, it is often desirable to estimate the quantity of energy necessary to heat the structure under typical weather conditions and with typical inputs from internal heat sources. This is a distinctly different procedure from design heat load calculations which are usually made for one set of design conditions neglecting solar effects and internal heat sources. It was mentioned earlier that the building with its air-conditioning equipment and variable indoor and outdoor environmental conditions constitute a complex dynamic system. Simulation of such a system requires a digital computer.

Many computer codes are available that model building systems quite well and they may be used with all types of structures ranging from residential to heavy commercial. This approach will be discussed in Chapter 8, because the building simulation is usually carried out for a whole year considering heating, cooling, and other energy requirements.

There are cases, however, where computer simulation is not possible or cannot be justified. Residential buildings fall into this category. Reasonable results can be obtained in this case using hand calculation methods such as the degree day or bin method. These methods will be discussed briefly here. Reference 2 discusses energy estimating in detail.

The Degree Day Procedure

The traditional degree day procedure (2) for computing fuel requirements is based on the assumption that on a long-term basis, solar and internal gains will offset heat loss when the mean daily outdoor temperature is 65 F (18 C). It is further assumed that fuel consumption will be proportional to the difference between the mean daily temperature and 65 F or 18 C.

For cities in the United States and Canada, Table C-2 lists the average number of degree days that have occurred over a period of many years; they are listed by month and the yearly totals of these averages are given. Degree days are defined by the relationship

$$DD = \frac{(t - t_a)N}{24} \tag{6-24}$$

where N is the number of hours for which the average temperature t_a is computed and t is 65 F or 18 C. Studies made by the American Gas Association (3) and the National District Heating Association (4) utilized the 65 F base. Residential insulation practices have improved dramatically over the last 40 years, however, and internal heat gains have increased. These changes indicate that a temperature less than 65 F should be used for the base; nevertheless the data now available are based on 65 F. Another factor, which is not accounted for in traditional methods, is the decrease in efficiency of fuel-fired furnaces and heat pumps under partial load. Until methods now under development are completed, a modified degree day procedure that accounts for these factors in an approximate way will be in current use.

The general relation for fuel calculations using this procedure is

$$F = \frac{24DD\dot{q}}{\eta(t_i - t_o)H} C_D \tag{6-25}$$

where

   F = the quantity of fuel required for the period desired; the units depend on H

   DD = the degree days for period desired, F-day or C-day

   $\dot{q}$ = the total calculated heat loss based on design conditions, t_i and t_o, Btu/ hr or W

   η = an efficiency factor which includes the effects of rated full load efficiency, part load performance, oversizing, and energy conservation devices

   H = the heating value of fuel, Btu or kW-hr per unit volume or mass

   C_D = the interim correction factor for degree days based on 65 F or 18 C, Fig. 6-10

Figure 6-10 gives values for the correction factor, C_D, as a function of yearly degree days. These values were calculated using typical modern single-family construction, and generally agree with electric utility experience (6).

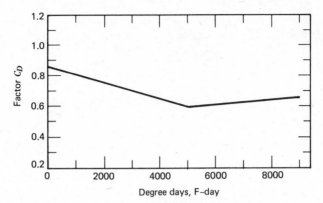

Figure 6-10. Correction factor for use in Eq. (6-25). (Reprinted by permission from *ASH-RAE Handbook and Product Directory-Systems*, 1980.)

The efficiency factor η of Eq. (6-25) is empirical. In an analysis of 140 gas heated homes by the American Gas Association (9), a mean value of 0.55 was indicated. There are other indications that a value of 0.65 is more appropriate for newer energy conserving homes. Oversizing of the heating equipment or special energy conservation features will produce variations. For most applications, however, a value between 0.55 and 0.65 is recommended. For electric resistance heat, η has a value of 1.0.

Example 6-8 Estimate the amount of natural gas required to heat a residence in Stillwater, Oklahoma, using the modified degree day method. The heating value of the fuel is 1000 Btu/std ft³. The calculated heat loss from the house is 80,000 Btu/hr with indoor and outdoor design temperatures of 70 F and 0 F, respectively. The furnace efficiency factor is approximately 0.55.

Solution The average winter temperature and the degree days for Stillwater are estimated to be 48 F and 3860 from Table C-2. Equation (6-17) will give an estimate of the fuel required by the prescribed method.
The correction factor C_D is 0.66 from Fig. 6-9 for 3860 degree days.

$$F = \frac{24(3860)80,000(0.661)}{0.55(70 - 0)1,000} = 127,050 \text{ std ft}^3$$

or $F = 127$ mcf of natural gas

Example 6-9 Suppose the house of Example 6-8 is equipped with an electric resistance furnace. Compute the energy requirement in kW-hr, and the source energy in mcf of natural gas assuming a power plant efficiency of 33 percent and no transmission or distribution loss.

Solution   Here the efficiency factor η is set equal to 1.0. The heating value of the fuel H becomes the conversion factor between the Btu and kW-hr.

$$F = \frac{24(3860)80,000(0.66)}{1.0(70 - 0)3413} = 20,474 \text{ kW-hr}$$

The equivalent amount of energy required to generate the electric energy for heating is

$$E = F/\eta_P = 20,474/0.33 = 62,042 \text{ kW-hr}$$

or

$$E = \frac{62,042(3413)}{(1000)(1000)} = 212 \text{ mcf}$$

Bin Method

The energy estimating method discussed above is based on average conditions and does not take into account actual day-to-day weather variations and the effect of temperature on equipment performance. The *bin method* is a hand calculation procedure where energy requirements are determined at many outdoor temperature conditions. The bins are usually 5 F (about 3 C) in size and the day may be divided into three daily eight-hour shifts. The bin method requires hourly weather data as well as equipment characteristics for the temperature range of interest. Reference (6) describes this method in detail and gives some representative weather and equipment performance data.

REFERENCES

1. *ASHRAE Handbook of Fundamentals,* American Society of Heating, Refrigerating and Air-Conditioning Engineers, New York, 1977.

2. *ASHRAE Handbook and Product Directory—Systems,* American Society of Heating, Refrigerating and Air-Conditioning Engineers, New York, 1972.

3. *House Heating,* American Gas Association, Industrial Gas Series, 3rd Ed., Chicago, Illinois.

4. "Report of Commercial Relations Committee," *National District Heating Association Proceedings,* 1932.

5. ASHRAE Standard 90 A-80, "Energy Conservation in New Building Design," American Society of Heating Refrigerating and Air-Conditioning Engineers, New York, 1980.

6. *ASHRAE Handbook and Product Directory—Systems,* American Society of Heating, Refrigerating and Air-Conditioning Engineers, New York, 1980.

7. *ASHRAE Cooling and Heating Load Calculation Manual* GRP 158, American Society of Heating, Refrigerating, and Air Conditioning Engineers, New York, 1979.

8. Janssen, P. E., et al., "Calculating Infiltration: An Examination of Handbook Models," *ASHRAE Transactions,* Vol. 86, Part 2, 1980.

9. Kelnhofer, W. J., *Evaluation of the ASHRAE Modified Degree Day Procedure for Predicting Energy Usage by Residential Gas Heating Systems,* American Gas Association, 1979.

PROBLEMS

6-1 Estimate the temperature in the crawl space of a house with floor plan dimensions of 30 × 60 ft. The concrete foundation has an average height of 2 ft and the wall is 6 in. thick. Use winter design conditions for Oklahoma City, Oklahoma.

6-2 Estimate the temperature in an unheated connecting garage. The side of the garage connected to the house is 24 ft long and the width is 20 ft. There is a 9 ft ceiling and an attic-ceiling combination with 6 in. of insulation similar to that in Example 4-3. The walls are of the same construction as those in Table 4-7 and have $3\frac{1}{2}$ in. of insulation. There is one 32 × 80 in. slab door, $1\frac{1}{2}$ in. thick and an overhead door 7 × 16 ft whose thickness is equivalent to $\frac{3}{4}$ in. plywood. The partition between house and garage is $\frac{3}{8}$ in. gypsum on each side of 2 × 4 studs with 3 in. of insulation; the floor is a slab. Indoor and outdoor design conditions are 75 F and 0 F, respectively.

6-3 Estimate the temperature in the knee space shown in Fig. 6-11. The roof is equivalent to 25 mm of wood on 2 × 6 rafters on 0.6 m centers. The walls are 2 × 4 studs on 0.4 m centers with 90 mm of insulation. The joist spaces all have 150 mm of insulation. Inside and outside temperatures are 21 and −12 C.

6-4 A large single-story business office is fitted with nine loose fitting, double-hung wood-sash windows 3 ft wide by 6 ft high. If the outside wind is 15 mph at a temperature of −10 F, what is the reduction in sensible heat loss if the windows are weather-

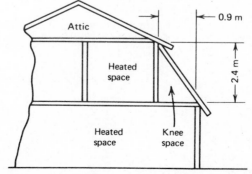

Figure 6-11.   Sketch of building for problem 6-3.

stripped? Assume an inside temperature of 70 F. Base your solution on a minimum condition in which half of the crackage undergoes infiltration and half undergoes exfiltration, with a quartering wind.

6-5 Using the crack method compute the infiltration for a tight fitting swinging door that is used occasionally. The door has dimensions of 0.9 × 2.0 m and is on the windward side of a house exposed to a 9 m/s wind. Neglect internal pressurization and stack effect.

6-6 A room has two 3 × 5 ft wood, double-hung windows of average fit that are not weather-stripped. The wind is 20 mph and normal to the wall with negligible pressurization of the room. (a) Find the infiltration rate assuming that the entire crack is admitting air. (b) The room has dimensions of 14 × 16 ft with the windows on one side. Estimate the infiltration by the air change method.

6-7 A small commercial building has a computed heat load of 200,000 Btu/hr sensible and 25,000 Btu/hr latent. Assuming a 50 F temperature rise for the heating unit, compute the quantity of air to be supplied by the unit with the following methods: (a) Use a psychrometric chart with room conditions of 70 F and 30 percent relative humidity. (b) Calculate the air quantity based on the sensible heat transfer.

6-8 Estimate the quantity of natural gas required to heat the building of Problem 6-7. The location is Stillwater, Oklahoma. Design conditions are 75 F indoor and 12 F outdoor temperatures. Assume a furnace efficiency of 60 percent. The heating value of the fuel is 1000 Btu/std ft³.

6-9 If electric resistance heat were used to heat the building mentioned in Problem 6-8, how much energy would be required in kW-hr assuming a 100 percent furnace efficiency? If the electrical energy costs five cents per kW-hr and natural gas costs $3 per mcf, what are the relative heating costs? Assuming a power plant efficiency of 33 percent, compare the total amounts of energy in terms of mcf of gas required to heat the building using a gas furnace and an electric furnace.

6-10 Compute the heating load for the structure described on the plans and specifications furnished by the instructor.

6-11 Compute the transmission heat loss for the structure described below. Use design conditions recommmended by ASHRAE Standard 90 A.

Location:	Kansas City, Missouri
Walls:	Figure 4-7 with 3½ in. fibrous glass insulation
Floor:	Concrete slab with 1 in. edge insulation, both horizontal and vertical, with heating ducts under the slab
Windows:	Double-insulating glass, ¼ in. air space, 3 × 3 ft, double hung, 3 on each side
Doors:	Wood, 1¾ in. with wood storm doors, 3 each, 3 × 6.75 ft
Roof-Ceiling:	Attic same as Example 4-3
House Plan:	Single story, 40 × 55 ft

6-12 A homeowner in Kansas City wants an evaluation concerning the wisdom of installing storm windows on his home. The existing windows are single glass, wood, double-hung, nonweather-stripped with an average fit. The house has 1700 ft of crack. (a)

Estimate the annual energy saving if the house is equipped with electric resistance heating equipment. (b) Assuming that the storm windows cost $5/ft² and electric power costs $0.04/kw-hr, estimate the number of years required to recover the cost of the storm windows. Assume 15 mph quartering wind.

6-13 The ceiling-attic-roof combination of a house in Chicago, Illinois, has $5\frac{1}{2}$ in. of fibrous glass insulation above the ceiling. Evaluate the merit of installing an additional 4 in. of fibrous glass batts with a reflective surface facing upward. The house has 2000 ft² of floor area. The additional insulation will cost $0.2/ft² and fuel costs $3/10⁶ Btu of input to the furnace. Assume a furnace efficiency of 50 percent.

6-14 Consider the knee space shown in Fig. 6-11. The vertical dimension is 8 ft, the horizontal dimension is 3 ft, and the space is 20 ft long. The walls and roof surrounding the space all have an overall heat transfer coefficient of about 0.09 Btu/(hr-ft²-F). Assuming an outdoor temperature of 2 F and an indoor temperature of 68 F, make a recommendation concerning the placement of water pipes in the knee space.

6-15 Estimate the temperature in an unheated basement which is completely below ground level with heated space above. The basement is located in Lincoln, Nebraska. Use standard design conditions.

6-16 A 20-story office building has plan dimensions of 100 ft by 60 ft and is oriented at 45 degrees to a 20 mph wind. All windows are fixed in place. There are double vestibule-type swinging doors on the 60 ft walls. The walls are average curtain wall construction and the doors have about $\frac{1}{8}$ in. cracks. (a) Compute the pressure differences for each wall due to wind and stack effect for the first, fifth, fifteenth, and twentieth floors. (b) Plot pressure difference versus height for each wall and estimate which surfaces have infiltration and exfiltration. (c) Compute the total infiltration rate for the first floor assuming a very low traffic rate. (d) Compute the net infiltration rate for the fifteenth floor. (e) Compute the net infiltration rate for the twentieth floor. Neglect any leakage through the roof. Assume winterdesign conditions for Dallas, Texas.

6-17 Compute the design infiltration rate for the house described in Problem 6-11 assuming an orientation normal to a 15 mph wind. The windows and doors are tight fitting.

6-18 Refer to Example 6-3. (a) Estimate the total pressure difference for each wall for the third and ninth floors. (b) Using design conditions for Denver, Colorado, estimate the heat load due to infiltration for floors 3 and 9.

6-19 Refer to Examples 6-3 and 6-4. (a) Estimate the infiltration rates for the windward and side doors for a low traffic rate. (b) Estimate the curtain wall infiltration for the first floor. (c) Compute the heat load due to infiltration for the first floor if the building is located in Cleveland, Ohio.

6-20 Refer to Problem 6-16. (a) Compute the heat gain due to infiltration for the first floor with the building located in Dallas, Texas. (b) Compute the heat gain due to infiltration for the fifteenth floor. (c) What is the heat gain due to infiltration for the twentieth floor?

6-21 Compute the heat gain due to infiltration for the house of Problem 6-17.

6-22 A light commercial building, located in Sioux City, Iowa, has construction and use characteristics much like a residence and a design heat load of 120,000 Btu/hr. The structure is heated with a natural gas warm air furnace and is considered an energy conserving house. Assuming standard design conditions, estimate the yearly fuel requirements.

6-23   Consider a building located in Syracuse, New York. Using Standard 90 A recommendations, compute the transmission loss for (a) a wall like Table 4-7, case 2, which is 3 m by 4 m with a window of one m² area, (b) a single pane window, (c) a roof-ceiling combination that is like Table 4-7e, case 1 with an area of 16 m².

CHAPTER SEVEN

The Cooling Load

A larger number of variables are considered in making cooling load calculations than in heating load calculations. In both situations the actual heat loss or gain is a transient one. As was explained in Chapter 6, however, for design purposes heat loss is usually based on steady state heat transfer, and the results obtained are usually quite adequate. In design for cooling, however, transient analysis must be used if satisfactory results are to be obtained. This is because the instantaneous heat gain into a conditioned space is quite variable with time primarily because of the strong transient effect created by the hourly variation in solar radiation. There may be an appreciable difference between the heat gain of the structure and the heat removed by the cooling equipment at a particular time. This difference is caused by the storage and subsequent transfer of energy from the structure and contents to the circulated air. If this is not taken into account, the cooling and dehumidifying equipment will usually be grossly oversized and estimates of energy requirements meaningless.

In determining seasonal energy requirements for either heating or cooling both the solar inputs and the transient effects due to storage must be considered for accurate results. These thermal simulations are usually made using computer programs with some type of assumed or historical weather data. These computations are often performed on an hourly or fraction of an hour basis over the period of interest.

7-1 HEAT GAIN, COOLING LOAD, AND HEAT EXTRACTION RATE

It is important to differentiate between *heat gain, cooling load,* and *heat extraction rate.* Heat gain is the rate at which energy is transferred to or generated within a space. It has two components, sensible heat and latent heat, which must be computed and tabulated separately. Heat gains usually occur in the following forms:

1. Solar radiation through openings
2. Heat conduction through boundaries with convection and radiation from the inner surface into the space

185

3. Sensible heat convection and radiation from internal objects
4. Ventilation (outside) and infiltration air
5. Latent heat gains generated within the space

The *cooling load* is the rate at which energy must be removed from a space to maintain the temperature and humidity at the design values. The cooling load will generally differ from the heat gain at any instant of time. This is because the radiation from the inside surface of walls and interior objects as well as the solar radiation coming directly into the space through openings does not heat the air within the space directly. This radiant energy is mostly absorbed by floors, interior walls, and furniture, which are then cooled primarily by convection as they attain temperatures higher than that of the room air. Only when the room air receives the energy by convection does this energy became part of the cooling load. The heat storage characteristics of the structure and interior objects determine the thermal lag and therefore the relationship between heat gain and cooling load. For this reason the thermal mass (product of mass and specific heat) of the structure and its contents must be considered in such cases. Figure 7-1 illustrates the phenomenon. The reduction in peak cooling load because of the thermal lag can be quite important in sizing the cooling equipment.

The *heat extraction rate* is the rate at which energy is removed from the space by the cooling and dehumidifying equipment. This rate is equal to the cooling load when the space conditions are constant and the equipment is operating. However, this is rarely true because some fluctuation in room temperature is necessary for the control system to operate. Because the cooling load is also below the peak or design value most of the time intermittent or variable operation of the cooling equipment is required.

Figure 7-2 shows the relation between heat gain and cooling load and the effect of the mass of the structure. The attenuation and delay of the peak heat gain is very evident especially for heavy construction. Figure 7-3 shows the cooling load for fluorescent lights that are used only part of the time. The sensible heat component from people acts in a similar way. The part of the energy produced by the lights or people that is radiant energy is momentarily stored in the surroundings. The energy convected directly to the air by the lights and people, and later by the surrounding

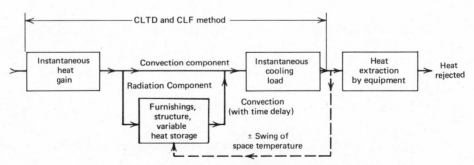

Figure 7-1. Schematic relation of heat gain to cooling load.

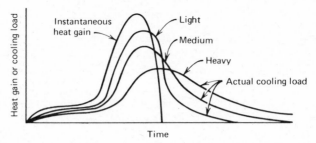

Figure 7-2.   Actual cooling load and solar heat gain for light, medium, and heavy construction.

objects, goes into cooling load. The area under the curves of Fig. 7-2 or Fig. 7-3 are all approximately equal. This means that about the same total amount of energy must be removed from the structure during the day; however, a larger portion is removed during the evening hours for the heavier constructions..

To explain the principles of heat gain and cooling load from the computational viewpoint consider a room enclosed by walls, windows, roof, and floor with infiltration and internal heat sources. The heat gain for the interior space at a given time is

$$\dot{q}_\theta = \dot{q}_{i,\theta} + \dot{q}_{r,\theta} + \dot{q}_{s,\theta} + \dot{q}_{L,\theta} \tag{7-1}$$

where

$\dot{q}_{i,\theta}$ = convective heat transfer from inside surfaces of boundaries at time θ
$\dot{q}_{r,\theta}$ = radiation heat transfer between the inside surfaces of boundaries and other interior surfaces at time θ
$\dot{q}_{s_\theta}$ = rate of solar energy entering through the windows at time θ
$\dot{q}_{L_\theta}$ = rate of heat generation by lights, people, and other internal sources at time θ

The sensible cooling load for the space at a given time may be expressed as:

$$\dot{q}_{c,\theta} = \dot{q}_{i,\theta} + \dot{q}_{I,\theta} + \dot{q}_{v,\theta} + \dot{q}_{sc,\theta} + \dot{q}_{Lc,\theta} \tag{7-2}$$

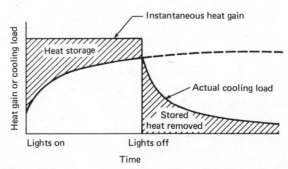

Figure 7-3.   Actual cooling load from fluorescent lights.

where

$\dot{q}_{I,\theta}$ = rate of heat gain due to infiltration at time θ

$\dot{q}_{v,\theta}$ = rate of heat gain due to outdoor ventilation air at time θ

$\dot{q}_{sc,\theta}$ = rate of energy convected from the inner window and other interior surfaces to the room air at time θ

$\dot{q}_{Lc,\theta}$ = rate of heat generated by lights, people, and other internal sources
which is convected into the room air at time θ

Equation (7-2) is actually the relation of most interest. In order to solve either Eqs.
(7-1) or (7-2), the equations describing the conduction heat transfer in the boundary
walls, floors, and ceilings must be solved simultaneously because they are coupled
through the inside surface temperature. The convective components of the solar and
internal heat generation are related to the room air temperature also.

Consider the heat conduction problem. Calculation of heat transfer through a
wall or roof section with a variable temperature in the outdoor environment and with
a variable solar radiation input on the outside surface requires the heat conduction
equation with nonlinear, time-dependent boundary conditions. In addition, the wall
is a complex assembly of materials and may be two-dimensional. However, if the
wall or roof is a single homogeneous slab, the governing differential equation is

$$\frac{\partial t}{\partial \theta} = \frac{k}{\rho c} \frac{\partial^2 t}{\partial x^2} \tag{7.3}$$

where

t = local temperature at a point in the slab, F or C

θ = time, hour

$k/\rho c$ = thermal diffusivity of the slab, ft^2/hr or m^2/hr

x = length, ft or m

At $x = 0$, the outside surface, the boundary condition is

$$-k_w \left(\frac{\partial t}{\partial x} \right)_{x=0} = h_o(t_o(\theta) - t_{wo}) + \dot{q}_r(\theta) \tag{7-4}$$

where $\dot{q}_r$ is the net solar radiation heat transfer to the exterior at a particular time.
Both t_o and $\dot{q}_r$ are functions of time. At $x = L$, the inside surface, the boundary
condition is

$$-k_w \left(\frac{\partial t}{\partial x} \right)_{x=L} = h_i(t_{wi} - t_i) \tag{7-5}$$

The initial condition for Eq. (7-3) is some specified temperature distribution in the
wall at a time equal to zero.

The nonlinear, time dependent boundary condition at the outside surface, Eq.
(7-4), is the primary obstacle in obtaining a solution to Eq. (7-3). To circumvent this
difficulty the concept of the sol-air temperature is introduced. The *sol-air temperature* t_e is the fictitious temperature that in the absence of all radiation exchanges

gives the same rate of heat transfer to the exterior surface as actually occurs by solar radiation and convection. In terms of the sol-air temperature the heat transfer to the outer surface is

$$\dot{q}_o = h_o A(t_e - t_{wo}) \tag{7-6}$$

In terms of the actual outdoor temperature t_o the heat transfer rate is

$$\dot{q}_o = h_o A(t_o - t_{wo}) + \alpha A G_t - \varepsilon \Delta R A \tag{7-7}$$

where

h_o = coefficient of heat transfer, Btu/(hr-ft²-F) or W/(m²-C)
ε = emittance of the surface
ΔR = difference between the long wavelength radiation incident on the surface from the sky and the radiation emitted from a black body at outdoor air temperature, Btu/(hr-ft²) or W/m²
α = absorptance of the wall surface
G_t = total incident solar radiation upon the surface, Btu/(hr-ft²) or W/m²
A = surface area, ft² or m²

Equations (7-6) and (7-7) may be combined to obtain

$$t_e = t_o + \alpha \frac{G_t}{h_o} - \frac{\varepsilon \Delta R}{h_o} \tag{7-8}$$

The term $\varepsilon \Delta R / h_o$ varies from about zero for a vertical surface to about 7 F or 4 C for horizontal surfaces. The ratio α / h_o varies from about 0.15 (hr-ft²-F)/Btu or 0.026 (m²-C)/W for a light-colored surface up to a maximum of about 0.30 (hr-ft²-F)/Btu or 0.053 (m²-C)/W. The total incident solar radiation G_t is computed as shown in Chapter 5.

Note that the sol-air temperature given by Eq. (7-8) is time dependent because t_o and G_t depend on time. However, the variation is approximately harmonic. The boundary condition at the outer surface given by Eq. (7-4) is greatly simplified by the sol-air temperature concept and becomes

$$-k_w \left(\frac{\partial t}{\partial x} \right)_{x=0} = h_o(t_e(\theta) - t_{wo}) \tag{7-9}$$

With this simplification and the assumption that t_e may be represented by a harmonic function, Fourier series solutions of Eq. (7-3) are possible using the digital computer. The solution for a particular wall yields the inside surface temperature as a function of time. The rigorous approach outlined above requires the most refined digital computers. Even so, very few computer programs are in use where space cooling load is computed in this manner. The number of surfaces and separate spaces to be considered adds even more complications.

To simplify and reduce the time required for computations, transform methods have been applied to this problem (1). A digital computer is still necessary but the speed of calculations is rapid even for small computers. Design cooling loads for one day as well as long-term energy calculations may be done using the transfer function approach. Details of this method are discussed in section 7-11.

It is not always practical to compute the cooling load using the transfer function method; therefore, a hand calculation method has been developed from the transfer function procedure (1, 4) and is referred to as the *Cooling Load Temperature Difference* (*CLTD*) method. The method involves extensive use of tables and charts and various factors to express the dynamic nature of the problem and predicts cooling loads within about 5 percent of the values given by the transfer function method. The following sections describe the procedure beginning with design conditions which are the same for all calculation methods.

7-2   OUTDOOR AND INDOOR DESIGN CONDITIONS

The problem of selecting outdoor design conditions for calculation of heat gain is similar to that for heat loss. Again it is not reasonable to design for the worst conditions on record because a great excess of capacity will result. The heat storage capacity of the structure also plays an important role in this regard. A massive structure will reduce the effect of overload from short intervals of outdoor temperature above the design value. The *ASHRAE Handbook of Fundamentals* (1) gives extensive outdoor design data. Tabulation of dry bulb and mean coincident wet bulb temperatures that are equaled or exceeded 1, $2\frac{1}{2}$, and 5 percent of the total hours during June through September (2928 hours) are given. For example a normal summer in Stillwater, Oklahoma, will have about 30 hours at 100 F dry bulb or greater, about 75 hours at 96 F or greater, and about 150 hours at 93 F or greater. Table C-1 gives the $2\frac{1}{2}$ percent values that are recommended for design purposes by ASHRAE Standard 90 A (3). The daily range of temperature given in Table C-1 is the difference between the average maximum and average minimum for the warmest month. The daily range is usually larger for the higher elevations where temperatures may be quite low late at night and during the early morning hours. The daily range has an effect on the energy stored by the structure. The variation in dry bulb temperature for a typical design day may be computed using the peak outdoor dry bulb temperature and the daily range. This is discussed in Article 7-10 in connection with the transfer function method.

The local wind velocity for summer conditions is usually taken to be about one half the winter design value but not less than about $7\frac{1}{2}$ mph or 3.4 m/s.

The indoor design conditions are governed by principles outlined in Chapter 3. For the average job in the United States and Canada, however, a condition of 75 F or 24 C dry bulb and relative humidity of 50 percent is typical when activity and dress of the occupants are light. Reference 3 sets the indoor design temperature at 78 F or 25 C with relative humidity within the comfort envelope defined in Figure 3-3. The designer should be alert for unusual circumstances that may lead to uncomfortable conditions. Certain activities may require occupants to engage in active work or require heavy protective clothing, both of which would require lower design temperatures.

7-3 THE *CLTD* METHOD

The *CLTD* method makes use of a temperature difference in the case of walls and roofs and *Cooling Load Factors* (*CLF*) in the case of solar gain through windows and internal heat sources. The *CLTD* and *CLF* vary with time and are a function of environmental conditions and building parameters. They have been derived from computer solutions using the transfer function procedure. A great deal of care has been taken to sample a wide variety of conditions in order to obtain reasonable accuracy. These factors have been derived for a fixed set of surface and environmental conditions; therefore, correction factors must often be applied. In general, calculations proceed as follows:

For walls and roofs

$$\dot{q}_\theta = UA(CLTD)_\theta \qquad (7\text{-}10)$$

where

U = overall heat transfer coefficient, Btu/(hr-ft^2-F) or W/(m^2-C)
A = area, ft^2 or m^2
$CLTD$ = temperature difference which gives the cooling load at time θ, F or C

The *CLTD* accounts for the thermal response (lag) in the heat transfer through the wall or roof, as well as the response (lag) due to radiation of part of the energy from the interior surface of the wall to objects within the space.

For solar gain through glass

$$\dot{q}_\theta = A(SC)SHGF(CLF)_\theta \qquad (7\text{-}11)$$

where

A = area, ft^2 or m^2
SC = shading coefficient (internal shade)
$SHGF$ = solar heat gain factor, Btu/(hr-ft^2) or W/m^2
CLF = cooling load factor for time θ

The *SHGF* is the maximum for a particular month, orientation, and latitude. The *CLF* accounts for the variation of the *SHGF* with time, the massiveness of the structure, and internal shade. Again the *CLF* accounts for the thermal response (lag) of the radiant part of the solar input.

For internal heat sources

$$\dot{q}_\theta = \dot{q}_i(CLF)_\theta \qquad (7\text{-}12)$$

where

$\dot{q}_i$ = instantaneous heat gain from lights, people, and equipment, Btu/hr or W
CLF_θ = cooling load factor for time θ

The *CLF* accounts for the thermal response of the space to the various internal heat gains and is slightly different for each.

The time of day when the peak cooling load will occur must be estimated. In fact, two different types of peaks need to be determined. First, the time of the peak load for each room is needed in order to compute the air quantity for that room. Second, the time of the peak load for a zone served by a central unit is required to size the unit. It is at these peak times that cooling load calculations should be made. The estimated times when the peak load will occur are determined from the tables of *CLTD* and *CLF* values together with the orientation and physical characteristics of the room or space. These tables are described in the next section. The times of the peak cooling load for walls, roofs, windows, and so on, is obvious in the tables and the most dominant cooling load components will then determine the peak time for the entire room or zone. For example, rooms facing west with no exposed roof will experience a peak load in the late afternoon or early evening. East-facing rooms tend to peak during the morning hours. A zone made up of east and west rooms with no exposed roofs will tend to peak when the west rooms peak. If there is a roof, the zone will tend to peak when the roof peaks. High internal loads may dominate the cooling load in some cases and cause an almost uniform load throughout the day.

The details of computing the various cooling load components will be discussed in the following articles. Most of the tables to follow have been abridged and many of the footnotes have been included in the narrative to conserve space. Reference 1 should be consulted for complete data and an original source of information.

7-4   COOLING LOAD—EXTERNAL SURFACES

The calculation procedure is similar for walls, roofs, and conduction through glass. A different procedure is used for the glass solar gain.

Walls and Roofs

Tables 7-1 and 7-2 give the *CLTD* values in degrees F which were computed for the following conditions:

1. Dark surface for solar radiation absorption.
2. Inside temperature of 78 F or 26 C.
3. Outdoor maximum temperature of 95 F or 35 C and an outdoor daily range of 21 F or 12 C. This corresponds to a mean outdoor temperature of 85 F or 29 C.
4. Solar radiation for 40 degree North latitude on July 21.
5. Outside convective film coefficient of 3.0 Btu/(hr-ft²-F) or 17 W/m²-C.
6. Inside convective film coefficient of 1.46 Btu/(hr-ft²-F) or 8.3 W/(m²-C).
7. No forced ventilation or air ducts in the ceiling space.

To convert *CLTD* values from degree F to C simply multiply by 5/9. When conditions differ from the above, the *CLTD* (in degrees F) should be adjusted according to the following relation:

$$CLTD_{cor} = (CLTD + LM)K + (78 - t_i) + (t_{om} - 85) \qquad (7\text{-}13)$$

where

LM = is a correction for latitude and month from Table 7-3, F or C

K = color adjustment factor

t_i = room design temperature, F or C

t_{om} = outdoor mean temperature, $t_{om} = t_o - DR/2$, F or C

The color correction factor K should be used with caution especially for roofs. It has a value of 1.0 for dark surfaces and 0.5 for permanently light-colored surfaces. The designer must be confident that a light-colored surface will remain in that condition before using a value of K less than 1.0. Reference 1 gives a multiplying factor for Eq. (7-13) to account for attic ventilation; however, recent studies of attics and ceiling spaces have shown attic ventilation to be ineffective in reducing the cooling load. Use of this factor is not recommended for attics or ceiling air spaces.

Tables 7-4, 7-5, and 7-6 describe the roofs and walls given in Tables 7-1 and 7-2 in detail. Note, for example, in Table 7-4 the roof numbers and descriptions correspond to those in Table 7-1, and each roof is further described by use of code numbers for each layer. The details of each layer are given in Table 7-6. This same general procedure is used in Table 7-5 for walls, except a capital letter is used to denote construction groups that are thermally similar. Again the layers are described in Table 7-6. The overall heat transfer coefficients given in Table 7-1 for roofs and Table 7-5 for walls are to be used as guides. The actual value of U calculated as described in Chapter 4 should be used in calculations. When the actual roof or wall cannot be found in Tables 7-1 or 7-5, a thermally similar one should be selected. Thermally similar walls or roofs have similar mass and heat capacity.

A wall or roof may have more insulation than those given in Tables 7-1 and 7-2. For each R-7 increase in insulation, choose a roof of similar mass and heat capacity but with a peak $CLTD$ which occurs 2 hours later. For walls, simply move upward one letter in Table 7-2 to the next group for each R-7 increase in insulation. To reiterate, the actual overall heat transfer coefficient for the wall or roof should be used.

Example 7-1.   A building located in Tallahassee, Florida, has a roof described as number 1 in Table 7-4 with a suspended ceiling. There are $2\frac{1}{2}$ in. fibrous glass batts laid on top of the ceiling. Compute the cooling load per ft^2 at 10:00 A.M. and 5:00 P.M. using standard design conditions for August.

Solution.   The cooling load is calculated using Eq. (7-10) where U is computed for the actual roof. Data can be obtained from Chapter 4 or Table 7-6. The overall heat transfer coefficient including the ceiling air space and acoustic tile ceiling is 0.134 Btu/(hr-ft^2-F) as given in Table 7-1. Addition of the $3\frac{1}{2}$ in. batts which have a thermal resistance of 7 (hr-ft^2-F)/Btu will change U to 0.07 Btu/(hr-ft^2-F). The $CLTD$ should then be obtained from Table 7-1 for a thermally

Table 7-1   COOLING LOAD TEMPERATURE DIFFERENCES

Roof No	Description of Construction	Weight lb/ft²	U Btu/(hr-ft²-F)	1	2	3	4	5
						Without Suspended		
1	Steel sheet with 1-in. (or 2-in.) insulation	7 (8)	0.213 (0.124)	1	−2	−3	−3	−5
2	1-in. wood with 1-in. insulation	8	0.170	6	3	0	−1	−3
3	4-in. l.w. concrete	18	0.213	9	5	2	0	−2
4	2-in. h.w. concrete with 1-in. (or 2-in.) insulation	29	0.206 (0.122)	12	8	5	3	0
5	1-in. wood with 2-in. insulation	19	0.109	3	0	−3	−4	−5
6	6-in. l.w. concrete	24	0.158	22	17	13	9	6
7	2.5-in. wood with 1- insulation	13	0.130	29	24	20	16	13
8	8-in. l.w. concrete	31	0.126	35	30	26	22	18
9	4-in. h.w. concrete with 1-in. (or 2-in.) insulation	52 (52)	0.200 (0.120)	25	22	18	15	12
10	2.5-in. wood with 2-in. insulation	13	0.093	30	26	23	19	16
11	Roof terrace system	75	0.106	34	31	28	25	22
12	6-in. h.w. concrete with 1-in. (or 2-in.) insulation	(75) 75	0.192 (0.117)	31	28	25	22	20
13	4-in. wood with 1-in. (or 2-in) insulation	17 (18)	0.106 (0.078)	38	36	33	30	28
						With Suspended		
1	Steel sheet with 1-in. (or 2-in.) insulation	9 (10)	0.134 (0.092)	2	0	−2	−3	−4
2	1-in. wood with 1-in. insulation	10	0.115	20	15	11	8	5
3	4-in. l.w. concrete	20	0.134	19	14	10	7	4
4	2-in. h.w. concrete with 1-in. insulation	30	0.131	28	25	23	20	17
5	1-in. wood with 2-in. insulation	10	0.083	25	20	16	13	10
6	6-in. l.w. concrete	26	0.109	32	28	23	19	16
7	2.5 in. wood with 1-in. insulation	15	0.096	34	31	29	26	23
8	8-in. l.w. concrete	33	0.093	39	36	33	29	26
9	4-in. h.w. concrete with 1-in. (or 2-in.) insulation	53 (54)	0.128 (0.090)	30	29	27	26	24
10	2.5-in. wood with 2-in. insulation	15	0.072	35	33	30	28	26
11	Roof terrace system	77	0.082	30	29	28	27	26
12	6-in. h.w. concrete with 1-in. (or 2-in) insulation	77 (77)	0.125 (0.088)	29	28	27	26	25
13	4-in. wood with 1-in (or 2-in.)insulation	19 (20)	0.083 (0.064)	35	34	33	32	31

*Reprinted by permission from *ASHRAE Handbook of Fundamentals*, 1977.
[a]*CLTD* may be converted to degrees C by multiplying by 5/9.

FOR CALCULATING COOLING LOAD FROM FLAT ROOFS[a,*]

				Solar Time, hr														
6	7	8	9	10	11	12	13	14	15	16	17	18	19	20	21	22	23	24

Ceiling

6	7	8	9	10	11	12	13	14	15	16	17	18	19	20	21	22	23	24
−3	6	19	34	49	61	71	78	79	77	70	59	45	30	18	12	8	5	3
−3	−2	4	14	27	39	52	62	70	74	74	70	62	51	38	28	20	14	9
−3	−3	1	9	20	32	44	55	64	70	73	71	66	57	45	34	25	18	13
−1	−1	3	11	20	30	41	51	59	65	66	66	62	54	45	36	29	22	17
−7	−6	−3	5	16	27	39	49	57	63	64	62	57	48	37	26	18	11	7
3	1	1	3	7	15	23	33	43	51	58	62	64	62	57	50	42	35	28
10	7	6	6	9	13	20	27	34	42	48	53	55	56	54	49	44	39	34
14	11	9	7	7	9	13	19	25	33	39	46	50	53	54	53	49	45	40
9	8	8	10	14	20	26	33	40	46	50	53	53	52	48	43	38	34	30
13	10	9	8	9	13	17	23	29	36	41	46	49	51	50	47	43	39	35
19	16	14	13	13	15	18	22	26	31	36	40	44	45	46	45	43	40	37
17	15	14	14	16	18	22	26	31	36	40	43	45	45	44	42	40	37	34
25	22	20	18	17	16	17	18	21	24	28	32	36	39	41	43	43	42	40

Ceiling

6	7	8	9	10	11	12	13	14	15	16	17	18	19	20	21	22	23	24
−4	−1	9	23	37	50	62	71	77	78	74	67	56	42	28	18	12	8	5
3	2	3	7	13	21	30	40	48	55	60	62	61	58	51	44	37	30	25
2	0	0	4	10	19	29	39	48	56	62	65	64	61	54	46	38	30	24
15	13	13	14	16	20	25	30	35	39	43	46	47	46	44	41	38	35	32
7	5	5	7	12	18	25	33	41	48	53	57	57	56	52	46	40	34	29
13	10	8	7	8	11	16	22	29	36	42	48	52	54	54	51	47	42	37
21	18	16	15	15	16	18	21	25	30	34	38	41	43	44	44	42	40	37
23	20	18	15	14	14	15	17	20	25	29	34	38	42	45	46	45	44	42
22	21	20	20	21	22	24	27	29	32	34	36	38	38	38	37	36	34	33
24	22	20	18	18	18	20	22	25	28	32	35	38	40	41	41	40	39	37
25	24	23	22	22	22	23	23	25	26	28	29	31	32	33	33	33	33	32
24	23	22	21	21	22	23	25	26	28	30	32	33	34	34	34	33	32	31
29	27	26	24	23	22	21	22	22	24	25	27	30	32	34	35	36	37	36

Table 7-2   COOLING LOAD TEMPERATURE DIFFERENCES

	1	2	3	4	5	6	7	8	9	10
North Latitude										
Wall Facing										
N	14	14	14	13	13	13	12	12	11	11
E	24	24	23	23	22	21	20	19	19	18
S	20	20	19	19	18	18	17	16	16	15
W	27	27	26	26	25	24	24	23	22	21
N	15	14	14	13	12	11	11	10	9	9
E	23	22	21	20	18	17	16	15	15	15
S	21	20	19	18	17	15	14	13	12	11
W	29	28	27	26	24	23	21	19	18	17
N	15	14	13	12	11	10	9	8	8	7
E	22	21	19	17	15	14	12	12	14	16
S	21	19	18	16	15	13	12	10	9	9
W	31	29	27	25	22	20	18	16	14	13
N	15	13	12	10	9	7	6	6	6	6
E	19	17	15	13	11	9	8	9	12	17
S	19	17	15	13	11	9	8	7	6	6
W	31	27	24	21	18	15	13	11	10	9
N	12	10	8	7	5	4	3	4	5	6
E	14	12	10	8	6	5	6	11	18	26
S	15	12	10	8	7	5	4	3	4	5
W	25	21	17	14	11	9	7	6	6	6
N	8	6	5	3	2	1	2	4	6	7
E	10	7	6	4	3	2	6	17	28	38
S	10	8	6	4	3	2	1	1	3	7
W	17	13	10	7	5	4	3	3	4	6
N	3	2	1	0	−1	2	7	8	9	12
E	4	2	1	0	−1	11	31	47	54	55
S	4	2	1	0	−1	0	1	5	12	22
W	6	5	3	2	1	1	2	5	8	11

*Reprinted by permission from *ASHRAE Handbook of Fundamentals*, 1977.
^a*CLTD* may be converted to degrees C by multiplying by 5/9.

FOR CALCULATING COOLING LOAD FROM SUNLIT WALLS*

Solar Time, hr													
11	12	13	14	15	16	17	18	19	20	21	22	23	24
Group A Walls													
10	10	10	10	10	10	11	11	12	12	13	13	14	14
19	19	20	21	22	23	24	24	25	25	25	25	25	25
14	14	14	14	14	15	16	17	18	19	19	20	20	20
20	19	19	18	18	18	18	19	20	22	23	25	26	26
Group B Walls													
9	8	9	9	9	10	11	12	13	14	14	15	15	15
17	19	21	22	24	25	26	26	27	27	26	26	25	24
11	11	11	12	14	15	17	19	20	21	22	22	22	21
16	15	14	14	14	15	17	19	22	25	27	29	29	30
Group C Walls													
7	8	8	9	10	12	13	14	15	16	17	17	17	16
19	22	25	27	29	29	30	30	30	29	28	27	26	24
9	10	11	14	17	20	22	24	25	26	25	25	24	22
12	12	12	13	14	16	20	24	29	32	35	35	35	33
Group D Walls													
6	7	8	10	12	13	15	17	18	19	19	19	18	16
22	27	30	32	33	33	32	32	31	30	28	26	24	22
7	9	12	16	20	24	27	29	29	29	27	26	24	22
9	9	10	11	14	18	24	30	36	40	41	40	38	34
Group E Walls													
7	9	11	13	15	17	19	20	21	23	20	18	16	14
33	36	38	37	36	34	33	32	30	28	25	22	20	17
9	13	19	24	29	32	34	33	31	29	26	23	20	17
7	9	11	14	20	27	36	43	49	49	45	40	34	29
Group F Walls													
9	11	14	17	19	21	22	23	24	23	20	16	13	11
44	45	43	39	36	34	32	30	27	24	21	17	15	12
13	20	27	34	38	39	38	35	31	26	22	18	15	12
8	11	14	20	28	39	49	57	60	54	43	34	27	21
Group G Walls													
15	18	21	23	24	24	25	26	22	15	11	9	7	5
50	40	33	31	30	29	27	24	19	15	12	10	8	6
31	39	45	46	43	37	31	25	20	15	12	10	8	5
15	19	27	41	56	67	72	67	48	29	20	15	11	8

Table 7-3   *CLTD* CORRECTION FOR LATITUDE AND MONTH APPLIED TO
WALLS AND ROOFS, NORTH LATITUDES*,ᵃ

Lat.	Month	N	NE NNW	NE NW	ENE WNW	E W	ESE WSW	SE SW	SSE SSW	S	HOR
24	Dec	−5	−7	−9	−10	−7	−3	3	9	13	−13
	Jan/Nov	−4	−6	−8	−9	−6	−3	3	9	13	−11
	Feb/Oct	−4	−5	−6	−6	−3	−1	3	7	10	−7
	Mar/Sept	−3	−4	−3	−3	−1	−1	1	2	4	−3
	Apr/Aug	−2	−1	0	−1	−1	−2	−1	−2	−3	0
	May/Jul	1	2	2	0	0	−3	−3	−5	−6	1
	Jun	3	3	3	1	0	−3	−4	−6	−6	1
32	Dec	−5	−7	−10	−11	−8	−5	2	9	12	−17
	Jan/Nov	−5	−7	−9	−11	−8	−4	2	9	12	−15
	Feb/Oct	−4	−6	−7	−8	−4	−2	4	8	11	−10
	Mar/Sep	−3	−4	−4	−4	−2	−1	3	5	7	−5
	Apr/Aug	−2	−2	−1	−2	0	−1	0	1	1	−1
	May/Jul	1	1	1	0	0	−1	−1	−3	−3	1
	Jun	1	2	2	1	0	−2	−2	−4	−4	2
40	Dec	−6	−8	−10	−13	−10	−7	0	7	10	−21
	Jan/Nov	−5	−7	−10	−12	−9	−6	1	8	11	−19
	Feb/Oct	−5	−7	−8	−9	−6	−3	3	8	12	−14
	Mar/Sep	−4	−5	−5	−6	−3	−1	4	7	10	−8
	Apr/Aug	−2	−3	−2	−2	0	0	2	3	4	−3
	May/Jul	0	0	0	0	0	0	0	0	0	0
	Jun	1	1	1	0	1	0	0	−1	−1	2
48	Dec	−6	−8	−11	−14	−13	−10	−3	2	6	−25
	Jan/Nov	−6	−8	−11	−13	−11	−8	−1	5	8	−24
	Feb/Oct	−5	−7	−10	−11	−8	−5	1	8	11	−18
	Mar/Sep	−4	−6	−6	−7	−4	−1	4	8	11	−11
	Apr/Aug	−3	−3	−3	−3	−1	0	4	6	7	−5
	May/Jul	0	−1	0	0	1	1	3	3	4	0
	Jun	1	1	2	1	2	1	2	2	3	2

*Reprinted by permission from the *ASHRAE Cooling and Heating Load Calculation Manual*, 1979.
ᵃ*CLTD* correction may be converted to degrees C by multiplying by 5/9.

similar roof with a peak *CLTD* occurring 2 hours later than roof number 1. It
appears that roof number 2 is the best choice. Then the uncorrected *CLTD*
values for 10:00 A.M. and 5:00 P.M. are 13 F and 62 F, respectively. Equation
(7-13) is used to correct these values as follows:

$$LM = 0.5 \text{ F}; \text{ Table 7-3 at 30 deg. N. lat.}$$
$$K = 1.0; \text{ dark surface assumed}$$
$$t_i = 78 \text{ F}; \text{ Standard 90 A}$$
$$t_{om} = 92 - (19/2) = 82.5 \text{ F}; \text{ Table C-1}$$

for 10:00 A.M.

$$CLTD_{10} = (13 - 0.5)(1.0) + (78 - 78) + (82.5 - 85) = 10 \text{ F}$$

for 5:00 P.M.

$$CLTD_{17} = (62 - 0.5)(1.0) + (78 - 78) + (82.5 - 85) = 59 \text{ F}$$

then the cooling load at 10:00 A.M. is

$$\dot{q}_c/A = U(CLTD_{10}) = 0.07(10) = 0.7 \text{ Btu/(hr-ft}^2)$$

and at 5:00 P.M.

$$\dot{q}_c/A = U(CLTD_{17}) = 0.07(59) = 4.13 \text{ Btu/(hr-ft}^2)$$

Example 7-2.   The building of Example 7-1 has walls consisting of 4 in. face brick plus 6 in. H.W. concrete block with 1 in. insulation. Compute the cooling load per ft^2 at 10:00 A.M. and 5:00 P.M. for the south wall using standard design conditions for August.

Solution.   The wall is exactly as described in Table 7-5 as Group C. Because there is no variation from the table, the overall heat transfer coefficient will be the same as given, 0.275 Btu/(hr-ft^2-F). Then for a south, Group C wall in Table 7-2, the uncorrected *CLTD* values are 9 and 22 F, respectively. Using Eq. (7-13) the corrected values become

$$CLTD_{10} = 6 \text{ F}$$
$$CLTD_{17} = 19 \text{ F}$$

Table 7-4   ROOF CONSTRUCTION CODE*

Roof No.	Description	Code Number of Layers (see Table 7-6)
1	Steel sheet with 1-in. insulation	A0,E2,E3,B5,A3,E0
2	1-in. wood with 1-in. insulation	A0,E2,E3,B5,B7,E0
3	4-in. l.w. concrete	A0,E2,E3,C14,E0
4	2-in. h.w. concrete with 1-in. insulation	A0,E2,E3,B5,C12,E0
5	1-in. wood with 2-in. insulation	A0,E2,E3,B6,B7,E0
6	6-in. l.w. concrete	A0,E2,E3,C15,E0
7	2.5-in. wood with 1-in. insulation	A0,E2,E3,B5,B8,E0
8	8-in. l.w. concrete	A0,E2,E3,C16,E0
9	4-in. h.w. concrete with 1-in. insulation	A0,E2,E3,B5,C5,E0
10	2.5-in. wood with 2-in. insulation	A0,E2,E3,B6,B8,E0
11	Roof terrace system	A0,C12,B1,B6,E2,E3,C5,E0
12	6-in. h.w. concrete with 1-in. insulation	A0,E2,E3,B5,C13,E0
13	4-in. wood with 1-in. insulation	A0,E2,E3,B5,B9,E0

*Reprinted by permission from *ASHRAE Handbook of Fundamentals*, 1977.

Table 7-5 WALL CONSTRUCTION GROUP DESCRIPTION*

Group No.	Description of Construction	Weight (lb/ft²)	U-value Btu/(hr·ft²·F)	Code Numbers of Layers (see Table 7-4)
4-in. Face Brick + (Brick)				
D	4-in. Common Brick	90	0.415	A0,A2,C4,E1,E0
C	1-in. Insulation or Air space+4-in. Common Brick	90	0.174–0.301	A0,A2,C4,B1/B2,E1,E0
B	2-in. Insulation+4-in. Common Brick	88	0.111	A0,A2,B3,C4,E1,E0
A	Insulation or Air space+8-in. Common Brick	130	0.154–0.243	A0,A2,C9,B1/B2,E1,E0
4-in. Face Brick + (H.W. Concrete)				
B	2-in. Insulation+4-in. concrete	97	0.116	A0,A2,B3,C5,E1,E0
A	Air Space or Insulation+8-in. or more Concrete	143–190	0.110–0.112	A0,A2,B1,C10/11,E1,E0
4-in. Face Brick + (L.W. or H.W. Concrete Block)				
E	4-in. Block	62	0.319	A0,A2,C2,E1,E0
D	Air Space or Insulation+4-in. Block	62	0.153–0.246	A0,A2,C2,B1/B2,E1,E0
C	Air Space or 1-in. Insulation+6-in. or 8-in. Block	73–89	0.221–0.275	A0,A2,B1,C7/C8,E1,E0
B	2-in. Insulation+8-in. Block	89	0.096–0.107	A0,A2,B3,C7/C8,E1,E0

H.W. Concrete Wall + *(Finish)*			
E	4-in. Concrete	63	0.585	A0,A1,C5,E1,E0
D	4-in. Concrete+1-in. or 2-in. Insulation	63	0.119–0.200	A0,A1,C5,B2/B3,E1,E0
C	2-in. Insulation+4-in. Concrete	63	0.119	A0,A1,B6,C5,E1,E0
C	8-in. Concrete	109	0.490	A0,A1,C10,E1,E0
B	8-in. Concrete+1-in. or 2-in. Insulation	110	0.115–0.187	A0,A1,C10,B5/B6,E1,E0
A	2-in. Insulation+8-in. Concrete	110	0.115	A0,A1,B3,C10,E1,E0
L.W. and H.W. Concrete Block + *(Finish)*			
F	4-in. Block+Air Space/Insulation	29	0.161–0.263	A0,A1,C2,B1/B2,E1,E0
E	2-in. Insulation+4-in. Block	29–37	0.105–0.114	A0,A1,B3,C2/C3,E1,E0
E	8-in. Block	47–51	0.294–0.402	A0,A1,C7/C8,E1,E0
D	8-in. Block+Air Space/Insulation	41–57	0.149–0.173	A0,A1,C7/C8,B1/B2,E1,E0
Metal Curtain Wall			
G	With/without Air Space+1-in./2-in./3-in. Insulation	5–6	0.091–0.230	A0,A3,B5/B6/B12,A3,E0
Frame Wall			
G	1-in. to 3-in. Insulation	16	0.081–0.178	A0,A1,B1,B2/B3/B4,E1,E0

*Reprinted by permission from *ASHRAE Handbook of Fundamentals*, 1977.

Table 7-6   THERMAL PROPERTIES AND CODE NUMBERS OF LAYERS USED IN CALCULATIONS OF COEFFICIENTS FOR ROOF AND WALL[a]*

Description	Code Number	Thickness and Thermal Properties				R
		x	k	ρ	c	
Outside surface resistance	A0					0.333
1-in. Stucco (asbestos cement or wood siding plaster, etc.)	A1	0.0833	0.4	116	0.20	0.208
4-in. face brick (dense concrete)	A2	0.333	0.75	130	0.22	0.444
Steel siding (aluminum or other lightweight cladding)	A3	0.005	26.0	480	0.10	0.0002
Finish	A6	0.0417	0.24	78	0.26	0.174
Air space resistance	B1					0.91
1-in. insulation	B2	0.083	0.025	2.0	0.2	3.32
2-in. insulation	B3	0.167	0.025	2.0	0.2	6.68
3-in. insulation	B4	0.25	0.025	2.0	0.2	10.03
1-in. insulation	B5	0.0833	0.025	5.7	0.2	3.33
2-in. insulation	B6	0.167	0.025	5.7	0.2	6.68
1-in. wood	B7	0.0833	0.07	37.0	0.6	1.19
2.5-in. wood	B8	0.2083	0.07	37.0	0.6	2.98
4-in. wood	B9	0.333	0.07	37.0	0.6	4.76
2-in. wood	B10	0.167	0.07	37.0	0.6	2.39
3-in. wood	B11	0.25	0.07	37.0	0.6	3.58
3-in. insulation	B12	0.25	0.025	5.7	0.2	10.0

Code	Description	x	k	ρ	c	R
C1	4-in. clay tile	0.333	0.33	70.0	0.2	1.01
C2	4-in. l.w. concrete block	0.333	0.22	38.0	0.2	1.51
C3	4-in. h.w. concrete block	0.333	0.47	61.0	0.2	0.71
C4	4-in. common brick	0.333	0.42	120	0.2	0.79
C5	4-in. h.w. concrete	0.333	1.0	140	0.2	0.333
C6	8-in. clay tile	0.667	0.33	70	0.2	2.02
C7	8-in. l.w. concrete block	0.667	0.33	38.0	0.2	2.02
C8	8-in. h.w. concrete block	0.667	0.6	61.0	0.2	1.11
C9	8-in. common brick	0.667	0.42	120	0.2	1.59
C10	8-in. h.w. concrete	0.667	1.0	140	0.2	0.667
C11	12-in. h.w. concrete	1.0	1.0	140	0.2	1.00
C12	2-in. h.w. concrete	0.167	1.0	140	0.2	0.167
C13	6-in. h.w. concrete	0.5	1.0	140	0.2	0.50
C14	4-in. l.w. concrete	0.333	0.1	40	0.2	3.33
C15	6-in. l.w. concrete	0.5	0.1	40	0.2	5.0
C16	8-in. l.w. concrete	0.667	0.1	40	0.2	6.67
E0	Inside surface resistance					0.685
E1	0.75-in. plaster; 0.75-in. gypsum or other similar finishing layer	0.0625	0.42	100	0.2	0.149
E2	0.5-in. slag or stone	0.0417	0.83	55	0.40	0.050
E3	0.375-in. felt membrane	0.0313	0.11	70	0.40	0.285
E4	Ceiling air space					1.0
E5	Acoustic tile	0.0625	0.035	30	0.20	1.786

*Reprinted by permission from *ASHRAE Handbook of Fundamentals*, 1977.

^{a}Units: x = ft; c = Btu/(lb-F); k = Btu/(hr-ft-F); R = (hr-ft²-F)/Btu; ρ = lb/ft³.

the cooling load at 10:00 A.M. is then

$$\dot{q}_c/A = 0.275(6) = 1.65 \text{ Btu}/(\text{hr-ft}^2)$$

and at 5:00 P.M.

$$\dot{q}_c/A = 0.275(19) = 5.23 \text{ Btu}/(\text{hr-ft}^2)$$

Heat admission or loss through fenestration areas is affected by many factors of which the following are the most significant.

1. Solar radiation intensity and incident angle.
2. Difference between outdoor and indoor temperature.
3. Velocity and direction of air flow across the exterior and interior surfaces.
4. Low temperature radiation exchange between the surfaces of the glass and the surroundings.
5. Exterior or interior shading.

When solar radiation strikes an unshaded window, Fig. 7-4, about 8 percent of the radiant energy is typically reflected back outdoors, from 5 to 50 percent is absorbed within the glass, depending on the composition and thickness of the glass, and the remainder is transmitted directly indoors, to become part of the cooling load. The solar gain is the sum of the transmitted radiation and the portion of the absorbed radiation that flows inward. Because heat is also conducted through the glass whenever there is an outdoor-indoor temperature difference, the total rate of heat admission is

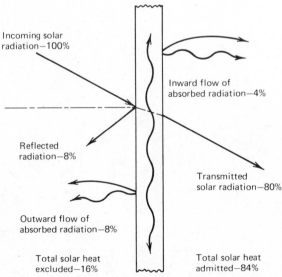

Figure 7-4. Distribution of solar radiation falling on clear plate glass.

Total heat admission through glass = Radiation transmitted through glass
     + Inward flow of absorbed solar radiation + Conduction heat gain   (7-14)

The first two quantities are related to the amount of solar radiation falling on the glass while the third quantity occurs whether or not the sun is shining. In winter the conduction heat flow may well be outward rather than inward.
    We can rewrite Eq. (7-14) to read

Total heat gain = Solar heat gain + Conduction heat gain   (7-14a)

The conduction heat gain per unit area is the product of the overall coefficient of heat transfer U for the existing fenestration and the outdoor-indoor temperature difference $(t_o - t_i)$. Values of U for a number of widely used glazing systems are given in Table 4-8. The cooling load is computed using a *CLTD* much the same as a wall or roof. The glass represents a much more simple situation, however, with small thermal capacity. Table 7-7 gives *CLTD* values for glass. The table is for 78 F indoor temperature, 95 F maximum outdoor temperature, and 21 F daily range. Corrections may be made as shown for walls and roof. The cooling load due to conduction heat gain through window glass is then given by

$$\dot{q}_c = UA(CLTD)$$ (7-15)

The solar heat gain is much more complex because the sun's apparent motion across the sky causes the irradiation of a surface to change minute by minute. As discussed in Chapter 5, the solar heat gain may be calculated when the type of glass, latitude and location, date and time, and orientation are known. It was mentioned earlier that the cooling load due to solar heat gain is expressed in terms of a shading coefficient, a solar heat gain factor, and a cooling load factor, Eq. (7-11). We will first consider the shading coefficient.
    The ASHRAE procedure for estimating solar heat gain assumes that there is a constant ratio between the solar heat gain through any given type of fenestration and the solar heat gain (under exactly the same solar conditions) through unshaded clear sheet glass (i.e., the *reference* glass). This ratio, called the *shading coefficient,* is unique for each type of fenestration or each combination of glazing and shading device.

$$SC = \frac{\text{solar heat gain of fenestration}}{\text{solar heat gain of double-strength glass}}$$ (7-16)

The shading coefficients for several types and combinations of glass are given in Table 7-8. The shading coefficient for any fenestration will rise above the tabulated values when the inner surface coefficient is increased and when the outer surface coefficient is decreased. The converse is also true. Table 7-8 shows the effect for outer film coefficients of 3 and 4 Btu/(hr-ft²-F) or 17 and 23 W/(m²-C) where the larger value is for a 7.5 mph or 3.4 m/s wind, and the smaller value is for a lower wind velocity. The effect of the film coefficient on the shading coefficient is related to the energy absorbed by the glass and then transferred away by convection.

Table 7-7   COOLING LOAD TEMPERATURE DIFFERENCE FOR CONDUCTION THROUGH GLASS AND
CONDUCTION THROUGH DOORS^{a}*

Solar Time, hr																							
1	2	3	4	5	6	7	8	9	10	11	12	13	14	15	16	17	18	19	20	21	22	23	24
CLTD, F																							
1	0	-1	-2	-2	-2	-2	0	2	4	7	9	12	13	14	14	13	12	10	8	6	4	3	2

Corrections: The values in the table were calculated for an inside temperature of 78 F and an outdoor maximum temperature of 95 F with an outdoor daily range of 21 F. The table remains approximately correct for other outdoor maximums (93 − 102 F) and other outdoor daily ranges (16 − 34 F), provided the outdoor daily average temperature remains approximately 85 F. If the room air temperature is different from 78 F, and/or the outdoor daily average temperature is different from 85 F, correct as shown in Eq. (7-13).

*Reprinted by permission from *ASHRAE Handbook of Fundamentals*, 1977.

^{a}The *CLTD* may be converted to deg C by multiplying by 5/9.

Table 7-8   SHADING COEFFICIENTS FOR SINGLE GLASS AND INSULATING GLASS*

A. Single Glass

Type of Glass	Nominal Thickness		Solar Trans.[b]	Shading Coefficient	
	in.	mm		$h_o = 4.0$ Btu/(hr-ft²-F)	$h_o = 3.0$ Btu/(hr-ft²-F)
Clear	$\frac{1}{8}$	3.2	0.84	1.00	1.00
	$\frac{1}{4}$	6.4	0.78	0.94	0.95
	$\frac{3}{8}$	9.5	0.72	0.90	0.92
	$\frac{1}{2}$	12.7	0.67	0.87	0.88
Heat Absorbing	$\frac{1}{8}$	3.2	0.64	0.83	0.85
	$\frac{1}{4}$	6.4	0.46	0.69	0.73
	$\frac{3}{8}$	9.5	0.33	0.60	0.64
	$\frac{1}{2}$	12.7	0.24	0.53	0.58

B. Insulating Glass[a]

Type of Glass			Solar Trans.	Shading Coefficient	
Clear Out, Clear In	$\frac{1}{8}$[c]	3.2	0.71[e]	0.88	0.88
Clear Out, Clear In	$\frac{1}{4}$	6.4	0.61	0.81	0.82
Heat Absorbing[d] Out, Clear In	$\frac{1}{4}$	6.4	0.36	0.55	0.58

*Reprinted by permission from *ASHRAE Handbook of Fundamentals,* 1977.
[a]Refers to factory-fabricated units with $\frac{3}{16}$, $\frac{1}{4}$, or $\frac{1}{2}$-in. air space or to prime windows plus storm sash.
[b]Refer to manufacturer's literature for values.
[c]Thickness of each pane of glass, not thicknes of assembled unit.
[d]Refers to gray, bronze, and green tinted heat-absorbing float glass.
[e]Combined transmittance for assembled unit.

The reference glass is double-strength glass with a transmittance of 0.87, a reflectance of 0.08, and an 0.05 absorptance.

Blinds, shades, and drapes or curtains that are often installed on the inside next to windows decrease the solar heat gain. The shading coefficient is also used to express this effect. Tables 7-9 and 7-10 give representative data for single sheet and insulating glass with indoor shading by blinds and shades. Note that the shading coefficient applies to the combination of glass and shading device.

Shading coefficients for draperies are a complex function of color and the weave of the fabric. Although other variables also have an effect, reasonable correlation has

Table 7-9   SHADING COEFFICIENTS FOR SINGLE GLASS WITH INDOOR SHADING BY VENETIAN BLINDS AND ROLLER SHADES*

Type of Glass	Nominal Thickness in.	mm	Solar Transmittance	Venetian Blinds Medium	Venetian Blinds Light	Roller Shade Opaque Dark	Roller Shade Opaque White	Roller Shade Translucent Light
Regular sheet	$\frac{3}{32}$ to $\frac{1}{4}$	2–6	0.87–0.80					
Regular plate/float	$\frac{1}{4}$ to $\frac{1}{2}$	6–13	0.80–0.71	0.64	0.55	0.59	0.25	0.39
Regular pattern	$\frac{1}{8}$ to $\frac{3}{32}$	3–6	0.87–0.79					
Heat-absorbing pattern	$\frac{1}{8}$	3	—					
Gray sheet	$\frac{3}{16}, \frac{7}{32}$	5–6	0.74, 0.71					
Heat-absorbing plate/float	$\frac{3}{16}, \frac{1}{4}$	5–6	0.46					
Heat-absorbing pattern	$\frac{3}{16}, \frac{1}{4}$	5–6	—	0.57	0.53	0.45	0.30	0.36
Gray sheet	$\frac{1}{8}, \frac{7}{32}$	3–6	0.59, 0.45					
Heat-absorbing plate/float or pattern	—		0.44–0.30					
Heat-absorbing plate/float	$\frac{3}{8}$	10	0.34	0.54	0.52	0.40	0.28	0.32
Heat-absorbing plate			0.29–0.15	0.42	0.40	0.36	0.28	0.31
or pattern	—		0.24					
Reflective coated glass (no inside shade)								
Shading coefficient = 0.30				0.25	0.23			
Shading coefficient = 0.40				0.33	0.29			
Shading coefficient = 0.50				0.42	0.38			
Shading coefficient = 0.60				0.50	0.44			

*Adapted by permission from *ASHRAE Handbook of Fundamentals*, 1977

Table 7-10 SHADING COEFFICIENTS FOR INSULATING GLASS WITH INDOOR SHADING BY VENETIAN BLINDS AND ROLLER SHADES*

Type of Glass	Nominal Thickness, Each light in.	mm	Solar Transmittance Outer Pane	Inner Pane	Type of Shading Venetian Blinds Medium	Light	Roller Shade Opaque Dark	White	Translucent Light
Regular sheet out	$\frac{3}{32}, \frac{1}{8}$	2–3	0.87	0.87	0.57	0.51	0.60	0.25	0.37
Regular sheet in									
Regular plate/float out	$\frac{1}{4}$	6	0.80	0.80	0.39	0.36	0.40	0.22	0.30
Regular plate/float in									
Heat-absorbing plate/float out	$\frac{1}{4}$	6	0.46	0.80					
Regular plate/float in									
Reflective coated glass (no inside shade)									
Shading coefficient = 0.20					0.19	0.18			
Shading coefficient = 0.30					0.27	0.26			
Shading coefficient = 0.40					0.34	0.33			

*Adapted by permission from *ASHRAE Handbook Fundamentals,* 1977.

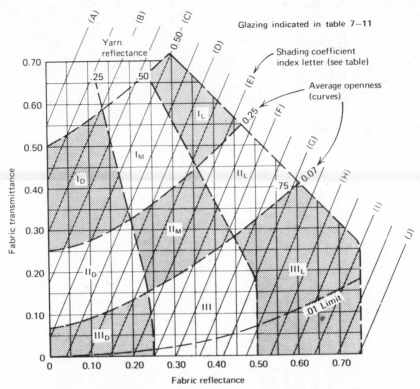

Figure 7-5. Indoor shading properties of drapery fabrics. (Reprinted by permission from *ASHRAE Handbook of Fundamentals*, 1977.)

been obtained using only color and openness of the weave. Figure 7-5 and Table 7-11 are a brief summary of information given in the *ASHRAE Handbook of Fundamentals* (1). Many manufacturers of drapery materials furnish data on the radiation properties of their products. This is usually in the form of the reflectance and transmittance, the abscissa and ordinate for Fig. 7-5. The shading coefficient index letter is read from Fig. 7-5 (*A* to *J*) and used in Table 7-11 to determine the shading coefficient for the combination glass and drapery system. The reflectance and transmittance of the drapery are often not known. In this case the index letter may be estimated from the openness of the weave and color of the material. Openness is classified as: open, I; semiopen, II; and closed, III. Color is classified as: dark, D; medium, M; and light, L. A light-colored, closed-weave material would then be classified III$_L$. From Fig. 7-5 the index letter varies from A to D for this classification, and judgment is required in making a final selection.

There are other data available for shading coefficients of such things as shade screen, domed skylights, and hollow glass blocks. The *ASHRAE Handbook* or manufacturers' catalogs may be consulted for these cases.

The solar heat gain factor of Eq. (7-11) is the maximum value for the month

Table 7-11   SHADING COEFFICIENTS FOR SINGLE AND INSULATING GLASS WITH DRAPERIES*

Glazing	Glass Transmittance	Glass Shading Coefficient	J	I	H	G	F	E	D	C	B	A
Single Glass												
¼ in. or 6 mm Regular	0.80	0.95	0.35	0.40	0.45	0.50	0.55	0.60	0.65	0.70	0.75	0.80
⅛ in. or 13 mm Regular	0.71	0.88	0.35	0.39	0.43	0.48	0.52	0.56	0.61	0.66	0.70	0.74
¼ in. or 6 mm heat absorbing	0.46	0.67	0.33	0.36	0.38	0.41	0.44	0.46	0.49	0.52	0.54	0.57
½ in. or 13 mm heat absorbing	0.24	0.50	0.30	0.32	0.33	0.34	0.36	0.38	0.39	0.40	0.42	0.43
Reflective coated	—	0.60	0.33	0.36	0.38	0.41	0.43	0.46	0.49	0.51	0.54	0.57
(See manufacturers'	—	0.50	0.31	0.33	0.34	0.36	0.38	0.39	0.41	0.42	0.44	0.46
literature for exact	—	0.40	0.26	0.27	0.28	0.29	0.30	0.32	0.33	0.34	0.35	0.36
values)	—	0.30	0.20	0.21	0.21	0.22	0.23	0.23	0.23	0.24	0.24	0.25
Insulating glass (½ in. or 13 mm air space) Regular out and regular in	0.64	0.83	0.35	0.37	0.42	0.45	0.48	0.52	0.56	0.58	0.62	0.66
Heat absorbing out and regular in	0.37	0.56	0.32	0.33	0.35	0.37	0.39	0.41	0.43	0.45	0.47	0.49
Reflective coated	—	0.40	0.28	0.28	0.29	0.31	0.32	0.34	0.36	0.37	0.37	0.38
(See manufacturers'	—	0.30	0.24	0.24	0.25	0.25	0.26	0.26	0.27	0.27	0.28	0.29
literature for exact values.)	—	0.20	0.15	0.15	0.16	0.16	0.17	0.17	0.18	0.18	0.19	0.19

*Adapted by permission from *ASHRAE Handbook of Fundamentals*, 1977.

for a given orientation and latitude. The cooling load factor then represents the ratio of the actual solar heat gain, which becomes cooling load to the maximum solar heat gain. This approach reduces the amount of tabulated data. Table 7-12 gives maximum solar heat gain factors for 32, 40, and 48 degrees north latitude. Table 7-13 gives maximum solar heat gain factors for a special case of external shade: This will be discussed later.

The cooling load factors depend on the actual solar heat gain for a particular time, the internal shade, and the building construction when there is no internal shade. Table 7-14 gives *CLF* values for glass with no interior shading, whereas Table 7-15 is for glass with interior shading. Notice that the type of construction is not a variable when interior shading is present, Table 7-15. The interior shading has two effects. First, it reflects some of the solar radiation so that it never enters the space.

Table 7-12   MAXIMUM SOLAR HEAT GAIN FACTOR,

32 Deg

	N (Shade)	NNE/ NNW	NE/ NW	ENE/ WNW	E/ W	ESE/ WSW	SE/ SW	SSE/ SSW	S	HOR
Jan.	24	24	29	105	175	229	249	250	246	176
Feb.	27	27	65	149	205	242	248	232	221	217
Mar.	32	37	107	183	227	237	227	195	176	252
Apr.	36	80	146	200	227	219	187	141	115	271
May	38	111	170	208	220	199	155	99	74	277
June	44	122	176	208	214	189	139	83	60	276
July	40	111	167	204	215	194	150	96	72	273
Aug.	37	79	141	195	219	210	181	136	111	265
Sep.	33	35	103	173	215	227	218	189	171	244
Oct.	28	28	63	143	195	234	239	225	215	213
Nov.	24	24	29	103	173	225	245	246	243	175
Dec.	22	22	22	84	162	218	246	252	252	158

40 Deg

	N (Shade)	NNE/ NNW	NE/ NW	ENE/ WNW	E/ W	ESE/ WSW	SE/ SW	SSE/ SSW	S	HOR
Jan.	20	20	20	74	154	205	241	252	254	133
Feb.	24	24	50	129	186	234	246	244	241	180
Mar.	29	29	93	169	218	238	236	216	206	223
Apr.	34	71	140	190	224	223	203	170	154	252
May	37	102	165	202	220	208	175	133	113	265
June	48	113	172	205	216	199	161	116	95	267
July	38	102	163	198	216	203	170	129	109	262

Second, the shading device prevents the solar radiation from being absorbed by the floor, interior walls, and furnishings. Instead the shading device absorbs the radiation which is subsequently convected to the room air.

Example 7-3.   The building of Example 7-1 has a 4 ft × 5 ft single glass window in the south wall. The window has light-colored venetian blinds. Compute the cooling load due to the window at 10:00 A.M. and 5:00 P.M. solar time for August using standard design conditions.

Solution.   There are two components of cooling load for the window. One is due to heat conduction, the other to solar radiation. Equation (7-15) gives the cooling load due to conduction where the overall coefficient is obtained from

BTU/(HR-FT²) FOR SUNLIT GLASS, NORTH LATITUDES*,a

40 Deg

	N (Shade)	NNE/ NNW	NE/ NW	ENE/ WNW	E/ W	ESE/ WSW	SE/ SW	SSE/ SSW	S	HOR
Aug.	35	71	135	185	216	214	196	165	149	247
Sep.	30	30	87	160	203	227	226	209	200	215
Oct.	25	25	49	123	180	225	238	236	234	177
Nov.	20	20	20	73	151	201	237	248	250	132
Dec.	18	18	18	60	135	188	232	249	253	113

48 Deg

	N (Shade)	NNE/ NNW	NE/ NW	ENE/ WNW	E/ W	ESE/ WSW	SE/ SW	SSE/ SSW	S	HOR
Jan.	15	15	15	53	118	175	216	239	245	85
Feb.	20	20	36	103	168	216	242	249	250	138
Mar.	26	26	80	154	204	234	239	232	228	188
Apr.	31	61	132	180	219	225	215	194	186	226
May	35	97	158	200	218	214	192	163	150	247
June	46	110	165	204	215	206	180	148	134	252
July	37	96	156	196	214	209	187	158	146	244
Aug.	33	61	128	174	211	216	208	188	180	223
Sep.	27	27	72	144	191	223	228	223	220	182
Oct.	21	21	35	96	161	207	233	241	242	136
Nov.	15	15	15	52	115	172	212	234	240	85
Dec.	13	13	13	36	91	156	195	225	233	65

*Reprinted by permission from *ASHRAE Handbook of Fundamentals*, 1977.

ᵃTo convert to W/m² multiply by 3.155

Table 7-13   MAXIMUM SOLAR HEAT GAIN FACTOR FOR
EXTERNALLY SHADED GLASS, BTU/(HR-FT^2) (BASED
ON GROUND REFLECTANCE OF 0.2)*,^{a}

Use for latitudes 0–24 deg.
For latitudes greater than 24, use north orientation, Table 7-12
For horizontal glass in shade, use the tabulated values for all latitudes

	N	NNE/ NNW	NE/ NW	ENE/ WNW	E/ W	ESE/ WSW	SE/ SW	SSE/ SSW	S	(All Latit.) HOR
Jan.	31	31	31	32	34	36	37	37	38	16
Feb.	34	34	34	35	36	37	38	38	39	16
Mar.	36	36	37	38	39	40	40	39	39	19
Apr.	40	40	41	42	42	42	41	40	40	24
May	43	44	45	46	45	43	41	40	40	28
June	45	46	47	47	46	44	41	40	40	31
July	45	45	46	47	47	45	42	41	41	31
Aug.	42	42	43	45	46	45	43	42	42	28
Sept.	37	37	38	40	41	42	42	41	41	23
Oct.	34	34	34	36	38	39	40	40	40	19
Nov.	32	32	32	32	34	36	38	38	39	17
Dec.	30	30	30	31	32	34	36	37	37	15

*Reprinted by permission from *ASHRAE Handbook of Fundamentals,* 1977.
^{a}To convert to W/m^2 multiply by 3.155

Chapter 4 as 0.81 Btu/(hr-ft^2-F). The *CLTD* values are given in Table 7-7 as
4 F and 13 F for 10:00 A.M. and 5:00 P.M., respectively. Because the daily average temperature is about 83 F, the table values should be reduced by 2 F. The
cooling load due to conduction at 10:00 A.M. is

$$\dot{q}_c = 20(0.81)(4 - 2) = 32 \text{ Btu/hr}$$

and for 5:00 P.M.

$$\dot{q}_c = 20(0.81)(13 - 2) = 178.0 \text{ Btu/hr}$$

Considering the solar radiation and using Eq. (7-11) with a shading coefficient
of 0.55 from Table 7-9; a maximum solar heat gain factor of 111 Btu/(hr-ft^2)
is obtained from Table 7-12 for 32 deg north latitude. The cooling load factors
from Table 7-15 are 0.58 and 0.27. The cooling load for 10:00 A.M. is

$$\dot{q}_c = (20)(0.55)(111)0.58 = 708 \text{ Btu/hr}$$

and for 5:00 P.M. is

$$\dot{q}_c = (20)(0.55)(111)(0.27) = 330 \text{ Btu/hr}$$

The total cooling load for the window at 10:00 A.M. is then

$$\dot{q}_c = 32 + 708 = 740 \text{ Btu/hr}$$

and for 5:00 P.M. is

$$\dot{q}_c = 178 + 330 = 508 \text{ Btu/hr}$$

Common situations are overhangs, side projections, or setback. The shaded portion of a window is easily estimated using the methods and data presented in Chapter 5. Separate calculations are then made for the sunlit and shaded portions of the glass to obtain the cooling load; however, the cooling load due to conduction is the same for both parts.

At latitudes greater than 24 deg, the shaded portion of the glass is treated as a north-facing window with the SHGF and CLF read from Tables 7-12, 7-14, or 7-15 for the north orientation. The shading coefficient is the same for the sunlit and shaded parts.

For the special case of shaded horizontal glass and shaded glass for latitudes less than 24 deg, the SHGF is given in Table 7-13. CLF values are read from Tables 7-14 and 7-15 for the north orientation.

Example 7-4.   Compute the cooling load for a 4 ft by 6 ft regular plate glass window facing south at 40 deg north latitude. The solar time is 3:00 P.M. on 21 July. A drape of semiopen weave and medium color covers the window on the inside. There is a 2 ft overhang at the top of the glass which is much wider than the window. Assume standard design conditions.

Solution.   The cooling load due to heat conduction is not affected by any external shade that may exist. Therefore, using an overall coefficient of 0.81 for single glass with internal shade and a CLTD of 14 F from Table 7-7

$$\dot{q}_c = 24(0.81)14 = 272 \text{ Btu/hr}$$

The cooling load due to solar heat gain must be done separately for the sunlit and shaded portions; however, the shading coefficient will be the same for each calculation. The drapery material has a II_M classification. Using Fig. 7-5, the index letter ranges from about D to F. To be conservative, D will be used. Then from Table 7-11 the shading coefficient is 0.65.

To find the shaded portion of the glass, the distance the shadow extends downward is computed using Eqs. (5-6) and (5-7). The solar altitude and azimuth angles are 47.2 and 76.7 deg, respectively, and the wall azimuth is 45 deg. Then

$$\gamma = |\phi - \psi| = |76.7 - 45| = 31.7 \text{ deg}$$

and

$$y = b \tan \beta / \cos \gamma$$
$$y = 2 \tan (47.2)/\cos (31.7) = 2.53 \text{ ft}$$

Table 7-14   COOLING LOAD FACTORS FOR GLASS

Glass Facing	Room Construction	1	2	3	4	5	6	7	8	9	10
N (Shaded)	L	0.17	0.14	0.11	0.09	0.08	0.33	0.42	0.48	0.56	0.63
	M	0.23	0.20	0.18	0.16	0.14	0.34	0.41	0.46	0.53	0.59
	H	0.25	0.23	0.21	0.20	0.19	0.38	0.45	0.49	0.55	0.60
NE	L	0.04	0.04	0.03	0.02	0.02	0.23	0.41	0.51	0.51	0.45
	M	0.07	0.06	0.06	0.05	0.04	0.21	0.36	0.44	0.45	0.40
	H	0.09	0.08	0.08	0.07	0.07	0.23	0.37	0.44	0.44	0.39
E	L	0.04	0.03	0.03	0.02	0.02	0.19	0.37	0.51	0.57	0.57
	M	0.07	0.06	0.06	0.05	0.05	0.18	0.33	0.44	0.50	0.51
	H	0.09	0.09	0.08	0.08	0.07	0.20	0.34	0.45	0.49	0.49
SE	L	0.05	0.04	0.04	0.03	0.03	0.13	0.28	0.43	0.55	0.62
	M	0.09	0.08	0.07	0.06	0.05	0.14	0.26	0.38	0.48	0.54
	H	0.11	0.10	0.10	0.09	0.08	0.17	0.28	0.40	0.49	0.53
S	L	0.08	0.07	0.05	0.04	0.04	0.06	0.09	0.14	0.22	0.34
	M	0.12	0.11	0.09	0.08	0.07	0.08	0.11	0.14	0.21	0.31
	H	0.13	0.12	0.12	0.11	0.10	0.11	0.14	0.17	0.24	0.33
SW	L	0.12	0.10	0.08	0.06	0.05	0.06	0.08	0.10	0.12	0.14
	M	0.15	0.14	0.12	0.10	0.09	0.09	0.10	0.12	0.13	0.15
	H	0.15	0.14	0.13	0.12	0.11	0.12	0.13	0.14	0.16	0.17
W	L	0.12	0.10	0.08	0.06	0.05	0.06	0.07	0.08	0.10	0.11
	M	0.15	0.13	0.11	0.10	0.09	0.09	0.09	0.10	0.11	0.12
	H	0.14	0.13	0.12	0.11	0.10	0.11	0.12	0.13	0.14	0.14
NW	L	0.11	0.09	0.08	0.06	0.05	0.06	0.08	0.10	0.12	0.14
	M	0.14	0.12	0.11	0.09	0.08	0.09	0.10	0.11	0.13	0.15
	H	0.14	0.12	0.11	0.10	0.10	0.10	0.12	0.13	0.15	0.16
HOR.	L	0.11	0.09	0.07	0.06	0.05	0.07	0.14	0.24	0.36	0.48
	M	0.16	0.14	0.12	0.11	0.09	0.11	0.16	0.24	0.33	0.43
	H	0.17	0.16	0.15	0.14	0.13	0.15	0.20	0.28	0.36	0.45

*Reprinted by permission from *ASHRAE Handbook of Fundamentals*, 1977.

Table 7-15   COOLING LOAD FACTORS FOR GLASS WITH (ALL ROOM CONSTRUCTIONS)

Glass Facing	Solar 1	2	3	4	5	6	7	8	9	10	11	12
N	0.08	0.07	0.06	0.06	0.07	0.73	0.66	0.65	0.73	0.80	0.86	0.89
NE	0.03	0.02	0.02	0.02	0.02	0.56	0.76	0.74	0.58	0.37	0.29	0.27
E	0.03	0.02	0.02	0.02	0.02	0.47	0.72	0.80	0.76	0.62	0.41	0.27
SE	0.03	0.03	0.02	0.02	0.02	0.30	0.57	0.74	0.81	0.79	0.68	0.49
S	0.04	0.04	0.03	0.03	0.03	0.09	0.16	0.23	0.38	0.58	0.75	0.83
SW	0.05	0.05	0.04	0.04	0.03	0.07	0.11	0.14	0.16	0.19	0.22	0.38
W	0.05	0.05	0.04	0.04	0.03	0.06	0.09	0.11	0.13	0.15	0.16	0.17
NW	0.05	0.04	0.04	0.03	0.03	0.07	0.11	0.14	0.17	0.19	0.20	0.21
HOR.	0.06	0.05	0.04	0.04	0.03	0.12	0.27	0.44	0.59	0.72	0.81	0.85

*Reprinted by permission from *ASHRAE Handbook of Fundamentals*, 1977.

Solar Time, hr													
11	12	13	14	15	16	17	18	19	20	21	22	23	24
0.71	0.76	0.80	0.82	0.82	0.79	0.79	0.84	0.61	0.48	0.38	0.31	0.25	0.20
0.65	0.70	0.74	0.75	0.76	0.74	0.75	0.79	0.61	0.50	0.42	0.36	0.31	0.27
0.65	0.69	0.72	0.72	0.72	0.70	0.70	0.75	0.57	0.46	0.39	0.34	0.31	0.28
0.39	0.36	0.33	0.31	0.28	0.26	0.23	0.19	0.15	0.12	0.10	0.08	0.06	0.05
0.36	0.33	0.31	0.30	0.28	0.26	0.24	0.21	0.17	0.15	0.13	0.11	0.09	0.08
0.34	0.31	0.29	0.27	0.26	0.24	0.22	0.20	0.17	0.14	0.13	0.12	0.11	0.10
0.50	0.42	0.37	0.32	0.29	0.25	0.22	0.19	0.15	0.12	0.10	0.08	0.06	0.05
0.46	0.39	0.35	0.31	0.29	0.26	0.23	0.21	0.17	0.15	0.13	0.11	0.10	0.08
0.43	0.36	0.32	0.29	0.26	0.24	0.22	0.19	0.17	0.15	0.13	0.12	0.11	0.10
0.63	0.57	0.48	0.42	0.37	0.33	0.28	0.24	0.19	0.15	0.12	0.10	0.08	0.07
0.56	0.51	0.45	0.40	0.36	0.33	0.29	0.25	0.21	0.18	0.16	0.14	0.12	0.10
0.53	0.48	0.41	0.36	0.33	0.30	0.27	0.24	0.20	0.18	0.16	0.14	0.13	0.12
0.48	0.59	0.65	0.65	0.59	0.50	0.43	0.36	0.28	0.22	0.18	0.15	0.12	0.10
0.42	0.52	0.57	0.58	0.53	0.47	0.41	0.35	0.29	0.25	0.21	0.18	0.16	0.14
0.43	0.51	0.56	0.55	0.50	0.43	0.37	0.32	0.26	0.22	0.20	0.18	0.16	0.15
0.16	0.24	0.36	0.49	0.60	0.66	0.66	0.58	0.43	0.33	0.27	0.22	0.18	0.14
0.17	0.23	0.33	0.44	0.53	0.58	0.59	0.53	0.41	0.33	0.28	0.24	0.21	0.18
0.19	0.25	0.34	0.44	0.52	0.56	0.56	0.49	0.37	0.30	0.25	0.21	0.19	0.17
0.12	0.14	0.20	0.32	0.45	0.57	0.64	0.61	0.44	0.34	0.27	0.22	0.18	0.14
0.13	0.14	0.19	0.29	0.40	0.50	0.56	0.55	0.41	0.33	0.27	0.23	0.20	0.17
0.15	0.16	0.21	0.30	0.40	0.49	0.54	0.52	0.38	0.30	0.24	0.21	0.18	0.16
0.16	0.17	0.19	0.23	0.33	0.47	0.59	0.60	0.42	0.33	0.26	0.21	0.17	0.14
0.16	0.17	0.18	0.21	0.30	0.42	0.51	0.54	0.39	0.32	0.26	0.22	0.19	0.16
0.18	0.18	0.19	0.22	0.30	0.41	0.50	0.51	0.36	0.29	0.23	0.20	0.17	0.15
0.58	0.66	0.72	0.74	0.73	0.67	0.59	0.47	0.37	0.29	0.24	0.19	0.16	0.13
0.52	0.59	0.64	0.67	0.66	0.62	0.56	0.47	0.38	0.32	0.28	0.24	0.21	0.18
0.52	0.59	0.62	0.64	0.62	0.58	0.51	0.42	0.35	0.29	0.26	0.23	0.21	0.19

INTERIOR SHADING, NORTH LATITUDES*

Time, hr											
13	14	15	16	17	18	19	20	21	22	23	24
0.89	0.86	0.82	0.75	0.78	0.91	0.24	0.18	0.15	0.13	0.11	0.10
0.26	0.24	0.22	0.20	0.16	0.12	0.06	0.05	0.04	0.04	0.03	0.03
0.24	0.22	0.20	0.17	0.14	0.11	0.06	0.05	0.05	0.04	0.03	0.03
0.33	0.28	0.25	0.22	0.18	0.13	0.08	0.07	0.06	0.05	0.04	0.04
0.80	0.68	0.50	0.35	0.27	0.19	0.11	0.09	0.08	0.07	0.06	0.05
0.59	0.75	0.83	0.81	0.69	0.45	0.16	0.12	0.10	0.09	0.07	0.06
0.31	0.53	0.72	0.82	0.81	0.61	0.16	0.12	0.10	0.08	0.07	0.06
0.22	0.30	0.52	0.73	0.82	0.69	0.16	0.12	0.10	0.08	0.07	0.06
0.85	0.81	0.71	0.58	0.42	0.25	0.14	0.12	0.10	0.08	0.07	0.06

The shaded area is then

$$A_{sh} = 4 \times (2.53) = 9.12 \text{ ft}^2$$

and the sunlit area is

$$A_{sl} = A - A_{sh} = 24 - 9.12 = 14.88 \text{ ft}^2$$

The cooling load for each part may now be computed. For the sunlit portion, the *SHGF* is 170 Btu/(hr-ft²) from Table 7-12 and the *CLF* is 0.83 from Table 7-15. Then

$$\dot{q}_{sl} = (14.88)(0.65)170(0.83) = 1365 \text{ Btu/hr}$$

For the shaded portion, the *SHGF* is 38 Btu/(hr-ft²) from Table 7-12 and the *CLF* is 0.82 from Table 7-15. Then

$$\dot{q}_{sh} = (9.12)(0.65)(38)(0.82) = 185 \text{ Btu/hr}$$

To summarize, the total cooling load for the window is

$$\dot{q}_t = \dot{q}_c + \dot{q}_{sl} + \dot{q}_{sh}$$
$$\dot{q}_t = 272 + 1365 + 185 = 1820 \text{ Btu/hr}$$

7-5 COOLING LOAD—INTERNAL SOURCES

Internal sources of heat energy may contribute significantly to the total cooling load of a structure, and poor judgment in the estimation of their magnitude can lead to unsatisfactory operation and/or high costs when part of the capacity is unneeded. These internal sources fall into the general categories of people, lights, and miscellaneous equipment.

People

Chapter 3 contains detailed information concerning the rates at which heat and moisture are given up by occupants engaged in different levels of activity and Table 7-16 summarizes the data needed for heat gain calculations. Although the data of Table 7-16 are quite accurate, large errors are often made in the computation of heat gain from occupants because of poor estimates of the periods of occupancy or the number of occupants. Great care should be taken to be realistic about the allowance for the number of people in a structure. It should be kept in mind that rarely will a complete office staff be present or a classroom be full. On the other hand a theater may often be completely occupied and sometimes may contain more occupants than it is designed for. Each design problem must be judged on its own merits. With the exception of theaters and other high occupancy spaces, most spaces are designed with too large an allowance for its occupants. One should not allow for more than the equivalent full-time occupants.

The heat gain from people has two components: sensible and latent. Latent heat

Table 7-16   RATES OF HEAT GAIN FROM OCCUPANTS OF CONDITIONED SPACES[a]*

Degree of Activity	Typical Application	Total Heat Adults, Male		Total Heat Adjusted[b]		Sensible Heat		Latent Heat	
		Watts	Btu/hr	Watts	Btu/hr	Watts	Btu/hr	Watts	Btu/hr
Seated at rest	Theater, movie	115	400	100	350	60	210	40	140
Seated, very light work writing	Offices, hotels, apts	140	480	120	420	65	230	55	190
Seated, eating	Restaurant[c]	150	520	170	580[c]	75	255	95	325
Seated, light work, typing	Offices, hotels, apts	185	640	150	510	75	255	75	255
Standing, light work or walking slowly	Retail Store, Bank	235	800	185	640	90	315	95	325
Light bench work	Factory	255	880	230	780	100	345	130	435
Walking, 3 mph, light machine work	Factory	305	1040	305	1040	100	345	205	695
Bowling[d]	Bowling alley	350	1200	280	960	100	345	180	615
Moderate dancing	Dance hall	400	1360	375	1280	120	405	255	875
Heavy work, heavy machine work, lifting	Factory	470	1600	470	1600	165	565	300	1035
Heavy work, athletics	Gymnasium	585	2000	525	1800	185	635	340	1165

*Reprinted by permission from *ASHRAE Handbook of Fundamentals*, 1977.

[a]Note: Tabulated values are based on 78 F room dry-bulb temperature. For 80 F room dry-bulb, the total heat remains the same, but the sensible heat value should be decreased by approximately 8% and the latent heat values increased accordingly.

[b]Adjusted total heat gain is based on normal percentage of men, women, and children for the application listed, with the postulate that the gain from an adult female is 85% of that for an adult male, and that the gain from a child is 75% of that for an adult male.

[c]Adjusted total heat value for eating in a restaurant, includes 60 Btu/hr for food per individual (30 Btu sensible and 30 Btu latent).

[d]For bowling figure one person per alley actually bowling, and all others as sitting (400 Btu/hr) or standing and walking slowly (790 Btu/hr).

Table 7-17   SENSIBLE HEAT COOL

Total Hours in Space	Hours after Each										
	1	**2**	**3**	**4**	**5**	**6**	**7**	**8**	**9**	**10**	**11**
2	0.49	0.58	0.17	0.13	0.10	0.08	0.07	0.06	0.05	0.04	0.04
4	0.49	0.59	0.66	0.71	0.27	0.21	0.16	0.14	0.11	0.10	0.08
6	0.50	0.60	0.67	0.72	0.76	0.79	0.34	0.26	0.21	0.18	0.15
8	0.51	0.61	0.67	0.72	0.76	0.80	0.82	0.84	0.38	0.30	0.25
10	0.53	0.62	0.69	0.74	0.77	0.80	0.83	0.85	0.87	0.89	0.42
12	0.55	0.64	0.70	0.75	0.79	0.81	0.84	0.86	0.88	0.89	0.91
14	0.58	0.66	0.72	0.77	0.80	0.83	0.85	0.87	0.89	0.90	0.91
16	0.62	0.70	0.75	0.79	0.82	0.85	0.87	0.88	0.90	0.91	0.92
18	0.66	0.74	0.79	0.82	0.85	0.87	0.89	0.90	0.92	0.93	0.94

*Reprinted by permission from *ASHRAE Handbook of Fundamentals,* 1977.

gain goes directly into the air in the space; therefore, this component immediately becomes cooling load with no delay. However, the sensible component from a person is delayed due to storage of a part of this energy in the room and furnishings. The cooling load factor is used to express this delayed effect. The *CLF* depends on the total hours the occupants are in the space and varies from the time of entry. Table 7-17 gives *CLF* values for people. The data are for continuously operating cooling equipment. When the cooling equipment is turned on, and off when the occupants arrive and leave, the *CLF* is 1.0. This is because any energy stored in the building at closing time is still essentially there the next morning. When the cooling equipment is operated for several hours after occupancy, however, the *CLF* values are about the same as for continuous equipment operation. When the density of people is large such as in a theater, a *CLF* of 1.0 should be used.

Example 7-5. An office suite is designed with 10 private offices, a secretarial area with space for four secretaries, a reception area and waiting room, and an executive office with a connecting office for a secretary. Estimate the cooling load from the occupants at 3:00 P.M.

Solution. In the absence of additional data the following approach seems reasonable. For the private offices assume that an average of 7 out of the 10 will be occupied between 8:00 A.M. and 5:00 P.M. Assume that 3 of the 4 secretaries are always present and a receptionist is always there. The waiting room will have a transient occupancy, assume 2 people. The executive office will probably experience variable occupancy. However, an average of one occupant is about right. Assume the secretary is always present. The total number of people for which the heat gain is to be based is then 15. We will assume sedentary, very light work and use data from Tables 7-16 and 7-17. Assuming an adjusted group of males and females and very light work, the sensible and latent heat

ING LOAD FACTORS FOR PEOPLE*

Entry into Space

12	13	14	15	16	17	18	19	20	21	22	23	24
0.03	0.03	0.02	0.02	0.02	0.02	0.01	0.01	0.01	0.01	0.01	0.01	0.01
0.07	0.06	0.06	0.05	0.04	0.04	0.03	0.03	0.03	0.02	0.02	0.02	0.01
0.13	0.11	0.10	0.08	0.07	0.06	0.06	0.05	0.04	0.04	0.03	0.03	0.03
0.21	0.18	0.15	0.13	0.12	0.10	0.09	0.08	0.07	0.06	0.05	0.05	0.04
0.34	0.28	0.23	0.20	0.17	0.15	0.13	0.11	0.10	0.09	0.08	0.07	0.06
0.92	0.45	0.36	0.30	0.25	0.21	0.19	0.16	0.14	0.12	0.11	0.09	0.08
0.92	0.93	0.94	0.47	0.38	0.31	0.26	0.23	0.20	0.17	0.15	0.13	0.11
0.93	0.94	0.95	0.95	0.96	0.49	0.39	0.33	0.28	0.24	0.20	0.18	0.16
0.94	0.95	0.96	0.96	0.97	0.97	0.97	0.50	0.40	0.33	0.28	0.24	0.21

gains per person are 230 and 190 Btu/hr, respectively. The latent cooling load due to people is

$$\dot{q}_l = 15(190) = 2850 \text{ Btu/hr}$$

There is a question about where the occupants go for lunch. Assume they stay in the space, then the total hours in the space are 9 and 3:00 P.M. represents the seventh hour after entry. From Table 7-17, CLF is between 0.82 and 0.83. The cooling load due to sensible heat is

$$\dot{q}_s = 15(230)0.825 = 2850 \text{ Btu/hr}$$

and the total cooling load at 3:00 P.M. is

$$\dot{q} = \dot{q}_l + \dot{q}_s = 2850 + 2850 = 5700 \text{ Btu/hr}$$

The cooling load due to lighting is often the major component of the space load and an accurate estimate is essential. A number of factors need to be considered because the heat gain to the air may differ significantly from the power supplied to the lights. Some of the energy emitted by the lights is in the form of radiation which is absorbed in the space. The absorbed energy is later transferred to the air by convection. The manner in which the lights are installed, the type of air distribution system, and the mass of the structure are important. Obviously a recessed light fixture will tend to transfer heat to the surrounding structure, whereas a hanging fixture will transfer more heat directly to the air. Some light fixtures are designed so that air returns through them absorbing heat that would otherwise go into the space. Lights are often turned off and on to save energy, which makes accurate computations difficult. Lights left on 24 hours a day approach an equilibrium condition where the cooling load equals the power input.

The designer should be careful to use the correct wattage in making calculations. It is not good practice to assume nominal wattage per unit area. Additionally,

Table 7-18   COOLING LOAD FACTORS

"a" Classi- fication	"b" Classi- fication	0	1	2	3	4	5	6	7	8	Number of 9
	A	0.03	0.47	0.58	0.66	0.73	0.78	0.82	0.86	0.88	0.91
	B	0.10	0.54	0.59	0.63	0.66	0.70	0.73	0.76	0.78	0.80
0.45	C	0.15	0.59	0.61	0.64	0.66	0.68	0.70	0.72	0.73	0.75
	D	0.18	0.62	0.63	0.64	0.66	0.67	0.68	0.69	0.69	0.70
	A	0.02	0.57	0.65	0.72	0.78	0.82	0.85	0.88	0.91	0.92
	B	0.08	0.62	0.66	0.69	0.73	0.75	0.78	0.80	0.82	0.84
0.55	C	0.12	0.66	0.68	0.70	0.72	0.74	0.75	0.77	0.78	0.79
	D	0.15	0.69	0.70	0.71	0.72	0.73	0.73	0.74	0.75	0.76
	A	0.02	0.66	0.73	0.78	0.83	0.86	0.89	0.91	0.93	0.94
	B	0.06	0.71	0.74	0.76	0.79	0.81	0.83	0.84	0.86	0.87
0.65	C	0.09	0.74	0.75	0.77	0.78	0.80	0.81	0.82	0.83	0.84
	D	0.11	0.76	0.77	0.77	0.78	0.79	0.79	0.80	0.81	0.81
	A	0.01	0.76	0.81	0.84	0.88	0.90	0.92	0.93	0.95	0.96
	B	0.04	0.79	0.81	0.83	0.85	0.86	0.88	0.89	0.90	0.91
0.75	C	0.07	0.81	0.82	0.83	0.84	0.85	0.86	0.87	0.88	0.89
	D	0.08	0.83	0.83	0.84	0.84	0.85	0.85	0.86	0.86	0.87

*Reprinted by permission from *ASHRAE Handbook of Fundamentals*, 1977.

all of the installed lights may not be used all the time. The instantaneous heat gain for lights may be expressed as

$$q_i = 3.412 \ W \ F_u F_s \qquad (7\text{-}17)$$

where

W = summation of all installed light wattage, Watts
F_u = use factor—ratio of wattage in use to that installed
F_s = special allowance factor for lights requiring more power than their rated wattage. For typical 40 W fluorescent lamps, $F_s = 1.20$

The cooling load is then given by

$$q = q_i(CLF) \qquad (7\text{-}18)$$

The cooling load factor is a function of the building mass, air circulation rate, type of fixture, and time. Table 7-18 gives cooling load factors as a function of time for lights which are on for 10 hours. The "a" and "b" classifications are given in Tables 7-19 and 7-20. The "a" classification depends on the nature of the light fixture and the return air system, whereas the "b" classification depends on the construction of

WHEN LIGHTS ARE ON FOR 10 HOURS*

Hours After Lights Are Turned on

10	11	12	13	14	15	16	17	18	19	20	21	22	23
0.93	0.49	0.39	0.32	0.26	0.21	0.17	0.13	0.11	0.09	0.07	0.06	0.05	0.04
0.82	0.39	0.35	0.32	0.28	0.26	0.23	0.21	0.19	0.17	0.15	0.14	0.12	0.11
0.76	0.33	0.31	0.29	0.27	0.26	0.24	0.23	0.21	0.20	0.19	0.18	0.17	0.16
0.71	0.27	0.26	0.26	0.25	0.24	0.23	0.23	0.22	0.21	0.21	0.20	0.19	0.19
0.94	0.40	0.32	0.26	0.21	0.17	0.14	0.11	0.09	0.07	0.06	0.05	0.04	0.03
0.85	0.32	0.29	0.26	0.23	0.21	0.19	0.17	0.15	0.14	0.12	0.11	0.10	0.09
0.81	0.27	0.25	0.24	0.22	0.21	0.20	0.19	0.17	0.16	0.15	0.14	0.14	0.13
0.76	0.22	0.22	0.21	0.20	0.20	0.19	0.18	0.18	0.17	0.17	0.16	0.16	0.15
0.95	0.31	0.25	0.20	0.16	0.13	0.11	0.08	0.07	0.05	0.04	0.04	0.03	0.02
0.89	0.25	0.22	0.20	0.18	0.16	0.15	0.13	0.12	0.11	0.10	0.09	0.08	0.07
0.85	0.21	0.20	0.18	0.17	0.16	0.15	0.14	0.14	0.13	0.12	0.11	0.11	0.10
0.82	0.17	0.17	0.16	0.16	0.15	0.15	0.14	0.14	0.14	0.13	0.13	0.12	0.12
0.97	0.22	0.18	0.14	0.12	0.09	0.08	0.06	0.05	0.04	0.03	0.03	0.02	0.02
0.92	0.18	0.16	0.14	0.13	0.12	0.10	0.09	0.08	0.08	0.07	0.06	0.06	0.05
0.89	0.15	0.14	0.13	0.12	0.12	0.11	0.10	0.10	0.09	0.09	0.08	0.08	0.07
0.87	0.12	0.12	0.12	0.11	0.11	0.11	0.10	0.10	0.10	0.09	0.09	0.09	0.09

the building and the type of supply and return air system. The *CLF* values of Table 7-18 are for the case of continuously operating cooling equipment. If the cooling equipment is turned on and off on the same schedule as the lights, the *CLF* is equal to 1.0 because this condition is similar to the case when lights and cooling equipment are on continuously. When the cooling equipment is operated for a few hours after the lights are used, however, the system behaves as if the cooling equipment were operating continuously.

In the case of vented light fixtures which are cooled by the return air, the cooling load given by Eq. (7-18) using the specified "a" coefficient is actually the cooling load imposed on the cooling coil and not the space. The return air carries part of the load directly back to the coil. The amount of load absorbed by the return air is a function of many variables but can be up to 50 percent of the light power.

Example 7-6.   The office suite of Example 7-5 has total installed light wattage of 8400 W. The fluorescent light fixtures are recessed and unvented with 40 W lamps. Supply and return air is through the ceiling. The lights are turned on at 8:00 A.M. and turned off at 6:00 P.M. Estimate the cooling load at 10:00 A.M. and 4:00 P.M. The floor is 6 in. concrete.

Solution.   Assuming that about 15 percent of the lights are off all the time due to unoccupied offices, the use factor F_u is 0.85. The allowance factor F_s is 1.2. Then from Eq. (7-17)

$$\dot{q}_i = 3.412(8400)0.85(1.2) = 29,234 \text{ Btu/hr}$$

The "a" and "b" classifications are 0.55 and C, respectively. Then from Table 7-18, CLF_{10} is 0.68 and CLF_{16} is 0.78. Then

$$\dot{q}_{10} = 29,234(0.68) = 19,880 \text{ Btu/hr}$$
$$\dot{q}_{15} = 29,234(0.78) = 22,800 \text{ Btu/hr}$$

Miscellaneous Equipment

There is an infinite variety of equipment and appliances that will provide a heat gain when installed in the conditioned space. If the equipment is completely enclosed, the heat gain equals the input to the equipment. However, this does not often happen. An electric motor may be within the space, but the device powered by the motor may be outside it. In this case the heat gain results from the inefficiency of the motor and will be only about 10 percent of the rated power of the motor. Kitchen appliances

Table 7-19   "a" CLASSIFICATION FOR LIGHTS[a,*]

"a"	Light Fixture and Ventilation Arrangements
0.45	Recessed lights which are not vented Low air supply rate—less than 0.5 cfm/ft² of floor area Supply and return diffusers below ceiling
0.55	Recessed lights which are not vented Medium to high air supply rate—more than 0.5 cfm/ft² of floor area Supply and return diffusers below ceiling or through ceiling space and grill
0.65	Vented light fixtures Medium to high air supply rate—more than 0.5 cfm/ft² of floor area Supply air through ceiling or wall but return air flows around light fixtures and through ceiling space
0.75	Vented or free hanging lights Supply air through ceiling or wall but return air flows around light fixtures and through a ducted return

*Reprinted by permission from *ASHRAE Handbook of Fundamentals,* 1977.
[a]Based on rooms having an average amount of furnishings.

Table 7-20   "b" CLASSIFICATION FOR LIGHTS*[a]

Room Air Circulation and Type of Supply and Return	Floor Construction and Floor Weight in Pounds Per Square Foot of Floor Area				
	2 in. Wooden Floor 10 lb/ft²	3 in. Concrete Floor 40 lb/ft²	6 in. Concrete Floor 75 lb/ft²	8 in. Concrete Floor 120 lb/ft²	12 in. Concrete Floor 160 lb/ft²
Low ventilation rate—minimum required to handle cooling load. Supply through floor, wall or ceiling diffuser. Ceiling space not vented.	B	B	C	D	D
Medium ventilation rate. Supply through floor, wall or ceiling diffuser. Ceiling space not vented.	A	B	C	D	D
High room air circulation induced by primary air of induction unit or by fan coil unit. Return through ceiling space.	A	B	C	C	D
Very high room air circulation used to minimize room temperature gradients. Return through ceiling space.	A	A	B	C	D

*Reprinted by permission from *ASHRAE Handbook of Fundamentals*, 1977.
[a]Based on floor covered with carpet and rubber pad. For floor covered with floor tile use letter designation in next row down with the same floor weight.

usually have a hood through which air is exhausted, which reduces the heat gain appreciably. Remember, however, that the exhausted air must be replaced by outdoor air. The *ASHRAE Handbook* (1) gives extensive data for commercial cooking equipment. Just as with people and lights, care must be taken to properly estimate the actual periods of use, so that an unreasonably large heat gain is not obtained.

After the heat gain is determined, the cooling load is computed using a cooling load factor in a fashion identical to that for lights. Table 7-21 gives *CLF* values for the case of unhooded equipment. Typical office equipment, small computers and motors, and small electric appliances fall in this category. Similar data are given in reference 1 for hooded appliances which are mainly commercial kitchen equipment.

Example 7-7.   Suppose the office suite previously described has a collection of typewriters, duplicating machines, electric coffee pot, and the like, with a total nameplate rating of 3.7 kW. It is estimated that only about one-half of the equipment is in use on a continuous basis. Estimate the cooling load at 5:00 P.M.

Table 7-21   SENSIBLE HEAT COOLING LOAD

Total Operational Hours									Hours	
	1	2	3	4	5	6	7	8	9	10
2	0.56	0.64	0.15	0.11	0.08	0.07	0.06	0.05	0.04	0.04
4	0.57	0.65	0.71	0.75	0.23	0.18	0.14	0.12	0.10	0.08
6	0.57	0.65	0.71	0.76	0.79	0.82	0.29	0.22	0.18	0.15
8	0.58	0.66	0.72	0.76	0.80	0.82	0.85	0.87	0.33	0.26
10	0.60	0.68	0.73	0.77	0.81	0.83	0.85	0.87	0.89	0.90
12	0.62	0.69	0.75	0.79	0.82	0.84	0.86	0.88	0.89	0.91
14	0.64	0.71	0.76	0.80	0.83	0.85	0.87	0.89	0.90	0.92
16	0.67	0.74	0.79	0.82	0.85	0.87	0.89	0.90	0.91	0.92
18	0.71	0.78	0.82	0.85	0.87	0.89	0.90	0.92	0.93	0.94

*Reprinted by permission from *ASHRAE Handbook of Fundamentals*, 1977.

Solution.   Assume the equipment is turned on at 8:00 A.M. and operates until 4:00 P.M. From Table 7-21 the *CLF* is 0.33 because 5:00 P.M. is 9 hours after startup time. The cooling load is

$$\dot{q} = 3.7(3412)(0.33)/2 = 2080 \text{ Btu/hr}$$

Duct Heat Gain

Strictly speaking, heat gain to the air while it is moving through ducts is not an internal component; however, any heat transfer to the air in the supply system is a cooling load imposed on the space. Heat transfer to the air in the return system is additional load imposed on the central cooling unit. These load components are estimated as discussed in Chapter 6; however, extra attention must be given to thermal radiation effects.

7-6 HEAT GAIN FROM INFILTRATION AND OUTDOOR VENTILATION AIR

The necessity of introducing outdoor air was discussed in Chapter 3 and current recommendations were given. Outdoor air should be held to a minimum consistent with the health and comfort of the occupants and energy conservation considerations. The outdoor air should be reduced to allow for the expected infiltration. Computation of the heat gain from outdoor air is properly a part of the air quantity calculation as discussed in Chapter 2; however, it may be computed separately to obtain an estimate of the total load. In this case the method given below for infiltration may be utilized.

The methods used to estimate the quantity of infiltration air were discussed in Chapter 6 when the heating load was considered. The same methods apply to heat

FACTORS FOR APPLIANCES—UNHOODED

after	Appliances	Are	On										
11	12	13	14	15	16	17	18	19	20	21	22	23	24
0.03	0.03	0.02	0.02	0.02	0.02	0.01	0.01	0.01	0.01	0.01	0.01	0.01	0.01
0.07	0.06	0.05	0.05	0.04	0.04	0.03	0.03	0.02	0.02	0.02	0.02	0.01	0.01
0.13	0.11	0.10	0.08	0.07	0.06	0.06	0.05	0.04	0.04	0.03	0.03	0.03	0.02
0.21	0.18	0.15	0.13	0.11	0.10	0.09	0.08	0.07	0.06	0.05	0.04	0.04	0.03
0.36	0.29	0.24	0.20	0.17	0.15	0.13	0.11	0.10	0.08	0.07	0.07	0.06	0.05
0.92	0.93	0.38	0.31	0.25	0.21	0.18	0.16	0.14	0.12	0.11	0.09	0.08	0.07
0.93	0.93	0.94	0.95	0.40	0.32	0.27	0.23	0.19	0.17	0.15	0.13	0.11	0.10
0.93	0.94	0.95	0.96	0.96	0.97	0.42	0.34	0.28	0.24	0.20	0.18	0.15	0.13
0.94	0.95	0.96	0.96	0.97	0.97	0.97	0.98	0.43	0.35	0.29	0.24	0.21	0.18

gain calculations. In this case both a sensible and latent heat gain will result and are computed as follows:

$$\dot{q}_s = \dot{m}_a c_p (t_o - t_i) = \frac{\dot{Q}_o c_p}{v_o} (t_o - t_i) \qquad (7\text{-}19)$$

$$\dot{q}_l = \dot{m}_a (W_o - W_i) i_{fg} = \frac{\dot{Q}_o}{v_o} (W_o - W_i) i_{fg} \qquad (7\text{-}20)$$

These heat gains may also be computed using the psychrometric chart as described in Chapter 2. Wind velocities are lower in the summer, which makes an appreciable difference in the computed infiltration rate.

The direction of the prevailing winds usually changes from winter to summer. This should be considered in making infiltration estimates because the load will be imposed mainly in the space where the air enters. It should also be remembered that during the summer infiltration will enter the upper floors of high-rise buildings instead of the lower floors.

7-7 SUMMATION OF HEAT GAIN AND COOLING LOAD

It is important to adopt a systematic approach to the calculation and summation of the cooling load for a structure. If this is not done, some important parts of the problem may be overlooked. The computer is exceptionally good for this purpose. However, many jobs are too small to justify the use of a computer, and some companies cannot afford a data processing system. In these situations a calculation form or worksheet is a necessity. Figure 7-6 is an example of such a worksheet designed to use the *CLTD* method. The worksheet has a number of advantages over a page-by-page analysis of the problem. A worksheet provides a concise record of the load estimate that can be easily checked, and a person without engineering training can be taught to make the calculation with a limited amount of supervision.

		Direction	U or SC	CLTD or SHGF			CLF or Δt			Area, cfm, etc.	Heat Load Btuh	Cooling Load-Btuh		
1	Identification of Room			1										
2	Length of Exposed Wall – Feet													
3	Room Dimensions – Feet													
4	Comments													
	Load Component			Hour			Hour					Hour		
5	Gross Exposed Walls	N												
		S												
		E												
		W												
6	Windows and Doors– Convective	N												
		S												
		E												
		W												
7	Windows and Glass Doors– Solar	N												
		S												
		E												
		W												
8	Net Exposed Walls and Partitions	N												
		S												
		E												
		W												
9	Roof													
10	Floors													
11	Infiltration, Sensible			1.1										
12	Infiltration, Latent			4840										
13	People, Sensible													
14	People, Latent													
15	Internal, Sensible													
16	Internal, Latent													
17	Lights													
18	Duct Loss													
19	Total Btuh													

Figure 7-6.   Cooling load worksheet.

2					3				
Area, cfm, etc.	Heat Load Btuh	Cooling Load–Btuh			Area, cfm, etc.	Heat Load Btuh	Cooling Load–Btuh		
							Hour		

It is emphasized that the total space cooling load does not generally equal the load imposed on the central cooling unit or cooling coil. The outdoor ventilation air is usually mixed with return air and conditioned before it is supplied to the space. The air circulating fan may be upstream of the coil, in which case the fan power input is a load on the coil. In the case of vented light fixtures, the heat absorbed by the return air is imposed on the coil and not the room.

The next steps are to determine the air quantities and to select the equipment. These steps may be reversed depending on the type of equipment to be utilized.

7-8   AIR QUANTITIES

The preferred method of computing air quantity for cooling and dehumidification has been described in section 2-6 of Chapter 2. That method should always be used when the conditions and size of the cooling load warrant specification of special equipment. This means that the cooling and dehumidifying coil is designed to match the sensible and latent heat requirements of a particular job and that the fan is sized to handle the required volume of air. The fan, cooling coil, control dampers, and the enclosure for these components are referred to as an *air handler*. These units are assembled at the factory in a wide variety of coil and fan models to suit almost any requirement. The design engineer usually specifies the entering and leaving moist air conditions, the volume flow rate of the air, and the total pressure the fan must produce.

Specially constructed equipment cannot be justified for small commercial and residential applications. Furthermore these applications generally have a higher sensible heat factor, and dehumidification is not as critical as it is in large commercial buildings. Therefore, the equipment is manufactured to operate at or near one particular set of conditions. For example typical residential and light commercial cooling equipment operates with a coil SHF of 0.75 to 0.8 with the air entering the coil at about 80 F or 27 C dry bulb and 67 F or 19 C wet bulb temperature. This equipment usually has a capacity of less than 10 tons or 35 kW. When the peak cooling load and latent heat requirements are appropriate, this less expensive type of equipment is used. In this case the air quantity is determined in a different way. The peak cooling load is first computed as 1.3 times the peak sensible cooling load for the structure to match the coil *SHF*. The equipment is then selected to match the peak cooling load as closely as possible. The air quantity is specified by the manufacturer for each unit and is about 400 cfm/ton or 0.0537 $m^3/(s\text{-}kW)$. The total air quantity is then divided among the various rooms according to the cooling load of each room. Table 7-22 shows typical performance data for residential and light commercial cooling equipment.

7-9   ENERGY REQUIREMENTS

The only reliable methods available for estimating cooling equipment energy requirements require hour by hour predictions of the cooling load and must be done using a computer and representative weather data. This is mainly because of the great importance of thermal energy storage in the structure and the complexity of

Table 7-22   TYPICAL PERFORMANCE DATA FOR UNIT TYPE HEATING AND COOLING EQUIPMENT

Rated Cooling Btu/hr	Cooling CFM	Heating		Range Temperature Rise, F	Heating CFM	Temperature Rise, F	External Static Pressure, in. of water	
		Input Btu/hr	Output Btu/hr				Factory Setting	Maximum
36,000	1200	120,000	90,000	70–100	1200	72	0.25	0.50
		138,000	103,500	45–75		60		
48,000	1750	160,000	120,000	55–85	1750	70	0.25	0.50
54,000	2000	200,000	150,000	70–100	2000	72	0.25	0.50
88,000	3200	320,000	240,000	70–100	3200	72	0.25	0.50
116,000	4000	400,000	300,000	70–100	4000	72	0.25	0.50

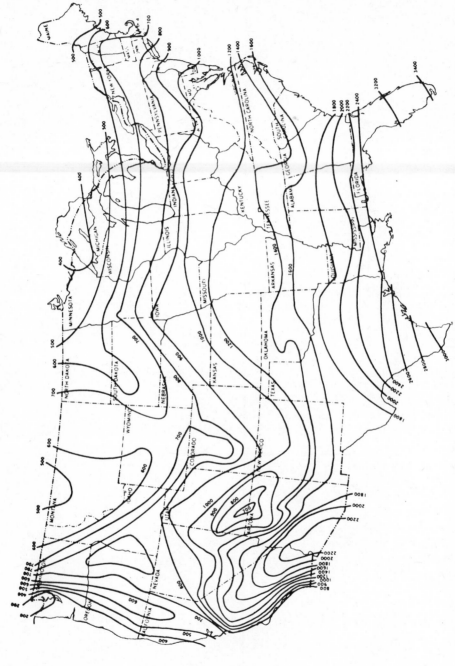

Figure 7-7. Hours of compressor operation for residential systems. (Reprinted by permission from *ASHRAE Journal*, 16, Part I, No. 2, 1974.)

Table 7-23   APPROXIMATE POWER INPUT FOR
RESIDENTIAL AND LIGHT
COMMERCIAL COOLING EQUIPMENT

	kW/ton	kW/W × 10⁴
Refrigerant compressor	1.25 to 1.75	3.56 to 4.98
Circulating fan	0.10	0.28
Condenser fan	0.15	0.43
Total	1.5 to 2.0	4.27 to 5.69

the equipment used. This approach is becoming much easier due to the development of desktop minicomputers. This complex problem is discussed in a later section.

There has been recent work related to residential and light commercial applications that is adaptable to hand calculations (2). Analog computer studies were made using one house as a source of data to verify the computer model. Calculations were made for 217 stations or cities throughout the United States using Air Force, Army, and Navy weather data. The indoor design temperature was 75 F or 24 C dry bulb with allowances made for moisture from infiltration and occupants. The analysis assumes a correctly sized system. Figure 7-7 summarizes the results of the study of compressor operating time for all locations inside the contiguous 48 states. With the compressor operating time it is possible to make an estimate of the energy consumed by the equipment for an average cooling season. The Air-Conditioning and Refrigeration Institute (ARI) publishes data concerning the power requirements of cooling and dehumidifying equipment and most manufacturers can furnish the same data. Table 7-23 shows representative data for residential and light commercial equipment. For residential systems it is generally best to cycle the circulating fan with the compressor. In this case all the equipment in Table 7-23 operates at the same time. However, for light commercial applications the circulating fan will probably operate continuously, and this should be taken into account.

Example 7-8.   Estimate the energy required in kW-hr and the cost at 5 cents per kW-hr to cool a home in Dallas, Texas, during an average season. Assume an indoor design temperature of 75 F dry bulb and that the circulating fan cycles with the compressor. The cooling equipment has a capacity of 36,000 Btu/hr.

Solution.   The required energy is expressed as

$$\text{energy} = [\text{hours of operation}][\text{kW/ton}][\text{tons capacity}]$$

From Fig. 7-7 the estimated hours of compressor operation is 1800 and from Table 7-23 the power required per ton is 1.75, using an average value. Then

$$\text{energy} = 1800(1.75)36,000/12,000 = 9450 \text{ kW-hr}$$

and the cost is

$$\text{Cost} = 9450(0.05) = \$472.50$$

7-10   COOLING LOAD BY THE TRANSFER FUNCTION METHOD

The *CLTD* method for predicting the cooling load was discussed in the previous sections. This method is not as precise as one would like and does not supply sufficient data for detailed energy studies of large structures. Through use of the computer it is possible to solve the heat gain-cooling load problem, taking into account more variables to obtain solutions for every hour of the day with greater accuracy than hand calculations. This approach also makes possible detailed studies of the energy requirements of a structure. The general procedure is referred to as the *transfer function method* and was developed by the ASHRAE Task Group on Energy Requirements (1).

A transfer function is a set of coefficients that relate an output function at some specific time to the value of one or more driving functions at that time and to previous values of both the input and output functions. For example, the heat transfer from the inside surface of a wall at a particular time is related through use of appropriate coefficients to the sol-air temperature at the same time and previous times, to the space temperature, and to the heat transfer rate at previous times. This may be written as

$$\dot{q}_{i,\theta} = A\left[\sum_{n=0} b_n(t_{e,\theta-n\Delta}) - \sum_{n=1} d_n\left(\frac{\dot{q}_{i,\theta-n\Delta}}{A} \right) - t_i \sum_{n=0} c_n \right] \qquad (7\text{-}21)$$

where b_n, d_n, and c_n are the transfer function coefficients that depend on the construction of the wall or roof section. The heat gains from solar radiation on windows, heat sources within the space, ventilation, infiltration, and latent heat gains are computed as previously discussed.

The cooling load for a space at a particular time may be related through the use of appropriate coefficients to the heat gain at that and previous times and to the cooling load at previous times. This is written as

$$\dot{q}_{c,\theta} = \sum_{i=1} (v_0\dot{q}_{i,\theta} + v_1\dot{q}_{i,\theta-\Delta} + v_2\dot{q}_{i,\theta-2\Delta} + \cdots)$$

$$- w_1\dot{q}_{c,\theta-\Delta} - w_2\dot{q}_{c,\theta-2\Delta} - w_3\dot{q}_{c,\theta-3\Delta} - \cdots \qquad (7\text{-}22)$$

where the coefficients depend on the size of the time element Δ, the nature of the heat gain, and the heat storage capacity of the space.

Heat Gain Through Exterior Walls and Roofs

Transfer function coefficients have been computed for a wide range of wall and roof constructions. These calculations involve the solution of the heat conduction equation, Eq. (7-3), with the boundary conditions given by Eqs. (7-5) and (7-9) to obtain the coefficients b_n, d_n, and c_n. A sampling of the transfer function coefficients given in reference 1 are shown in Tables 7-24 and 7-25. The outside surface film coefficient used for the calculations was equal to 3.0 Btu/(hr-ft²-F) or 17 W/(m²-C) while the inside surface film coefficient was 1.46 Btu/(hr-ft²-F) or 8.3 W/(m²-C). The time interval Δ was one hour. Sol-air temperatures used in Eq. (7-21) should also be based on a film coefficient of 3.0 Btu/(hr-ft²-F) or 17 W/(m²-C). Table 7-26

Table 7-24 TRANSFER FUNCTION COEFFICIENTS FOR EXTERIOR WALLS (TIME INTERVAL = 1.0 hr)

Construction Description	Code Numbers of Layers			$n=0$	$n=1$	$n=2$	$n=3$	$n=4$	$n=5$	$n=6$	U	$\sum_{n=0} c_n$
4 in or 100 mm face brick, 2 in. or 50 mm insulation, and 4 in. or 100 mm lightweight concrete block	A0, A2, B3, C2, E1, E0	b	0.00000	0.00046	0.00225	0.00150	0.00016				0.102	0.00437
		d	1.00000	−1.73771	0.90936	−0.11373	0.00496	−0.00001				
4 in. or 100 mm face brick, air space, and 4 in. or 100 mm common brick	A0, A2, B1, C4, E1, E0	b	0.00000	0.00086	0.00485	0.00378	0.00050	0.00001			0.301	0.01000
		d	1.00000	−1.79201	0.98014	−0.16102	0.00609	−0.00003				
4 in. or 100 mm face brick, air space, and 4 in. or 100 mm lightweight concrete block	A0, A2, B1, C2, E1, E0	b	0.00003	0.00286	0.01029	0.00504	0.00037				0.248	0.01859
		d	1.00000	−1.50943	0.65654	−0.07415	0.00212					
12 in. or 300 mm heavyweight concrete	A0, A1, C11, E1, E0	b	0.00000	0.00029	0.00303	0.00412	0.00105	0.00005			0.421	0.00854
		d	1.00000	−1.86853	1.09284	−0.21487	0.01094	−0.00009				
Frame wall with 4 in. or 100 mm brick veneer	A0, A2, B6, A6, E0	b	0.00037	0.00823	0.00983	0.00125	0.00001				0.121	0.01969
		d	1.00000	−1.03045	0.20108	−0.00726						
Frame wall	A0, A6, B6, A6, E0	b	0.01977	0.06317	0.01064	0.00006					0.124	0.09364
		d	1.00000	−0.25848	0.01072							
Frame wall with 3 in. or 75 mm insulation	A0, A1, B1, B4, E1, E0	b	0.00509	0.02644	0.00838	0.00010					0.081	0.04001
		d	1.00000	−0.59602	0.08757	−0.00002						

Table 7-24 TRANSFER FUNCTION COEFFICIENTS FOR EXTERIOR WALLS (TIME INTERVAL = 1.0 hr)
(*continued*)

Construction Description	Code Numbers of Layers		Coefficients[a] b_n and d_n						$n = 6$ U	$\sum\limits_{n=0} c_n$
			$n = 0$	$n = 1$	$n = 2$	$n = 3$	$n = 4$	$n = 5$		
4 in. or 100 mm face brick, 2 in. or 50 mm insulation and 4 in. or 100 mm common brick	A0, A2, B3, C4, E1, E0	b	0.00000	0.00012	0.00084	0.00082	0.00014		0.111	0.00192
		d	1.00000	−1.96722	1.20279	−0.22850	0.01033	−0.00006		
8 in. or 200 mm heavyweight concrete block	A0, A1, C8, E1, E0	b	0.0004	0.0171	0.0310	0.0065	0.0001		0.402	0.0551
		d	1.0000	−1.1621	0.3132	−0.0139				
4 in. or 100 mm face brick, 8 in. or 200 mm clay tile and airspace	A0, A2, C6, B1, E1, E0	b	0.0000	0.0000	0.0007	0.0019	0.0011	0.0001	0.221	0.0038
		d	1.0000	−2.1290	1.5667	−0.4781	0.0605	−0.0029		
4 in. or 100 mm heavyweight concrete with 1 in. or 25 mm insulation	A0, A1, C5, B2, E1, E0	b	0.0005	0.0094	0.0106	0.0013			0.200	0.0218
		d	1.0000	−1.1763	0.3011	−0.0157				
4 in. or 100 mm heavyweight concrete	A0, A1, C5, E1, E0	b	0.0078	0.0705	0.0355	0.0011			0.585	0.1149
		d	1.0000	−0.8789	0.0753	−0.0001				
Sheet metal with 1 in. or 25 mm insulation	A0, A3, B2, B1, A3, E0	b	0.1424	0.0479					0.191	0.1903
		d	1.0000	−0.0013						

*Adapted by permission from *ASHRAE Handbook of Fundamentals*, 1977.

[a]U, b's, and c's are in Btu/(hr-ft²-F), and d is dimensionless. To convert U, b's, and c's to W/(m²-C) multiply by 5.6783.

Table 7-25 TRANSFER FUNCTION COEFFICIENTS FOR ROOFS (TIME INTERVAL = 1.0 hr)*

Construction Description	Code Numbers of Layers	b, d	$n=0$	$n=1$	$n=2$	$n=3$	$n=4$	$n=5$	$n=6$	U	$\sum\limits_{n=0} c_n$
Roof terrace system	A0, C12, B1, B6, E2, E3, C5, E4, E5, E0	b	0.00000	0.00008	0.00048	0.00039	0.00006			0.082	0.00101
		d	1.0000	−1.7304	0.8564	−0.1611	0.0024				
1 in. or 25 mm wood with 1 in. or 25 mm insulation	A0, E2, E3, B5, B7, E4, E5, E0	b	0.0003	0.0082	0.0103	0.0014				0.115	0.0202
		d	1.0000	−1.0046	0.1845	−0.0046					
8 in. or 200 mm lightweight concrete	A0, E2, E3, C16, E4, E5, E0	b	0.00000	0.00002	0.00046	0.00133	0.00079	0.00011		0.092	0.00271
		d	1.00000	−1.91091	1.22135	−0.31019	0.03001	−0.00095	0.00001		
4 in. or 100 mm lightweight concrete	A0, E2, E3, C14, E4, E5, E0	b	0.0001	0.0055	0.0141	0.0045	0.0002			0.134	0.0244
		d	1.0000	−1.0698	0.2665	−0.0143	0.0001				
Steel sheet with 1 in. or 25 mm insulation	A0, E2, E3, B5, A3, E4, E5, E0	b	0.0085	0.0505	0.0179	0.0004				0.134	0.0773
		d	1.0000	−0.4700	0.0476						
2.5 in. or 64 mm wood with 1 in. or 25 mm insulation	A0, E2, E3, B5, B8, E0	b	0.0000	0.0017	0.0068	0.0035	0.0003			0.130	0.0123
		d	1.0000	−1.3557	0.5121	−0.0634	0.0015				
6 in. or 150 mm heavyweight concrete with 1 in. or 25 mm insulation	A0, E2, E3, B5, C13, E0	b	0.0001	0.0036	0.0068	0.0016	0.0001			0.192	0.121
		d	1.0000	−1.3001	0.3991	−0.0361					
4 in. or 100 mm heavyweight concrete with 1 in. or 25 mm insulation	A0, E2, E3, B5, C5, E0	b	0.0008	0.0117	0.0100	0.0008				0.200	0.0233
		d	1.0000	−1.0800	0.2015	−0.0051					

*Adapted by permission from *ASHRAE Handbook of Fundamentals*, 1977.

^{a}U, b's and c's are in Btu/(hr-ft²-F), and d is dimensionless. To convert U, b's, and c's to W/(m²-C), multiply by 5.6783.

Table 7-26   TRANSFER FUNCTION COEFFICIENTS FOR INTERIOR PARTITIONS, FLOORS, AND CEILINGS (TIME INTERVAL = 1.0 HR)

Construction Description	Code Numbers of Layers		Coefficients[a] b_n and d_n							U	$\sum\limits_{n=0} c_n$
			$n=0$	$n=1$	$n=2$	$n=3$	$n=4$	$n=5$	$n=6$		
4 in. or 100 mm lightweight concrete block with $\frac{3}{4}$ in. or 19 mm plaster	E0, E1, C2, E1, E0	b	0.0048	0.0514	0.0339	0.0018				0.314	0.0919
		d	1.0000	−0.8456	0.1397	−0.0015					
4 in. or 100 mm heavyweight concrete block with $\frac{3}{4}$ in. or 19 mm plaster	E0, E1, C3, E1, E0	b	0.0092	0.0705	0.0318	0.0008				0.421	0.1123
		d	1.0000	−0.8203	0.0874	−0.0001					
8 in. or 200 mm lightweight concrete block, plastered both sides	E0, E1, C7, E1, E0	b	0.0002	0.0106	0.0214	0.0054	0.0001			0.271	0.0377
		d	1.0000	−1.2098	0.3736	−0.0248	0.0001				
8 in. or 200 mm heavyweight concrete block, plastered both sides	E0, E1, C8, E1, E0	b	0.0004	0.0141	0.0231	0.0043	0.0001			0.360	0.0420
		d	1.0000	−1.1995	0.3293	−0.0132					
12 in. or 300 mm heavyweight concrete plastered both sides	E0, E1, C11, E1, E0	b	0.0000	0.0002	0.0023	0.0029	0.0007			0.372	0.0061
		d	1.0000	−1.8959	1.1220	−0.2203	0.0107	−0.0001			

Construction		b, d						
Frame, partition with $\frac{3}{4}$ in. or 19 mm gypsum board	E0, E1, B1, E1, E0	b	0.0729	0.1526	0.0159		0.388	0.2414
		d	1.0000	-0.3986	0.0208			
2 in. or 50 mm heavyweight concrete floor deck	E0, A5, C12, E0	b	0.0505	0.0691	0.0011		0.362	0.1207
		d	1.0000	-0.6662				
4 in. or 100 mm heavyweight concrete floor deck	E0, A5, C5, E0	b	0.0111	0.0405	0.0072		0.341	0.0588
		d	1.0000	-0.8507	0.0229			
4 in. or 100 mm lightweight concrete floor deck	E0, A5, C2, E0	b	0.0200	0.0710	0.0120		0.243	0.1030
		d	1.0000	-0.5878	0.0116			
2 in. or 50 mm wood deck	E0, A5, B10, E0	b	0.0020	0.0245	0.0174	0.0010	0.201	0.0449
		d	1.0000	-0.9025	0.1271	-0.0008		
4 in. or 100 mm lightweight concrete deck with false ceiling	E0, A5, C2, E4, E5, E0	b	0.0020	0.0195	0.0104	0.0004	0.144	0.0323
		d	1.0000	-0.8295	0.0534	-0.0003		
2 in. or 50 mm wood deck with false ceiling	E0, A5, B10, E4, E5, E0	b	0.0001	0.0048	0.0078	0.0014	0.129	0.0141
		d	1.0000	-1.1372	0.2530	-0.0061		
Steel deck with false ceiling	E0, A5, A3, E4, E5, E0	b	0.0749	0.0853	0.0022		0.186	0.1624
		d	1.0000	-0.1257	0.0001			

*Adapted by permission from *ASHRAE Handbook of Fundamentals*, 1977.

[a] U, b's, and c's are in Btu/(hr-ft^2-F), and d is dimensionless. To convert U, b's, and c's to W/(m^2-C) multiply by 5.6783.

Table 7-27 PERCENTAGE OF THE DAILY RANGE*

Time, hr	Percent	Time, hr	Percent	Time, hr	Percent	Time, hr	Percent
1	87	7	93	13	11	19	34
2	92	8	84	14	3	20	47
3	96	9	71	15	0	21	58
4	99	10	56	16	3	22	68
5	100	11	39	17	10	23	76
6	98	12	23	18	21	24	82

*Reprinted by permission from *ASHRAE Handbook of Fundamentals*, 1977.
Hour 1 is 1:00 A.M. solar time.

gives some coefficients for interior partitions and Table 7-6 gives data to more adequately describe the walls, roofs, and partitions given in Tables 7-24, 7-25, and 7-26.

The sol-air temperature as a function of time is required to solve Eq. (7-21) for
the heat gain through a wall or roof. Equation (7-8) expresses the sol-air temperature
as a function of the dry bulb temperature, emittance and absorptance of the surface,
total solar intensity, and the convective heat transfer coefficient for the surface.

$$t_e = t_o + \alpha \frac{G_t}{h_o} - \frac{\varepsilon \, \Delta R}{h_o} \tag{7-8}$$

As mentioned in section 7-1 the term α/h_o varies from about 0.15 (hr-ft^2-F)/Btu or
0.026 (m^2-C)/W for a light-colored surface to a maximum of about 0.30 (hr-ft^2-F)/
Btu or 0.053 (m^2-C)/W, whereas the term $\varepsilon \, \Delta R/h_o$ varies from about zero for a
vertical surface to about 7 F or 4 C for a horizontal surface. The total radiation
incident on the surface is computed by the methods of Chapter 5.

The dry bulb temperature required in Eq. (7-8) is computed from the design
dry bulb temperature and the daily range of temperature given in Table C-1 with
the percentage of daily range given in Table 7-27 as

$$t_o = t_d - DR(X) \tag{7-23}$$

where

t_d = design dry bulb temperature, F or C
DR = daily range, F or C
X = percentage of daily range divided by 100

The last term in Eq. (7-21) is a constant because the inside design temperature
is assumed to be constant. Equation (7-21) can now be solved for a particular wall
or roof using Eq. (7-8) and transfer coefficient data from Tables 7-24, 7-25, or 7-26.
The only practical way of doing this is by digital computer; however some of the
equations will be set up to demonstrate the procedure. With a time interval of one
hour, the value for which coefficients are available, Eq. (7-21) may be expanded to

give 24 separate equations for θ ranging from 1 to 24. The first equation for 1:00 A.M. is

$$
\dot{q}_{i,1} = \begin{bmatrix} b_0 t_{e,1} \\ + b_1 t_{e,24} \\ + b_2 t_{e,23} \\ + b_3 t_{e,22} \\ + \cdots \end{bmatrix} - \begin{bmatrix} d_1 \dot{q}_{i,24} \\ + d_2 \dot{q}_{i,23} \\ + d_3 \dot{q}_{i,22} \\ + d_4 \dot{q}_{i,21} \\ + \cdots \end{bmatrix} - \left[t_i \sum_{n=0} c_n \right] \qquad (7\text{-}21a)
$$

The equation for 2:00 A.M. is

$$
\dot{q}_{i,2} = \begin{bmatrix} + b_0 t_{e,2} \\ + b_1 t_{e,1} \\ + b_2 t_{e,24} \\ + b_3 t_{e,23} \\ + \cdots \end{bmatrix} - \begin{bmatrix} + d_1 \dot{q}_{i,1} \\ + d_2 \dot{q}_{i,24} \\ + d_3 \dot{q}_{i,23} \\ + d_4 \dot{q}_{i,22} \\ + \cdots \end{bmatrix} - \left[t_i \sum_{n=0} c_n \right] \qquad (7\text{-}21b)
$$

and the remaining 22 equations could be written in a similar way. The complete set of equations is solved sequentially starting with Eq. (7-21a) and continued until the solution to each set does not change more than a specified amount. Three or four calculation cycles are usually required for an acceptable solution. To start the calculation the unknown heat gains $\dot{q}_{i,\theta}$ are set equal to zero. The effect of this assumption is negligible after successive calculation cycles. The final result of these calculations is the heat gain to a space for a particular wall, roof, or partition as a function of time.

When Eq. (7-21) is used to compute the heat gain for a partition, the sol-air temperation t_e must be changed to the temperature in the adjacent uncooled space t_a and its variation with time must be known. If the temperature in the uncooled space is constant, Eq. (7-21) reduces to the steady state situation given by

$$
\dot{q}_i = UA(t_a - t_i) \qquad (7\text{-}24)
$$

Cooling Load by the Transfer Function Method

The cooling load depends on both the magnitude and nature of the sensible heat gain. Each component of the space heat gain gives rise to a distinct component of cooling load. For example, the cooling load resulting from heat absorbed by the floor is quite different from that absorbed by the window glass or drapes. However, the sum of all the cooling load components for one time is equal to the total cooling load at that time.

The latent heat gain component of the cooling load may or may not be a part of the space cooling load. The latent portion of the infiltration heat gain is a part of the room cooling load, and outdoor ventilation air may be dehumidified before it is distributed to the space. In any case the latent heat gain at a particular time results in cooling load at the same time.

Different room transfer functions are used to convert the various heat gains to cooling load. In other words, the v coefficients in Eq. (7-22) depend on the nature of the heat gain. These coefficients as well as the w coefficients also depend on the mass of the structure. The coefficients for the room transfer function have been computed using the digital computer. The process is rather complex; however, a simple energy balance on the space forms the basis for such calculations.

$$\text{heat gain} = \text{heat loss} + \text{energy storage}$$

or

$$\Sigma \dot{q}_i = \dot{q}_c + \dot{q}_o + mc \frac{dt}{d\theta} \tag{7-25}$$

where $\dot{q}_o$ accounts for any heat transfer from the space to the environment. Rearrangement of Eq. (7-25) yields

$$\dot{q}_c = \Sigma \dot{q}_i - \dot{q}_o - mc \frac{dt}{d\theta} \tag{7-25a}$$

The room transfer coefficients may be derived from Eq. (7-25a) and have been placed in four main groups according to the nature of the heat gain as shown in Table 7-28. Note that the coefficients also depend on the mass of the structure. The coefficients in Table 7-28 apply when all of the heat gains eventually appear as cooling load. In most cases at least a small part of the heat gain is lost to the surroundings. This depends on the thermal conductance of the enclosure. Equation (7-26) has been prepared to estimate the fraction of the heat gain F_c that results in the cooling load.

$$F_c = 1 - 0.02K_t \tag{7-26}$$

The v coefficients are then modified by multiplying by F_c. The unit length conductance K_t is given by

$$K_t = \frac{1}{L} [(UA)_w + (UA)_{ow} + (UA)_c] \tag{7-27}$$

where

L = length of exterior wall, ft or m
U = overall heat transfer coefficient of room element, Btu/(hr-ft²-F) or W/(m²-C) (w for window, ow for outside wall, and c for corridor)
A = area of the room element, ft² or m²

The solution of Eq. (7-22) for the cooling load for a particular space is carried out in a manner analogous to that of Eq. (7-21) previously discussed. It is important to note that Eq. (7-22) must be applied to each different class of sensible heat gain shown in Table 7-28. This means that it must be solved three or four times for each space. The sum of the cooling load components at each time then equals the total cooling load at that time.

Table 7-28   COEFFICIENTS OF ROOM TRANSFER FUNCTIONS*

Heat Gain Component	Room Envelope Construction	v_0	v_1 Dimensionless	v_2
Solar heat gain through glass with no interior shading and heat generated by equipment and people which is dissipated by radiation	Light	0.224	$= 1 + w_1 - v_0$	0.0
	Medium	0.197	$= 1 + w_1 - v_0$	0.0
	Heavy	0.187	$= 1 + w_1 - v_0$	0.0
Conduction heat gain through exterior walls, roofs, partitions and doors, and windows with blinds or drapes	Light	0.703	$= 1 + w_1 - v_0$	0.0
	Medium	0.681	$= 1 + w_1 - v_0$	0.0
	Heavy	0.676	$= 1 + w_1 - v_0$	0.0
Heat generated by lights	Light	0.0	$=$ "a" in Table 7-19	$= 1 + w_1 - v_1$
	Medium	0.0	$=$ "a" in Table 7-19	$= 1 + w_1 - v_1$
	Heavy	0.0	$=$ "a" in Table 7-19	$= 1 + w_1 - v_1$
Heat generated by equipment and people which is dissipated by convection and energy gain due to ventilation and infiltration air	Light	1.0	0.0	0.0
	Medium	1.0	0.0	0.0
	Heavy	1.0	0.0	0.0

*Reprinted by permission from *ASHRAE Handbook of Fundamentals*, 1977.

Table 7-28a   THE VALUE OF w_1 FOR DIFFERENT ROOM AIR CIRCULATION RATES AND ENVELOPE CONSTRUCTION*

Room Envelope Construction[b]	2-in. Wood Floor	3-in. Concrete Floor	6-in. Concrete Floor	8-in. Concrete Floor	12-in. Concrete Floor	Room Air[a] Circulation & Type of Supply and Return
Specific Mass, lb/ft² of floor area	10	40	75	120	160	
	−0.88	−0.92	−0.95	−0.97	−0.98	Low
	−0.84	−0.90	−0.94	−0.96	−0.97	Medium
	−0.81	−0.88	−0.93	−0.95	−0.97	High
	−0.77	−0.85	−0.92	−0.95	−0.97	Very high
	−0.73	−0.83	−0.91	−0.94	−0.96	

*Reprinted by permission from *ASHRAE Handbook of Fundamentals*, 1977.

[a] *Low:* Low ventilation rate-minimum required to cope with cooling load due to lights and occupants in interior zone. Supply through floor, wall, or ceiling diffuser. Ceiling space not vented.

Medium: Medium ventilation rate, supply through floor, wall, or ceiling diffuser. Ceiling space not vented.

High: Room air circulation induced by primary air of induction unit or by fan coil unit. Return through ceiling space.

Very high: High room circulation used to minimize temperature gradient in a room. Return through ceiling space.

[b] Floor covered with carpet and rubber pad; for a floor covered only with floor tile take next w_1 value down the column.

Summary

The heat gain and cooling load calculations using the transfer function method should be carried out by a digital computer. The general procedure is outlined as follows:

1. Derive hourly values of the outdoor dry bulb temperature, Eq. (7-23). This may be part of the computer program.
2. Compute the sol-air temperatures for each surface and each hour using Eq. (7-8).
3. Compute the instantaneous sensible heat gain for each wall, partition, and roof using the procedure discussed above and Eq. (7-21). There is a heat gain for each hour of the day for each surface.
4. Compute the instantaneous sensible heat gain for the doors. Doors are usually assumed to have negligible energy storage. Then

$$\dot{q}_{i,\theta} = (UA)_D(t_{e,\theta} - t_i) \tag{7-28}$$

5. Compute the convective heat gain for the windows.

$$\dot{q}_{i,\theta} = (UA)_w(t_{o,\theta} - t_i) \tag{7-29}$$

6. Compute the solar radiation heat gain for the windows.

$$\dot{q}_{i,\theta} = A_w(SC)(SHGF)_\theta \tag{7-30}$$

7. Compute the heat gain due to the lights, which is simply power input to the lights for the times they are on.
8. Compute the sensible heat gain due to people for the hours the space is occupied.
9. Compute the sensible heat gain due to infiltration for each hour. This may not be constant for every hour.
10. Compute the latent heat gain due to infiltration and people for each hour.
11. Sum the instantaneous heat gains that will appear immediately as cooling load. This includes lights that are on continuously (item 7 above), sensible heat gain due to infiltration (item 9 above), and the sensible heat gain from people (item 8 above). This total corresponds to line 4 in Table 7-28.
12. Sum the solar heat gains for glass that have no interior shading (item 6 above). This total corresponds to line 1 in Table 7-28.
13. Sum the conduction-convection heat gains through roof, walls, windows, and solar heat gain for windows with inside blinds or drapes (3, 4, 5, 6 above). This total corresponds to line 2 of Table 7-28.
14. Sum the heat gain due to lights that are not on all the time (item 7 above and line 3 of Table 7-28).
15. Sum the heat gains due to equipment and people and dissipated by radiation (item 9 above). This total corresponds to line 1 of Table 7-28.

16. Transform the separate heat gain totals of items 12, 13, 14, and 15 above to cooling load by using Eq. (7-22) and the appropriate transfer function coefficients from Table 7-28. Adjust the coefficients using Eq. 7-26 as required.

17. Obtain the total space cooling load for each hour by summing items 10, 11, and 16 above.

7-11 HEAT EXTRACTION RATE AND ROOM TEMPERATURE

The cooling load calculation procedure discussed in the previous section produces data that may be used to compute the heat extraction rate and the room temperature obtained with a particular terminal unit. This is accomplished by the use of a transfer function in a somewhat analogous manner to that used to find the cooling load. When the cooling equipment removes heat energy from the space air at a rate equal to the cooling load, the space air temperature will remain constant and the heat extraction rate $\dot{q}_x$ is equal to the cooling load $\dot{q}_c$. However this is seldom true. A simple energy balance on the space air yields

$$\dot{q}_x - \dot{q}_c = (mc)_{air} \frac{dt_r}{d\theta} \tag{7-28}$$

Equation (7-28) shows the general behavior of the system; however the actual process is complex because of the dependence of $\dot{q}_x$ and $\dot{q}_c$ on the air temperature and time. Therefore a transfer function has been devised, which has an obvious similarity to Eq. (7-28), to describe the process and the digital computer has been utilized to compute the transfer function coefficients. The room air transfer function is

$$\sum_{i=0}^{1} p_i(\dot{q}_{x,\theta-i\Delta} - \dot{q}_{c,\theta-i\Delta}) = \sum_{i=0}^{2} g_i(t_i - t_{r,\theta-i\Delta}) \tag{7-29}$$

where

p_i, g_i = transfer function coefficients
$\dot{q}_c$ = cooling load at the various times
t_i = room temperature used for cooling load calculations
t_r = actual room temperature at the various times

The room air transfer function coefficients are given in Table 7-29. Note that the g coefficients have been normalized to unit floor area. The coefficients g_o and g_1 depend on the heat conductance to the surroundings UA and the infiltration and ventilation rate to the space.

$$g_0 = g_0^* A_{fl} + [UA + \rho c_p(\dot{Q}_i + \dot{Q}_v)]p_0 \tag{7-30}$$
$$g_1 = g_1^* A_{fl} + [UA + \rho c_p(\dot{Q}_i + \dot{Q}_v)]p_1 \tag{7-30a}$$

where

g^* = normalized g coefficients, Btu/(hr-ft²-F) or W/(m²-C)
A_{fl} = floor area, ft² or m²

UA = conductance between the space and the surrounding (average), Btu/(hr-F) or W/C

$\dot{Q}_i$ = infiltration rate, ft^3/hr or m^3/s

$\dot{Q}_v$ = ventilation rate, ft^3/hr or m^3/s

p_i = coefficients from Table 7-29, dimensionless

The remaining g coefficient g_2 is given by

$$g_2 = g_2^* A_{fl} \tag{7-31}$$

The next step is to describe the characteristic of the terminal unit. This can often be done by a linear expression of the form

$$\dot{q}_{\dot{x},\theta} = W + St_{r,\theta} \tag{7-32}$$

where W and S are parameters that characterize the equipment at time θ. The equipment being modeled here is actually the cooling coil and the associated control system which matches the coil load to the space load. A face and bypass system would be typical.

The ability of the cooling unit to extract heat energy from the air varies from some minimum to some maximum value over the throttling range of the control system (thermostat). Equation (7-32) is valid only in the throttling range. The throttling range Δt_r may be described as the temperature at which the thermostat calls for maximum cooling minus the temperature at which the thermostat calls for minimum cooling. Figure 7-8 shows the relationship of the heat extraction rate $\dot{q}_x$ and the room air temperature according to Eq. (7-32). When the room air temperature lies outside the throttling range, the heat extraction rate is either $\dot{q}_{x,\min}$ or $\dot{q}_{x,\max}$ depending on whether t_r is above or below the set-point temperature t_r^*. The value of S is the slope of the characteristic in the throttling range:

$$S = [\dot{q}_{x,\max} - \dot{q}_{x,\min}]/\Delta t_r \tag{7-33}$$

The value of W is the intercept of Eq. (7-32) with the $\dot{q}_x$ axis and is defined by

$$W = [\dot{q}_{x,\max} + \dot{q}_{x,\min}]/2 - St_{r,\theta}^* \tag{7-34}$$

Table 7-29   NORMALIZED COEFFICIENTS OF THE ROOM AIR TRANSFER FUNCTION*,[a]

Room Envelope Construction	Btu/(hr-ft²-F)			Dimensionless	
	g_0^*	g_1^*	g_2^*	p_0	p_1
Light	+1.68	−1.73	+0.05	1.0	−0.82
Medium	+1.81	−1.89	+0.08	1.0	−0.87
Heavy	+1.85	−1.95	+0.10	1.0	−0.93

*Reprinted with permission from *ASHRAE Handbook of Fundamentals,* 1977.

[a]For all cases, the room is assumed furnished.

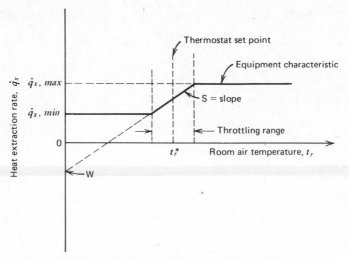

Figure 7-8. Cooling equipment characteristic.

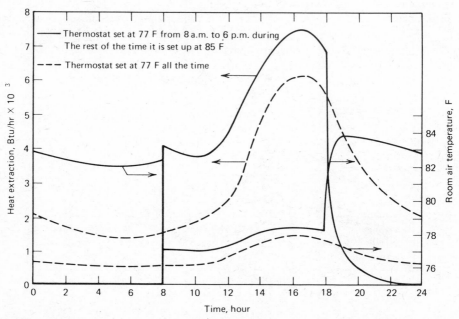

Figure 7-9. Room air temperature and heat extraction rates for continuous and intermittent operation.

where $t_{r,\theta}^*$ the set-point temperature is assumed to be in the center of the throttling range. The set-point temperature may be changed during the day.

Equations (7-29) and (7-32) may be combined and solved for $\dot{q}_{x,\theta}$

$$\dot{q}_{x,\theta} = \frac{g_o W}{S + g_o} + \frac{S}{S + g_o} G_\theta \tag{7-35}$$

where

$$G_\theta = t_i \sum_{i=0}^{2} g_i - \sum_{i=1}^{2} g_i(t_{i,\theta - i\Delta}) + \sum_{i=0}^{1} p_i(\dot{q}_{c,\theta - i\Delta}) - \sum_{i=1}^{1} p_i(\dot{q}_{x,\theta - i\Delta}) \tag{7-36}$$

When the value of $\dot{q}_{x,\theta}$ computed by Eq. (7-35) is greater than $\dot{q}_{x,max}$ it is made equal to $\dot{q}_{x,max}$, and when it is less than $\dot{q}_{x,min}$ it is made equal to $\dot{q}_{x,min}$. Finally, Eqs. (7-32) and (7-35) can be combined and solved for $t_{r,\theta}$;

$$t_{r,\theta} = \frac{(G_\theta - \dot{q}_{x,\theta})}{g_o} \tag{7-37}$$

The solution of Eq. (7-35) is carried out in the same manner as Eqs. (7-21) and (7-22) for the heat gain and cooling load discussed in Section 7-10. To summarize, Eq. (7-35) is written for each hour of the day and solved successively beginning with hour one. Cyclic calculations are carried on until there is no significant change in $\dot{q}_{x,\theta}$ or $t_{r,\theta}$ for each hour. The transfer function coefficients are all constant for a given space unless the infiltration and ventilation rates change with time. In this case the values of g_o and g_1 must be recalculated for each set of conditions.

Figure 7-9 illustrates results obtained by using the above method. Both continuous and intermittent operation are shown using somewhat undersized cooling equipment. Notice that only a small sacrifice in space temperature results from intermittent operation, but the total heat energy removed seems to be less, which would reduce operating costs.

The transfer function method just described is useful in finding the space cooling load and the coil load on an hour-to-hour basis for a design day. The same method may be used to find design heating loads. In such a case the method is identical, except that the solar heat gain and the internal heat gains should be set equal to zero.

7-12   BUILDING AND EQUIPMENT THERMAL SIMULATION

Following the design of the environmental control system for a building, it is often desirable to determine the anticipated energy requirements of the structure for heating, cooling, lighting, and other powered equipment. This same information is often required in energy conservation studies involving existing buildings. Simulation implies that the complete system configuration is already determined; therefore, this type analysis is distinctly different from design.

The transfer function method is readily adaptable to simulation analysis and works equally well for heating or cooling. The main differences as compared with the design computation are:

1. Hourly historical or assumed temperature data are used instead of the design maximum temperature and the daily range.
2. Historical or assumed hourly solar radiation data are used instead of computed clear sky data so that the effect of cloudiness, haze, and so on, are taken into account.
3. More realistic occupancy and lighting schedules may be used.
4. Central cooling equipment such as chillers and boilers is modeled to compute the energy required to meet the heating or cooling demand (heat extraction rate).
5. The simulation is carried out for periods of time greater than one day and generally up to a year.

Various weather data are available on magnetic tape. One type, known as the Typical Meteorological Year (TMY), is in popular use. As the name implies, this weather tape contains data which are typical of the weather patterns for several cities in the United States for a complete year on an hourly basis.

It often happens that a building is not used or the environmental equipment operated under the same conditions for which it was designed. Therefore, these changes should be reflected in the simulation analysis.

Modeling of the central cooling and heating plant can become quite complex; however, this doesn't have to be true. The model should take into account the effect of environmental conditions and load on the operating efficiency. For example, the coefficient of performance of a water chiller depends on the chilled water temperature and the condensing water temperature. The chilled water temperature may be relatively constant but the condensing water temperature may depend on the outdoor

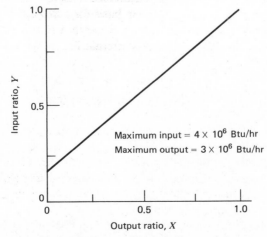

Figure 7-10.   A simple boiler model.

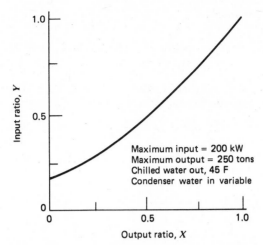

Figure 7-11.   A simple centrifugal chiller model.

wet bulb temperature and the load on the chiller. Boiler performance does not depend so much on environmental conditions but does lose efficiency rapidly with decreasing load.

A useful and simple way of modeling all types of heating and cooling equipment is to normalize the energy input and the capacity with the rated full load input and capacity. Then the normalized input is

$$Y = E/E_{max} \tag{7-38}$$

and the normalized capacity is

$$X = \dot{q}_x/\dot{q}_{x,max} \tag{7-38}$$

These quantities may then be plotted and a curve fitted which comprises a model. Figure 7-10 is an example of such a model for a hot water boiler and Fig. 7-11 is for a centrifugal chiller. To construct the curves it is necessary to have performance data for partial load conditions. Most manufacturers can furnish such data. The plots of Y versus X tend to be straight lines. When this is true, it is possible to develop a model with a limited amount of performance data.

The model for the central plant must also include pumps, fans, cooling towers, and any auxiliary equipment that uses energy. The energy consumed by the lights is also often included in the overall equipment model. An estimate of the total energy consumption of the building is the overall objective. For existing buildings, the total predicted energy usage may be compared with the actual utility data.

Many computer programs are available to carry out both the design and simulation analysis described above. Among these are: TRACE, developed by the Trane Company; E-CUBE, developed by the American Gas Association; DOE-2, developed by the federal government; TRNSYS, developed at the University of Wisconsin; and CHLOAD and CHLSYM, developed at Oklahoma State University.

REFERENCES

1. *ASHRAE Handbook of Fundamentals,* American Society of Heating, Refrigerating and Air-Conditioning Engineers, New York, 1977.

2. L. W. Nelson and J. R. Tobias, "Energy Savings in Residential Buildings," *ASHRAE Journal, 16,* No. 2, February 1974.

3. *ASHRAE* Standard 90-A-80, "Energy Conservation in New Building Design," American Society of Heating, Refrigerating and Air-Conditioning Engineers, New York, 1980.

4. *ASHRAE Cooling and Heating Load Calculation Manual,* GRP 158, American Society of Heating, Refrigerating, and Air Conditioning Engineers, New York, 1979.

PROBLEMS

7-1 Compute the cooling load for a window facing southeast at 40 degrees north latitude at 10:00 A.M. central daylight time in August. The window is regular insulating glass with a $\frac{1}{2}$ in. air space. Inside drapes have a transmittance of 0.35 and reflectance of 0.35. The indoor design temperature is 75 F and the outdoor temperature is 100 F with a 24 F daily range. Window dimensions are 2 ft wide by 8 ft high.

7-2 The window in Problem 7-1 has an 18 in. overhang at the top. How far will the shadow extend downward?

7-3 Compute the cooling load for the window in Problem 7-1 with the external shade in Problem 7-2.

7-4 Compute the cooling load for a window facing south at 40 degrees north latitude at 1:00 P.M. sun time in October. The glass is 5 mm gray sheet with venetian blinds of medium color. The inside temperature is 25 C while the outdoor temperature is 35 C with a 13 C daily range. The window is 1 by 2 m.

7-5 The window in Problem 7-4 has an overhang at the top of 0.6 m. What percentage of the window is shaded assuming that the overhang is much wider than the window?

7-6 Compute the cooling load for the window in Problem 7-4 with the shade in Problem 7-5.

7-7 Compute the cooling load for the south windows of an office building that has no external shading. The windows are the insulating type with regular plate glass inside and out. Drapes with a reflectance of 0.2 and transmittance of 0.5 are fully closed. Make the calculation for 12:00 noon in (a) July, and (b) December at 40 degrees north latitude. The total window area is 400 ft². Assume the indoor temperature is 78 F and the outdoor temperatures are 100 and 40 F, respectively with a 22 F daily range.

7-8 A 6 by 6 ft regular plate glass window faces west. There is a 2 ft overhang at the top. Compute the cooling load at 3:00 P.M. in August at 40 degrees north latitude. Assume that the outdoor-indoor temperature differential is 25 F and the outdoor temperature is 100 F.

7-9   A 1 m² regular plate glass window has inside venetian blinds of medium color. The window is located at 40 degrees north latitude. Calculate the cooling load for (a) the window facing east at 10:00 A.M., June with a mean outdoor temperature of 29 C and (b) the window facing west at 2:00 P.M. in June with a mean outdoor temperature of 32 C. Assume an indoor temperature of 25 C.

7-10   A window is set back similar to that shown in Fig. 5-6. If the window faces southwest, at what time on June 21 will the shading on the sides be zero?

7-11   Compute the cooling load for the south wall of a building at 40 degrees north latitude in July. The time is 4:00 P.M. sun time. The wall is frame with an overall heat transfer coefficient of 0.08 Btu/(hr-ft²-F). The inside design temperature is 75 F and the outside temperature is 101 F. The daily range of temperature is 22 F. The wall is 8 × 16 ft with a 4 × 5 ft window. (Do not make a calculation for the window.)

7-12   The building of Problem 7-11 has a roof of light construction (No. 1, Table 7-1) with $U = 0.12$ Btu/(hr-ft²-F). Compute the cooling load for a 16 × 20 ft room using data from Problem 7-11 (roof only).

7-13   Compute the peak cooling load per square meter for a west wall similar to that of Table 4-7b. Assume that the wall is a dark color and located at 40 degrees north latitude. The date is July 21. What time of day does the peak occur? The outdoor and indoor design temperatures are 38 and 25 C, respectively, and the daily range is 12 C.

7-14   Compute the cooling load for a 8 × 40 m roof-ceiling combination similar to that of Table 4-7d at 6:00 P.M. in July. Assume 40 degrees north latitude, 39 and 24 C outdoor and indoor temperatures, and a daily range of 13 C.

7-15   A ceiling-attic combination has a ceiling thermal resistance of 9.5 (hr-ft²-F)/Btu. Estimate the overall heat transfer coefficient for the combination. Assume a dark-colored roof and time of 3:00 P.M. August 21. (There are no reflective surfaces.)

7-16   Assuming that the ceiling-roof combination in Problem 7-15 has a reflective surface facing the air space in the attic, compute the overall heat transfer coefficient for the combination.

7-17   Compute the cooling load for a frame wall like that shown in Table 4-7 except 3½ in. insulation batts have been added in the stud space. The wall is on the south side of a house in Shreveport, Louisiana. Make calculations for 10:00 A.M., 1:00 A.M., and 6:00 P.M. Daylight Savings Time in June, using ASHRAE Standard 90 A conditions.

7-18   The house of Problem 7-17 has a typical attic ceiling combination which can be approximated by roof number 1 of Table 7-1 with suspended ceiling and 3½ in. insulation batts. Compute the cooling loads for the times and conditions given in Problem 7-17.

7-19   A large office space has an average occupancy of 20 people from 8:00 A.M. to 5:00 P.M. Lighting is 2.5 W/ft² of recessed, unvented fluorescent fixtures. Assume a low ventilation rate and 6 in. concrete floor. Miscellaneous equipment operates intermittently and amounts to an average of 3 horsepower. Compute the cooling load at 4:00 P.M. for the space assuming a floor area of 4000 ft².

7-20   A space has an occupancy of 40 people engaged in sedentary activity from 8:00 A.M. to 5:00 P.M. The average light level is 20 W/m² of free hanging fluorescent fixtures.

Assume a medium ventilation rate and a 3 in. concrete floor. Equipment amounts to 6 kW. Estimate the cooling load for a floor area of 750 m^2.

7-21 A large office complex has a variable occupancy pattern. Twenty people arrive at 8:00 A.M. and leave at 4:00 P.M. Forty people arrive at 10:00 A.M. and leave at 4:00 P.M. Ten people arrive at 1:00 P.M. and leave at 5:00 P.M. Assume seated, light activity and compute the cooling load at 4:00 P.M. and 6:00 P.M.

7-22 A space contains 90 people at 8:00 A.M. The people start leaving at 10:00 A.M. at the rate of 10 people per hour. What is the cooling load at 6:00 P.M. assuming seated, very light work?

7-23 A large room has 5000 watts of vented fluorescent light fixtures. The floor construction is 6 in. concrete. The air flows from the lights through a ducted return.
(a) Assuming the lights have been on for 10 hours, compute the cooling load at 5:00 P.M.
(b) Assuming that 20 percent of the heat from the lights is convected to the return air, what is the actual cooling load for the space?

7-24 Compute the total cooling load for the structure described by the plans and specifications furnished by the instructor using (a) the *CLTD* method and (b) the transfer function method.

7-25 Estimate the power requirements for Problem 7-24a, assuming that the vapor compression cooling equipment is properly sized for the load.

7-26 Compute the heat extraction rate and space temperature as a function of time for Problem 7-24b.

7-27 A house located in Atlanta, Georgia, has a 12 kW design cooling load and the same-size cooling equipment. (a) Estimate the energy required to cool the house for a typical summer. (b) If energy costs an average of 5 cents per kW-hr, what is the annual cost to cool the house?

7-28 Estimate the energy requirements for the structure of Problem 7-24b.

CHAPTER EIGHT

Air-Conditioning Systems

Those aspects of the HVAC system design problem treated in Chapters 1 to 7 do not require consideration of the particular type of equipment to be used. After the heating and cooling loads and the air quantities have been computed, the required conditioning equipment must be selected and located in the spaces so that the refrigerant and water piping, the air distribution system, controls, services, and auxiliary devices may be sized and integrated into a complete air conditioning system. Therefore, various systems that meet the requirements of different building types and uses, load variations, and economic considerations are discussed in the following sections.

The basic system of air conditioning was the forced warm air heating and ventilating systems with centrally located equipment and distributed tempered air through ducts. It was found that cooling and dehumidification equipment added to these systems produced satisfactory air-conditioning in spaces where heat gains were relatively uniform throughout the conditioned area.

The use of this system was made complicated by the presence of variable heat sources within the space served. When this occurred, the area had to be divided into sections or zones and the central system supplemented by additional equipment and controls.

As the science of air conditioning progressed, variations in the basic designs were required to meet the functional and economic demands of individual buildings. Now air-conditioning systems are categorized according to the means by which the controllable heating and cooling is accomplished in the conditioned area. They are further classified to accomplish specific purposes by special equipment arrangements.

8-1 THE BASIC CENTRAL SYSTEM

The basic central system is an all-air, single-zone system that is used as part of most systems. It can be designed for low, medium, and high pressure air distribution. Normally the equipment is located outside the conditioned area in a basement, penthouse, or in a service area at the core of a commercial building. It can be installed in any convenient area of a factory, particularly in the roof truss area or on the roof. The equipment can be located adjacent to the heating and refrigerating equipment or at a considerable distance from it by using a circulating refrigerant, chilled water, hot water, or steam for energy transfer sources. It is most important that the heat gains and losses within the area conditioned by a central system be uniformly distributed if a single-zone constant air volume duct system is to be used.

Spaces with Uniform Loads

In spaces with relatively large open areas and small external loads such as theaters, auditoriums, department stores, and the public spaces of most buildings air-conditioning loads are fairly uniform throughout. Adjustments for minor variations can be made by supplying more or less air, by changing the supply air temperatures in the original design and by balance of the system. In commercial buildings the interior areas generally meet these criteria when local heat sources such as computers are treated separately. These interior areas usually require year-round cooling, and any isolated spaces of limited occupancy may require special attention.

Spaces Requiring Precision Control

Spaces with stringent requirements for cleanliness, humidity, and temperature control, and/or air distribution are usually isolated within the larger building and require precision control. The components of a central system can be selected and assembled to meet the exact requirements of the area. Locating the equipment outside the conditioned space permits routine inspection and maintenance without interference.

Multiple Systems for Large Areas

In large spaces such as hangars, factories, and some large stores, practical considerations require multiple installation of central systems. The size of the individual system is usually limited only by the physical limitations of the structure. In multiple systems the equipment is often located in the truss space, against the outside wall, or on the roof where it least interferes with the operations within the conditioned space and outdoor air is readily available.

Primary Source of Conditioned Air for Other Systems

There are various systems for controlling conditions in individual zones in which a constant supply of conditioned outdoor air is used for ventilation and for some of the air-conditioning load. This reduces the amount of conditioned air handled by the central system and, consequently, the space required for ductwork. By utilizing high velocities and designing for the resultant higher pressure and sound levels, the ductwork can be further reduced in size.

Systems for Environmental Control

For applications requiring close aseptic or contamination control of the environment all-air type systems generally are used to provide the necessary air supply to sustain adequate dilution of the controlled space. These applications are usually combinations of supply systems and scavenging exhaust systems to circulate the diluting air through the controlled environment space.

8-2 CENTRAL SYSTEM COMPONENTS

When the air handling requirements of the system have been determined from the use and physical characteristics of the space to be conditioned, and capacities for both air volumes and thermal exchange, the designer can select and arrange the various components. It is important that equipment be adequate, accessible for easy maintenance, and no more complex in arrangement and control than is necessary to produce the conditions required to meet the design criteria.

Figure 8-1 shows a general arrangement of the component parts of an all-air, central system for year-round conditioning. Although Fig. 8-1 indicates a built-up system, most of the components are available in subassembled sections ready for bolting together in the field or completely assembled by the manufacturer. Large preassembled sections are available but, because of the inflexibility of their dimensions, one should determine if a more efficient and economical use of equipment space could be obtained with a built-up system. Figure 8-2 shows a single zone central air handler that contains the fan, heating and cooling coil, and the filter section.

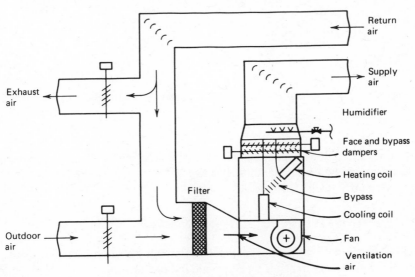

Figure 8-1. Typical central air system.

Figure 8-2.   Single zone central air handler. (Courtesy of Thermal Corp., Houston, Texas.)

The fan is located downstream of the coil and is referred to as a draw-through configuration. A multizone air handler is shown in Fig. 8-3, which has a hot deck, cold deck, and neutral deck. This is called a blow-through configuration.

8-3   CENTRAL MECHANICAL EQUIPMENT

The selection of central mechanical equipment for air-conditioning systems in large buildings chiefly depends on economic factors once the total required capacity has been determined. The decision as to the choice of components depends on the type of fuel available, environmental protection required, structural support, and available space, among other considerations.

In recent years the cost of energy has led to many designs that recover the internal heat from lights, people, and equipment to reduce the size of the heating plant.

Among the largest installations of central mechanical equipment are the central cooling and heating plants serving groups of large buildings. These provide higher diversity, greater efficiency, less maintenance cost, and lower labor costs compared to individual plants. The economics of these systems requires maximum use of heat recovery.

Most large buildings, however, have their own central equipment plant and the choice of equipment depends mainly on owning and operating costs.

Heating Equipment

Steam and hot water boilers for heating are manufactured for high or low pressure, using coal, oil, electricity, gas, or waste material for fuel.

Low pressure boilers are rated for a working pressure of 15 psig or 103 kPa for

Figure 8-3. Central air handler with hot deck, cold deck, and neutral deck. (Courtesy of Thermal Corporation, Houston, Texas.)

Figure 8-4. Packaged fire tube hot water boiler. (Courtesy of Federal Corporation, Oklahoma City, Oklahoma.)

steam, 160 psig or 1.1 MPa for water, with a maximum temperature limitation of 250 F or 121 C. High-rise buildings usually use high pressure boilers. However, a low pressure steam boiler may be used with a heat exchanger, called a converter, to heat water which is then pumped to the various floors in a high-rise building. This eliminates the need for an expensive high pressure boiler. Packaged boilers with all components and controls assembled as a unit are available. Figure 8-4 shows such a unit.

Refrigeration Compressors

The three major types used in large systems are:

1. Reciprocating: $\frac{1}{16}$ to 150 hp or 50 W to 112 kW.
2. Helical rotary: 100 to 1000 tons or 350 to 3500 kW.
3. Centrifugal: 100 tons or 350 kW to an upper limit of capacity determined only by physical size.

Compressors come with many types of drives including electric, gas, and diesel engines, and gas and steam turbines. Many compressors are purchased as part of a condensing unit; they consist of compressor, drive, cooler, condenser, and all necessary safety and operating controls. Reciprocating condensing units are available with air-cooled condenser as part of the unit or arranged for remote installation. Remote air-cooled condensers are available up to 500 tons or 1.8 MW. A typical commercial size, air-cooled condensing unit is shown in Fig. 8-5.

Absorption Chillers

The absorption chiller most commonly used for air conditioning is the indirect fired lithium bromide-water cycle unit. These are available in large units from 50 to 1500 tons or 176 kW to 5 MW capacity. Their generator sections are heated with low pressure steam, hot water, or other hot liquids. Small, direct gas fired units are also available. In large installations the absorption chiller is frequently combined with centrifugal compressors driven by steam turbines. Steam from the noncondensing turbine is taken to the generator of the absorption machine. When the centrifugal unit is driven by a gas turbine or an engine, the absorption machine generator is heated with the exhaust jacket water. Because of rising energy costs many absorption chillers which use primary energy are being replaced with more energy efficient reciprocating or centrifugal equipment.

Figure 8-5.   An air cooled condensing unit. (Courtesy of Technical Systems Inc., Pryor, Oklahoma.)

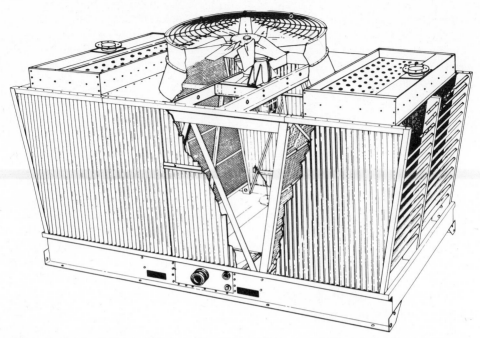

Figure 8-6. A mechanical draft cooling tower. (Courtesy of The Marley Company, Mission, Kansas.)

Cooling Towers

To remove the heat from the water-cooled condensers of air-conditioning systems the water is usually cooled by contact with the atmosphere. This is accomplished by natural draft or mechanical draft cooling towers or by spray ponds. Of these the mechanical draft tower can be designed for more exacting conditions because it is independent of the wind.

Air-conditioning systems use towers ranging from small package towers of 5 to 500 tons or 17.5 to 1760 kW or intermediate-size towers of 2000 to 4000 tons or 7 to 14 MW. A mechanical draft cooling tower is shown in Fig. 8-6.

Water treatment is a definite requirement for satisfactory operation. Units that operate during the entire year must be protected against freezing.

Pumps

The pumps used in air-conditioning systems are usually single inlet centrifugal pumps. The pumps for the larger system and for heavy duty have a horizontal split case with double suction impeller for easier maintenance and high efficiency. End suction pumps either close coupled or flexible connected are used for smaller tasks. Figure 8-7 shows a typical medium-sized centrifugal pump. The major applications for pumps in the equipment room are: primary and secondary chilled water, hot water, condenser water, steam condensate return, boiler feed water, and fuel oil.

It is common to provide multiple but identical pumps operating in parallel, one of which is a spare to maintain system continuity in case of a pump failure. Sometimes the chilled water and condenser water system characteristics permit using one spare pump for both systems with valved connections to the manifolds of each system.

Fuels and Energy

Consideration of the type of fuels must be made at the same time as the selection of the boiler to assure maximum efficiency. Chapter 14 of the 1977 *Handbook of Fundamentals* (2) gives the types and properties of fuels and factors in their proper combustion.

Figure 8-7.   A typical single inlet direct coupled centrifugal pump. (Courtesy of Pacific Pump Company, Oakland, California.)

The resource impact of the type of fuel or other energy to be used is an important consideration. Obviously, the selection should result in the least use of our natural resources. This will usually also result in least cost. ASHRAE Standard 90C provides a method for calculating and reporting the anticipated quantities of fuel and energy resources to be consumed.

Piping

Air-conditioning piping systems can be divided into two parts, the piping in the main equipment room and the piping required for the air handling systems throughout the building. The air handling system piping follows procedures developed in detail in Chapter 9. The major piping in the main equipment room consists of fuel lines, refrigerant piping, steam, and water connections. Chapter 9 gives details for sizing of pipes.

Instrumentation

At the time of installation all equipment should be provided with adequate gages, thermometers, flow meters, and balancing devices so that system performance is properly established. In addition, capped thermometer wells, gage cocks, capped duct openings, and volume dampers should be provided at strategic points for system balancing. A central control center to monitor and control a large number of control points should be considered for any large and complex air-conditioning system.

8-4   ALL-AIR SYSTEMS

An all-air system provides complete sensible heating and cooling and latent cooling by supplying only air to the conditioned space. In such systems there may be piping between the refrigerating and heat producing devices and the air handling device. No additional cooling is required at the zone. Heating may be accomplished by the air stream of the central system or at a particular zone. In some applications heating is accomplished by a separate air, water, steam, or electric heating system. The term *zone* implies a provision or the need for separate thermostatic control, whereas the term *room* implies a partitioned area that may or may not require separate control.

All-air systems may be briefly classified in two basic categories: (1) single-path systems and (2) dual-path systems. Single-path systems contain the main heating and cooling coils in a series flow air path using a common duct distribution system at a common air temperature to feed all terminal apparatus. Dual-path systems contain the main heating and cooling coils in a parallel flow or series-parallel flow air path using either (1) a separate cold and warm air duct distribution system that is blended at the terminal apparatus (dual-duct system), or (2) a single supply duct to each zone with a blending of warm and cold air at the main supply fan.

The all-air system may be adapted to all types of air-conditioning systems for comfort or process work. It is applied in buildings requiring individual control of conditions and having a multiplicity of zones such as office buildings, schools and universities, laboratories, hospitals, stores, hotels, and ships. Air systems are also used for many special applications where a need exists for close control of temperature and humidity, including clean rooms, computer rooms, hospital operating rooms, textile and tobacco factories.

Reheat Systems

The reheat system is a modification of the single-zone system. Its purpose is to permit zone or space control for areas of unequal loading, or to provide heating or cooling of perimeter areas with different exposures, or for process or comfort applications where close control of space conditions is desired. As the word *reheat* implies, the application of heat is a secondary process, being applied to either preconditioned primary air or recirculated room air. A single low pressure reheat system is produced when a heating coil is inserted in the duct system. The more sophisticated systems utilize higher pressure duct designs and pressure reduction devices to permit system balancing at the reheat zone. The medium for heating may be hot water, steam, or electricity.

Conditioned air is supplied from a central unit at a fixed cold air temperature designed to offset the maximum cooling load in the space. The control thermostat activates the reheat unit when the temperature falls below the upper limit of the controlling instrument's setting. A schematic arrangement of the components for a typical reheat system is shown in Fig. 8-8. To conserve energy reheat should not be used unless absolutely necessary. At the very least, reset control should be provided to maintain the cold air at the highest possible temperature to satisfy the space cooling requirement (4).

Variable Volume System

The variable volume system compensates for varying load by regulating the volume of air supplied through a single duct. Special zoning is not required because each space supplied by a controlled outlet is a separate zone. Figure 8-9 is a schematic of a true variable air volume (VAV) system.

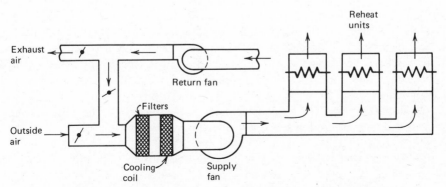

Figure 8-8.   Arrangement of components for a reheat system.

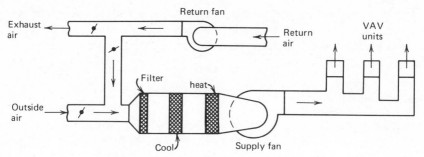

Figure 8-9. Variable air volume system.

Significant advantages of the variable volume system are low initial cost and low operating costs. The first cost of the system is far lower in comparison with other systems that provide individual space control because it requires only single runs of duct and a simple control at the air terminal. Where diversity of loading occurs, smaller equipment can be used and operating costs are generally the lowest among all the air systems. Because the volume of air is reduced with a reduction in load, the refrigeration and fan horsepower follow closely the actual air-conditioning load of the building. During intermediate and cold seasons outdoor air can be used for economy in cooling. In addition the system is virtually self-balancing.

Until recently there were two reasons why variable volume systems were not recommended for applications with loads varying more than 20 percent. First, throttling of conventional outlets down to 50 or 60 percent of their maximum design volume flow might result in the loss of control of room air motion with noticeable drafts resulting. Second, the use of mechanical throttling dampers produces noise, which increases proportionally with the amount of throttling.

Improvements in volume-throttling devices and aerodynamically designed outlets have helped overcome these problems and extended the potential application of variable volume systems. This system can now handle interior areas as well as building perimeter areas where load variations are greatest, and where throttling to 10 percent of design volume flow is often necessary.

Although some heating may be done with a variable volume system, it is primarily a cooling system and should be applied only where cooling is required the major part of the year. Buildings with internal spaces with large internal loads are the best candidates. A secondary heating system should be provided for boundary surfaces during the heating season. Baseboard perimeter heat is often used. During the heating season, the VAV system simply provides tempered ventilation air to the exterior spaces.

An important aspect of VAV system design is fan control. There are significant fan power savings where fan speed is reduced in relation to the volume of air being circulated. This topic is discussed in detail in Chapter 11.

Single duct variable volume systems should be considered in applications where full advantage can be taken of their low cost of installation and operation. Applications exist for office buildings, hotels, hospitals, apartments, and schools.

Dual- or Double-Duct System

In the dual-duct system the central station equipment supplies warm air through one duct run and cold air through the other. The temperature in an individual space is controlled by a thermostat that mixes the warm and cool air in proper proportions. Variations of the dual-duct system are possible with one form shown in Fig. 8-10.

For best performance some form of constant volume regulation should be incorporated into the system to maintain a constant flow of air. Without this the system is difficult to control because of the wide variations in system static pressure that occur from the normal demand from loading changes.

Many double-duct systems are installed in office buildings, hotels, hospitals, schools, and large laboratories. A common characteristic of these multiroom buildings is their highly variable sensible heat load. This system provides great flexibility in satisfying multiple loads and in providing prompt and opposite temperature response as required.

Space or zone thermostats may be set once to control year-round temperature conditions. All outdoor air can be used when the outdoor temperature is low enough to handle the cooling load. A dual-duct system should be provided with control that will automatically reset the cold air supply to the highest temperature acceptable and the hot air supply to the lowest temperature acceptable (4).

From the energy conservation viewpoint the dual-duct system has the same disadvantage as reheat. Although many of these systems are in operation, few are now being designed and installed.

Multizone System

The multizone central station units provide a single supply duct for each zone and obtain zone control by mixing hot and cold air at the central unit in response to room or zone thermostats. For a comparable number of zones this system provides greater flexibility than the single duct and involves lower cost than the dual-duct system, but it is physically limited by the number of zones that may be provided at each central unit.

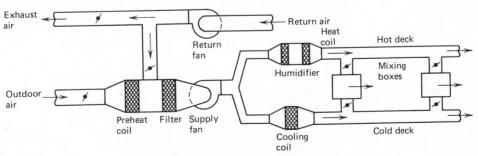

Figure 8-10. Dual-duct system.

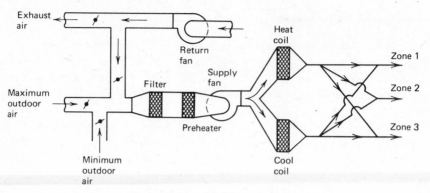

Figure 8-11. Multizone system.

Typical multizone equipment is similar in some respects to the dual-duct system but the two air streams are proportioned within the equipment instead of being mixed at each space served, and the proper temperature air is provided as it leaves the equipment. Figure 8-11 shows a sketch of a multizone system. The central station apparatus includes dampers that premix the proper amounts of cold and warm air for each duct and is controlled by room thermostats. The system conditions groups of rooms or zones by means of a blow-through central apparatus having heating and cooling coils in parallel downstream from the fan.

The multizone, blow-through system is applicable to locations and areas having high sensible heat loads and limited ventilation requirements. The use of many duct runs and control systems can make initial costs of this system high compared to other all-air systems. Also to obtain very fine control this system might require larger refrigeration and air handling equipment, which should be considered in estimating both initial and operating costs.

The use of these systems with simultaneous heating and cooling is now discouraged for energy conservation (3). However, through the use of outdoor air and either heating or cooling satisfactory control may be attained in many applications. The air handler shown in Fig. 8-3 was developed to achieve this type of operation.

8-5 AIR AND WATER SYSTEMS

In the all-air systems discussed in the previous section the spaces within the building are cooled solely by air supplied to them from the central air-conditioning equipment. In contrast, in an air and water system both air and water are distributed to each space to perform the cooling function. In virtually all air-water systems both cooling and heating functions are carried out by changing the air or water temperatures (or both) to permit control of space temperature during all seasons of the year.

There are several basic reasons for the use of this type of system. Because of the greater specific heat and much greater density of water compared to air the cross-sectional area required for the distribution pipes is markedly less than that required for ductwork to accomplish the same cooling task. Consequently, the quantity of air supplied can be low compared to an all-air system, and less building space need be allocated for the cooling distribution system.

The reduced quantity of air is usually combined with a high velocity method of air distribution to minimize the space required. If the system is designed so that the air supply is equal to the air needed to meet outside air requirements or that required to balance exhaust (including exfiltration) or both, the return air system can be eliminated for the areas conditioned in this manner.

The pumping horsepower necessary to circulate the water throughout the building is usually significantly less than the fan horsepower to deliver and return the air. Thus not only space but also operating cost savings can be realized.

Systems of this type have been commonly applied to office buildings, hospitals, hotels, schools, better apartment houses, research laboratories, and other buildings. Space saving has made these systems beneficial in high-rise structures.

The air side of air and water systems is comprised of central air-conditioning equipment, a duct distribution system, and a room terminal. The air is supplied at constant volume and is often referred to as primary air to distinguish it from room air that is recirculated over the room coil.

The water side in its basic form consists of a pump and piping to convey water to the heat transfer surface within each conditioned space. The heat exchange surface may be a coil which is an integral part of the air terminal (as with induction units), a completely separate component within the conditioned space (radiant panel), or either (as is true of fan-coil units).

Individual room temperature control is obtained by varying the capacity of the coil (or coils) within the room by regulation of either the water flow through it or the air flow over it. The coil may be converted to heating service during the winter or a second coil or a heating device within the space may provide heating capacity depending on the type of system.

Air and water systems are categorized as two-pipe, three-pipe, and four-pipe systems. They are basically similar in function and all incorporate both cooling and heating capabilities for all season air conditioning. However, arrangements of the secondary water circuits and control systems differ greatly.

Air-Water Induction System

The basic arrangement for air-water induction units is shown in Fig. 8-12. Centrally conditioned primary air is supplied to the unit plenum at high pressure. The plenum is acoustically treated to attenuate part of the noise generated in the duct system and in the unit. A balancing damper is used to adjust the primary air quantity within limits.

The high pressure air flows through the induction nozzles and induces secondary air from the room and over the secondary coil. This secondary air is either heated or cooled at the coil depending on the season, the room requirement, or both. Ordinarily no latent cooling is accomplished at the room coil, but a drain pan is provided to collect condensed moisture resulting from unusual latent loads of short duration. The primary and secondary air are mixed and discharged to the room.

Induction units are usually installed at a perimeter wall under a window, but

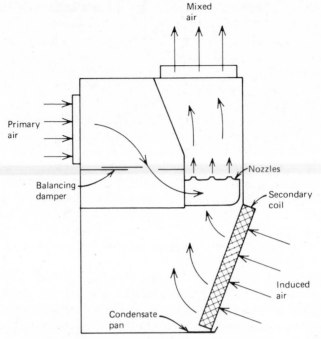

Figure 8-12.   Air-water induction unit.

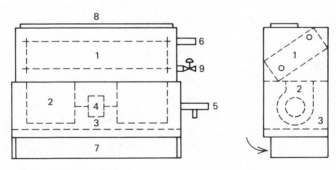

1. Finned tube coil
2. Fan scrolls
3. Filter
4. Fan motor
5. Auxiliary condensate pan
6. Coil connections
7. Return air opening
8. Discharge air opening
9. Water control valve

Figure 8-13.   Typical fan-coil unit.

units designed for overhead installation are available. During the heating season the floor-mounted induction unit can function as a convector during off hours with hot water to the coil and without a primary air supply.

Fan-Coil Conditioner System

The fan-coil conditioner unit is a versatile room terminal that is applied to both air-water and water-only systems. Despite shortcomings in the quality of air conditioning achieved with water-only systems the fan-coil units have been more commonly associated with that class of system than with air-water. Many of the standard features of the units are accordingly incorporated into it for water-only applications.

The basic elements of fan-coil units are a finned-tube coil and a fan section, Fig. 8-13. The fan section recirculates air continuously from within the perimeter space through the coil, which is supplied with either hot or chilled water. In addition, the unit may contain an auxiliary heating coil, which is usually of the electric-resistance type but which can be of the steam or hot-water type. Thus the recirculated room air is either heated or cooled.

Fan-coil unit capacity can be controlled by regulation of coil water flow, air bypass, fan speed, or a combination of these. Water flow can be thermostatically controlled by either return air or wall thermostats. Because of their lower cost two-position valves are often used for fan-coil applications instead of modulating valves. Water valves should not be used for control where outdoor intakes are used, unless freezing of the coils is prevented. Bypass dampers are available on some conditioners. Capacity control is achieved by modulation of a damper to bypass all or part of the air around the unit coil. Fan speed control may be automatic or manual. Automatic control is usually on-off with manual speed selection.

8-6 ALL-WATER SYSTEMS

All-water systems are those with fan-coil, unit ventilator, or valance-type room terminals, with unconditioned ventilation air supplied by an opening through the wall or by infiltration. Cooling and humidification are provided by circulating chilled water or brine through a finned coil in the unit. Heating is provided by supplying hot water through the same or a separate coil using two-, three-, or four-pipe water distribution from central equipment. Electric heating or a separate steam coil may also be used. Humidification is not practical in all-water systems unless a separate package humidifier is provided in each room.

The greatest advantage of the all-water system is its flexibility for adaptation to many building module requirements.

A fan-coil system applied without provision for positive ventilation or one taking ventilation air through an aperture is one of the lowest first-cost central station type perimeter systems in use today. It requires no ventilation air ducts, is comparatively easy to install in existing structures, and as with any central station perimeter system utilizing water in pipes instead of air ducts, its use results in considerable space savings throughout the building.

All-water systems have individual room control with quick response to thermostat settings and freedom from recirculation of air from other conditioned space. These systems have remote chilling and heating equipment. When fan-coil units are used with the three- or four-pipe water arrangements, each is its own zone with a choice of heating or cooling at all times and no seasonal changeover is required. All-water systems can be installed in existing buildings with a minimum of interference in the use of occupied space.

There is no positive ventilation unless wall openings are used, and these are affected by wind pressures and stack action on the building. Special precautions are required at each unit with an outside air opening to prevent freezing of coil and potential water damage from rain.

Seasonal changeover is required in most climates with a two-pipe system, and zoning and piping are required to reduce operating difficulties during intermediate seasons when a sun-exposed zone may need cooling while other zones need heat. When heating, units cannot be used as convectors in unoccupied rooms and fans must be kept running.

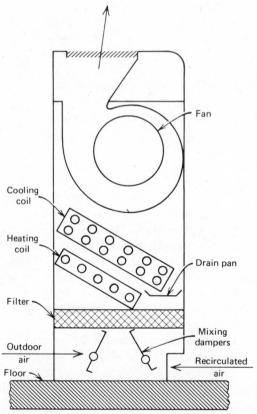

Figure 8-14.   Typical air-conditioning unit ventilator with separate coils.

If a two-pipe system has only one pump, the same quantity of hot and cold water is circulated even though the requirements for each may be different. With three- and four-pipe systems hot and cold water may be furnished throughout the year.

Maintenance and service work has to be done in the occupied areas; as the units become older the fan noise can become objectionable. Each unit requires a condensate drain line. It is difficult to limit bacterial growth in the unit. In extremely cold weather it is often necessary to close the outside air dampers to prevent freezing of coils, reducing ventilation air to that obtained by infiltration. Filters are small and inefficient and require freqent changing to maintain air volume.

Unit Ventilators

Figure 8-14 illustrates a typical air-conditioning unit ventilator with two separate coils, one used for heating and the other for cooling with a four-pipe system. In some cases the unit ventilator may have only one coil, such as the fan coil of Fig. 8-13.

The heating coil may use hot water, steam, or electricity. The cooling coil can be either a chilled water coil or a direct expansion refrigerant coil. Heating and cooling coils are sometimes combined in a single coil by providing separate tube circuits for each function. In such cases the effect is the same as having two separate coils.

Unit ventilator capacity control is essentially the same as described for fan coils in the previous section. In most cases, however, the control system will be slightly more sophisticated.

8-7   UNITARY AND ROOM AIR CONDITIONERS

Unitary Air Conditioners

Unitary air-conditioning equipment consists of factory matched refrigerant cycle components for inclusion in air-conditioning systems that are field designed to meet the needs of the user.

The following list of variations is indicative of the vast number of types of unitary air conditioners available.

1. Arrangement: single or split (evaporator connected in the field)
2. Heat rejection: air cooled, evaporative condenser, water cooled
3. Unit exterior: decorative for in-space application, functional for equipment room and ducts, weatherproofed for outdoors
4. Placement: floor standing, wall mounted, ceiling suspended
5. Indoor air: vertical upflow, counterflow, horizontal, 90 and 180 degree turns, with fan, or for use with forced air furnace
6. Locations: *indoor*—exposed with plenums or furred in ductwork, concealed in closets, attics, crawl spaces, basements, garages, utility rooms, or equipment rooms; *wall*—built in, window, transom; *outdoor*—rooftop, wall mounted, or on ground

7. Heat: intended for use with upflow, horizontal, or counterflow forced air furnace, combined with furnace, combined with electrical heat, combined with hot water or steam coil

Unitary air conditioners as contrasted to room air conditioners are designed with fan capability for duct work, although some units may be applied with plenums.

Heat pumps are also offered in many of the same types and capacities as unitary air conditioners.

Packaged reciprocating and centrifugal water chillers can be considered as unitary air conditioners particularly when applied with unitary-type chilled water blower coil units.

Figure 8-15 depicts a single package air conditioner while Fig. 8-16 shows a typical residential split system.

The nearly infinite combination of coil configurations, evaporator temperatures, air handling arrangements, refrigerating capacities, and other variations that are available in central systems are rarely possible with unitary systems. Consequently in many respects a higher level of design ingenuity and performance is required to develop superior system performance using unitary equipment than for central systems.

The need for zoning is widely recognized because it is desirable to give the occupant some control over his or her environment both during normal occupancy and outside normal operating hours. Unitary equipment tends to fall automatically into a zoned system with each zone served by its own unit.

For large single spaces where central systems work best the use of multiple units is often an advantage because of the movement of load sources within the larger space, giving flexibility to many smaller independent systems instead of one large central system.

Room Air Conditioners

A room air conditioner is an encased assembly designed as a unit primarily for mounting in a window, through a wall, or as a console. The basic function of a room air conditioner is to provide comfort by cooling, dehumidifying, filtering or cleaning, and circulating the room air. It may also provide ventilation by introducing outdoor air into the room, and by exhausting the room air to the outside. The conditioner may also be designed to provide heating by reverse cycle (heat pump) operation or by electric resistance elements.

Figure 8-17 shows a schematic view of a typical room air conditioner.

Through-the-Wall Conditioner Systems

A through-the-wall system uses an air-cooled room air conditioner designed for mounting through the wall and in many cases is capable of providing both the heating and cooling function. Design and manufacture parameters vary widely. Specification grades range from appliance grade through heavy duty commercial grade. The latter is called a packaged terminal air conditioner.

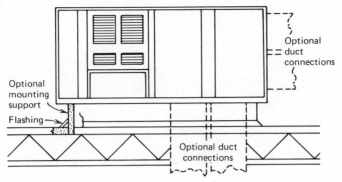

Figure 8-15. Rooftop installation of air-cooled single package or single package year-round unit.

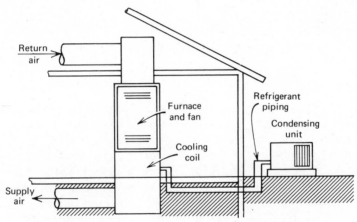

Figure 8-16. Outdoor installation of split system air-cooled condensing unit with downflow furnace.

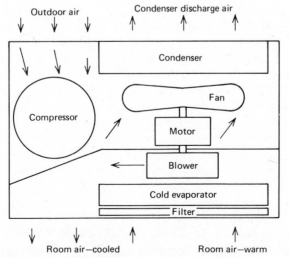

Figure 8-17. Schematic view of room air conditioner.

Figure 8-18. Packaged terminal air conditioning with combination heating and cooling chassis.

The through-the-wall concept incorporates a complete air-cooled refrigeration and air handling system in an individual package; it utilizes space normally occupied by the building wall for equipment with the remainder projecting inside the room.

Each packaged terminal air conditioner has a self-contained, direct-expansion cooling system, heating coil (electric, hot water, or steam), and packaged controls. Figure 8-18 shows a typical unit; it consists of a combination wall sleeve and room cabinet, combination heating and cooling chassis, and outdoor louver.

Through-the-wall systems are best applied in multizone applications. The initial cost of the through-the-wall system of air conditioning in multiroom applications is considerably less than central systems adapted to provide simultaneous functions of either heating or cooling in each room under control of the room occupants. With a through-the-wall system of heating and air conditioning, considerable space is saved by eliminating both duct work and equipment rooms.

The through-the-wall system is limited to multizone systems and generally cannot be used economically in large spaces. However, where the packaged terminal air conditioner system is coordinated with a well-designed core system, it can be economically used for large office areas while allowing for maximum flexibility in moving partitions. Limitation of through-the-wall units is often governed by the ability of the unit to throw the air across the room. One of the most popular applications has been in hotels and motels. This system is also applicable for renovation of existing buildings because there is less disruption and construction-forced sacrifice of rentable space than with alternative systems.

8-8 HEAT PUMP SYSTEMS

The term *heat pump* as applied to HVAC systems is a system in which refrigeration equipment is used such that heat is taken from a heat source and given up to the conditioned space when heating service is wanted and is removed from the space and discharged to a heat sink when cooling and dehumidification are desired. The thermal cycle is identical with that of ordinary refrigeration, but the application is equally concerned with the cooling effect produced at the evaporator and the heating effect produced at the condenser. In some applications both the heating and cooling effects obtained in the cycle are utilized.

Unitary heat pumps (as opposed to applied heat pumps) are shipped from the factory as a complete preassembled unit including internal wiring, controls, and piping. Only the ductwork, external power wiring, and condensate piping are required to complete the installation. For the split unit is is also necessary to connect the refrigerant piping between the indoor and outdoor sections. In appearance and dimensions, casings of unitary heat pumps closely resemble those of conventional air-conditioning units having equal capacity.

Capacities of unitary heat pumps range from about $1\frac{1}{2}$ to 25 tons or 5 to 90 kW although there is no specific limitation. This equipment is almost universally used in residential and the smaller commercial and industrial installations. The multiunit type of installation with a number of individual units of 2 to 20 tons or 7 to 70 kW of cooling capacity is particularly advantageous to obtain zoning and to provide simultaneous heating and cooling. It may also be used for heat reclaiming to conserve energy by connecting the units to a common water circuit.

Large central heat pumps of modern design within the capacity range of about 30 to 1000 horsepower or 20 to 750 kW of compressor-motor rating are now operating in a substantial number of buildings. A single or central system is generally used throughout the building but in some instances the total capacity is divided among several separate heat pump systems to facilitate zoning.

Heat Pump Types

The air-to-air heat pump is the most common type. It is particularly suitable for factory-built unitary heat pumps and has been widely used for residential and commercial applications. Figure 8-19 is typical of the refrigeration circuit employed. Outdoor air offers a universal heat-source, heat-sink medium for the heat pump. Extended-surface, forced-convection heat transfer coils are normally employed to transfer the heat between the air and the refrigerant.

In some air-to-air heat pump systems, the air circuit may be interchanged by means of dampers (motor driven or manually operated) to obtain either heated or cooled air for the conditioned space. With this system one heat exchanger coil is always the evaporator and the other is always the condenser. The conditioned air will pass over the evaporator during the cooling cycle and the outdoor air will pass over the condenser. The change from cooling to heating is accomplished by positioning the dampers.

Figure 8-20 shows typical curves of heat pump capacity versus outdoor dry bulb temperature. Imposed on the figure are approximate heating and cooling load curves for a building. In the heating mode it can be seen that the heat pump capacity decreases and the building load increases as the temperature drops. In the cooling mode the opposite trends are apparent. If the cooling load and heat pump capacity are matched at the cooling design temperature, then the balance point, where heating

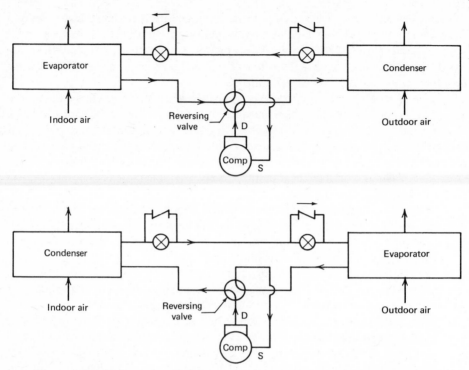

Figure 8-19. Schematics of a simple heat pump in cooling and heating modes.

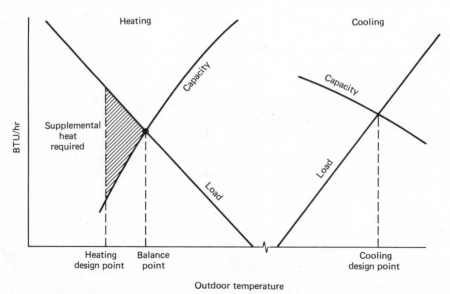

Figure 8-20. Comparison of building heat loads with heat pump capacities.

load and capacity match, is then fixed. This balance point will quite often be above the heating design temperature. In such cases supplemental heat must be furnished to maintain the desired indoor condition.

The most common type of supplemental heat for heat pumps in the United States is electrical resistance heat. This is usually installed in the air handler unit and is designed to turn on automatically, sometimes in stages, as the indoor temperature drops. In some systems the supplemental heat is turned on when the outdoor temperature drops below some preset value. It is also a common feature for the compressor to shut off when the supplemental heat is turned on and thus for the supplemental heat to carry the entire load at these low temperatures. Heat pumps which have fossil fuel-fired supplemental heat are referred to as hybrid or bivalent heat pumps. These are more common in Europe than in the United States.

If the heat pump capacity is sized to match the heating load, care must be taken that there is not excessive cooling capacity for summer operation. This excessive capacity could lead to poor summer performance, particularly in dehumidification of the air.

Air-to-water heat pumps are commonly used in large buildings where zone control is necessary and are also sometimes employed for the production of hot or cold water in industrial applications as well as heat reclaiming. Heat pumps for hot water heating are commercially available in residential sizes.

A water-to-air heat pump uses water as a heat source and sink and uses air to transmit heat to or from the conditioned space.

A water-to-water heat pump uses water as the heat source and sink for both cooling and heating operation. Heating-cooling changeover may be accomplished in the refrigerant circuit, but in many cases it is more convenient to perform the switching in the water circuits.

Water may represent a satisfactory and in many cases an ideal heat source. Well water is particularly attractive because of its relatively high and nearly constant temperature, generally about 50 F or 10 C in northern areas and 60 F or 16 C and higher in the south. However, abundant sources of suitable water are not always available and the application of this type of system is limited. Frequently sufficient water may be available from wells, but the condition of the water may cause corrosion in heat exchangers or it may induce scale formation. Other considerations to be made are the costs of drilling, piping and pumping, and the means for disposing of used water.

Surface or stream water may be utilized but under reduced winter temperatures the cooling spread between inlet and outlet must be limited to prevent freeze-up in the water chiller, which is absorbing the heat.

Under certain industrial circumstances waste process water such as spent warm water in laundries and warm condenser water may be a source for specialized heat pump operations.

Water-refrigerant heat exchangers generally take the form of direct expansion water coolers either of the shell-and-coil type or of the shell-and-tube type. They are circuited so that they can be used as a refrigerant condenser during the heating cycle and as a refrigerant evaporator during the cooling cycle.

Closed Loop Systems

In many cases a building may require cooling in interior zones while needing heat in exterior zones. The needs of the north zones of a building may also be different from those of the south. In many cases a closed loop heat pump system is a good choice. Closed loop systems may be solar assisted. A closed loop system is shown in Fig. 8-21.

Individual water-to-air heat pumps in each room or zone accept energy from or reject energy to a common water loop, depending on whether that area has a call for heating or for cooling. In the ideal case the loads will balance and there will be no surplus or deficiency of energy in the loop. If cooling demand is such that more energy is rejected to the loop than is required for heating, the surplus is rejected to the atmosphere by a cooling tower. In the other case an auxiliary furnace furnishes any deficiency.

Earth as a heat source and sink by heat transfer through buried coils has not been extensively used. This may be attributed to the high installation expense, the ground area requirements, and the difficulty and uncertainty of predicting performance.

The ground has been used successfully as a source-sink for heat pumps. The most common installation uses the vertical heat exchanger shown in Fig. 8-22. Water from the heat pump is pumped to the bottom of the "well" and slowly rises back up through the annulus. It exchanges heat with the surrounding earth before being returned back to the heat pump. Tests and analysis have shown rapid recovery in earth temperature around the well after the heat pump cycles off. Proper sizing of the well depends upon the nature of the earth surrounding the well, the water table level, and the efficiency of the heat pump.

Although still largely in the research stage, the use of solar energy as a heat source either on a primary basis or in combination with other sources is attracting

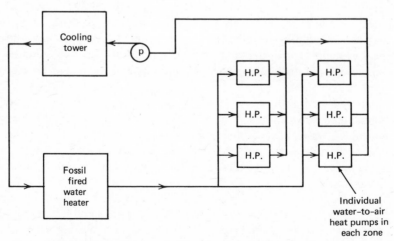

Figure 8-21.   Schematic of a closed loop heat pump system.

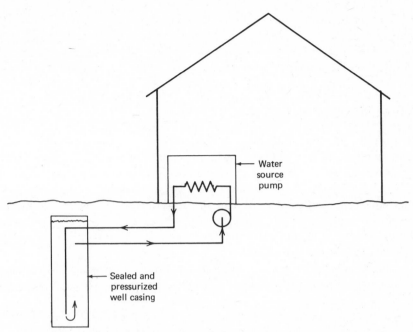

Figure 8-22.   A typical ground source heat pump system.

increasing interest. Solar energy is discussed in detail in Chapter 16. Heat pumps may be used with solar systems either in a series or a parallel arrangement, or a combination of both. In the parallel arrangement, either the heat pump or the solar system is operating to meet the needs of a building. In this system the solar energy might be used first. If the storage tank temperature drops below a level that can maintain comfort for the space, the heat pump comes on automatically. This has the advantage that an air-to-air heat pump can be used and heat is easily rejected during the cooling cycle. A disadvantage is that this system cannot utilize the solar energy available and stored at temperatures below approximately 100 F.

If the solar system and the heat pump are connected in series, the heat pump in the heating mode gets its source energy from the solar storage tank, which is usually at a higher temperature than the outdoor temperature. In addition, the solar storage tank does not have to be maintained at a high temperature in order to furnish useful energy. With a moderate source temperature the heat pump and the collectors can operate with a high efficiency. If the tank temperature drops below some specified amount, energy must be added to the tank from supplemental sources to prevent lowered performance and eventual freezing of the fluid in the tank.

The series arrangement requires a water source heat pump, a type which some major manufacturers have been slow to develop in residential sizes. A water source heat pump must reject heat to water when operating in the cooling mode. The tank must be capable of rejecting heat to the environment if it is to remain at an acceptable temperature. This requires a cooling tower or a source of cooling water such as a well, a stream, or a pond. Thermal simulations of series systems indicate that large storage is required to avoid extensive use of back-up fuels. This large storage in turn requires large collector areas for significant solar input. This large collector area creates an expensive system that is difficult to justify on an economic basis.

8-9   HEAT RECOVERY SYSTEMS

In large commercial applications considerable heat energy is generated internally and may require removal even during the coldest weather. This condition usually occurs within the central spaces, which do not have exterior walls. It is necessary to exhaust considerable quantities of air from large commercial structures because of the introduction of outdoor ventilation air. Considerable savings in energy can be realized if the heat energy from the interior spaces and the exhaust air can be recovered and used in heating the exterior parts of the structure. Heat energy may also be recovered from waste water.

Redistribution of heat energy within a structure can be accomplished through the use of heat pumps of the air-to-air or water-to-water type. Another approach is the use of the dual path systems described earlier.

Recovery of heat energy from exhaust air is accomplished through the use of air-to-air heat exchangers, rotating (periodic type) heat exchangers, and air-to-water heat exchangers connected by a circulating water loop. Sometimes spray systems are used; they may contain desiccants to enhance latent heat transfer. Figures 8-23 and 8-24 illustrate the air-to-water and rotating heat recovery systems while Fig. 8-25

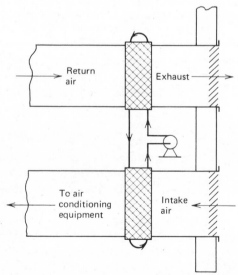

Figure 8-23.   Air-to-water type of heat recovery system.

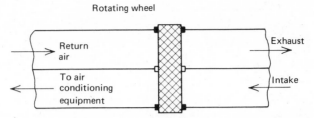

Figure 8-24.   Rotating-type heat exchanger used for heat recovery.

shows how an air-to-air system might be arranged. The air-to-air and rotating type systems are the most effective in recovering energy but require that the intake and exhaust to the building be at the same location, whereas the air-to-water system may have the exhaust and intake at widely separated locations; however, the air-to-water system is not as effective because of the added thermal resistance of the water. To prevent freezing, glycol must also be introduced, which further reduces the effectiveness of the air-to-water system.

All of the above described systems may also be effective during the cooling season when they function to cool and dehumidify outdoor ventilation air.

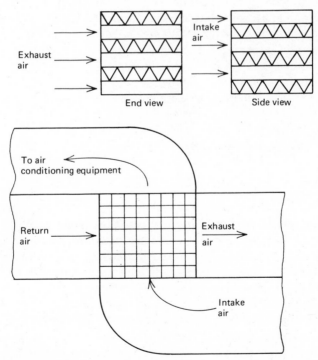

Figure 8-25.   Air-to-air type of heat recovery system.

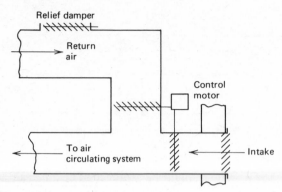

Figure 8-26.   Arrangement of components in an economizer system.

8-10   ECONOMIZER SYSTEMS

It is often possible to cool a space either totally or partially using outdoor air. This will be true during the spring and fall of the year and at night in northern regions and at high altitudes. Outdoor air can also be used to cool interior spaces, which may otherwise require operation of the cooling equipment. Because this type of cooling results in economies of energy and money, the general term *economizer* has been adopted to describe these systems.

The design of economizers are quite varied but generally have provisions for introducing outdoor air and exhausting indoor air (Fig. 8-26). There are a great variety of control arrangements, and the economizer may be combined with the heat recovery system. It is important to sense both outdoor temperature and humidity so that hot humid air is not introduced to the space. This would defeat the purpose of the economizer. Adequate precautions must be taken to prevent contamination of the space by dirty or odorous outdoor air.

8-11   SUMMARY

It is apparent from the preceding sections that the entire subject of HVAC systems is complex and that the types and combinations of equipment are numerous. The design engineer must constantly keep acquainted with the literature from the various manufacturers to stay abreast with new innovations. It is the business of the sales or application engineer to be particularly well versed on available equipment and to give the best advice possible to the design engineer.

It is obvious that the flow of liquids and gases (primarily water and air) are of great importance in all HVAC applications. Therefore, we shall consider the flow of liquids and the design of piping systems in Chapter 9 and in Chapters 10 and 11 consider air distribution and duct design.

REFERENCES

1. *ASHRAE Handbook and Product Directory—Systems*, American Society of Heating, Refrigerating and Air-Conditioning Engineers, New York, 1980.

2. *ASHRAE Handbook of Fundamentals,* American Society of Heating, Refrigerating and Air-Conditioning Engineers, New York, 1977.

3. "ASHRAE Standard 90C–75R, Energy Conservation in New Building Design," American Society of Heating, Refrigerating and Air Conditioning Engineers, New York, 1979.

4. "ASHRAE Standard 90A–80, Energy Conservation in New Building Design," American Society of Heating, Refrigerating and Air Conditioning Engineers, New York, 1980.

PROBLEMS

8-1 Consider the small single-story office building shown in Fig. 8-27. Lay out an all-air central system using an air handler with two zones. There is space between the ceiling and roof for ducts. The air handler is equipped with a direct expansion cooling coil and a hot water heating coil. Show all associated equipment schematically.

8-2 Suppose the building of Problem 8-1 is to use a combination air-water system where fan coil units in each room are used for heating. Schematically lay out this part of the system with related equipment. Discuss the general method of control for (a) the supplied air, (b) the fan coil units.

8-3 Lay out a year-round all-water system for the building of Problem 8-1. Show all equipment schematically. Discuss the control and operation of the system in the summer, winter, and between seasons.

8-4 Apply single package year-round rooftop type unit(s) to the single story building in Figure 8-27.

8-5 Suppose a variable air volume (*VAV*) all-air system is to be used to condition the space shown in Figure 8-28. Assume the space is the ground floor of a multistory

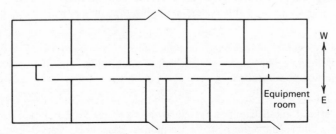

Figure 8-27. Floor plan of small office building.

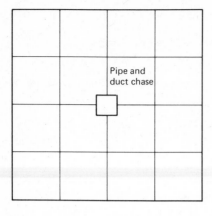

Figure 8-28. Schematic floor plan of one floor of a large building.

office building. Describe the system using a schematic diagram. The lighting and occupant load are variable. Discuss the general operation of the system during the (a) colder months, (b) the warmer months.

8-6 Devise a central equipment arrangement for the system of Problem 8-5 that would save energy during the winter months. Sketch the system schematically.

8-7 Suppose an air-to-water heat pump is used to condition each space of Figure 8-28 where the water side of each heat pump is connected to a common water circuit. Sketch this system schematically showing all necessary additional equipment. Discuss the operation of this system during the (a) colder months, (b) warmer months, and (c) intermediate months.

8-8 A building, such as that shown in Figure 8-28, requires some outdoor air. Explain and show schematically how this may be done with the system of Problem 8-5. Incorporate some sort of heat recovery device in the system. What controls would be necessary?

8-9 How can an economizer be used to advantage during the (a) winter months, (b) the summer months, and (c) the intermediate seasons?

8-10 The system proposed in Problem 8-7 requires the distribution of outdoor ventilation air to each space. Sketch a central air handler system for this purpose that has energy recovery equipment and an economizer. Do not sketch the air distribution system. Discuss the control of this system assuming that the air will always be delivered at 72 F.

8-11 Make a single line, block diagram of an all water cooling system. The system has unit ventilators in each room with a packaged water chiller, and pumps. Explain how the system will be controlled.

8-12 Sketch a diagram of an air-water system that uses fan coils around the perimeter and an overhead air distribution system from a central air handler. Show a hot water boiler, chiller, and water distribution pumps. Explain the operation of the system in the summer and in the winter. What kind of controls does the system need?

8-13 Make a sketch of a variable volume system with a secondary perimeter heating system. Discuss the operation and control of the system for the different seasons of the year.

8-14 Diagram a combination air-to-air heat recovery and economizer system. Describe the operation and control of the system for various times of the year.

8-15 Water enters the condenser of a water to air heat pump at a temperature of 60 F with a flow rate of 6 gpm. The heat pump is removing 27,000 Btu/hr from the conditioned space and rejecting it plus the compressor energy to the water. The Energy Efficiency Ratio (EER) is assumed to be 10 Btu/W. Calculate the temperature of the water exiting from the condenser.

8-16 Determine the air flow rate needed (in cfm) through the evaporator coil of an air source heat pump if the heat pump is furnishing 30,000 Btu/hr to a conditioned space. The COP of the heat pump is assumed to be 2.5 and the desired decrease in the bulk temperature of the air is 10 F.

8-17 A house has a heat loss rate of approximately 600 Btu/hr-F temperature difference between inside and outside temperature. The balance point of the house and its heat pump is 30 F. How much supplementary resistance heat (kw) is required to maintain the indoor design temperature when the temperature outside is 10 F? Assume that the compressor shuts off below the balance point.

CHAPTER NINE

Flow, Pumps, and Piping Design

The distribution of fluids by pipes, ducts, and conduits is essential to all heating and cooling systems. The fluids encountered are gases, vapors, liquids, and mixtures of liquid and vapor (two-phase flow). From the standpoint of overall design of the building system, water and air are of greatest importance. This chapter deals with the fundamentals of incompressible flow of fluids like air and water in conduits, considers the basics of centrifugal pumps, and develops simple design procedures for water piping systems.

9-1 THE MECHANICAL ENERGY EQUATION

The incompressible adiabatic, steady flow of a fluid in a pipe or conduit is governed by the first law of thermodynamics, which may be written

$$\frac{P_1}{\rho} + \frac{\overline{V}_1^2}{2g_c} + \frac{gz_1}{g_c} = \frac{P_2}{\rho} + \frac{\overline{V}_2^2}{2g_c} + \frac{gz_2}{g_c} + w + \frac{g}{g_c}l_f \qquad (9\text{-}1)$$

where

P = static pressure, lbf/ft^2 or N/m^2
ρ = mass density, lbm/ft^3 or kg/m^3
$\overline{V}$ = average velocity, ft/sec or m/s
g = local acceleration due to gravity, ft/sec^2 or m/s^2
g_c = constant, 32.17 $(lbm\text{-}ft)/(lbf\text{-}sec^2)$ or 1.0 $(kg\text{-}m)/(N\text{-}s^2)$
z = elevation, ft or m
w = work, (ft-lbf)/lbm or J/kg
l_f = lost head, ft or m

The last term on the right in Eq. (9-1) is the internal conversion of energy due to friction. Each term of Eq. (9-1) has the units of energy per unit mass or specific energy. The first three terms on each side of the equality are the pressure energy, kinetic energy, and potential energy, respectively. A sign convention has been selected for the work such that work done on the fluid is negative.

Another governing relation for steady flow in a conduit is the conservation of mass. For flow along a single conduit the mass rate of flow at any two cross sections 1 and 2 is given by

$$\dot{m} = \rho_1 \overline{V}_1 A_1 = \rho_2 \overline{V}_2 A_2 \tag{9-2}$$

where

$\dot{m}$ = mass flow rate, lbm/sec or kg/s
A = cross-sectional area, ft^2 or m^2

When the fluid is incompressible, Eq. (9-2) becomes

$$\dot{Q} = \overline{V}_1 A_1 = \overline{V}_2 A_2 \tag{9-3}$$

where

$$\dot{Q} = \text{volume flow rate, ft}^3/\text{sec or m}^3/\text{s}$$

Equations (9-1) and (9-3) assume that the velocity and the mass density are constant over the cross-sectional area or that a suitable average value is used.

Equation (9-1) has other useful forms. If it is multiplied by the mass density, an equation is obtained where each term has the units of pressure.

$$P_1 + \frac{\rho \overline{V}_1^2}{2g_c} + \frac{\rho g z_1}{g_c} = P_2 + \frac{\rho \overline{V}_2^2}{2g_c} + \frac{\rho g z_2}{g_c} + \rho w + \frac{\rho g l_f}{g_c}. \tag{9-1a}$$

In this form the first three terms on each side of the equality are the static pressure, the velocity pressure, and the elevation pressure, respectively. The work term now has units of pressure and the last term on the right is the pressure lost due to friction.

Finally, if Eq. (9-1) is multiplied by g_c/g, an equation results where each term has the units of length, commonly referred to as *head*.

$$\frac{g_c}{g} \frac{P_1}{\rho} + \frac{\overline{V}_1^2}{2g} + z_1 = \frac{g_c}{g} \frac{P_2}{\rho} + \frac{\overline{V}_2^2}{2g} + z_2 + \frac{g_c w}{g} + l_f \tag{9-1b}$$

The first three terms on each side of the equality are the static head, velocity head, and elevation head, respectively. The work term is now in terms of head and the last term is the lost head due to friction.

Equations (9-1) and (9-2) are complementary because they have the common variables of velocity and density. When Eq. (9-1) is multiplied by the mass flow rate $\dot{m}$, another useful form of the energy equation results.

$$\dot{W} = \dot{m} \left[\frac{P_1 - P_2}{\rho} + \frac{\overline{V}_1^2 - \overline{V}_2^2}{2g_c} + \frac{g(z_1 - z_2)}{g_c} - \frac{g}{g_c} l_f \right] \tag{9-4}$$

where

$$\dot{W} = \text{power, work per unit time}$$

It is well to note that all terms on the right-hand side of the equality may be positive or negative except the lost energy, which must always be positive.

Some of the terms in Eqs. (9-1) and (9-4) may be zero or in some cases negligibly small. When the fluid flowing is a liquid such as water, the velocity terms are usually rather small and can be neglected. In the case of flowing gases such as air, the potential energy terms are usually very small and can be neglected; however, the kinetic energy terms may be quite important. Obviously the work term will be zero when no pump, turbine, or fan is present.

The *total pressure*, a very important concept, is the sum of the static pressure and the velocity pressure.

$$P_0 = P + \frac{\rho \overline{V}^2}{2g_c} \tag{9-5}$$

In terms of head Eq. (9-5) is written

$$\frac{g_c P_0}{g\rho} = \frac{g_c P}{g\rho} + \frac{\overline{V}^2}{2g} \tag{9-5a}$$

or

$$H_0 = H + H_v \tag{9-5b}$$

Equations (9-1) and (9-4) may be written in terms of total head and with rearrangement of terms become

$$\frac{g_c}{g} \frac{(P_{01} - P_{02})}{\rho} + (z_1 - z_2) = \frac{g_c w}{g} + l_f \tag{9-1c}$$

This form of the equation is much simpler to use with gases because the term $(z_1 - z_2)$ is negligible and when no fan is in the system the lost head equals the loss in total pressure head.

Lost Head

For incompressible flow in pipes and ducts the lost head is expressed as

$$l_f = f \frac{L}{D} \frac{\overline{V}^2}{2g} \tag{9-6}$$

where

f = Moody friction factor
L = length of the pipe or duct, ft or m
D = diameter of the pipe or duct, ft or m
$\overline{V}$ = average velocity in the conduit, ft/sec or m/s
g = acceleration due to gravity, ft/sec^2 or m/s^2

The lost head then has the units of feet or meters of the fluid flowing. For conduits of noncircular cross section the diameter may be replaced by the hydraulic diameter D_h given by

$$D_h = \frac{4(\text{cross-sectional area})}{\text{wetted perimeter}} \qquad (9\text{-}7)$$

Use of the hydraulic diameter is restricted to turbulent flow and cross-sectional geometries without extremely sharp corners.

Figure 9-1 shows friction data correlated by Moody (1), which is commonly referred to as the Moody diagram. Figure 9-2 is an example of relative roughness data for various types of pipe and tubing. The friction factor is a function of the Reynolds number Re, and the relative roughness e/D of the conduit in the transition zone; is a function of only the Reynolds number for laminar flow; and is a function of only relative roughness in the complete turbulence zone. Notice that the friction factor can be read directly from Fig. 9-2 if the Reynolds number and relative roughness are sufficiently large for the flow to be considered as complete turbulence.

The Reynolds number is defined as

$$\text{Re} = \frac{\rho \overline{V} D}{\mu} = \frac{\overline{V} D}{\nu} \qquad (9\text{-}8)$$

where

ρ = mass density of the flowing fluid, lbm/ft³ or kg/m³
μ = dynamic viscosity, lbm/(ft-sec) or (N-s)/m²
ν = kinematic viscosity, ft²/sec or m²/s

The hydraulic diameter is used to calculate Re when the conduit is noncircular. The *ASHRAE Handbook of Fundamentals* (2) has viscosity data on a wide variety of fluids. Appendix B contains data for water, air, and refrigerants.

In water systems it is often necessary to use an antifreeze solution that is usually a mixture of ethylene glycol and water. Figure 9-3 gives specific gravity and viscosity data for water and various solutions of ethylene glycol and water. Note that the viscosity is given in units of centipoise (1 lbm/(ft-sec) = 1490 centipoises and 10^{-3} centipoise = 1 (N-s)/m²). The following example demonstrates calculation of lost head for pipe flow.

Example 9-1.   Compare the lost head for water and a 30 percent ethylene glycol solution flowing at the rate of 110 gallons per minute (gpm) in a 3 in. standard (Schedule 40) commercial steel pipe, 200 ft in length. The temperature of the water is 50 F.

Solution.   Equation (9-6) will yield the desired information. From Table D-1 the inside diameter of 3 in. nominal diameter Schedule 40 pipe is seen to be 3.068 in. The Reynolds number is given by Eq. (9-8) and the average velocity in the pipe is

$$\overline{V} = \dot{Q}/A = \frac{110}{7.48} \times \frac{1}{\left(\dfrac{\pi}{4}\right)\left(\dfrac{3.068}{12}\right)^2} = 286 \text{ ft/min} = 4.77 \text{ ft/sec}$$

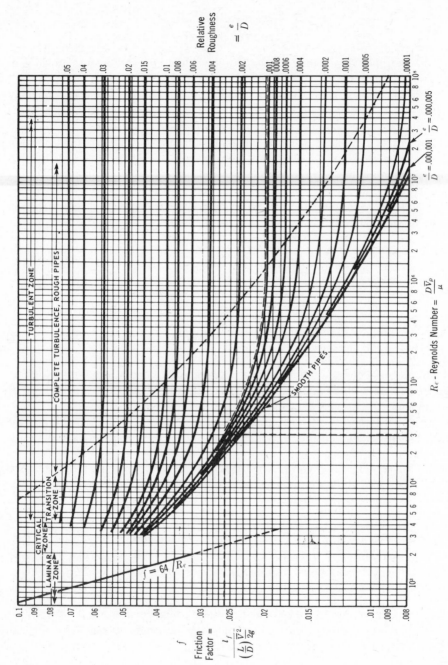

Figure 9-1. Friction factors for pipe flow.

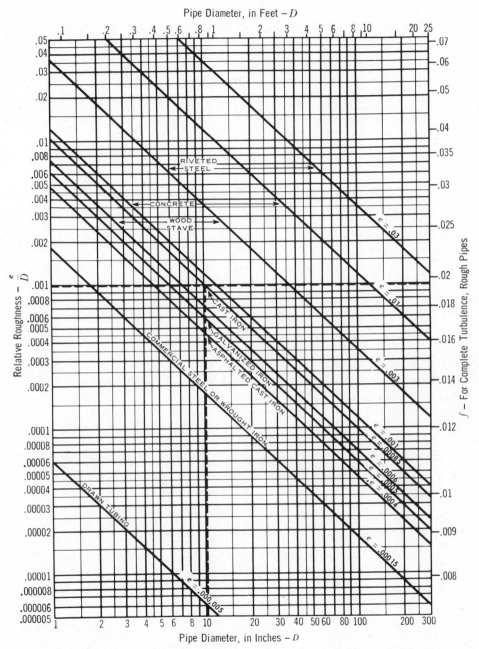

Figure 9-2. Relative roughness of pipe materials and friction factors for complete turbulence.

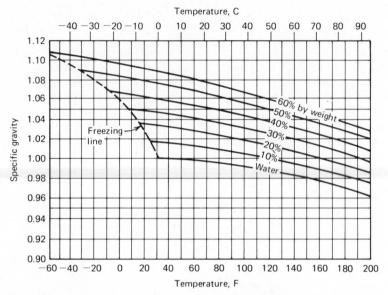

Figure 9-3a. Specific gravity of aqueous ethylene glycol solutions. (Adapted by permission from *ASHRAE Handbook and Product Directory-Systems*, 1980.)

The absolute viscosity of pure water at 50 F is 1.4 centipoises or 9.4×10^{-4} lbm/(ft-sec) from Fig. 9-3 b. Then

$$\text{Re} = \frac{62.4(4.77)(3.068/12)}{9.4 \times 10^{-4}} = 8.1 \times 10^4$$

From Fig. 9-2 the relative roughness e/D is 0.00058 for commercial steel pipe. Then the flow is in the transition zone and the friction factor f is 0.021 from Fig. 9-1. The lost head for pure water is then computed as

$$l_{fw} = 0.021 \times \frac{200}{(3.068/12)} \frac{(4.77)^2}{2(32.2)} = 5.80 \text{ ft of water}$$

The absolute viscosity of the 30 percent ethylene glycol solution is 3.1 centipoises and its specific gravity is 1.042 from Fig. 9-3 a. The Reynolds number for this case is

$$\text{Re} = \frac{1.042(62.4)(4.77)(3.068/12)}{(3.1)/1490} = 3.8 \times 10^4$$

and the friction factor is 0.024 from Fig. 9-1. Then

$$l_{fe} = 0.024 \times \frac{200}{(3.068/12)} \frac{(4.77)^2}{2(32.2)} = 6.63 \text{ ft of E.G.S.}$$
$$= 6.91 \text{ ft of water}$$

The increase in lost head when the antifreeze is added is

$$\text{Percent increase} = \frac{100(6.91 - 5.80)}{5.80} = 19\%$$

System Characteristic

The behavior of a piping system may be conveniently represented by plotting total head versus volume flow rate. Equation (9-1c) becomes

$$H_p = \frac{g_c(P_{01} - P_{02})}{g\rho} + (z_1 - z_2) - l_f \tag{9-1d}$$

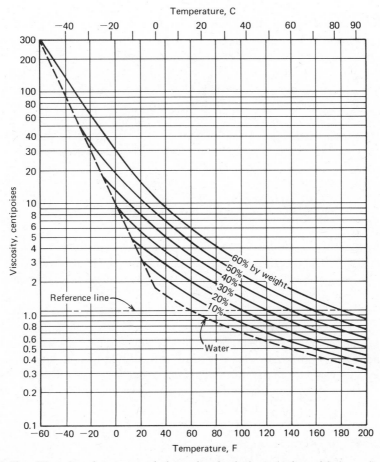

Figure 9-3b.   Viscosity of aqueous ethylene glycol solutions. (Adapted by permission from *ASHRAE Handbook and Product Directory-Systems*, 1980.)

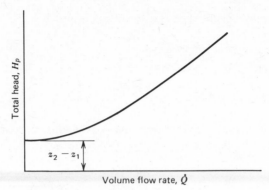

Figure 9-4. Typical system characteristic.

where H_p represents the total head required to produce the change in static, velocity, and elevation head and to offset the lost head. If a pump were present in the system H_p is the total head it must produce for a given volume flow rate. Since the lost head and velocity head are proportional to the square of the velocity, the plot of total head versus flow rate is approximately parabolic as shown in Fig. 9-4. Note that the elevation head is the same regardless of the flow rate. System characteristics are useful in analyzing complex circuits such as the parallel arrangement of Fig. 9-5. Circuit 1a2 and 1b2 each have a characteristic as shown in Fig. 9-6. The total flow rate is equal to the sum of $\dot{Q}_a$ and $\dot{Q}_b$ and the total head is the same for both circuits; therefore the characteristics are summed for various values of H_p to obtain the curve for the complete system shown as $(a + b)$. Series circuits have a common flow rate and the total heads are additive (Fig. 9-7).

Flow Measurement

It is important to make provisions for the measurement of flow rate in piping and duct systems. The most common devices for making these measurements are the pitot tube, orifice meter or other restrictive device, and anemometers, which are generally direct velocity measuring instruments. The pitot tube and orifice meter will be discussed here because they have greatest application in duct and piping systems. Figure 9-8 shows a pitot tube installed in a duct. The pitot tube senses both total and static head. The difference, the velocity head, is measured with a manometer. If the pitot tube is very small relative to the duct size, then local measurements and traverses may be made. When Eq. (9-1) is applied to a streamline between the tip of

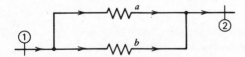

Figure 9-5. Arbitrary parallel flow circuit.

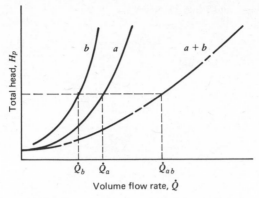

Figure 9-6. System characteristic for parallel circuits.

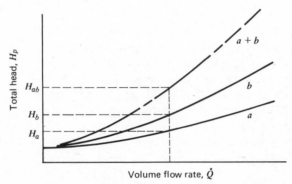

Figure 9-7. System characteristic for series circuits.

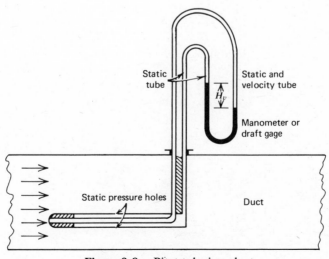

Figure 9-8. Pitot tube in a duct.

the pitot tube and a point a short distance upstream the following equation results (the lost energy is negligibly small):

$$\frac{P_1}{\rho} + \frac{\overline{V}_1^2}{2g_c} = \frac{P_2}{\rho} = \frac{P_{02}}{\rho} \tag{9-9}$$

or

$$\frac{P_{02} - P_1}{\rho} = \frac{\overline{V}_1^2}{2g_c} \tag{9-9a}$$

Solving for $\overline{V}_1$, we get

$$\overline{V}_1 = \left[2g_c \frac{(P_{02} - P_1)}{\rho} \right]^{1/2} \tag{9-10}$$

Equation (9-10) yields the velocity upstream of the pitot tube. It should be emphasized that the velocity may be different at other transverse locations in the pipe. For laminar flow the velocity distribution is parabolic while turbulent flow exhibits a more nearly uniform velocity distribution except near the wall.

Figure 9-9 shows a profile that might be typical of turbulent flow. It is generally necessary to traverse the pipe or duct and to integrate either graphically or numerically to find the average velocity in the duct. Equations (9-2) and (9-3) are then used to find the mass or volume flow rate.

The average velocity for incompressible steady flow is expressed by

$$\overline{V}_{avg} A = \int_A \overline{V}(x) \, dA \tag{9-11}$$

which is simply an adaptation of Eq. (9-3). For a circular duct the cross-sectional area A is given by

$$A = \pi x^2 \quad \text{and} \quad dA = 2\pi x \, dx \tag{9-12}$$

Then, letting x equal r at the wall, Eq. (9-11) becomes

$$\overline{V}_{avg} = \frac{2}{r^2} \int_0^r \overline{V}(x) x \, dx \tag{9-13}$$

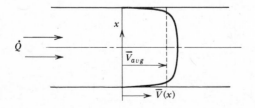

Figure 9-9.   Velocity profile in a pipe or duct.

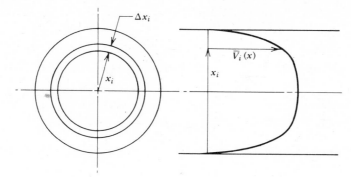

Figure 9-10.   Integrating the velocity profile over the pipe cross-sectional area.

In fully developed turbulent flow the velocity profile can be well represented by

$$\frac{\overline{V}}{\overline{V}_{max}} = \left(\frac{x}{r}\right)^{1/n} \tag{9-14}$$

where n varies from 6 to 8 over the Reynolds number range of 4×10^3 to 5×10^5 (6). For practical cases of air flow in ducts, $\overline{V}_{avg}/\overline{V}_{max}$ is about 0.82 for fully developed flow. In practice it is usually necessary to determine the velocity profile using the pitot tube and to then integrate graphically or numerically. In Fig. 9-10 and Eq. (9-13), the average velocity in the duct is given by

$$\overline{V}_{avg} = \frac{2}{r^2} \sum_{i=1}^{m} x_i \overline{V}_i(x) \, \Delta x_i \tag{9-13a}$$

The number of elements m required in Eq. (9-13a) depends on how $\overline{V}(x)$ varies with x. There is usually a central core over which the velocity is relatively constant. There are cases when the velocity profile is not symmetrical, that is, where the velocity is not constant throughout the circular element for a particular radius. In this case velocity profiles should be measured along at least two diameters, which are oriented at 90 degrees. Then for each circular element in the summation of Eq. (9-13a) the velocity $\overline{V}_i(x)$ is the average of the four measured values at the radius x_i.

   Pitot tubes are more frequently used to measure the velocity of gases than liquids. With gases such as air the available velocity head is often small, making precise measurements quite important. Micromanometers and inclined or Hook gages are used to measure $(P_{02} - P_1)$ in these cases.

   Example 9-2.   A pitot tube is installed in an air duct on the centerline. The velocity head as indicated by an inclined gage is 0.32 in. of water, the static head is 0.25 in. of water, the air temperature is 60 F, and barometric pressure is 29.92 in. of mercury. Assuming fully developed turbulent flow exists where the average velocity is approximately 82 percent of the centerline value, compute the volume and mass flow rates for a 10 in. diameter duct.

Solution. The centerline velocity is computed using Eq. (9-10). Using the English engineering system the term $(P_{02} - P_1)/\rho$ has the units of (ft-lbf)/lbm of air.

$$\frac{P_{02} - P_1}{\rho} = \left(\frac{g}{g_c}\right)\frac{0.32}{12}\left(\frac{\text{ft-lbf}}{\text{lbmw}}\right)\rho_w\left(\frac{\text{lbmw}}{\text{ft}^3}\right)\frac{1}{\rho_a}\left(\frac{\text{ft}^3}{\text{lbma}}\right); \frac{\text{ft-lbf}}{\text{lbma}}$$

If we assume an ideal gas,

$$\rho_a = \frac{P_a}{R_a T_a} = \frac{29.92(0.491)144}{53.35(60 + 460)} = 0.076 \frac{\text{lbma}}{\text{ft}^3}$$

which neglects the slight pressurization of the air in the duct. Finally

$$\frac{P_{02} - P_1}{\rho} = \frac{0.32}{12}\left(\frac{62.4}{0.076}\right)\frac{g}{g_c} = 21.9 \text{ (ft-lbf)/lbma}$$

The centerline velocity is given by

$$\overline{V}_{cl} = [2(32.2)(21.9)]^{1/2} = 37.6 \text{ ft/sec}$$

and the average velocity is

$$\overline{V}_{avg} = 0.82\overline{V}_{cl} = 0.82(37.6) = 30.8 \text{ ft/sec}$$

The mass flow rate is given by Eq. (9-2) with the area given by

$$A = \frac{\pi}{4}\left(\frac{10}{12}\right)^2 = 0.545 \text{ ft}^2$$

$$\dot{m} = \rho_a\overline{V}_{avg}A = 0.076(30.8)0.545 = 1.28 \text{ lbm/sec}$$

The volume flow rate is

$$\dot{Q} = \overline{V}_{avg}A = 30.8(0.545) = 16.8 \text{ ft}^3/\text{sec}$$

using Eq. (9-3).

Example 9-3. Air flows in a 0.3 m diameter circular duct. At the end of a long straight run the centerline velocity is measured to be 12 m/s. Estimate the average velocity and mass flow rate of the air assuming standard conditions for the air.

Solution. Because the measurement follows a long straight run of pipe, we shall assume that the flow is fully developed and that the velocity profile is symmetrical. It will be necessary to integrate Eq. (9-13) using Eq. (9-14) to describe the velocity profile. The exponent n should be about 7 for this case and can be checked later. Substitution of Eq. (9-14) in Eq. (9-13) and integration with $n = 7$ yields

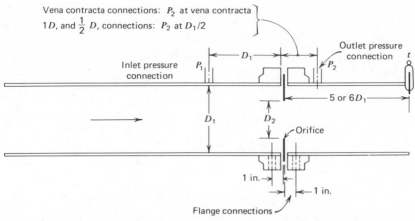

Figure 9-11.   Recommended location of pressure taps for use with concentric thin-plate, and square-edged orifices according to reference 3.

$$\overline{V}_{avg} = \frac{2\,\overline{V}_{max}}{r^{15/7}} \int_0^r x^{8/7}\, dx = \frac{14}{15}\,\overline{V}_{max}$$

and

$$\overline{V}_{avg} = \frac{14}{15}\,(12) = 11.2 \text{ m/s}$$

The mass flow rate is given by Eq. (9-2)

$$\dot{m} = \rho\overline{V}A = 1.2(11.2)\,\frac{\pi}{4}\,(0.3)^2 = 0.95 \text{ kg/s}$$

where the mass density of standard air, 1.2 kg/m³, has been used. To check the value of the exponent n, it is necessary to compute the Reynolds number.

$$\text{Re} = \frac{\rho\overline{V}D}{\mu} = \frac{1.2(11.2)0.3}{20.4 \times 10^{-6}} = 197{,}650$$

where the dynamic viscosity is from Table B-4a. Reference 6 indicates that n equal to 7 is valid for the Reynolds number computed above.

Flow measuring devices of the restrictive type use the pressure drop across an orifice, nozzle, or venturi to predict flow rate. The square-edged orifice is widely used because of its simplicity. Figure 9-11 shows such a meter with the location of the pressure taps (3). The flange-type pressure taps are widely used in HVAC piping systems and are standard fittings available commercially. The orifice plate may be fabricated locally or may be purchased. Reference 3 outlines the manufacturing procedure in detail.

The orifice meter is far from being an ideal flow device and introduces an appreciable loss in total pressure. An empirical *discharge coefficient* is

$$C = \frac{\dot{Q}_{actual}}{\dot{Q}_{ideal}} \tag{9-15}$$

The ideal flow rate may be derived from Eq. (9-1) with the lost energy equal to zero. Applying Eq. (9-1) between the cross sections defined by the pressure taps gives

$$\frac{P_1}{\rho} + \frac{\overline{V}_1^2}{2g_c} = \frac{P_2}{\rho} + \frac{\overline{V}_2^2}{2g_c} \tag{9-16}$$

To eliminate the velocity $\overline{V}_1$ from Eq. (9-12), Eq. (9-3) is recalled and

$$\overline{V}_1 = \overline{V}_2 \frac{A_2}{A_1} \tag{9-3a}$$

Substitution of Eq. (9-3a) into Eq. (9-16) and rearrangement yields

$$\overline{V}_2 = \frac{1}{[1 - (A_2/A_1)^2]^{1/2}} \left[2g_c \frac{(P_1 - P_2)}{\rho} \right]^{1/2} \tag{9-17}$$

Then by using Eq. (9-15) and (9-17) we get

$$\dot{Q}_{actual} = \frac{CA_2}{[1 - (A_2/A_1)^2]^{1/2}} \left[2g_c \frac{(P_1 - P_2)}{\rho} \right]^{1/2} \tag{9-18}$$

The term $[1 - (A_2/A_1)^2]^{1/2}$ is referred to as the *velocity of approach factor*. In practice the discharge coefficient and velocity of approach factor are often combined and called the *flow coefficient* C_d.

$$C_d = \frac{C}{[1 - (A_2/A_1)^2]^{1/2}} \tag{9-19}$$

This is merely a convenience. For precise measurements other corrections and factors may be applied, especially for compressible fluids (3). Figure 9-12 shows representative values of the flow coefficient C_d. The data apply to pipe diameters over a wide range (1 to 8 in.) and to flange or radius taps within about 5 percent. When precise flow measurement is required, reference 3 should be consulted for more accurate flow coefficients.

In complex piping systems it is necessary to install meters at strategic locations so that the system can be balanced when it is put into operation. The designer must specify an orifice which when the desired flow rate is attained, will have a reasonable pressure drop but one not large enough to be wasteful of pumping power. The following example illustrates an orifice design procedure.

Example 9-4.   Design an orifice to be used in a 4 in. standard commercial steel pipe that has a design flow rate of 200 gal per minute (gpm). The pipe will supply chilled water in summer at 45 F and in winter hot water at 150 F. A change in head, H_{12} of 5 to 10 in. of mercury is acceptable.

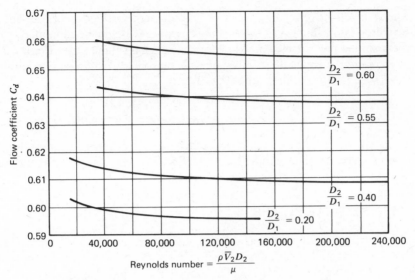

Figure 9-12. Flow coefficients for square-edge orifices.

Solution. Referring to Fig. 9-12, we note that the flow coefficient curves become independent of Reynolds number at high Reynolds numbers. It would therefore be desirable to operate in this region. It is also desirable to operate with a diameter ratio of 0.5 to 0.6 because the flow coefficient data seem to be most accurate in this range. Therefore, it is assumed that $D_2/D_1 = 0.55$ and $C_d = 0.637$. The change in head $\Delta P/\rho$ is calculated from Eq. (9-18).

$$\frac{P_1 - P_2}{\rho} = \left(\frac{\dot{Q}}{C_d A_2}\right)^2 \frac{1}{2g_c}$$

$$D_2 = 0.55D_1 = 0.55(4.026) = 2.21 \text{ in.}$$

where the inside diameter of 4 in. pipe is 4.026 in.

$$A_2 = \frac{\pi}{4}\left(\frac{2.21}{12}\right)^2 = 0.027 \text{ ft}^2$$

$$\dot{Q} = \frac{200}{60(7.48)} = 0.446 \text{ ft}^3/\text{sec}$$

where 7.48 gal equals 1 ft³.

$$\frac{P_1 - P_2}{\rho} = \left[\frac{0.446}{(0.637)0.027}\right]^2 \frac{1}{2(32.2)} = 10.38 \frac{\text{ft-lbf}}{\text{lbmw}}$$

or

$$H_{12} = \frac{10.38}{12.55} \times 12 = 9.92 \text{ in. of mercury}$$

where the effective specific gravity of the manometer fluid is 12.55. This pressure drop is satisfactory. The Reynolds number will now be computed to check the assumed flow coefficient. The largest expected value of the viscosity is used to compute the Reynolds number because this yields the lowest Re number. If we use Eq. (9-8) and data from Fig. 9-3,

$$\text{Re} = \frac{\rho \overline{V}_2 D_2}{\mu} = \frac{62.4(0.446/0.027)(2.21/12)}{(1.6/1490)} = 1.8 \times 10^5$$

From Fig. 9-12 note that the flow coefficient C_d is nearly constant at Reynolds numbers greater than about 2×10^5. For this particular case the Reynolds number is sufficiently large to be acceptable.

9-2 CENTRIFUGAL PUMPS

The centrifugal pump is by far the most frequently used type of pump in HVAC systems. The essential parts of a centrifugal pump are the rotating member or impeller and the surrounding case. The impeller is usually driven by an electric motor that may be close-coupled (on the same shaft as the impeller) or flexible coupled, (Fig. 9-13 a). The fluid enters the center of the rotating impeller, is thrown into

Figure 9-13a.   A single inlet, flexible coupled centrifugal pump. (Courtesy of ITT Bell and Gossett, Skokie, Ill.)

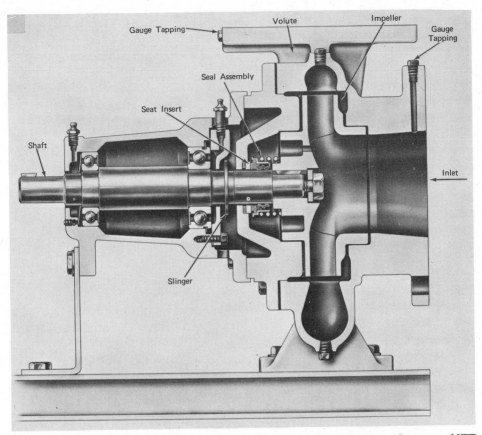

Figure 9-13b.   Cutaway of single inlet, flexible coupled centrifugal pump. (Courtesy of ITT Bell and Gossett, Skokie, Ill.)

the volute, and flows outward through the diffuser (Fig. 9-13 b). The fluid leaving the impeller has high kinetic energy that is converted to static pressure in the volute and diffuser as efficiently as possible. Although there are various types of impellers and casings (4), the principle of operation is the same. The pump shown in Fig. 9-13 is a single suction pump because the fluid enters the impeller from only one side. The double-suction type has fluid entering from both sides. A pump may be staged with more than one impeller on the same shaft with one casing. The fluid leaves the first stage and enters the impeller of the second stage before leaving the casing.

  Pump performance is most commonly given in the form of curves. Figure 9-14 is an example of such data for a pump that may be operated at two different speeds with four different impellers. For each speed a different curve is given for each impeller. These curves give the total dynamic head, efficiency, shaft power, and the net positive suction head as a function of capacity.

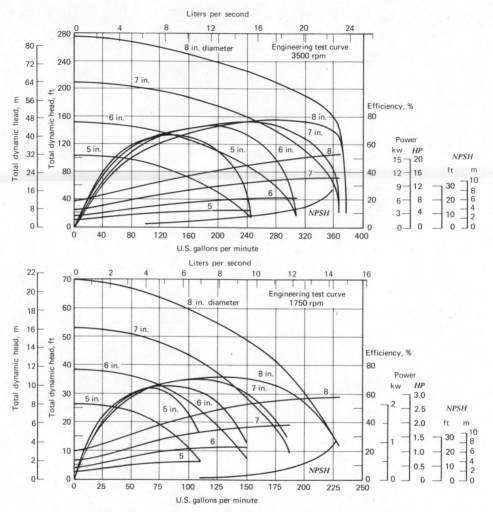

Figure 9-14. Typical centrifugal pump performance data.

The *total dynamic head* furnished by a pump can be understood by applying Eq. (9-1b) to the fluid entering and leaving the pump.

$$H_p = \frac{w g_c}{g} = \frac{g_c(P_1 - P_2)}{g\rho} + \frac{\overline{V}_1^2 - \overline{V}_2^2}{2g} + (z_1 - z_2) \qquad (9\text{-}20)$$

The lost head is unavailable as useful energy and is omitted from the equation. Losses are typically accounted for by the *efficiency*, defined as the ratio of the useful power actually imparted to the fluid to the shaft power input.

$$\eta_p = \frac{\dot{W}}{\dot{W}_s} = \frac{w\dot{m}}{\dot{W}_s} = \frac{\dot{Q}w}{\dot{W}_s} \qquad (9\text{-}21)$$

The shaft power may be obtained from Eq. (9-21)

$$\dot{W}_s = \frac{\dot{m}w}{\eta_p} = \frac{\rho \dot{Q}w}{\eta_p} \tag{9-22}$$

Therefore a definite relationship exists between the curves for total head, efficiency, and shaft power in Fig. 9-14.

   If the static pressure of the fluid entering a pump approaches the vapor pressure of the liquid too closely, vapor bubbles will form in the impeller passages. This condition is detrimental to pump performance and the collapse of the bubbles is noisy and may damage the pump. This phenomenon is known as cavitation. The amount of pressure in excess of the vapor pressure required to prevent cavitation is known as the *required net positive suction head* (*NPSHR*). *NPSHR* is a characteristic of a given pump and varies considerably with speed and capacity. *NPSHR* is determined by the actual testing of each model.

   Whereas each pump has its own *NPSHR*, each system has its own *available net positive suction head* (*NPSHA*).

$$NPSHA = \frac{P_s g_c}{\rho g} + \frac{\overline{V}_s^2}{2g} - \frac{P_v g_c}{\rho g} \tag{9-23}$$

where

$g_c P_s / pg$ = static head at the pump inlet, ft or m, absolute
$\overline{V}_s^2 / 2g$ = velocity head at the pump inlet, ft or m
$g_c P_v / \rho g$ = static vapor pressure head of the liquid at the pumping temperature, ft or m, absolute

The net positive suction head available must always be greater than the *NPSHR* or noise and cavitation will result.

   Example 9-5.   Suppose the pump of Fig. 9-14 is installed in a system as shown in Fig. 9-15. The pump is operating at 3500 rpm with the 6 in. impeller and delivering 200 gpm. The suction line is standard 4 in. pipe that has an inside diameter of 4.026 in. Compute the *NPSHA* and compare it with the *NPSHR*. The water temperature is 60 F.

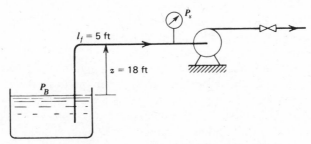

Figure 9-15.   Open system with suction lift.

Solution. From Fig. 9-14 the *NPSHR* is 5 ft of head. The available net positive suction head is computed from Eq. (9-23); however, the form will be changed slightly through the application of Eq. (9-1b) between the water surface and the pump inlet.

$$\frac{P_B g_c}{\rho g} = \frac{P_s g_c}{\rho g} + \frac{\overline{V}_s^2}{2g} + z_s + l_f$$

or

$$\frac{P_s g_c}{\rho g} + \frac{\overline{V}_s^2}{2g} = \frac{P_B g_c}{\rho g} - z_s - l_f$$

Then Eq. (9-23) becomes

$$NPSHA = \frac{P_B g_c}{\rho g} - z_s - l_f - \frac{P_v}{\rho}\frac{g_c}{g} \qquad (9\text{-}23a)$$

Assuming standard barometric pressure

$$\frac{P_B g_c}{\rho g} = \frac{29.92}{12} \times 13.55 = 33.78 \text{ ft of water}$$

$$\frac{P_v g_c}{\rho g} = \frac{0.2562 \times 144}{62.4} = 0.59 \text{ ft of water}$$

where P_v is read from Table A-1. Then from Eq. (9-23a)

$$NPSHA = 33.78 - 18 - 5 - 0.59 = 10.19 \text{ ft of water}$$

which is about twice as large as the *NPSHR*. If the water temperature is increased to 160 F and other factors remain constant, the *NPSHA* becomes

$$NPSHA = 33.78 - 18 - 5 - \left(\frac{4.74 \times 144}{61} \right) = -0.4 \text{ ft}$$

which is less than the *NSPHR* of 5 ft. Cavitation will undoubtedly result.

In an open system such as a cooling tower the pump suction (inlet) should be flooded if possible; that is, the inlet should be lower than the free water surface. This is particularly important if the fluid temperature is high such as in the case of condensate from a steam condenser. Long runs of suction piping should be eliminated whenever possible, and care should be taken to eliminate air traps on the suction side of the pump. In an open reservoir or sump the end of the suction pipe should be adequately covered with fluid to prevent entrainment of air from the vortex formed at the pipe entrance. An inlet velocity of less than 3 ft/sec or 1 m/s will minimize vortex formation. Care must be taken to locate the pump in a space where freezing will not occur and where maintenance may be easily performed.

The pump foundation, usually concrete, should be sufficiently rigid to support the pump base plate. This is particularly important for close-coupled pumps to maintain alignment between the pump and motor. The pump foundation should weigh from $1\frac{1}{2}$ to 3 times the total pump and motor weight for vibration and sound control.

It is extremely important to install the suction and discharge piping such that forces are not transmitted to the pump housing. Expansion joints are required on both the suction and discharge sides of the pump to isolate expansion and contraction forces, and the piping must be supported independent of the pump housing.

9-3   COMBINED SYSTEM AND PUMP CHARACTERISTICS

The combination of the system and pump characteristics (head versus capacity) is very useful in the analysis and design of piping systems. Figure 9-16 is an example of how a system with parallel circuits behaves with a pump installed. Recall that the total head H_p furnished by the pump is given by Eq. (9-20). Note that the combination operates at point t where the characteristics cross. The flow rate for each of the parallel circuits is quite obvious because the required change in total head from 1 to 2 is equal for both circuits.

Figure 9-17 illustrates a series-type circuit. When the valve is open, the operating point is at point a with flow rate $\dot{Q}_a$ and total head H_a. Partial closing of the valve introduces additional flow resistance (head loss) and is similar to adding series resistance in an electrical circuit. The new system characteristic crosses the pump curve at point c and the flow rate is $\dot{Q}_c$ with total head H_c. All piping systems should contain valves for control and adjustment purposes.

A typical design problem is one of pump selection. The following example illustrates the procedure.

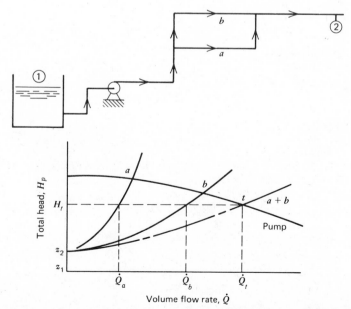

Figure 9-16.   Combination of system and pump characteristics-parallel circuits.

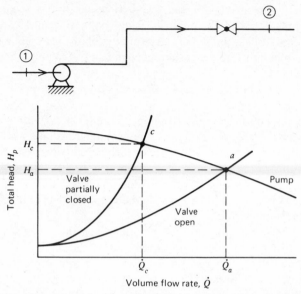

Figure 9-17. Combination of system and pump characteristics-series circuit.

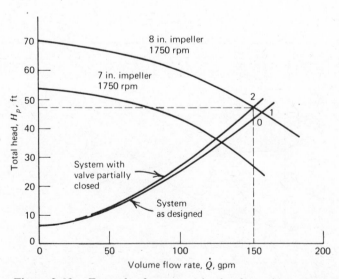

Figure 9-18. Example of a pump selection for a given system.

Example 9-6.   A water piping system has been designed to distribute 150 gpm and the total head requirement is 42 ft. Select a pump using the data of Fig. 9-14 and specify the power rating for the electric motor.

Solution.   Figure 9-18 shows the characteristic for the piping system as it was designed. Point 0 denotes the operating capacity desired. Examination of Fig. 9-14 indicates that the low-speed version of the given pump covers the desired range. The desired operating point lies between the curves for the 7 and 8 in. impellers. The curves are sketched on Fig. 9-18. Obviously, the pump with the 8 in. impeller must be selected, but the flow rate will be about 160 gpm as indicated by point 1. Therefore a valve must be adjusted (closed slightly) to modify the system characteristic as shown, to obtain 150 gpm at point 2. Again referring to Fig. 9-14 the shaft power requirement is read as 2.6 horsepower. Electric motors usually have an efficiency of 85 to 90 percent and a 3 horsepower motor should be specified.

9-4   PIPING SYSTEM DESIGN

There are many different types of piping systems used with HVAC components and there are many specialty items and refinements that make up these systems. Chapters 15, 16, 17, and 18 of reference 5 give a detailed description of various arrangements of the components making up the complete system. The main thrust of the discussion to follow is to develop methods for the design of basic piping systems used to distribute hot and chilled water. Chapter 32 of reference 2 pertains to the sizing of pipe. This aspect of the problem will be emphasized.

Open Loop System

A typical open loop piping system is shown in Fig. 9-19. Characteristically an open loop system will have some part of the circuit open to the atmosphere. The

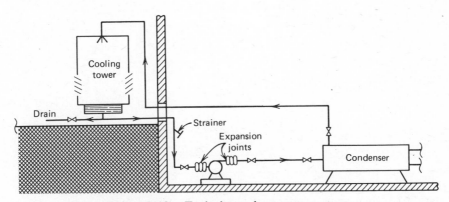

Figure 9-19.   Typical open-loop water system.

cooling tower circuit of Fig. 9-19 shows the usual valves, filters, and fittings installed in this type of circuit. It is important to protect the pump with a filter. The isolation valves provide for maintenance without complete drainage of the system, whereas a globe or plug valve should be provided at the pump outlet for control of the flow rate. Expansion joints and a rigid base support and isolate the pump as previously discussed.

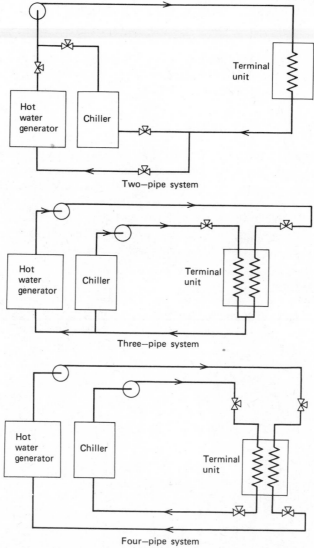

Figure 9-20. Schematic diagrams of two-pipe, three-pipe, and four-pipe constant flow systems.

Closed Loop System

Basic closed loop systems are shown in Fig. 9-20 and are known as two-pipe, three-pipe, and four-pipe systems. A single-pipe system is also possible but is not used extensively in modern times. Although it is not necessary to install a filter in a closed loop system, an expansion tank and air separator are required. The expansion tank protects the system from being damaged by a change in volume due to temperature variations. It also provides a collecting space for air removed by the separator. Closed loop systems are normally pressurized slightly so that the complete loop will be above atmospheric pressure. A pressure regulator is installed in the makeup water line and set at about 10 lbf/in.^{2} or 69 kPa gage. Isolation and control valves are required, and instrumentation to measure flow rates and temperatures at strategic locations is essential. This is especially important when parallel circuits are employed.

Design Criteria

Open loop systems should be designed for velocities of 5 to 10 ft/sec or 1.5 to 3 m/s. Noise generated by the flowing fluid is not the main consideration in this case, whereas minimizing the pipe sizes is an important economic factor. Velocities above 10 ft/sec or 3 m/s introduce large head losses and increase pump size and cost. When parallel circuits are employed, a reasonable effort should be made to balance the head loss in each circuit so that drastic valve adjustments are not required.

Closed loop piping systems often pass through or near occupied spaces where noise generated by the flowing fluid may be objectionable. Therefore, a velocity limit of 4 ft/sec or 1.2 m/s for two in. pipe and smaller is generally imposed. For larger sizes a limit on the head loss of 4 ft per 100 ft of pipe is imposed. This corresponds to about 0.4 kPa/m in SI Units. These criteria should not be treated as hard rules but rather as guides. Noise is caused by entrained air, abrupt pressure drops, and turbulence in general. If these factors can be minimized, the given criteria can be relaxed. A reasonable effort to design a balanced system will prevent drastic valve adjustments and will contribute to a quieter system.

The piping layout for a heating and air-conditioning system depends on the location of the central and terminal equipment and the type of system to be used. All of the piping may be located in the central equipment room or may run throughout the building to terminal units in every room. In this case the available space may be a controlling factor. It must be kept in mind that piping for domestic hot and cold water, sewage, and other services must be provided in addition to the heating and air-conditioning requirements. The designer must constantly check to make sure the piping will fit into the allowed space.

Pipe Sizing

After the piping layout has been completed, the problem of sizing the pipe consists mostly of applying the design criteria discussed above. Where possible the pipes should be sized so that drastic adjustments are not required. Often an ingenious layout helps in this respect. The system and pump characteristics are also useful in the design process.

To facilitate the actual pipe sizing and computation of head loss, charts such as those shown in Figs. 9-21, and 9-22 for pipe and copper tubing have been developed. Figures 9-21 and 9-22 are based on 60 F or 16 C water and give head losses that are about 10 percent high for hot water. Examination of Figs. 9-21, and 9-22 shows that head loss may be obtained directly from the flow rate and nominal pipe size or from flow rate and water velocity. When the head loss and flow rate are known, a pipe size and velocity may be obtained.

Pipe fittings and valves also introduce losses in head. These losses are usually accounted for by use of a *resistance coefficient K,* which is the number of velocity heads lost because of the valve or fitting. Thus,

$$l_f = K\frac{\overline{V}^2}{2g} \tag{9-24}$$

Comparing this definition with Eq. (9-6) it can be seen that

$$K = f\frac{L}{D} \tag{9-24a}$$

The ratio L/D is the equivalent length in pipe diameters of straight pipe that will cause the same pressure drop as the valve or fitting under the same flow conditions. This is a convenient concept to use when one is computing total pressure drop or head loss in a piping system. Representative values of resistance coefficients for some

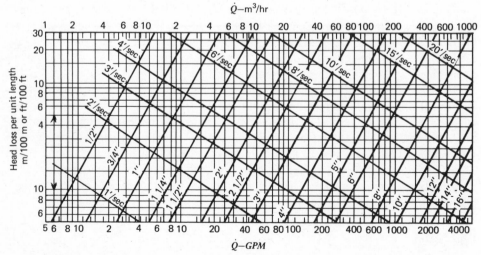

Figure 9-21.   Friction loss due to flow of water in commercial steel pipe (schedule 40). (Reprinted by permission from *ASHRAE Handbook of Fundamentals,* 1977.)

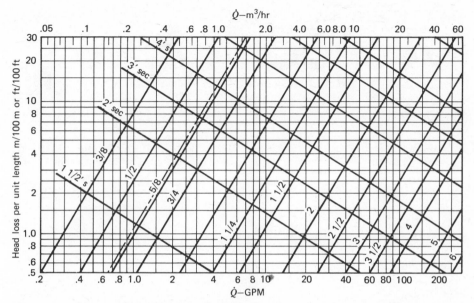

Figure 9-22. Friction loss due to flow of water in type L copper tube. (Reprinted by permission from *ASHRAE Handbook of Fundamentals*, 1977.)

common valves and fittings are given in Fig. 9-23 *a*. Conversions between K, L/D, and L can be obtained for various pipe sizes by the use of Fig. 9-23 *b*. When using SI units it is suggested that the L/D ratio be determined from Fig. 9-23 *b*, using the nominal pipe size. The equivalent length in meters may then be determined using the inside diameter D in meters. The lost head for a given length of pipe of constant diameter and containing fittings is computed as the product of the lost head per foot from Figs. 9-21 and 9-22 and the total equivalent length of the pipe and fittings.

Example 9-7. Compute the lost head for a 150 ft run of standard pipe, having a diameter of 3 in. The pipe run has 3 standard 90 degree elbows, a globe valve, and a gate valve. One hundred gpm of water flows in the pipe.

Solution. The equivalent length of the various fittings will first be determined by using Figs. 9-23 *a* and 9-23 *b*.

Globe Valve: $K_1 = 340 f_t, f_t = 0.018$; Fig. 9-23 *a* and Table 9-1
$K_1 = 340(0.018) = 6.1$
$L = 86$ ft; Fig. 9-23 *b*
Elbow: $K = 30 f_t, f_t = 0.018$
$K = 30(0.018) = 0.54$
$L = 8$ ft
Gate Valve: $K_1 = 8 f_t, f_t = 0.018$
$K_1 = 8(0.018) = 0.14$
$L = 2$ ft

GATE VALVES
Wedge Disc, Double Disc, or Plug Type

If: $\beta = 1$, $\theta = 0$ . . . . $K_1 = 8\,f_t$

$\beta < 1$ and $\theta \lesssim 45°$ . . . . $K_2 =$ Formula **1**

$\beta < 1$ and $\theta > 45° \lesssim 180°$ . . . $K_2 =$ Formula **2**

GLOBE AND ANGLE VALVES

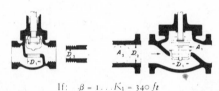

If: $\beta = 1$ . . . $K_1 = 340\,f_t$

90° PIPE BENDS AND
FLANGED OR BUTT-WELDING 90° ELBOWS

r/D	K	r/D	K
1	$20\,f_t$	10	$30\,f_t$
2	$12\,f_t$	12	$34\,f_t$
3	$12\,f_t$	14	$38\,f_t$
4	$14\,f_t$	16	$42\,f_t$
6	$17\,f_t$	18	$46\,f_t$
8	$24\,f_t$	20	$50\,f_t$

The resistance coefficient, K_B, for pipe bends other than 90° may be determined as follows:

$$K_B = (n - 1) \left(0.25\,\pi\,f_T \frac{r}{D} + 0.5\,K \right) + K$$

n = number of 90° bends
K = resistance coefficient for one 90° bend (per table)

CLOSE PATTERN RETURN BENDS

$K = 50\,f_t$

STANDARD ELBOWS

90°   45°

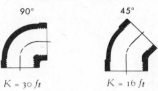

$K = 30\,f_t$ $K = 16\,f_t$

STANDARD TEES

Flow thru run . . . . $K = 20\,f_t$
Flow thru branch . . . . $K = 60\,f_t$

PIPE ENTRANCE

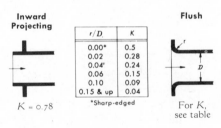

Inward Projecting

$K = 0.78$

Flush

r/D	K
0.00*	0.5
0.02	0.28
0.04*	0.24
0.06	0.15
0.10	0.09
0.15 & up	0.04

*Sharp-edged

For K, see table

PIPE EXIT

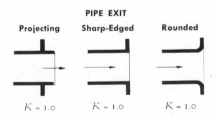

Projecting   Sharp-Edged   Rounded

$K = 1.0$ $K = 1.0$ $K = 1.0$

Figure 9-23a.   Resistance coefficients K for various valves and fittings. (Courtesy of the Crane Company, Technical Paper No. 410.)

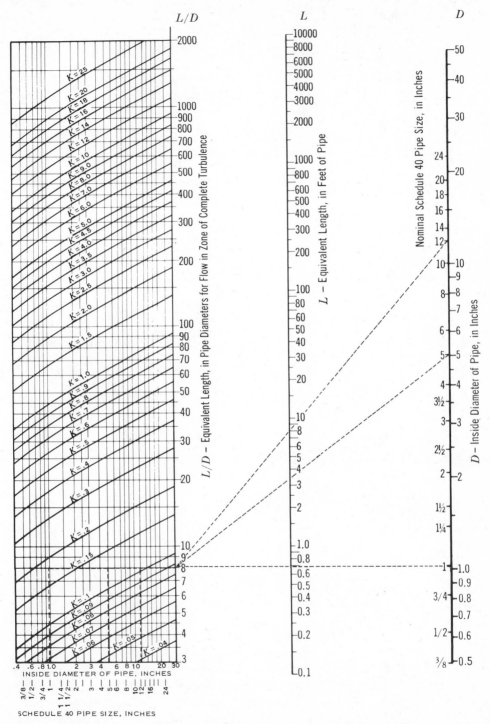

Figure 9-23b. Equivalent lengths L and L/D and resistance coefficient K. (Courtesy of the Crane Company, Technical Paper No. 410.)

Table 9-1   FORMULAS, DEFINITION OF TERMS, AND VALUES OF f_t FOR FIGURE 9-23

Formula 1 $K_2 = \dfrac{K_1 + \left(\sin\dfrac{\theta}{2}\right)0.8(1 - \beta^2) + 2.6(1 - \beta^2)^2}{\beta^4}$

Formula 2 $K_2 = \dfrac{K_1 + 0.5\left(\sin\dfrac{\theta}{2}\right)(1 - \beta^2) + (1 - \beta^2)^2}{\beta^4}$

$\beta = \dfrac{D_1}{D_2};\qquad \beta^2 = \left(\dfrac{D_1}{D_2}\right)^2 = \dfrac{A_1}{A_2};\qquad \begin{array}{l} D = \text{smaller diameter} \\ A_1 = \text{smaller area} \end{array}$

Nominal Size	$\frac{1}{2}$ in.	$\frac{3}{4}$ in.	1 in.	$1\frac{1}{4}$ in.	$1\frac{1}{2}$ in.	2 in.
Friction Factor (f_t)	0.027	0.025	0.023	0.022	0.021	0.019
$2\frac{1}{2}$, 3 in.	4 in.	5 in.	6 in.	8–10 in.	12–16 in.	18–24 in.
0.018	0.017	0.016	0.015	0.014	0.013	0.012

The total equivalent length is then

Actual length of pipe	150 ft
One globe valve	86 ft
Three elbows	24 ft
One gate valve	2 ft
Total	262 ft

From Fig. 9-21 the lost head l'_f is 2.3 ft per 100 ft of length or

$$l'_f = 2.3 \times 10^{-2} \text{ ft/ft of length}$$

The lost head for the complete pipe run is then given by

$$l_f = L_e l'_f = (262)2.3 \times 10^{-2} = 6.0 \text{ ft}$$

The pressure drop for control valves, check valves, strainers, and other such devices is often given in terms of a coefficient C_v. The coefficient is numerically equal to the flow rate of water at 60 F in gpm, which will give a pressure drop of one lbf/in.^{2} (2.31 ft of water). Because the head loss is proportional to the square of the velocity, the pressure drop or lost head may be computed at other flow rates.

$$\frac{l_{f1}}{l_{f2}} = \left(\frac{\dot{Q}_1}{\dot{Q}_2}\right)^2 \tag{9-25}$$

In terms of the coefficient C_v:

$$l_f = 2.31 \left(\frac{\dot{Q}}{C_v}\right)^2 \tag{9-26}$$

where $\dot{Q}$ and C_v are both in gpm and l_f is in feet of water.

It may be shown that the flow rate of any fluid is given by

$$\dot{Q} = C_v \left[\frac{\Delta P(62.4)}{\rho}\right]^{1/2} \tag{9-27}$$

where ΔP is in lbf/in.² and ρ is in lbm/ft³.

There is a relationship between C_v and the resistance coefficient K. By using Eqs. (9-3) and (9-6), we can show that

$$C_v = \frac{0.208 D^2}{\sqrt{K}}$$

where D is in ft. In SI units a flow coefficient C_{vs} is defined as the flow rate of water at 15 C in m³/s with a pressure loss of 1 kPa given by

$$C_{vs} = 1.11 \frac{D^2}{\sqrt{K}} \tag{9-28}$$

where D is in meters.

Example 9-8.   A strainer has a C_v rating of 60. It is to be used in a system to filter 50 gpm of water. What pressure drop and head loss can be expected?

Solution.   Equation (9-26) will yield the desired result.

$$l_f = 2.31 \left(\frac{50}{60}\right)^2 = 1.6 \text{ ft of water}$$

The pressure drop ΔP is given by

$$\Delta P = \frac{1.6(62.4)}{144} \frac{g}{g_c} = 0.7 \text{ lbf/in.}^2$$

Heating and cooling units and terminal devices usually have head loss information furnished by the manufacturer. The head loss is often used to indicate the flow rate for the adjustment of the system. Equation (9-25) may be used to estimate head loss at other than specified conditions.

There is no one set procedure for pipe sizing. The following examples will demonstrate some approaches to the problem.

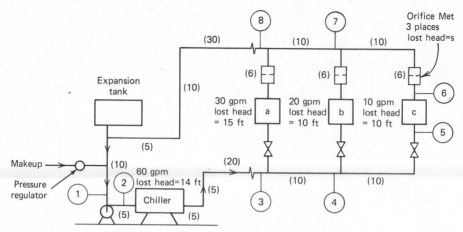

Figure 9-24. Two-pipe system-design example.

Example 9-9. Figure 9-24 shows a two-pipe water system such as might be found in a central equipment room. The terminal units a, b, and c are air handling units that contain air-to-water finned tube heat exchangers. An actual system would probably contain both a heater and a chiller; only one or the other is to be considered here. Size the piping and specify the pumping requirements.

Solution. The first step is to select criteria for sizing of the pipe. Because the complete system is confined to a central equipment room, the velocity and head loss criteria may be relaxed somewhat. Let the maximum velocity be 5 ft/sec and the maximum head loss be about 7 ft per 100 foot in the main run. Somewhat higher values may be used in the parallel circuits. By using Fig. 9-21 we select pipe sizes as follows:

$$D_{23} = D_{81} = 2\tfrac{1}{2} \text{ in.}; \quad \overline{V} = 4 \text{ ft/sec}; \quad l'_f = 2.6 \text{ ft/100 ft}$$
$$D_{34} = D_{78} = 1\tfrac{1}{2} \text{ in.}; \quad \overline{V} = 4\tfrac{3}{4} \text{ ft/sec}; \quad l'_f = 6.5 \text{ ft/100 ft}$$
$$D_{45} = D_{67} = 1 \text{ in.}; \quad \overline{V} = 3\tfrac{3}{4} \text{ ft/sec}; \quad l'_f = 7.4 \text{ ft/100 ft}$$

The lost head for the various sections is now computed. The equivalent length used for the various fittings is approximate in the following analysis. In actual practice the fittings should be treated as shown in Examples 9-7 and 9-8. A three-dimensional piping diagram would also be used to describe the piping layout in detail.

Section 8 to 3 (exclusive of the heater or chiller)

$$L_{83} = 90 + 50 = 140 \text{ ft}$$
$$l_{f83} = 2.6(140)/100 = 3.6 \text{ ft}$$

Sections 3 to 4 and 7 to 8

$$L_e = L_{43} + L_{78} = 2(15) = 30 \text{ ft}$$
$$l_{f34} + l_{f78} = 30(6.5)/100 = 2.0 \text{ ft}$$

Sections 4 to 5 and 6 to 7

$$L_e = L_{45} + L_{67} = 40 + 20 = 60 \text{ ft}$$
$$l_{f45} + l_{f67} = 60(7.4)/100 = 4.4 \text{ ft}$$

Chiller, Unit c, and Orifice

$$l_f = 14 + 10 + 5 = 29 \text{ ft}$$

The total lost energy for this pipe run including fittings and equipment is

$$\frac{(P_{02} - P_{01})}{\rho} = (l_f)_t \frac{g}{g_c} = 3.6 + 2.0 + 4.4 + 29 = 39.0 \text{ (ft-lbf)/lbm}$$

Applying Eq. (9-1c) to just the pump, we get

$$w = \frac{(P_{01} - P_{02})}{\rho}$$

Therefore,

$$w = -39.0 \text{ (ft-lbf)/lbm}$$

which is equivalent to stating that the pump must develop 39 ft of head.

The two remaining parallel circuits will now be sized to balance the system within reason.

Section 4 to 7

$$(l_{f47})_t = l_{f4567} = 4.4 + 10 + 5 = 19.4 \text{ ft}$$
$$L_{47} = 6 + 39 = 45 \text{ ft}$$
$$l_{f47} = 19.4 - 15 = 4.4 \text{ ft}$$
$$(l'_f)_{47} = \frac{4.4(100)}{45} = 9.8 \text{ ft}/100 \text{ ft}$$

From Fig. 9-21

$$D_{47} = 1\tfrac{1}{4} \text{ in.}; \qquad \overline{V} = 4\tfrac{1}{3} \text{ ft/sec}$$

Section 3 to 8

$$(l_{f38})_t = l_{f3478} = (l_{f47})_t + l_{f34} + l_{f78} = 19.4 + 2.0 = 21.4 \text{ ft}$$
$$L_{38} = 6 + 34 = 40 \text{ ft}$$
$$l_{f38} = 21.4 - 20 = 1.4 \text{ foot}$$
$$(l'_f)_{38} = \frac{1.4(100)}{40} = 3.5 \text{ ft}/100 \text{ ft}$$

From Fig. 9-21,

$$D_{38} = 2 \text{ in.}; \qquad \overline{V} = 3 \text{ ft/sec}$$

Only minor adjustments should be required when the system is put into operation.

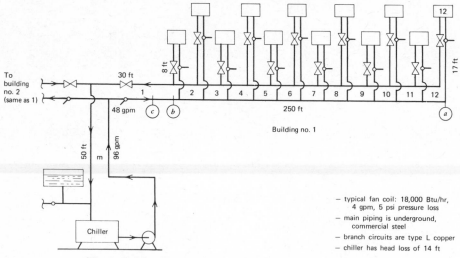

Figure 9-25.   Two-pipe system for apartment complex.

Example 9-10.   A typical piping system for two apartment buildings served by one central chiller is shown in Fig. 9-25. Pertinent data are given on the diagram. Size the piping and determine the pump requirements.

Solution.   Since the main parts of the system are underground, water velocities of 5 to 7 ft/sec can be tolerated without a noise problem because of the absorbing effect of the earth. Lower velocities must be used for the branch circuits to the fan coil units and near the ends of the main runs where the pipe will be small (4 ft/sec or less). Figures 9-21 and 9-22 are used to size the various pipe sections using the velocity criteria discussed above. The equivalent length of each section will then be estimated and the lost head computed. Table 9-2 summarizes this procedure.

   All of the branch circuits to the remaining fan coils should be $\frac{3}{4}$ in. copper. A smaller size will produce too high a velocity and noise. The flow rates must be balanced using the ball valves.

   To compute the pump requirement for complex circuits, it is helpful to use the system characteristics of each part to arrive at the total system characteristic. In this example the circuits to each building are parallel and in series with the main supply circuit coming from the chiller. The characteristic for the main supply circuits is obtained as follows and is shown on Fig. 9-26 as $(m + c)$.

$$l_{fmc} = l_{fm} + l_{fc} = 10 + 14 = 24 \text{ ft}$$

and the volume flow rate is 96 gpm. The total head for the circuit supplying building 1 is the head loss for the longest run of pipe or the sum of sections 1 through 12 plus fan coil 12 from Table 9-2.

$$l_{f1} = 32.3 \text{ ft}$$

Table 9-2   SUMMARY OF PIPE SIZES AND LOST HEAD—EXAMPLE
9-10

Section	Pipe Size inches	Velocity ft/sec	l_f' ft/100 ft	L_e feet	l_f feet	$\dot{Q}$ gpm
m	$2\frac{1}{2}$	$6\frac{3}{4}$	6.7	150	10.0	96
1	2	$4\frac{1}{2}$	4.8	75	4.0	48
2	2	$4\frac{1}{4}$	3.8	45	1.7	44
3	2	<4	3.3	45	1.5	40
4	2	<4	2.6	45	1.2	36
5	2	<4	2.0	45	0.9	32
6	2	<4	1.8	45	0.8	28
7	2	<4	1.3	45	0.6	24
8	$1\frac{1}{2}$	$3\frac{1}{2}$	3.1	45	1.4	20
9	$1\frac{1}{4}$	<3	3.8	45	1.7	16
10	$1\frac{1}{4}$	<3	2.4	45	1.0	12
11	1	$2\frac{1}{2}$	4.8	45	2.2	8
12	$\frac{3}{4}a$	$2\frac{3}{4}$	5.3	72	3.8	4
Chiller	—	—	—	—	14.0	96
Fan coil 12	—	—	—	—	11.5	4

[a]Copper tubing.

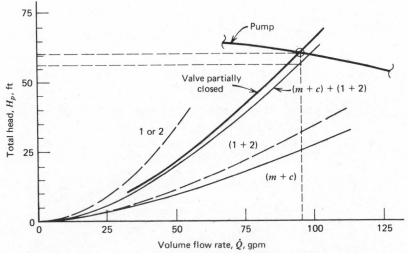

Figure 9-26. System and pump characteristics for Example 9-10.

and

$$Q_1 = 48 \text{ gpm}$$

This characteristic is sketched on Fig. 9-26. The characteristic for building 2 is identical to building 1. The characteristics for buildings 1 and 2 are summed as shown $(1 + 2)$ because those parts of the system are in parallel. Now the parts of the system represented by characteristics $(1 + 2)$ and $(m + c)$ are in series and are summed as shown in Fig. 9-26. The pump must produce at least 56 ft of head with a flow rate of 96 gpm. From the pump data given in Fig. 9-14 we see that the 1750 rpm pump with an 8 in. impeller will meet the system require-ments. A portion of the pump characteristic is plotted on Fig. 9-26. Because the pump is too large, a valve must be adjusted (closed slightly) to obtain the proper flow rate. From Fig. 9-14 the required power for the pump is about 2.2 horse-power and a $2\frac{1}{2}$ horsepower electric motor should be specified.

The system as shown in Fig. 9-25 is difficult to balance. This problem may be alleviated by connecting a separate supply line from points a to c in Fig. 9-25. The existing supply would be capped off at point b. With this arrangement all the parallel runs through the fan coils have about the same equivalent length.

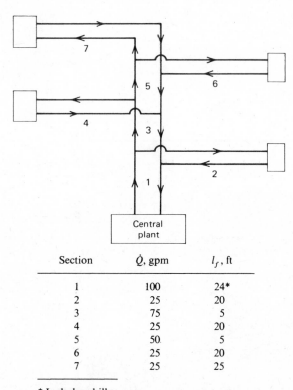

Section	$\dot{Q}$, gpm	l_f, ft
1	100	24*
2	25	20
3	75	5
4	25	20
5	50	5
6	25	20
7	25	25

* Includes chiller

Figure 9-27.   A complex branch circuit system.

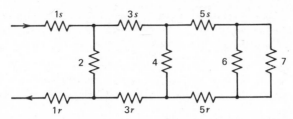

Figure 9-28. Equivalent electrical circuit for piping system of Example 9-11.

Note, however, that supply sections 2 through 12 would have to be resized. This arrangement is called a *reverse return system*.

In Example 9-10 the total system characteristic was obtained in a rather intuitive way. When more complicated systems are encountered an electrical circuit analogy is helpful.

Example 9-11. The system, shown in Fig. 9-27, has four branch circuits. Each of the branches serves a building or zone. Find the pump requirement.

Solution. An electrical circuit equivalent to the piping system is shown in Fig. 9-28. The objective is to reduce the circuit to one single resistor that will correspond to the total system characteristic. The electrical circuit is used as a guide and the resistances or characteristics are combined. This can be done quite easily if the resistors have a constant value; however, flow resistance depends on the square of the fluid velocity. For this reason the characteristics must be combined graphically. Note that resistors 6 and 7 are parallel and are combined as shown in Fig. 9-29 to obtain (6 + 7). The combination of (6 + 7)

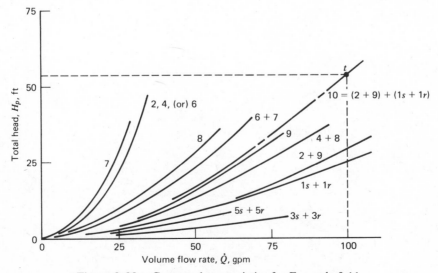

Figure 9-29. System characteristics for Example 9-11.

is then in series with $5s$ and $5r$ or $(5s + 5r)$. These characteristics are then summed as shown to obtain $(6 + 7) + (5s + 5r) = 8$. Resistor 8 is then parallel to resistor 4 and these are combined to obtain $(4 + 8)$. Characteristic $(4 + 8)$ is then in series with $(3s + 3r)$, $(4 + 8) + (3s + 3r) = 9$, and the summation is continued until only one resistor or characteristic 10 remains. The final result is shown in Fig. 9-29. Point t represents the minimum capacity and head requirements of the pump, 100 gpm at 53 ft of head.

9-5 VARIABLE FLOW SYSTEMS

Each of the water piping systems discussed in section 9-4 had a constant volume flow rate, which is common for smaller systems where pumping power is not very large. With systems that serve a very large building or a group of buildings, it is desirable to reduce the water flow rate when the load decreases. This is easily accomplished simply by throttling the water at each terminal unit; however, the effect on the remainder of the system must be considered. For example, a water chiller requires a constant flow rate to operate efficiently. If a single chiller is to be used, a bypass arrangement must be installed to maintain a constant flow rate through the chiller. This is not the best solution because some of the benefit of variable flow is lost.

An interesting scheme to accomplish variable flow is shown in Fig. 9-30. Such an arrangement would be used only with large sized chillers, approximately 100 tons

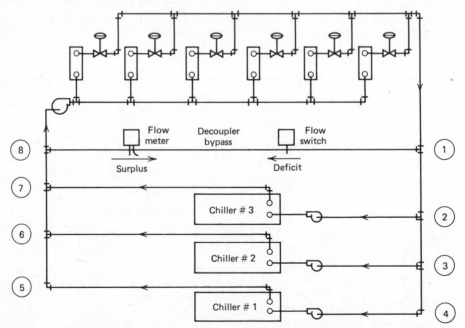

Figure 9-30. Schematic of a variable flow chilled water system.

capacity or above. This scheme decouples the chilled water production from the distribution. Each chiller operates independently with a constant flow rate while the distribution system has a variable flow rate that depends on the load.

The bypass line is the decoupler which automatically accepts any surplus water from the chillers or the distribution system. The flow meter senses surplus flow through the bypass from the chillers and cycles them off one by one as load decreases. The flow switch detects surplus flow from the distribution system and cycles on the chillers as load increases. Temperature control of each chiller is independent and a chiller can operate only when its pump is operating.

This type of system is very adaptable to expansion because it does not have to be disturbed during construction. The chiller capacity tends to be matched to the load so that the warmest water possible enters the chillers for best operating efficiency.

The single pump shown in the distribution system (Fig. 9-30) will usually be a combination of pumps that cycle off and on in response to the load.

The design and sizing of the piping and pumps for the variable flow system follow the same general procedures given for constant flow systems in Section 9-4. Each part of the variable flow system is designed for full load. Partial load operation is then controlled as described above. Note that the system of Fig. 9-30 balances itself and forces the water to flow where needed.

REFERENCES

1. L. F. Moody, "Friction Factors for Pipe Flow," *Trans. ASME, 66,* 1944.

2. *ASHRAE Handbook of Fundamentals,* American Society of Heating, Refrigerating and Air-Conditioning Engineers, New York, 1977.

3. *Fluid Meters, Their Theory and Application,* American Society of Mechanical Engineers, New York, 1959.

4. *ASHRAE Guide and Data Book—Equipment,* American Society of Heating, Refrigerating and Air-Conditioning Engineers, New York, 1979.

5. *ASHRAE Handbook and Product Directory—Systems,* American Society of Heating, Refrigerating and Air-Conditioning Engineers, New York, 1980.

6. H. Schlichting, *Boundary Layer Theory,* 4th Ed., McGraw-Hill, New York, 1960.

7. "Flow of Fluids Through Valves, Fittings, and Pipes," Technical Paper No. 410, The Crane Co., Chicago, Ill., 1976.

PROBLEMS

9-1 Consider the system shown in Fig. 9-31. Compute (a) the work done on the water in (ft-lbf)/lbm; ft of head, and lbf/in.^{2}, and (b) the power delivered to the water. (c) Sketch the system characteristic.

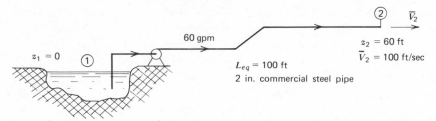

Figure 9-31.   Sketch for Problem 9-1.

9-2 The system shown in Fig. 9-32 transfers water to the tank at a rate of 0.015 m³/s through standard commercial steel 4 in. pipe (ID = 103 mm). The total equivalent length of the pipe is 100 m. The increase in elevation is 40 m. Compute (a) the work done on the water in kJ/kg, (b) the power delivered to the water in kW, and (c) sketch the system characteristic.

9-3 Consider the piping system shown in Fig. 9-33. Sketch the characteristics for each separate part of the system and combine them to obtain the characteristic for the complete system.

9-4 Compute the lost head for 100 gpm of 20 percent ethlylene glycol solution flowing through 300 ft of $2\frac{1}{2}$ in. commercial steel pipe. The temperature of the solution is 60 F.

9-5 Compute the lost head for 0.006 m³/s of 10 C water flowing in $2\frac{1}{2}$ in. (62 mm) standard commercial steel pipe. The total equivalent length is 100 m.

9-6 A pitot tube is being used to measure the flow rate of air in a 6 in. diameter duct. The pressure indicated by the inclined gage (velocity head) is 0.25 in. of water. The static pressure is essentially standard atmospheric and the air temperature is 120 F. Assume the average velocity is 80 percent of the centerline velocity. Compute the volume flow rate in cubic feet per minute.

9-7 Assume the measurement made in Problem 9-6 is on the centerline and the velocity profile can be described by Eq. (9-14) with n = 7.5. Find (a) the average velocity in the duct, (b) the volume flow rate, (c) the mass flow rate, and (d) the ratio of the average to the maximum velocity.

9-8 Saturated water vapor at 101.35 kPa flows in a standard 8 in. pipe (203 mm). A pitot tube located at the center of the pipe shows a velocity head of 10 mm of mercury.

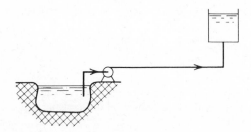

Figure 9-32.   Sketch for Problem 9-2.

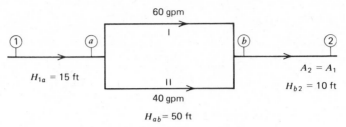

60 gpm

$H_{1a} = 15$ ft

$A_2 = A_1$

$H_{b2} = 10$ ft

40 gpm

$H_{ab} = 50$ ft

Figure 9-33. Schematic for Problem 9-3.

Find (a) the velocity of the water vapor at this location, (b) the mass flow rate assuming the average velocity is 82 percent of the maximum velocity.

9-9 Integrate the velocity profile of Problem 9-8, Eq. (9-14) assuming $n = 8$ and find the average velocity of the steam. Compute the Reynolds number. Does this agree with the assumed value of n?

9-10 Design an orifice to be used in 2 in. standard commercial steel pipe to measure 60 gpm of 30 percent ethylene glycol solution at 50 F. A pressure drop of 5 to 10 in. of mercury is required.

9-11 A square edged orifice is installed in standard 4 in. water pipe (103 mm I.D.). The orifice diameter is 50 mm and a head differential across the orifice of 98 mm of mercury is observed. Compute the volume flow rate of the water assuming a temperature of 10 C. What is the Reynolds number based on the orifice diameter? Does the Reynolds number agree with the flow coefficient?

9-12 The piping system of Problem 9-4 has an increase in elevation of 18 ft from inlet to outlet. (a) Select a pump using Fig. 9-14 and sketch both the system and pump characteristics. (b) How much power is delivered to the fluid in horsepower? (c) How much shaft power is required? (d) If the electric motor has an efficiency of 90 percent, what is the size of the motor required?

9-13 Two 5 in. 1750 rpm pumps as shown in Fig. 9-14 are used in parallel to deliver 140 gpm at 20 ft of head. (a) Sketch the system and pump characteristics. (b) What is the shaft power requirement of each pump? (c) If one pump fails, what is the flow rate and shaft power requirement of the pump still in operation? (d) Could this type of failure cause a problem in general?

9-14 An 8 in. 3500 rpm pump, shown in Fig. 9-14, is to be used to transfer lake water to a water treatment plant. The flow rate is to be 340 gpm. What is the maximum height that the pump can be located above the lake surface to prevent cavitation? Assume the water has a maximum temperature of 80 F, the lost head in the suction line is 2 ft of water, and barometric pressure is 28 in. of mercury.

9-15 The apartment complex of Example 9-10 is to be enlarged by adding two more buildings beyond buildings 1 and 2. The main supply piping is to be extended 100 ft and then will turn to the right and left into buildings 3 and 4. These buildings will have just one air handling unit, which requires 20 gpm. The total equivalent length of pipe to each building is 200 ft and the head loss through the cooling coil is 10 ft. The chiller, pump, and existing main run of pipe will be replaced. Assume the new chiller

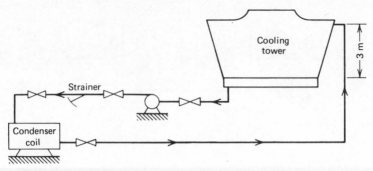

Figure 9-34. Sketch for Problem 9-16.

will have the same head loss as the existing unit. Size the new piping required and select a pump to replace the existing unit. Sketch the system and pump characteristics.

9-16 Size the piping for the cooling tower circuit shown in Fig. 9-34. The water flow rate is 0.03 m³/s and the total equivalent length of the pipe and fittings is 200 m. Pressure loss for the condenser coil is 35 kPa and the strainer has a C_{vs} of 1.14×10^{-2} m³/s per kPa pressure loss. What is the head requirement for the pump in m and kPa?

9-17 Size the piping for the layout shown in Fig. 9-35 and specify the pump requirements. Assume all the turns and fittings are as shown on the diagram. The pipe is commercial

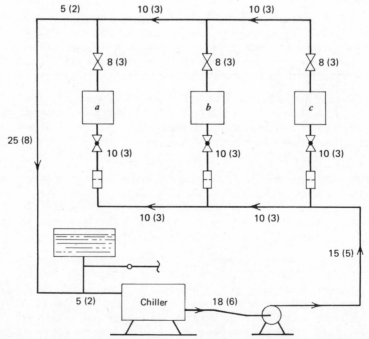

Figure 9-35. Sketch for Problem 9-17—lengths are in feet and meters in parentheses.

Table 9-3   DATA FOR PROBLEM
9-17

Unit	gpm	Head Loss in Feet	
		Coil	Orifice
a	30	15	6
b	40	12	6
c	50	10	6
Chiller	120	20	—

steel and the whole system is located in a basement equipment room. Table 9-3 gives
the required data.

9-18   Size the piping and specify pump requirements for a cooling tower installation similar
to that shown in Fig. 9-19. The volume flow rate of the water is 300 gpm. The piping
is commercial steel. Assume fittings are as shown. The head loss in the condenser is
20 ft of water. C_v for the strainer is 200. The horizontal distance from the condenser
to the cooling tower is 80 ft. The vertical distance from the pump to the top of the
tower is 30 ft. The tower sump is 12 ft above the pump.

9-19   Size the piping for the layout shown in Fig. 9-35. Assume that fittings are as shown
and the pipe is standard commercial steel. Table 9-4 gives pertinent data. Specify
pump requirements.

9-20   Refer to Fig. 9-30. (a) Sketch the system and pump characteristic for the water distribution part of the system. (b) Under partial load, the control valves throttle the
flow to match the load. Sketch the system characteristics for half load. (c) Suppose
there are two identical pumps in parallel instead of one. The pumps in parallel are
equivalent to the one above and at half load one pump cycles off. Sketch the system
and pump characteristics.

9-21   Suppose the water distribution part of the variable flow system in Fig. 9-30 uses three
3500 rpm pumps with 6 in. impellers as shown in Fig. 9-14, connected in parallel.

Table 9-4   DATA FOR PROBLEM
9-19

Unit	m³/s	Head Loss in Meters	
		Coil	Orifice
a	0.002	5	2
b	0.003	4	2
c	0.0033	4	2
Chiller	0.0083	10	—

Each pump delivers 240 gpm when the system is at full load with full flow. (a) Sketch the pump and system characteristics to scale. The use of log-log graph paper is suggested. (b) Determine the total shaft horsepower and the efficiency of each pump. (c) The total flow rate is reduced by one third and one pump cyles off. Show the condition on the graph of part (a) above and determine the total shaft horsepower and efficiency of each pump. (d) The total flow rate is reduced to 400 gpm. How many pumps should cycle off? Determine the total shaft horsepower and efficiency of each pump.

9-22 Consider the chilled water producing part of the system in Fig. 9-30. The pumps and chillers are identical. Each chiller requires a flow rate of 240 gpm. (a) Size the piping for this part of the system using commercial steel (schedule 40) pipe and full flow conditions. (b) Compute the head loss for each section of pipe assuming the distance from point 1 to point 8 is 50 ft and the distance from point 1 to point 4 is about 30 ft. Assume that the fittings are as shown. (c) How much head must each of the chiller

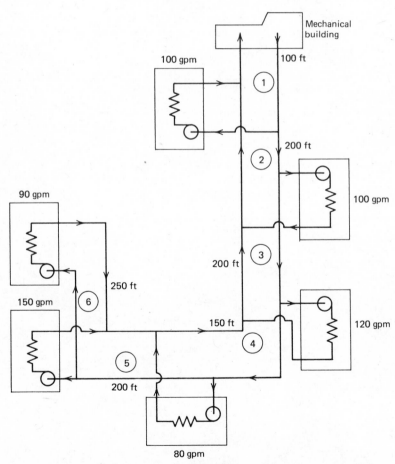

Figure 9-36.   Schematic of a central chilled water system.

pumps produce under full flow conditions if each chiller has a head loss of 40 ft of water? Assume the pressure at points 1 and 2 is zero.

9-23 Consider the water distribution part of the system in Fig. 9-30. (a) Size the various sections of pipe in this part of the system if the total flow rate is 720 gpm and each terminal unit is the same size. (b) For what maximum flow rate should the bypass be sized if each chiller requires 240 gpm? Size the bypass line.

9-24 The diagram in Fig. 9-36 shows a large chilled water distribution system. The water pump in each building is designed to match the head loss of the branch circuits into and out of that building. The water flow rates shown are the full load design values. (a) Size the main supply and return lines (sections 1 through 5). (b) Two pumps operating in parallel will circulate water in the variable flow circuit. What total head and flow rate should they develop as a unit? (c) The pumps are identical. Sketch approximate characteristics and estimate how much flow one pump will provide in the circuit (use Fig. 9-14 as a guide in sketching pump characteristics).

9-25 Devise a variable flow water system that utilizes one chiller and five air handlers. The air handlers have throttling type water control (Fig. 9-30). To maintain constant flow through the chiller use a differential pressure sensor that controls a three-way valve. Sketch the system and describe operation under the following conditions. (a) full load, (b) half load, (c) no load. (d) What could be done with respect to the pump in part (c)?

CHAPTER TEN

Room Air Distribution

The object of air distribution in warm air heating, ventilating, and air conditioning systems is to create the proper combination of temperature, humidity, and air motion in the occupied portion of the conditioned room. To obtain comfort conditions within this space, standard limits for an acceptable effective draft temperature have been established. This term comprises air temperature, air motion, relative humidity, and their physiological effect on the human body. Any variation from accepted standards of one of these elements may result in discomfort to the occupants. Discomfort also may be caused by lack of uniform conditions within the space or by excessive fluctuation of conditions in the same part of the space. Such discomfort may arise due to excessive room air temperature variations (horizontally, vertically, or both), excessive air motion (draft), failure to deliver or distribute the air according to the load requirements at the different locations, or rapid fluctuation of room temperature or air motion (gusts).

10-1   AIR MOTION AND COMFORT

A measure of the effective temperature difference between any point in the occupied space and the control conditions is called the effective draft temperature. It is defined by the equation proposed by Rydberg and Norback (1).

$$EDT = (t_x - t_r) - M(\overline{V}_x - \overline{V}_r) \qquad (10\text{-}1)$$

where

t_r = 76 F or 24 C
$\overline{V}_r$ = 30 ft/min or 0.15 m/s
t_x = local air stream dry bulb temperature, F or C
$\overline{V}_x$ = local air stream velocity, ft/min or m/s
M = 0.077 (F-min)/ft or 7.66 (C-s)/m

334

Equation (10-1) takes into account the feeling of coolness produced by air motion. It also shows that the effect of a one degree F temperature change is equivalent to a 15 ft/min velocity change. In summer the local air stream temperature t_x is usually below the control temperature. Hence both temperature and velocity terms are negative when the velocity $\overline{V}_x$ is greater than $\overline{V}_r$ and both of them add to the feeling of coolness. If in winter $\overline{V}_x$ is above $\overline{V}_n$ it will reduce the feeling of warmth produced by t_x. Therefore, it is usually possible to have zero difference in effective temperature between location x and the control point in winter but not in summer. Research indicates that a high percentage of people in sedentary occupations are comfortable where the effective draft temperature is between -3 F $(-1.7$ C) and $+2$ F (1.1 C) and the air velocity is less than 70 ft per minute (0.36 m/s). These conditions are used as criteria for developing the Air Distribution Performance Index (ADPI) to be discussed later.

Conditioned air is normally supplied to air outlets at velocities much higher than would be acceptable in the occupied space. The conditioned air temperature may be above, below, or equal to the temperature of the air in the occupied space. Proper air distribution therefore causes entrainment of room air by the primary air stream and reduces the temperature differences to acceptable limits before the air enters the occupied space. It also counteracts the natural convection and radiation effects within the room.

10-2 BEHAVIOR OF JETS FROM OUTLETS

The air projection from free round openings, grilles, perforated panels, ceiling diffusers, and other outlets is related to the average velocity at the face of the air supply opening. A free jet has four zones of expansion, and the centerline velocity in any of the zones is related to the initial velocity as shown in Fig. 10-1. Regardless of the type of opening the jet will tend to assume a circular shape. The effect of nearby surfaces will be considered later. In zone III, the most important zone from the point of view of room air distribution, the relation between the jet centerline velocity and the initial velocity is given by

$$\frac{\overline{V}_x}{\overline{V}_0} = K \frac{\sqrt{A_0}}{x} \tag{10-2}$$

or

$$\overline{V}_x = \frac{K\dot{Q}_0}{(\sqrt{A_0}x)} \tag{10-2a}$$

where

$\overline{V}_x$ = centerline velocity at any x, ft/min or m/s
$\overline{V}_0$ = initial velocity, ft/min or m/s
A_0 = area corresponding to initial velocity, ft^2 or m^2
x = distance from outlet to point of measurement of $\overline{V}_x$, ft or m
$\dot{Q}_0$ = air flow rate at outlet, cfm or m^3/s
K = constant of proportionality, dimensionless

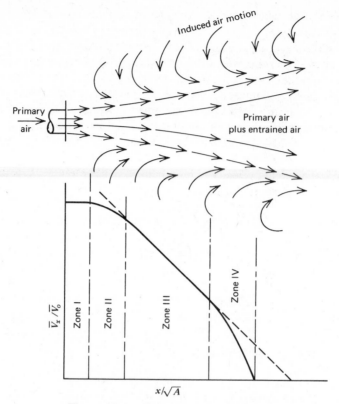

Figure 10-1.   Isothermal air jet behavior.

Equations (10-2) and (10-2a) strictly pertain to isothermal free jets, but with the proper A and K the equations define the throw for any type of outlet. The throw is the distance from the outlet to where the maximum velocity in the jet has decreased to some specified value such as 50, 100, or 150 ft/min. The constant K varies from about 6 for free jets to about 1 for ceiling diffusers.

The jet expands because of entrainment of room air; the air beyond zone II is a mixture of primary and induced air. The ratio of the total volume of the jet to the initial volume of the jet at a given distance from the origin depends mainly on the ratio of initial velocity $\overline{V}_0$ to the terminal velocity $\overline{V}_x$. The *induction ratio* is

$$\frac{\dot{Q}_x}{\dot{Q}_0} = C\frac{\overline{V}_0}{\overline{V}_x} \tag{10-3}$$

where

$\dot{Q}_x$ = total air mixture at distance x from the outlet, cfm or m³/s
C = entrainment coefficient, 2 for round, free jet, dimensionless

In zone IV where the terminal velocity is low, Eq. (10-3) will give values about 20 percent high.

When a jet is projected parallel to and within a few inches of a surface, the induction or entrainment is limited on the surface side of the jet. A low pressure region is created between the surface and the jet, and the jet attaches itself to the surface. This phenomenon results if the angle of discharge between the jet and the surface is less than about 40 degrees and if the jet is within about one foot of the surface. The jet from a floor outlet is drawn to the wall and the jet from a ceiling outlet is drawn to the ceiling. This surface effect increases the throw for all types of outlets and decreases the drop for horizontal jets. The drop is illustrated in Fig. 10-2.

In nonisothermal jets, buoyant forces cause the jet to rise when the air is warm and drop when cool. These conditions result in shorter throws for jet velocities less than 150 ft/min or 0.76 m/s. The bottom portion of Fig. 10-2 shows the drop for a cool jet of air.

The following general statements may be made concerning the characteristics of air jets.

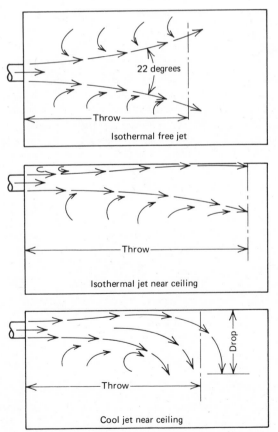

Figure 10-2.   Schematic showing entrainment, surface effect, and drop.

1. Surface effect increases the throw and decreases the drop compared to free space conditions.
2. Increased surface effect may be obtained by moving the outlet away from the surface somewhat so that the jet spreads over the surface after impact.
3. Increased surface effect may be obtained by spreading the jet when it is discharged.
4. Spreading the air stream reduces the throw and drop.
5. Drop primarily depends on the quantity of air and only partially on the outlet size or velocity. Thus the use of more outlets with less air per outlet reduces drop.

Room Air Motion

Room air near the jet is entrained and must then be replaced by other room air. The room air always moves toward the supply and thus sets all the room air into motion. Whenever the average room air velocity is less than about 50 ft/min or 0.25 m/s, buoyancy effects may be significant. In general, about 8 to 10 air changes per hour are required to prevent stagnant regions (velocity less than 15 ft/min or 0.08 m/s). However, stagnant regions are not necessarily a serious condition. The general approach is to supply air in such a way that the high velocity air from the outlet does not enter the occupied space. The region within 1 ft of the wall and above about 6 ft from the floor is out of the occupied space for practical purposes.

Figure 10-3 shows velocity envelopes for a high sidewall outlet. Equation (10-2) has been used to estimate the throw for the terminal velocities shown. In order to interpret the air motion shown in terms of comfort it is necessary to estimate the local air temperatures corresponding to the terminal velocities. The relationship between the centerline velocities and the temperature differences is given approximately (3) by

$$\Delta t_x = 0.8 \, \Delta t_0 \, \frac{\overline{V}_x}{\overline{V}_0} \tag{10-4}$$

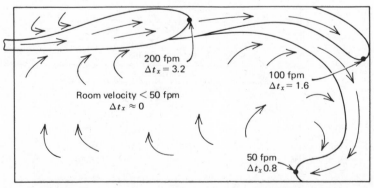

Figure 10-3. Jet and room air velocities and temperatures for $\overline{V}_0 = 1000$ ft/min and $\Delta t_0 = -20$ F.

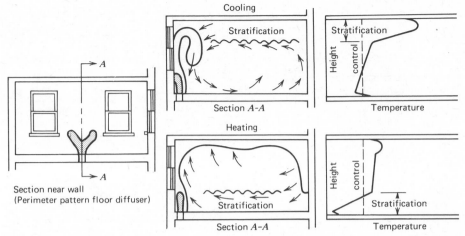

Figure 10-4. Room air distribution patterns for perimeter floor diffuser—spreading jet. (Courtesy of National Environmental Systems Contractors Association, 1501 Wilson Blvd., Arlington, Va. 22209.)

where Δt_x and Δt_0 are the differences in temperature between the local stream temperature and the room $(t_x - t_r)$ and the outlet air and the room $(t_0 - t_r)$. Equation (10-4) is similar to Eq. (10-1) but slightly more general. Temperatures calculated using Eq. (10-4) are shown in Fig. 10-3. On the opposite wall where the terminal velocity is 100 ft/min the air temperature is 1.6 F below the room temperature. The temperature difference for the 50 ft/min envelope shows that within nearly the entire occupied space the temperature is less than about 0.8 F below the room temperature and the room air motion is under 50 ft/min.

Basic Flow Patterns

The basic flow patterns for the most often used types of outlets are shown in Figs. 10-4 to 10-7 (4). The high velocity primary air is shown by the dark shading, and the total air is represented by the light shading. These areas represent the high momentum regions of the room air motion. Natural convection (buoyancy) effects are evident in all cases. Note that stagnant zones always have a large temperature gradient. When this occurs in the occupied space, air needs to be projected into the stagnant region to enhance mixing. An ideal condition would be uniform room temperature from the floor to about 6 ft above the floor. However, a gradient of about 4 F or 2 C should be acceptable to about 85 percent of the occupants.

The perimeter-type outlets shown in Fig. 10-4 are generally regarded as superior for heating applications. This is particularly true when the floor is over an unheated space or a slab and where considerable glass area exists in the wall. Diffusers with a wide spread are usually best for heating because buoyancy tends to increase the throw. For the same reason the spreading jet is not as good for cooling applications because the throw may not be adequate to mix the room air thoroughly.

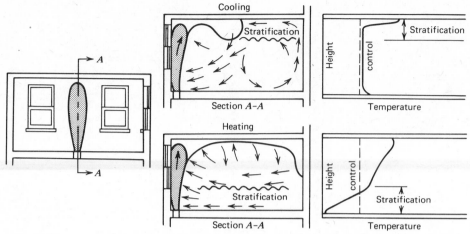

Figure 10-5.   Room air distribution patterns for floor register—nonspreading jet. (Courtesy of National Environmental Systems Contractors Association, 1501 Wilson Blvd., Arlington, Va. 22209.)

However, the perimeter outlet with a nonspreading jet is quite satisfactory for cooling. Figure 10-5 shows a typical cooling application of the nonspreading perimeter diffuser. It can be seen that the nonspreading jet is less desirable for heating because a larger stratified zone will usually result. Perimeter diffusers are usually selected to have a throw based on a 50 ft/min terminal velocity equal to the ceiling height plus about one third the room width. Diffusers are available which may be changed from the spreading to nonspreading type according to the season.

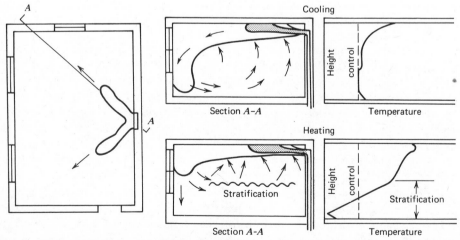

Figure 10-6.   Room air distribution patterns for high sidewall register. (Courtesy of National Environmental Systems Contractors Association, 1501 Wilson Blvd., Arlington, Va. 22209.)

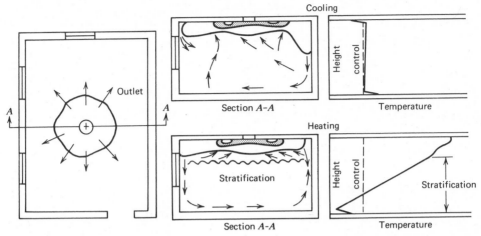

Figure 10-7. Room air distribution patterns for ceiling diffuser. (Courtesy of National Environmental Systems Contractors Association, 1501 Wilson Blvd., Arlington, Va. 22209.)

The high sidewall type of register shown in Fig. 10-6 is often used in mild climates and on the second and succeeding floors of multistory buildings. This type of outlet is not recommended for cold climates or with unheated floors. Figure 10-6 shows that a considerable temperature gradient may exist between floor and ceiling when heating; however, this type outlet gives good air motion and uniform temperatures in the occupied zone for cooling application. These registers are generally selected to project air from about three-fourths to full room width.

The ceiling diffuser shown in Fig. 10-7 is very popular in commercial applications and many variations of it are available. The air patterns shown in Fig. 10-7 are typical. Because the primary air is projected radially in all directions, the rate of entrainment is large, causing the high momentum jet to diffuse quickly. This feature enables the ceiling diffuser to handle larger quantities of air at higher velocities than most other types. Figure 10-7 shows that the ceiling diffuser is quite effective for cooling applications but generally poor for heating. However, satisfactory results may be obtained in commercial structures when the floor is heated.

The return air intake generally has very little effect on the room air motion. But the location may have a considerable effect on the performance of the heating and cooling equipment. Because it is desirable to return the coolest air to the furnace and the warmest air to the cooling coil, the return air intake should be located in a stagnant region.

Noise

Noise produced by the air diffuser and air can be annoying to the occupants of the conditioned space. Noise associated with air motion usually does not have distinguishable frequency characteristics and the level or loudness is basically a statistically representative sample of human reactions. Loudness contours or curves of equal loudness versus frequency can be established from such reactions.

A method of providing information on the spectrum content of noise is the use of the noise criteria (NC) curves and numbers. The NC curves are shown in Fig. 10-8. These are a series of curves constructed using loudness contours, and the speech-interfering properties of noise and are used as a simple means of specifying sound level limits for an environment by a simple, single number rating. They have been

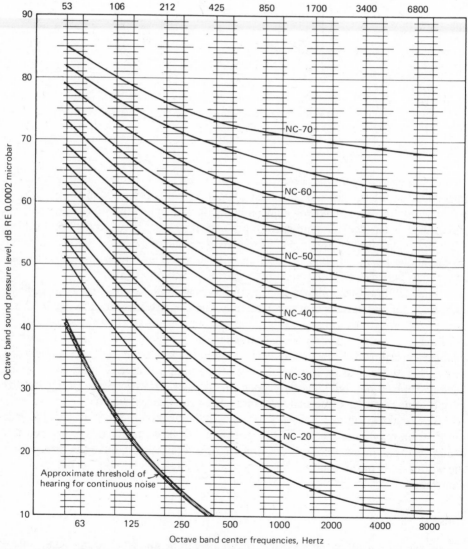

Figure 10-8.   Noise criteria (NC) curves. (Reprinted by permission from *ASHRAE Handbook of Fundamentals*, 1977.)

Table 10-1   RECOMMENDED NOISE CRITERIA FOR ROOMS*

Application	
Broadcasting studios	**
Concert halls	**
Legitimate theaters	**
Schoolrooms	NC 25–35
Apartments and hotels	NC 30–35
Assembly halls (amplification)	NC 30–35
Homes (sleeping area)	NC 25–30
Conference rooms	NC 25–30
Motion picture theaters	NC 30–35
Hospitals	NC 25–30
Churches	NC 20–30**
Courtrooms	NC 30–35
Libraries	NC 30–35
Small private offices	NC 30–35
Restaurants	NC 35–40
Coliseums for sports only	NC 30–40
Stenographic offices	NC 35–45

*Reprinted by permission from *ASHRAE Handbook and Product Directory—Systems*, 1980.
**An acoustics expert should be consulted.

found to be quite generally applicable for conditions of comfort. In general, levels below a NC of 30 are considered to be quiet, whereas levels above a NC of 50 or 55 are considered noisy. Of course the activity within the space is a major consideration in determining an acceptable level. Table 10-1 gives recommended noise criteria numbers for various applications (5). To determine the acceptability of a given space for a given specification, sound pressure level must be measured at several octave band center frequencies and compared with the specified NC curve of Fig. 10-8. To meet the particular NC rating, the actual octave band reading should lie on or below the NC curve.

Some manufacturers of air-diffusing equipment rate their products using the noise criteria concept. An example of these data is presented in the next section.

10-3   AIR DISTRIBUTION SYSTEM DESIGN

This section discusses the selection and placement of the air outlets. If this is done purely on the basis of comfort, the preceding discussions on room air motion dictate the type of system and the location of the air inlets. However, the architectural design and the functional requirements of the building often override comfort.

When the designer is free to select the type of air distribution system based on comfort, the perimeter type of system with vertical discharge of the supply air is to be preferred for exterior spaces when the heating requirements exceed 2000 degrees (F) days. This type system is excellent for heating and satisfactory for cooling when adequate throw is provided. When the floors are warmed and the degree (F) day requirement is between about 3500 and 2000, the high sidewall outlet with horizontal discharge toward the exterior wall is acceptable for heating and quite effective for cooling. When the heating requirement falls below about 2000 degree (F) days, the overhead ceiling outlet or high sidewall diffuser are recommended because cooling is the predominant mode. Interior spaces in commercial structures are usually provided with overhead systems because cooling is required most of the time.

Commercial structures often are constructed in such a way that ducts cannot be installed to serve the desired air distribution system. Floor space is very valuable and the floor area required for outlets may be covered by shelving or other fixtures, making a perimeter system impractical. In this case an overhead system must be used. In some cases the system may be a mixture of the perimeter and overhead type.

Renovation of commercial structures may represent a large portion of a design engineer's work. Compromises are almost always required in this case, and the air distribution system is often dictated by the nature of the existing structure.

In all cases where an ideal system cannot be used it is particularly important that the air-diffusing equipment be carefully selected and located. Although most manufacturers of air diffusers and grilles furnish extensive data on the performance of their products, there is no substitute for experience and good judgment in designing the air distribution system.

Table 10-2 gives performance data for a type of diffuser that may be used for perimeter systems having a vertical discharge from floor outlets or as a linear diffuser in the ceiling or sidewall. Note that the data pertain to the capacity, throw, total pressure loss, noise criteria, and free area as a function of the size. It is important to read the notes given with these catalog data. Notice that throw values for three different terminal velocities are given. The diffuser may be almost any length, but its capacity is based on a length of 1 ft, whereas the throw is based on a 4 ft active length, and the NC is based on 10 ft of length. Some corrections are also required when the diffuser is used as a return intake.

Performance data for one type of round ceiling diffuser are shown in Table 10-3, and Table 10-4 shows data for an adjustable diffuser that would generally be used for high sidewall applications. The same general data are given. Note that the diffuser of Table 10-4 has adjustable vanes and throw data are given for three different settings, 0, $22\frac{1}{2}$, and 45 degrees. Figure 10-9 shows a T-Bar type diffuser which is used extensively with modular ceilings. These diffusers are often associated with variable air volume systems and sometimes have automatic flow control built into the diffuser itself. The diffuser shown produces horizontal throw parallel to the ceiling in opposing directions. Table 10-5 gives performance data for the T-Bar diffuser.

Return grilles are quite varied in design. The construction of the grille has very little to do with the overall performance of the system except to introduce some loss in pressure and noise if not properly sized. The appearance of a return grille is important, and the louver design is usually selected on this basis. Table 10-6 gives data for one style of grille. Note that the capacity, pressure loss, and noise criteria are the main performance data given.

The Air Distribution Performance Index (ADPI) was mentioned in connection with Eq. (10-1). The ADPI is defined as the percentage of measurements taken at many locations in the occupied zone of a space which meet the -3 F to 2 F effective draft temperature criteria. The objective is to select and place the air diffusers so that an ADPI approaching 100 percent is achieved. Note that ADPI is based only on air velocity and effective draft temperature, a local temperature difference from the room average, and is not directly related to the level of dry bulb temperature or relative humidity. These effects and other factors such as mean radiant temperature must be accounted for as discussed in Chapter Three. The ADPI provides a means of selecting air diffusers in a rational way. There are no specific criteria for selection of a particular type of diffuser except as discussed above, but within a given type the ADPI is the basis for selecting the throw. The space cooling load per unit area is an important consideration. Heavy loading tends to lower the ADPI. However, loading does not influence design of the diffuser system significantly. Each type of diffuser has a characteristic room length as shown in Table 10-7. Table 10-8 is the ADPI selection guide. Table 10-8 gives the recommended ratio of throw to characteristic length which should maximize the ADPI. A range of throw-to-length ratios are also shown which should give a minimum ADPI. Note that the throw is based on a terminal velocity of 50 ft per minute for all diffusers except the ceiling slot type. The general procedure for use of Table 10-8 is as follows:

1. Determine the air flow requirements and the room size.
2. Select the type of diffuser to be used.
3. Determine the room characteristic length.
4. Select the recommended throw-to-length ratio from Table 10-8.
5. Calculate the throw.
6. Select the appropriate diffuser from catalog data such as that in Tables 10-2, 10-3, 10-4, or 10-5.
7. Make sure any other specifications are met (noise, total pressure, etc.).

To illustrate the use of diffuser and grille performance data to design air distribution systems, let us consider a few examples.

Example 10-1.   The room shown in Fig. 10-10 is part of a single story office building located in the central United States. A perimeter type of air distribution system is used since heating will be important. The air quantity required for the room is 250 cfm. Select diffusers for the room.

Solution.   Diffusers of the type shown in Table 10-2 should be used for this application. Because the room has two exposed walls, an air outlet should be placed under each window in the floor near the wall, Fig. 10-10c. This will help

Table 10-2   PERFORMANCE DATA FOR A TYPICAL LINEAR DIFFUSER*

Size/Area	Total Pressure	0.009	0.020	0.036	0.057	0.080	0.109	0.143	0.182	0.225
2 in.	Flow, cfm/ft	22	33	44	55	66	77	88	99	110
	NC	—	—	12	18	23	27	31	34	37
0.055	Throw, ft—sill or floor	1-1-1	4-4-4	7-7-7	9-9-10	11-11-12	13-14-16	14-16-18	15-17-20	17-19-21
3 in.	Flow, cfm/ft	38	58	77	96	115	134	154	173	192
	NC	—	—	11	17	22	26	30	33	36
0.096	Throw, ft—sill or floor	2-2-2	7-7-7	10-10-11	12-13-14	15-16-17	18-19-20	20-21-23	23-24-25	25-25-26
4 in.	Flow, cfm/ft	56	83	111	139	167	195	222	250	278
	NC	—	—	12	18	23	27	31	34	37
0.139	Throw, ft—sill or floor	3-3-3	9-9-9	13-13-13	16-16-17	20-20-21	22-23-24	24-25-26	27-27-27	30-30-30
5 in.	Flow, cfm/ft	72	107	143	179	215	250	286	322	358
	NC	—	—	12	18	23	27	31	34	37
0.179	Throw, ft—sill or floor	4-4-4	10-10-10	14-14-14	18-18-18	22-22-23	24-24-24	27-27-28	30-30-31	32-32-32

6 in.	88	133	177	221	265	310	354	398	442
Flow, cfm/ft									
Nc	—	—	13	19	24	28	32	35	38
0.221 Throw, ft—sill or floor	5-5-5	10-10-10	15-15-15	18-18-18	23-23-23	25-25-25	28-28-28	31-31-31	32-32-32

Area is given in square feet per foot of length.

1. All pressures are in inches of water.
2. Minimum throw values refer to a terminal velocity of 150 ft/min, middle to 100 ft/min and maximum to 50 ft/min, for a 4 ft. active section with a cooling temperature differential of 20 F. The multiplier factors listed in the table are applicable for other lengths.

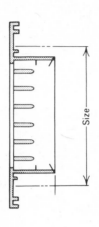

Size

Terminal Velocity

Active Length	150 ft/min	100 ft/min	50 ft/min
1 ft	0.5	0.6	0.7
10 ft or Continuous	1.6	1.4	1.2

3. The NC values are based on a room absorption of 80 Db, re 10^{-12} watts and a 10 ft active section.

NC Correction for Length

Active Length, ft	1	2	4	6	8	10	15	20	25	30
Correction	−10	−7	−4	−2	−1	0	+2	+3	+4	+5

*Reprinted by permission of Environmental Elements Corporation, Dallas, Texas.

Table 10-3 PERFORMANCE DATA FOR A TYPICAL ROUND CEILING DIFFUSER*

Size	Neck velocity, ft/min Velocity pressure	400 0.010	500 0.016	600 0.023	700 0.031	800 0.040	900 0.051	1000 0.063	1200 0.090
6 in.	Total pressure	0.026	0.041	0.059	0.079	0.102	0.130	0.161	0.230
	Flow rate, cfm	80	100	120	140	160	160	200	235
	Radius of diffusion, ft	2-2-4	2-3-5	2-4-6	3-4-7	3-5-8	4-5-9	4-6-10	5-7-11
	NC	—	—	14	19	23	26	30	35
8 in.	Total pressure	0.033	0.052	0.075	0.101	0.130	0.166	0.205	0.292
	Flow rate, cfm	140	175	210	245	280	315	350	420
	Radius of diffusion, ft	2-4-6	3-4-7	4-5-9	4-6-10	5-7-11	5-8-13	6-9-14	7-11-17
	NC	—	15	21	26	31	34	37	44
10 in.	Total pressure	0.027	0.043	0.062	0.084	0.108	0.138	0.170	0.243
	Flow rate, cfm	220	270	330	380	435	490	545	655
	Radius of diffusion, ft	3-4-7	3-5-8	4-6-10	5-7-11	5-8-13	6-9-15	7-10-16	8-12-20
	NC	—	11	17	21	26	30	33	39
12 in.	Total pressure	0.026	0.042	0.060	0.081	0.105	0.134	0.166	0.236
	Flow rate, cfm	315	390	470	550	630	705	785	940
	Radius of diffusion, ft	3-5-8	4-6-10	5-7-12	6-8-13	6-10-15	7-11-17	8-12-19	10-14-23
	NC	—	11	17	22	26	30	33	39

24 in.	Total pressure	0.024	0.038	0.054	0.073	0.094	0.120	0.148	0.211
	Flow rate, cfm	1260	1570	1880	2200	2510	2820	3140	3770
	Radius of diffusion, ft	6-9-15	8-12-19	9-14-22	11-16-26	12-19-30	14-21-34	16-23-37	19-28-45
	NC	—	13	19	24	28	32	35	41
36 in.	Total pressure	0.024	0.038	0.055	0.074	0.096	0.122	0.151	0.216
	Flow rate, cfm	2820	3520	4230	4930	5630	6340	7040	8450
	Radius of diffusion, ft	9-14-22	12-18-28	14-21-34	16-25-39	19-28-45	21-32-50	23-35-56	28-42-67
	NC	—	15	21	26	30	34	38	43

1. All pressures are inches of water.
2. Minimum radii of diffusion are to a terminal velocity of 150 ft/min, middle to 100 ft/min, and maximum to 50 ft/min.
3. The NC values are based on a room absorption of 18 Db, re 10^{-13} watts or 8 Db, re 10^{-12} watts.

Dimensions

		Size		
A	B	C	D	E
6	$6\frac{1}{2}$	$11\frac{1}{8}$	$1\frac{3}{4}$	$1\frac{1}{8}$
8	$8\frac{1}{2}$	$14\frac{3}{4}$	$2\frac{1}{8}$	$1\frac{1}{2}$
10	$10\frac{1}{2}$	$18\frac{1}{4}$	$2\frac{7}{8}$	$2\frac{1}{8}$
12	$12\frac{1}{2}$	22	$3\frac{1}{8}$	$2\frac{3}{8}$
			$7\frac{3}{4}$	
24	$24\frac{1}{2}$	$43\frac{1}{4}$		$6\frac{5}{8}$
36	$36\frac{1}{2}$	$64\frac{1}{2}$	$10\frac{1}{8}$	$8\frac{3}{8}$

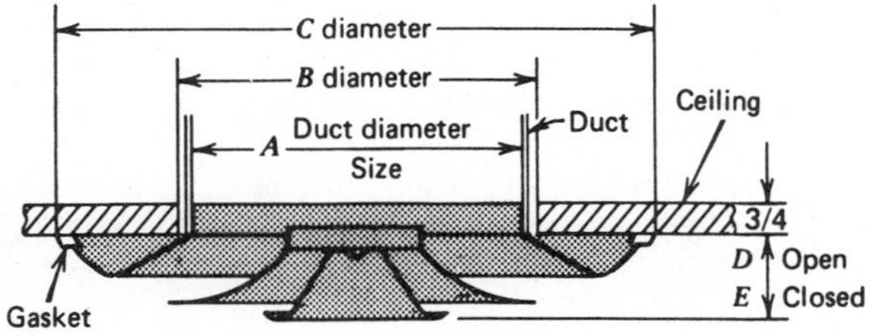

*Reprinted by permission of Environmental Elements Corporation, Dallas, Texas.

Table 10-4 PERFORMANCE DATA FOR AN ADJUSTABLE TYPE, HIGH SIDEWALL DIFFUSER*

	Velocity, ft/min Velocity pressure	300 0.006	400 0.010	500 0.016	600 0.022	700 0.030	800 0.040	1000 0.062	1200 0.090
Total Pressure	0	0.010	0.017	0.028	0.038	0.052	0.069	0.107	0.156
	$22\frac{1}{2}$	0.011	0.019	0.031	0.043	0.058	0.078	0.120	0.175
	45	0.016	0.029	0.047	0.064	0.088	0.117	0.181	0.263
Size inches									
8 × 4 7 × 5	cfm	55	70	90	110	125	145	180	215
	NC				10	15	19	25	31
6 × 6 $A_c = 0.18\ ft^2$	Throw, ft 0	4-7-13	6-8-15	7-11-17	9-13-19	10-15-20	11-16-22	14-17-24	15-19-26
	$22\frac{1}{2}$	3-6-10	5-6-12	6-9-14	7-10-15	8-12-16	9-13-18	11-14-19	12-15-21
	45	2-3-7	3-4-8	4-5-9	4-7-10	5-7-10	6-8-11	7-9-12	8-10-13
10 × 4 8 × 5	cfm	65	90	110	130	155	175	220	265
	NC				10	15	19	25	31
7 × 6 $A_c = 0.22\ ft^2$	Throw, ft 0	4-7-14	7-10-17	8-12-19	9-15-21	11-16-23	13-17-24	16-19-27	17-21-29
	$22\frac{1}{2}$	3-6-11	6-8-14	6-10-15	7-12-17	9-13-18	10-14-19	13-15-22	14-17-23
	45	2-4-7	3-5-9	4-6-10	5-7-10	6-8-11	6-9-12	8-10-13	9-11-15
12 × 4 10 × 5	cfm	80	105	130	155	180	210	260	310
	NC				11	16	20	26	32

8 × 6, $A_c = 0.26$ ft² (16 × 4, 12 × 5)

cfm	100	135	170	205	240	270	340	410
Throw, ft 0	5-8-16	7-11-19	9-13-21	10-16-23	12-17-24	14-19-26	17-21-29	19-23-32
22½	4-6-13	6-9-15	7-10-17	8-13-18	10-14-19	11-15-21	14-17-23	15-18-26
45	3-4-8	4-5-9	4-7-10	5-8-11	6-9-12	7-9-13	8-11-15	9-12-16
NC				12	17	21	27	33

10 × 6, $A_c = 0.34$ ft² (18 × 4, 14 × 5)

cfm	115	155	195	235	275	310	390	470
Throw, ft 0	5-9-18	8-12-21	10-15-24	12-19-26	14-20-28	16-22-30	20-24-33	22-26-37
22½	4-7-14	6-10-17	8-12-19	10-15-21	11-16-22	13-18-24	16-19-26	18-21-30
45	3-4-9	4-6-11	5-8-11	6-9-13	7-10-14	8-11-15	10-12-17	11-13-18
NC				13	18	22	28	34

12 × 6, 8 × 8, $A_c = 0.39$ ft²

Throw, ft 0	6-9-19	9-13-23	11-16-25	13-19-28	15-22-30	17-23-32	21-26-36	23-27-40
22½	5-7-15	7-10-18	9-13-20	10-15-22	12-18-24	14-18-26	17-21-29	18-22-32
45	3-5-10	4-6-11	5-18-13	7-10-14	8-11-15	9-12-16	11-13-18	12-14-20

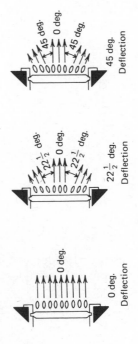

0 deg. Deflection

22½ deg. Deflection

45 deg. Deflection

*Reprinted by permission of Environmental Elements Corporation, Dallas, Texas.

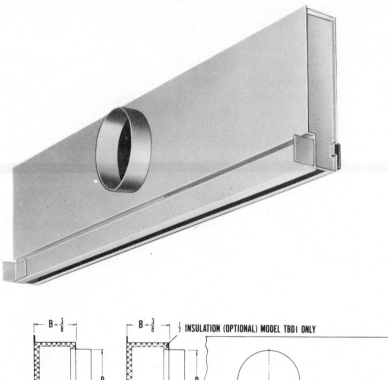

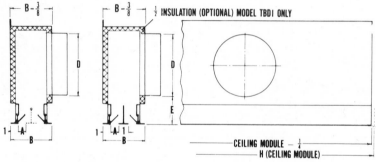

Model	A	H	B	C	D	E
27	½	24	4	12	5	5⅞
		48			7	3⅞
28	¾	24	4½	12	6	4⅞
		48			8	2⅞

Figure 10-9. A typical T-bar type diffuser assembly. (Courtesy of Environmental Corporation, Dallas, Tex.)

Table 10-5   PERFORMANCE DATA FOR THE T-BAR DIFFUSERS OF FIGURE 10-9*

Model 27

H-24	CFM	55	62	68	80	95	110	120	135	150
	Horiz. Proj.	2-3-4	2-3-4	2-3-5	2-4-6	3-5-7	3-6-8	3-6-9	4-7-10	4-8-11
	Total Press	.04	.06	.07	.10	.14	.18	.22	.28	.34
	NC	—	11	14	19	24	28	32	35	38
H-48	CFM	105	120	135	160	185	215	240	270	295
	Horiz. Proj.	2-4-5	2-4-6	3-5-7	3-6-8	4-6-9	4-7-10	5-8-12	5-9-13	6-10-14
	Total Press	.04	.05	.07	.10	.13	.18	.22	.28	.34
	NC		14	17	22	27	31	35	38	41

Model 28

H-24	CFM	80	90	100	120	140	160	180	200	215
	Horiz. Proj.	2-3-5	2-3-5	2-4-6	3-5-7	3-6-8	4-7-9	4-6-10	5-8-12	5-8-12
	Total Press	.05	.06	.08	.11	.15	.20	.25	.31	.36
	CFM	17	21	24	29	34	38	42	45	48
H-48	CFM	140	155	175	210	245	280	315	350	385
	Horiz Proj.	2-4-6	3-4-6	3-5-7	4-6-8	5-7-10	5-8-11	5-9-13	6-10-14	7-12-16
	Total Press	.04	.05	.06	.08	.11	.15	.19	.23	.28
	NC	15	19	22	27	32	36	40	43	46

*Reprinted by permission of Environmental Elements Corporation, Dallas, Texas.
1. All pressures are in inches of water.
2. Minimum projection is to a terminal velocity of 150 fpm, middle to 100 fpm and maximum to 50 fpm.
3. NC values are based on room absorption of 10 dB, re 10^{-12} watts.

to counteract the cold air moving downward from the window as a result of natural convection. The total air quantity is divided equally between the two diffusers. According to Table 10-1 the NC should be about 30 to 40. If we assume that the room has an 8 ft ceiling, the room characteristic length is 8 ft. Table 10-8 gives a throw-to-length ratio ranging from 1.7 to 0.9 for a straight vane diffuser. We will assume a room load of 40 Btu/(hr-ft²), then

$$x_{50}/L = 1.3$$

and

$$x_{50} = 1.3(8) = 10.4 \text{ ft}$$

From Table 10-2 a 4 in. by 12 in. diffuser with 125 cfm has a throw between

$$x_{50} = 13(0.7) = 9.1 \text{ ft}$$

and

$$x_{50} = 17(0.7) = 11.9 \text{ ft}$$

Table 10-6 PERFORMANCE DATA FOR ONE TYPE OF RETURN GRILLE

A_c	Size, inches		Core Velocity fpm	200	300	400	500	600	700	800
			Velocity pressure	0.002	0.006	0.010	0.016	0.023	0.031	0.040
			Negative static pressure	0.011	0.033	0.055	0.088	0.126	0.170	0.220
0.34 ft²	16 × 4 12 × 5		cfm NC	70	100	135 13	170 20	205 25	240 30	270 33
0.39 ft²	18 × 4 14 × 5	12 × 6 8 × 8	cfm NC	80	115	155 14	195 21	235 26	275 31	310 34
0.46 ft²	20 × 4 16 × 5	14 × 6 10 × 8	cfm NC	90	140	185 15	230 22	275 27	320 32	370 35
0.52 ft²	24 × 4 18 × 5	16 × 6	cfm NC	105	155	210 16	260 23	310 28	365 33	415 36
0.60 ft²	28 × 4 20 × 5	18 × 6 12 × 8	10 × 10	cfm NC	120	180	240 17	300 24	360 29	420 34	480 37
0.69 ft²	30 × 4 24 × 5	20 × 6 14 × 8	12 × 10	cfm NC	140	205	275 17	345 24	415 29	485 34	550 37
0.81 ft²	36 × 4 28 × 5	22 × 6 16 × 8	14 × 10	cfm NC	160	245 10	325 18	405 25	485 30	565 35	650 38
0.90 ft²	40 × 4 30 × 5	26 × 6 18 × 8	16 × 10 12 × 12	cfm NC	180	270 11	360 19	450 26	540 31	630 36	720 39
1.07 ft²	48 × 4 36 × 5	30 × 6 18 × 10	14 × 12	cfm NC	215	320 12	430 20	535 27	640 32	750 37	855 40

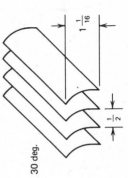

30 deg. (dimensions: $1\frac{1}{16}$, $\frac{1}{2}$)

Area	Duct sizes		1	2	3	4	5	6	7
1.18 ft²	34 × 6 / 24 × 8 ; 20 × 10 / 16 × 12 ; 14 × 14	cfm	235	355	470	590	710	825	945
		NC		13	21	28	33	38	41
1.34 ft²	60 × 4 / 48 × 5 ; 36 × 6 / 18 × 12 ; 16 × 14	cfm	270	400	535	670	805	940	1070
		NC		13	21	28	33	38	41
1.60 ft²	72 × 2 / 30 × 8 ; 24 × 10 / 22 × 12 ; 18 × 14 / 16 × 16	cfm	320	480	640	800	960	1120	1280
		NC		14	22	29	34	39	42
1.80 ft²	60 × 5 / 48 × 6 ; 36 × 12 / 30 × 10 ; 24 × 12 / 20 × 14 ; 18 × 16	cfm	360	540	720	900	1080	1260	1440
		NC		15	23	30	35	40	43
2.08 ft²	72 × 5 / 60 × 6 ; 40 × 8 / 36 × 10 ; 30 × 12 / 24 × 14 ; 20 × 16 / 18 × 18	cfm	415	625	830	1040	1250	1460	1660
		NC		16	24	31	36	41	44
2.45 ft²	72 × 6 / 48 × 8 ; 32 × 12 / 26 × 14 ; 24 × 16 / 20 × 18	cfm	490	735	980	1220	1470	1720	1960
		NC		17	25	32	37	42	45
2.78 ft²	36 × 12 / 30 × 14 ; 26 × 16 / 24 × 18 ; 22 × 20	cfm	555	835	1110	1390	1670	1950	2220
		NC		18	26	33	38	43	46
3.11 ft²	60 × 8 / 40 × 10 ; 40 × 12 / 36 × 14 ; 30 × 16 / 26 × 18 ; 24 × 20	cfm	620	935	1240	1560	1870	2180	2490
		NC		19	27	34	39	44	47
3.61 ft²	72 × 8 / 60 × 10 ; 48 × 12 ; 30 × 18 / 24 × 24 ; 36 × 16	cfm	720	1080	1440	1800	2170	2530	2890
		NC		20	28	35	40	45	48

*Reprinted by permission of Environmental Elements Corporation, Dallas, Texas.

1. All pressures are in inches of water.
2. The NC values are based on a room absorption of 8 Db, Re 10^{-12} watts and one return.

Table 10-7   CHARACTERISTIC ROOM LENGTH FOR
SEVERAL DIFFUSER TYPES*

Diffuser Type	Characteristic Length, L
High Sidewall Grille	Distance to wall perpendicular to jet
Circular Ceiling Diffuser	Distance to closest wall or intersecting jet
Sill Grille	Length of room in the direction of the jet
Ceiling Slot Diffuser	Distance to wall or midplane between outlets

*Reprinted by permission from ASHRAE Handbook of Fundamentals, 1977.

Table 10-8   ADPI SELECTION GUIDE*

Terminal Device	Room Load Btu/(hr-ft²)	x_{50}/L for Max. ADPI	Maximum ADPI	For ADPI Greater Than	Range of x_{50}/L
High	80	1.8	68	—	—
Sidewall	60	1.8	72	70	1.5–2.2
Grilles	40	1.6	78	70	1.2–2.3
	20	1.5	85	80	1.0–1.9
Circular	80	0.8	76	70	0.7–1.3
Ceiling	60	0.8	83	80	0.7–1.2
Diffusers	40	0.8	88	80	0.5–1.5
	20	0.8	93	90	0.7–1.3
Sill Grille	80	1.7	61	60	1.5–1.7
Straight	60	1.7	72	70	1.4–1.7
Vanes	40	1.3	86	80	1.2–1.8
	20	0.9	95	90	0.8–1.3
Sill Grille	80	0.7	94	90	0.8–1.5
Spread	60	0.7	94	80	0.6–1.7
Vanes	40	0.7	94	—	—
	20	0.7	94	—	—
Ceiling	80	0.3[a]	85	80	0.3–0.7
Slot	60	0.3[a]	88	80	0.3–0.8
Diffusers	40	0.3[a]	91	80	0.3–1.1
	20	0.3[a]	92	80	0.3–1.5

[a] x_{100}/L.

*Reprinted by permission from *ASHRAE Handbook of Fundamentals*, 1977.

because 125 cfm lies between 111 cfm and 139 cfm. The throw given in the table has been corrected for diffuser length. The NC is quite acceptable and is between 12 and 18 when corrected for length. The total pressure required by the diffuser is between 0.036 and 0.057 in. wat. and is about

$$\Delta P = (125/111)^2 \, (0.036) = 0.046 \text{ in. wat.}$$

An acceptable solution is listed as follows:

Size, in.	Capacity, cfm	Throw, ft	NC	ΔP_0, in. of water
4 × 12	125	10.5	5	0.046

The loss in total pressure for the diffuser is an important consideration. The value shown above would be acceptable for a light commercial system.

Example 10-2.   Suppose the room of Fig. 10-10 is located in the southern latitudes where overhead systems are recommended. Select a round ceiling diffuser system and a high sidewall system. Also select a return grille.

Solution.   The data of Table 10-3 with information from Tables 10-7 and 10-8 will be used to select a ceiling diffuser. The characteristic length is 7 or 8 ft and the throw-to-length ratio is 0.8 then,

$$x_{50} = 0.8(7.0) = 5.6 \text{ ft}$$

The best choice would be:

Size, in.	Throw, ft	NC	ΔP_0, in. of water
10	$7\frac{1}{2}$	10	0.035

The throw is larger than desired but the throw-to-length ratio is within the range to give a minimum ADPI of 80 percent. Figure 10-10 shows this application. A high sidewall diffuser may be selected from Table 10-4. In this case the throw-to-length ratio should be about 1.6 and the characteristic length is 14 ft, then

$$x_{50} = 1.6(14) = 22.4 \text{ ft}$$

The following units using the $22\frac{1}{2}$ degree spread would be acceptable:

Size, in.	Throw, ft	NC	ΔP_0, in. of water
16 ×4			
12 × 5	$22\frac{1}{2}$	18	0.063
10 × 6			

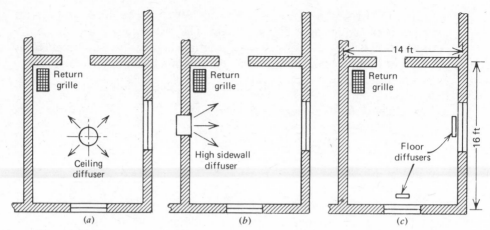

Figure 10-10. Plan view of a room showing location of different types of outlets.

Figure 10-10*b* shows the diffuser location. It would be desirable to locate the return air intake near the floor for heating purposes and in the ceiling for cooling. However, two different intakes are not generally used except in extreme cases, and the return will be located to favor the cooling case or to accommodate the building structure. For the room shown in Figure 10-10 it will be assumed that the building design prevents practical location of the return near the floor and the return is located in the ceiling as shown. We may select the following grilles from Table 10-5:

Size, in.	NC	ΔP_0, in. of water
24 × 4		
16 × 6		
18 × 5	23	0.072
12 × 6		
8 × 8	27	0.11

Example 10-3. Figure 10-11 shows a sketch of a recreational facility with pertinent data on ceiling height, air quantity, and building dimensions. The elevated seating rises 6 ft from the floor. The floor area and walls are not available for air outlets in the locker rooms. The structure is located in Topeka, Kansas. Select an air diffuser system for the complete structure.

Solution. It would be desirable to use a perimeter type of system throughout the structure; floor area is not available in all of the spaces, however, and air motion will be enhanced in the central part of the gymnasium by an overhead system.

The entry area is subject to large infiltration loads and has a great deal of glass area. Therefore, outlets should be located in the floor around the perimeter. There is 50 ft of perimeter wall with 12 ft taken up by doors. Then about 38 ft of linear diffuser could be used if required. Noise is not a limiting factor and the throw should be about 12 ft based on the ADPI, Table 10-8. If we refer to Table 10-2, the 2 in. size has a throw of 12 ft, total pressure loss of 0.08 in. of water, a NC of 23, and a capacity of 66 cfm/ft. The total length of the required diffusers would then be

$$L_d = \frac{1000}{66} = 15 \text{ or } 16 \text{ ft}$$

This total length should be divided into four equal sections and located as shown in Figure 10-11.

The office and classrooms should also be equipped with perimeter air inlets. The throw should be about 12 to 15 ft and a NC of about 30 would be acceptable. Referring to Table 10-2 a 3 in. size may be used with a capacity of 115 cfm/ft. The NC is 22 and the throw is 15 ft with a loss in total pressure of 0.08 in. of water. The total length of diffuser is then computed as

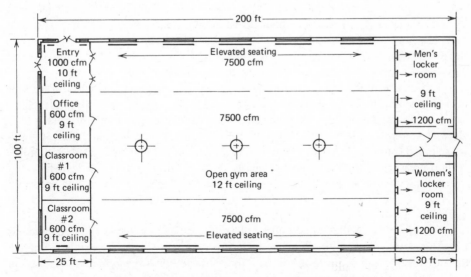

Figure 10-11.   A single story recreational facility.

$$L_d = \frac{600}{115} = 5.2 \text{ or } 5 \text{ ft}$$

The total length may be divided into two sections, or a single 5 ft length will function adequately as shown in Fig. 10-11. The corner classroom should have two outlets.

The elevated seating on each side of the gym should also be equipped with perimeter upflow air outlets because of the exposed walls and glass. A throw of 10 ft would be acceptable because the seating is elevated about 6 ft. Noise is not a major factor. There is about 145 ft of exposed wall on each side and 7500 cfm is required. Therefore, a capacity of at least 52 cfm/ft is required. From Table 10-2 a 2 in. size with a capacity of 55 cfm/ft will give a throw of 10 ft with a loss in total pressure of 0.057 in. of water. The total length of diffuser is computed as

$$L_d = \frac{7500}{55} = 136.4 \text{ or } 136 \text{ ft}$$

The total length should be divided into at least five sections and located beneath each window as shown in Fig. 10-11.

The central portion of the gymnasium should be equipped with round ceiling diffusers. Table 10-3 has data for this type of outlet. The total floor area is divided into imaginary squares and a diffuser selected with a capacity to serve that area with a throw just sufficient to reach the boundary of the area. It is better for the throw to be a little short than too long. If the total area is divided into 12 equal squares of about 25 × 25 ft, a 12 in. diffuser in each area with a capacity of about 630 cfm would be in the acceptable range, although the throw of 15 ft is slightly large. Since this is a gym area, some sacrifice of comfort is acceptable and a more economical system will result if larger diffusers are used. Imagine that the area is divided into three equal squares of about 50 × 50 ft. Then each diffuser should provide about 2500 cfm and have a throw of about 25 ft. A 24 in. size, which has a capacity of 2510 cfm and a throw of 30 ft, would be acceptable. The loss in total pressure is about 0.094 in. of water. The throw is slightly high, but is within the range given in Table 10-8. Three diffusers should be located as shown in Fig. 10-11.

The locker room areas may be equipped either with ceiling-type diffusers or with high sidewall outlets because the floor area is all covered near the walls. Since moisture will tend to collect near the ceiling and condense on ceiling diffusers, a high sidewall system will be used. If four 16 × 4 in. diffusers with capacity of 300 cfm are selected from Table 10-4, a throw of 31 ft (zero deflection) will result with a loss in total pressure of 0.085 in. of water with a NC of 24. This throw is less than that required by the ADPI of Table 10-8. A different diffuser type should be used in a critical situation. The diffusers should be equally spaced about 6 to 12 in. below the ceiling as shown in Fig. 10-11.

The air return grilles should all be placed in the ceiling unless the structure has a basement, which would make placement of grilles near the floor feasible if desired. Because cooling and ventilation will be important factors in the gym and locker room area, a ceiling type of return air system will be utilized. Return grilles may be selected from Table 10-5 as follows:

No.	Size, in.	Capacity, cfm	ΔP_0, in. of water	NC	Location
1	24 × 12	900	0.07	30	Entry
1	24 × 8	590	0.07	28	Office
1	24 × 8	590	0.07	28	Classroom 1
1	24 × 8	590	0.07	28	Classroom 2
10	30 × 24	2320	0.07	36	Gym
1	24 × 16	1220	0.07	32	Mens L.R.
1	24 × 16	1220	0.07	32	Womens L.R.

It has been assumed that all of the air will flow back through the air return before any of it is exhausted.

Variable air volume air distribution systems usually involve the use of linear or T-Bar diffusers and a thermostat-controlled metering device, referred to as a *VAV* terminal box. Figure 10-12 shows how such a device is used in relation to the main air supply and the diffusers. Figure 10-13 is a schematic of a typical terminal box. There are almost infinite variations in these devices depending on the manufacturer. Some are self-powered, using energy from the flowing air, whereas others use power from an external source. Many of the self-powered boxes require a relatively high

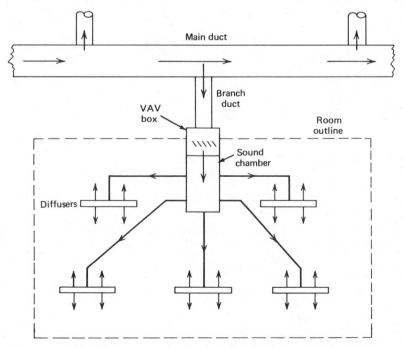

Figure 10-12.   Schematic of a *VAV* air distribution system for a room.

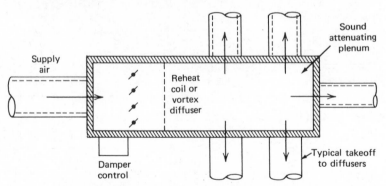

Supply
air

Reheat
coil or
vortex
diffuser

Sound
attenuating
plenum

Typical takeoff
to diffusers

Damper
control

Figure 10-13.   A single duct *VAV* terminal box.

static pressure and therefore are adaptable only to high velocity systems. However, there are models available that operate with pressures compatible with low velocity systems. The layout and selection of the diffusers follow the principles and methods previously discussed.

REFERENCES

1. J. Rydberg and P. Norback, "ASHVE Research Report No. 1362—Air Distribution and Draft," *ASHVE Trans.,65,* 1949.

2. F. C. Houghten, Carl Gutberlet, and Edward Witkowski, "Draft Temperatures and Velocities in Relation to Skin Temperatures and Feeling of Warmth," *ASHVE Trans., 44,* 1938.

3. Alfred Koestel, "Computing Temperature and Velocities in Vertical Jets of Hot or Cold Air," *ASHVE Trans., 60,* 1954.

4. H. E. Straub, "Principles of Room Air Distribution," *Heating, Piping, and Air Conditioning,* 1969.

5. *ASHRAE Handbook and Product Directory—Systems,* American Society of Heating, Refrigerating and Air-Conditioning Engineers, New York, 1980.

PROBLEMS

10-1 A free isothermal jet is discharged horizontally from a circular opening. There is no nearby surface. The initial velocity and volume flow rate are 700 ft/min and 200 cfm, respectively. Compute (a) the throw for terminal velocities of 50, 100, and 150 ft/min, (b) the total volume flow rate of the jet for each terminal velocity in (a) above.

10-2 Air is discharged from a circular pipe 100 mm in diameter to an open space. The discharge velocity is 5 m/s. Assuming standard air, compute (a) the throw for terminal velocities of 0.25 and 0.5 m/s, and (b) the total volume flow rate of the air at each terminal velocity in (a) above.

10-3 A free jet is discharged horizontally below a ceiling. The initial velocity and volume flow rate are 1000 ft/min and 300 cfm. The initial jet temperature is 100 F while the room is to be maintained at 75 F. If the coefficient K in Eq. (10-2) has a value of 9, compute the difference in temperature between the centerline of the jet and the room at terminal velocities of 50, 100, and 150 ft/min.

10-4 Air at 49 C is discharged into a room at 20 C with a velocity of 4 m/s. Estimate the stream temperature where the jet velocities are 0.25 and 0.5 m/s.

10-5 Suppose a given space requires a very large quantity of circulated air for cooling purposes. What type of diffuser system would be best? Why?

10-6 A space has a low but essentially constant occupancy with a moderate cooling load. What type of air diffuser system would be best for heating and cooling? Explain. Assume that the space is on the ground floor.

10-7 Consider a single story structure with many windows. What would be the best all-around air distribution system for (a) the northern part of the United States, and (b) the southern states? Explain.

10-8 Consider a relatively large open space with a small cooling load and low occupancy located in the southern part of the United States. What type of air distribution system would be best? Explain.

10-9 Select a perimeter type diffuser system for the building shown in Fig. 10-14. It is general office space. Dimensions given are in feet.

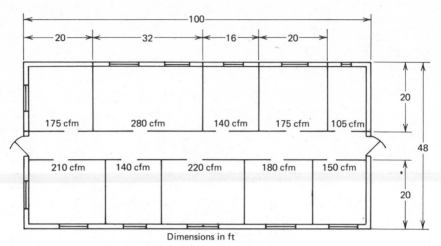

Figure 10-14.   Floor plan for Problem 10-9.

10-10   Select a round ceiling diffuser system for the building in Problem 10-9.

10-11   Select a high sidewall diffuser system for the building in Problem 10-9.

10-12   Select return air grilles for the building in Problem 10-9. Assume the return system must be placed in the attic and each room must have a return.

10-13   Select an air distribution system and diffusers and grilles for the structure described by plans and specifications furnished by the instructor.

10-14   A linear floor diffuser is required for a space with an air supply rate of 480 cfm. The room has a 12 ft ceiling and cooling load of 40 Btu/(hr-ft²). (a) Select a diffuser from Table 10-2 for this application (b) Determine the total pressure and NC for your selection.

10-15   Suppose a round ceiling diffuser is to be used in the situation described in problem 10-14. The room has plan dimensions of 16 ft × 18 ft. (a) Select a diffuser from Table 10-3 for this application. (b) Determine the total pressure and NC for the diffuser.

10-16   Assume that two high side wall diffusers are to be used for the room described in Problems 10-14 and 10-15, and they are to be installed in the wall with the longest dimension. (a) Select suitable diffusers from Table 10-4 (b) Determine the total pressure and NC for the diffusers.

10-17   Select a suitable return grille from Table 10-5 for the room described in Problem 10-14. Total pressure for the grille should be less than 0.10 in. water and one dimension should be 12 in.

10-18   Consider a room with a 20 ft exposed wall which has two windows. The other dimension is 30 ft. The room is part of a variable air volume system. (a) Lay out and select T-Bar diffusers from Table 10-6 if the room requires a total air quantity of 800 cfm and the maximum total pressure available is 0.10 in. of water (b) Note the total pressure, the throw to where the maximum velocity has decreased to 100 ft/min, and the NC for each diffuser.

10-19   Consider a 18 ft × 30 ft room in the southwest corner of a zone. There are windows on both exterior walls and the peak air quantity for the room is 1000 cfm. (a) Lay out and select T-Bar diffusers from Table 10-6 using a maximum total pressure of 0.15 in. of water (b) Note the total pressure, the throw to where the maximum velocity has decreased to 100 ft/min, and the NC for each diffuser.

CHAPTER ELEVEN

Fans and Building Air Distribution

The previous chapter considered the distribution and movement of the air within the conditioned space. It was assumed that the proper amount of air at the required total pressure was delivered to each diffuser. This chapter discusses the details of distributing the air to the various spaces in the structure. Proper design of the duct system and the selection of appropriate fans and accessories are essential. A poorly designed system may be noisy, inefficient and lead to discomfort of occupants. Correction of faulty design is expensive and sometimes practically impossible.

11-1 FANS

The fan is an essential component of almost all heating and air-conditioning systems. Except in those cases where free convection creates air motion, a fan is used to move air through ducts and to induce air motion in the space. An understanding of the fan and its performance is necessary if one is to design a satisfactory duct system.

The *centrifugal fan* is the most widely used because it can efficiently move large or small quantities of air over a wide range of pressures. The principle of operation is similar to the centrifugal pump in that a rotating impeller mounted inside a scroll type of housing imparts energy to the air or gas being moved. Figure 11-1 shows the various components of a centrifugal fan. The impeller blades may be forward-curved, backward-curved, or radial. The blade design influences the fan characteristics and will be considered later.

The *vaneaxial fan* is mounted on the centerline of the duct and produces an axial flow of the air. Guide vanes are provided before and after the wheel to reduce rotation of the air stream.

The *tubeaxial fan* is quite similar to the vaneaxial fan but does not have the guide vanes. Figure 11-2 illustrates both types.

Axial flow fans are not capable of producing pressures as high as those of the

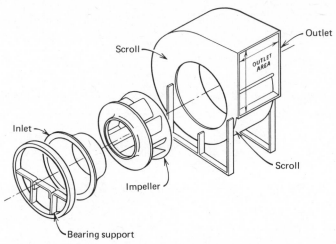

Figure 11-1.   Exploded view of a centrifugal fan. (Reprinted by permission from *ASHRAE Handbook and Product Directory—Equipment,* 1979.)

centrifugal fan but can move large quantities of air at low pressure. Axial flow fans generally produce higher noise levels than centrifugal fans.

11-2   FAN PERFORMANCE

The performance of fans is generally given in the form of a graph showing pressure, efficiency, and power as a function of capacity. The energy transferred to the air by the impeller results in an increase in static and velocity pressure; the sum of the two pressures gives the total pressure. Although these quantities are commonly referred to as pressure, they are usually expressed in terms of head such as inches or millimeters of water. When Eq. (9-1b) is applied to a fan, the following result is obtained:

$$\frac{g_c w}{g} = \frac{g_c}{g}\left(\frac{P_1}{\rho} - \frac{P_2}{\rho}\right) + \frac{1}{2g}(\overline{V}_1^2 - \overline{V}_2^2) = \frac{g_c}{g}\frac{(P_{01} - P_{02})}{\rho} \qquad (11\text{-}1)$$

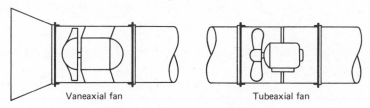

Vaneaxial fan   Tubeaxial fan

Figure 11-2.   Axial flow fans.

In this form the equation expresses the increase in total head of the air. Multiplying Eq. (11-1) by g/g_c gives

$$w = \frac{P_{01} - P_{02}}{\rho} \tag{11-1a}$$

which is an expression for the energy imparted to the air. Multiplication of Eq. (11-1a) by the mass flow rate of the air produces an expression for the *total power* imparted to the air.

$$\dot{W}_t = \frac{\dot{m}(P_{01} - P_{02})}{\rho} \tag{11-2}$$

The *static power* is the part of the total power that is used to produce the change in static head.

$$\dot{W}_s = \frac{\dot{m}(P_1 - P_2)}{\rho} \tag{11-3}$$

Fan efficiency may be expressed in two ways. The *total fan efficiency* is the ratio of total air power to the shaft power input.

$$\eta_t = \frac{\dot{W}_t}{\dot{W}_{sh}} = \frac{\dot{m}(P_{01} - P_{02})}{\rho \dot{W}_{sh}} \tag{11-4}$$

In terms of volume flow rate Eq. (11-4) becomes

$$\eta_t = \frac{\dot{Q}(P_{01} - P_{02})}{\dot{W}_{sh}} \tag{11-4a}$$

where

$$\dot{Q} = \text{volume flow rate, ft}^3/\text{min or m}^3/\text{s}$$
$$P_{01} - P_{02} = \text{change in total pressure, lbf/ft}^2 \text{ or Pa}$$
$$\dot{W}_{sh} = \text{shaft power, (ft-lbf)/min or W}$$

It has been common practice in the United States for $\dot{Q}$ to be in ft^3/min., ($P_{01} - P_{02}$) to be in inches of water, and for $\dot{W}_{sh}$ to be in horsepower. In this special case,

$$\eta_t = \frac{\dot{Q}(P_{01} - P_{02})}{6350 \dot{W}_{sh}} \tag{11-4b}$$

The *static fan efficiency* is the ratio of the static air power to the shaft power input.

$$\eta_s = \frac{\dot{W}_s}{\dot{W}_{sh}} = \frac{\dot{m}(P_1 - P_2)}{\rho \dot{W}_{sh}} = \frac{\dot{Q}(P_1 - P_2)}{\dot{W}_{sh}} \tag{11-5}$$

where the same consistent units are used as with Eq. (11-4a). Using the units of Eq. (11-4b), we get

$$\eta_s = \frac{\dot{Q}(P_1 - P_2)}{6350 \dot{W}_{sh}} \tag{11-5a}$$

Figures 11-3, 11-4, and 11-5 illustrate typical performance curves for centrifugal fans. Note the difference in the pressure characteristics for the different types of blade. Also note the point of maximum efficiency with respect to the point of maximum pressure. Fan characteristics are discussed in greater detail later.

A conventional representation of fan performance is shown in Fig. 11-6 for a specific backward-curved blade fan. In this case total pressure and total efficiency

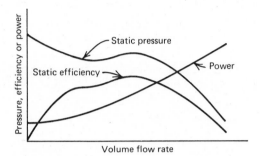

Figure 11-3. Forward-tip fan characteristics.

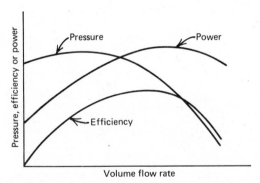

Figure 11-4. Backward-tip fan characteristics.

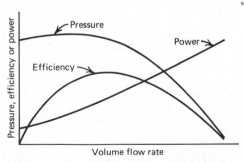

Figure 11-5. Radial-tip fan characteristics.

are also given. Note that the area for desired application is marked. When data from this zone are plotted on a logarithmic scale, the curves appear as shown in Fig. 11-7. This plot has some advantages over the conventional representation (6). Many different fan speeds can be conveniently shown, and the system characteristic is a straight line parallel to the efficiency lines.

Table 11-1 compares some of the more important characteristics of centrifugal fans.

The noise emitted by a fan is of great importance in many applications. For a given pressure the noise level is proportional to the tip speed of the impeller and to the air velocity leaving the wheel. Furthermore, fan noise is roughly proportional to the pressure developed regardless of the blade type. However, backward-curved fan blades are generally considered to have the better (lower) noise characteristics.

The pressure developed by a fan is limited by the maximum allowable speed. If noise is not a factor, the straight radial blade is superior. Fans may be operated in series to develop higher pressures, and multistage fans are also constructed. However, difficulties may arise when fans are used in parallel. Surging back and forth between fans may develop, particularly if the system demand is changing. Forward-curved blades are particularly unstable when operated at the point of maximum efficiency.

Combining both the system and fan characteristics on one plot is very useful in matching a fan to a system and to ensure fan operation at the desired conditions.

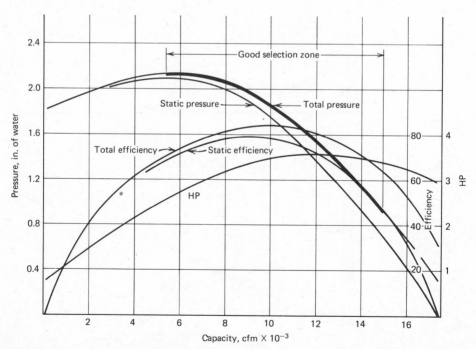

Figure 11-6. Conventional curves for backward-curved blade fan. (Reprinted by permission from *ASHRAE Journal, 14,* Part I, No. 1, 1972.)

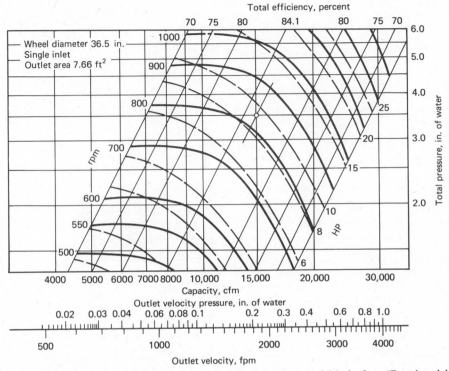

Figure 11-7. Performance data for a typical backward-curved blade fan. (Reprinted by permission from *ASHRAE Journal, 14,* Part I, No. 1, 1972.)

Figure 11-8 illustrates the desired operating range for a forward-curved blade fan. The range is to the right of the point of maximum efficiency. The backward-curved blade fan has a selection range that brackets the range of maximum efficiency and is not so critical to the point of operation; however this type should always be operated to the right of the point of maximum pressure. Figure 11-9 shows the combined system and fan characteristics using the logarithmic plot. The system characteristic

Table 11-1   COMPARISON OF CENTRIFUGAL FAN TYPES

Item	Forward-Curved Blades	Radial Blades	Backward-Curved Blades
Efficiency	Medium	Medium	High
Space required	Small	Medium	Medium
Speed for given pressure rise	Low	Medium	High
Noise	Poor	Fair	Good

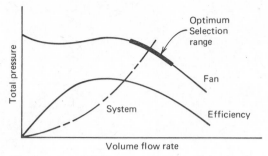

Figure 11-8.   Optimum match between system and forward-curved blade fan.

is line $S - S$. Note that for a given system the efficiency does not change with speed; however capacity, total pressure, and power all depend on the speed. Also note that changing the fan speed will not change the relative point of intersection between the system and fan characteristics. This can only be done by changing fans.

Figure 11-10 shows fan characteristics for a forward-curved blade fan using SI units except that capacity is in m^3/min instead of m^3/s.

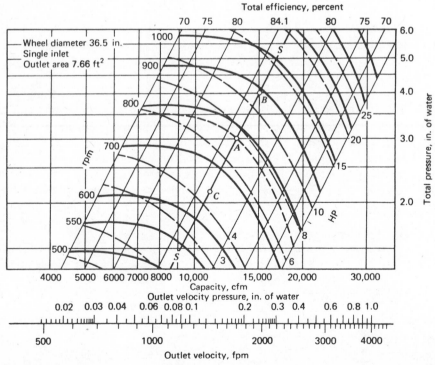

Figure 11-9.   Performance chart showing combination of fan and system. (Reprinted by permission from *ASHRAE Journal, 14*, Part I, No. 1, 1972.)

There are several simple relationships between fan capacity, pressure, speed, and power, which are referred to as the *fan laws*. The first three fan laws are the most useful and are stated as follows:

1. The capacity is directly proportional to the fan speed.
2. The pressure (static, total, or velocity) is proportional to the square of the fan speed.
3. The power required is proportional to the cube of the fan speed.

The last three fan laws are of less utility but are nevertheless useful.

4. The pressure and power are proportional to the density of the air at constant speed and capacity.
5. Speed, capacity, and power are inversely proportional to the square root of the density at constant pressure.

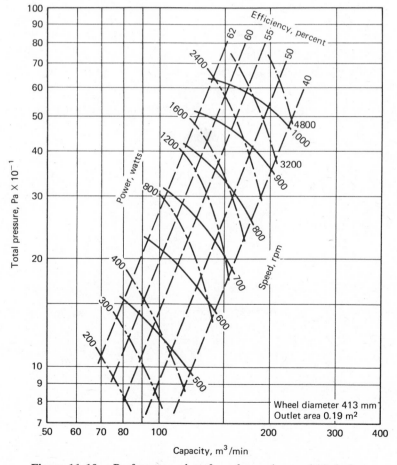

Figure 11-10.   Performance data for a forward-curved blade fan.

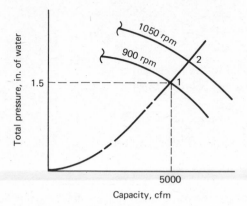

Figure 11-11. Fan and system characteris
tics for Example 11-1.

6. The capacity, speed, and pressure are inversely proportional to the density, and the power is inversely proportional to the square of the density at a constant mass flow rate.

Example 11-1. A centrifugal fan is operating as shown in Fig. 11-11 at point 1. Estimate the capacity, total pressure, and power requirement when the speed is increased to 1050 rpm. The initial power requirement is 2 horsepower.

Solution. The first three fan laws may be used to estimate the new capacity, total pressure, and power.

Capacity:

$$\frac{\dot{Q}_1}{\dot{Q}_2} = \frac{rpm_1}{rpm_2}$$

or

$$\dot{Q}_2 = \dot{Q}_1 \frac{rpm_2}{rpm_1} = 5000 \left(\frac{1050}{900} \right) = 5833 \ ft^3/min.$$

Total pressure:

$$\frac{P_{01}}{P_{02}} = \left(\frac{rpm_1}{rpm_2} \right)^2$$

$$P_{02} = P_{01} \left(\frac{rpm_2}{rpm_1} \right)^2 = 1.5 \left(\frac{1050}{900} \right)^2 = 2.04 \ \text{in. of water}$$

Power:

$$\frac{\dot{W}_1}{\dot{W}_2} = \left(\frac{rpm_1}{rpm_2} \right)^3$$

$$\dot{W}_2 = \dot{W}_1 \left(\frac{rpm_2}{rpm_1} \right)^3 = 2 \left(\frac{1050}{900} \right)^3 = 3.2 \ hp$$

We will see later when discussing the variable air volume system that it is desirable to reduce fan speed as air volume flow rate is reduced under part load conditions. This reduces the fan power considerably. The variable air volume fan and duct system will be discussed in Section 11-6.

11-3   FAN SELECTION

To select a fan for a given system it is necessary to know the capacity and total pressure requirement of the system. The type and arrangement of the prime mover, the possibility of fans in parallel or series, and nature of the load (variable or steady), and the noise constraints must be considered. After the system characteristics have been determined, the main considerations in the actual fan selection are efficiency, reliability, size and weight, speed, noise, and cost.

To assist in the actual fan selection, manufacturers furnish graphs such as those shown in Fig. 11-6 and 11-7 with the areas of preferred operation shown. The static pressure is often given but not the total pressure. The total pressure may be computed from the capacity and the fan outlet dimensions. Data pertaining to noise are also available from most manufacturers and are generally similar to those discussed in Chapter 10.

In many cases manufacturers present their fan performance data in the form of tables. Tables 11-2 and 11-2a are examples of such data for two forward-curved blade fans. Note that the static pressure is given instead of the total pressure; however, the outlet velocity is given, which makes it convenient to calculate the velocity pressure to find the total pressure.

It is important that the fan be efficient and quiet. Generally a fan will generate the least noise when operated near the peak efficiency. Operation considerably beyond the point of maximum efficiency will be noisy. Forward-curved blades operated at high speeds will be noisy and straight blades are generally noisy, especially at high speed. Backward-curved blades may be operated on both sides of the peak efficiency at relatively high speeds with less noise than the other types of fans.

Example 11-2   Comment on the suitability of using the fan described by Fig. 11-7 to move 15,000 cfm at 3.5 in. of water total pressure. Estimate the speed and power requirement.

Solution.   Examination of Fig. 11-7 shows that the fan would be quite suitable. The operating point would be just to the right of the maximum efficiency curve and the fan speed is between 800 and 900 rpm. Therefore, the fan will operate in a relatively quiet manner. The speed and power required may be estimated directly from the graph as 830 rpm and 9.5 horsepower, respectively.

Example 11-3.   Determine whether the fan given in Table 11-2 is suitable for use with a system requiring 1250 cfm at 1.8 in. of water total pressure.

Table 11-2 PRESSURE-CAPACITY TABLE FOR A FORWARD-CURVED BLADE FAN

cfm	Outlet Velocity	½ in. of water[a]		⅝ in. of water		¾ in. of water		1 in. of water		1¼ in. of water		1½ in. of water	
		rpm	bhp[b]	rpm	bhp	rpm	bhp	rpm	bhp	rpm	bhp	rpm	bhp
851	1200	848	0.13	933	0.16	1018	0.19	—	—	—	—	—	—
922	1300	866	0.15	945	0.18	1019	0.21	—	—	—	—	—	—
993	1400	884	0.17	957	0.20	1030	0.23	1175	0.30	—	—	—	—
1064	1500	901	0.19	973	0.22	1039	0.26	1182	0.32	—	—	—	—
1134	1600	926	0.22	997	0.24	1057	0.29	1190	0.35	1320	0.43	—	—
1205	1700	954	0.25	1020	0.27	1078	0.31	1200	0.38	1325	0.46	1436	0.55
1276	1800	983	0.28	1044	0.31	1100	0.34	1210	0.42	1330	0.50	1440	0.59
1347	1900	1011	0.31	1068	0.35	1126	0.38	1230	0.46	1341	0.54	1447	0.63
1418	2000	1039	0.35	1092	0.39	1152	0.42	1250	0.50	1352	0.59	1458	0.66
1489	2100	1068	0.39	1115	0.43	1178	0.47	1275	0.54	1370	0.62	1470	0.72
1560	2200	1096	0.44	1147	0.47	1204	0.51	1300	0.59	1390	0.67	1482	0.77
1631	2300	1124	0.48	1179	0.52	1230	0.56	1325	0.64	1420	0.73	1500	0.83
1702	2400	1152	0.53	1210	0.58	1256	0.62	1350	0.70	1448	0.78	1525	0.88

Note. Data are for a 9 in. wheel diameter and an outlet of 0.71 ft.²

[a] Static pressure.

[b] bhp = shaft power in horsepower.

Table 11-2a PRESSURE-CAPACITY TABLE FOR A FORWARD-CURVED BLADE FAN

Volume Flow Rate m³/s	Outlet Velocity m/s	0.7 kPa[a] rpm	kW	0.8 kPa rpm	kW	0.9 kPa rpm	kW	1.0 kPa rpm	kW	1.1 kPa rpm	kW	1.2 kPa rpm	kW
3.35	7	692	4.33										
3.83	8	688	4.79	737	5.44								
4.32	9	679	5.20	732	6.06	778	6.90						
4.78	10	664	5.48	721	6.48	770	7.46	825	7.68				
5.27	11	654	5.82	704	6.82	755	7.98	819	8.43	864	9.47	900	11.2
5.75	12	656	6.38	699	7.31	743	8.43	808	9.02	855	10.1	887	11.7
6.23	13	663	7.12	702	7.98	741	8.87	790	9.47	840	10.5	871	12.3
6.72	14	674	7.90	710	8.72	747	9.62	781	9.84	825	11.0	855	12.7
7.18	15	686	8.95	720	9.77	755	10.7	781	10.6	817	11.6	853	13.5
7.67	16	702	10.1	733	10.8	765	11.6	787	11.6	820	12.5	860	14.5
8.13	17			748	12.0	778	12.9	797	12.6	828	13.6	869	15.8
8.62	18					793	14.3	808	13.9	839	14.8	880	17.3
9.10	19							822	15.4	851	16.3	891	18.9

[a]Static pressure.
[b]Outlet area = 0.479 m². Wheel diameter = 660 mm. Tip speed = RPM × 2.07 m/s.

Solution.   There is a possibility that the fan could be used. At 1250 cfm the outlet velocity is about 1750 ft/min. Then the velocity pressure is

$$H_v = \frac{\overline{V}^2}{2g} = \frac{\overline{V}^2}{2g}\frac{\rho_a}{\rho_w}\frac{12}{3600} = \left(\frac{\overline{V}}{4005}\right)^2$$

where

$\overline{V}$ = average velocity, ft/min
H_v = velocity pressure, in. of water

Then

$$H_v = \left(\frac{1750}{4005}\right)^2 = 0.19 \text{ in. of water}$$

The static pressure at the fan outlet is then computed as

$$H_s = H_t - H_v = 1.8 - 0.19 = 1.61 \text{ in. of water}$$

From Table 11-2 the maximum static pressure recommended for the fan is 1.5 in. of water and the speed is relatively high even at that condition. If the fan speed were to be further increased, fan noise would probably be unacceptable. A larger fan should be selected. The fan shown in Table 11-2 would be suitable for use with 1250 cfm and total pressures of 0.75 to 1 in. of water. The speed would range from about 975 to 1100 rpm.

Example 11-4.   A duct system requires a fan that will deliver 6 m³/s of air at 1.2 kPa total pressure. Is the fan of Table 11-2a suitable? If so, determine the speed, shaft power, and total efficiency.

Solution.   The required volume flow rate falls between 5.75 and 6.23 m³/s in the left-hand column of Table 11-2a. The corresponding outlet velocities are 12 and 13 m/s and the velocity pressure for each case is

$$(P_v)_{5.75} = \rho_a \frac{\overline{V}^2}{2} = 1.2 \frac{(12)^2}{2} = 86.4 \text{ Pa}$$

$$(P_v)_{6.23} = \frac{1.2(13)^2}{2} = 101.4 \text{ Pa}$$

Moving across the table to the 1.1 kPa static pressure column, the total pressure at 5.75 m³/s is

$$(P_0)_{5.75} = 1100 + 86.4 = 1186.4 \text{ Pa}$$

and at 6.23 m³/s

$$(P_0)_{6.23} = 1100 + 101.4 = 1201.4 \text{ Pa}$$

By interpolation the total pressure at 6 m³/s is

$$(P_0)_{6.0} = 1186.4 + \frac{6 - 5.75}{6.23 - 5.75} (1201.4 - 1186.4)$$
$$= 1194 \text{ Pa or } 1.194 \text{ kPa}$$

Although the total pressure at 6 m³/s is barely adequate, the fan speed can be increased to obtain total pressures up to almost 1.3 kPa at a capacity of 5.75 to 6.23 m³/s.

The fan speed may be determined by interpolation to be

$$\text{rpm} = 840 - \frac{6 - 5.75}{6.23 - 5.75} (840 - 825) = 832$$

and the shaft power is likewise found to be

$$\dot{W}_{sh} = 10.5 + \frac{6 - 5.75}{6.23 - 5.75} (0.5) = 10.76 \text{ kW}$$

The total power imparted to the air is given by Eq. (11-2):

$$\dot{W}_t = \frac{\dot{m}}{\rho} (P_{01} - P_{02}) = \dot{Q}(P_{01} - P_{02}) \qquad (11\text{-}2)$$

where $\dot{Q}$ is in m³/s, $(P_{01} - P_{02})$ is in N/m² or Pa, and $\dot{W}_t$ is in watts. Then

$$\dot{W}_t = (6)(1.2)(1000)/(1000) = 7.2 \text{ kW}$$

The total efficiency is then given by

$$\eta_t = \frac{\dot{W}_t}{\dot{W}_{sh}} = 7.2/10.76 = 0.67$$

11-4   FAN INSTALLATION

The performance of a fan can be drastically reduced by improper connection to the duct system. In general, the duct connections should be such that the air may enter and leave the fan as uniformly as possible with no abrupt changes in direction or velocity. Figure 11-12 compares some good and poor practices. The designer must rely on good judgment and ingenuity in laying out the system. Space is often limited for the fan installation, and a less than optimum connection may have to be used. In this case the designer must be aware of the penalties (loss in total pressure and efficiency). Some manufacturers furnish application factors from which a modified fan curve can be computed. Figure 11-13 shows a modified fan curve with associated efficiencies. Note that for a given system the capacity, total pressure, and efficiency are all reduced by a poor installation and the power requirement is increased.

The Air Movement and Control Association, Inc. (AMCA) has published, *system effect factors* in their *Fan Application Manual* (8), which express the effect of various fan connections on system performance. These factors are in the form of flow resistances that are added to the computed system resistance.

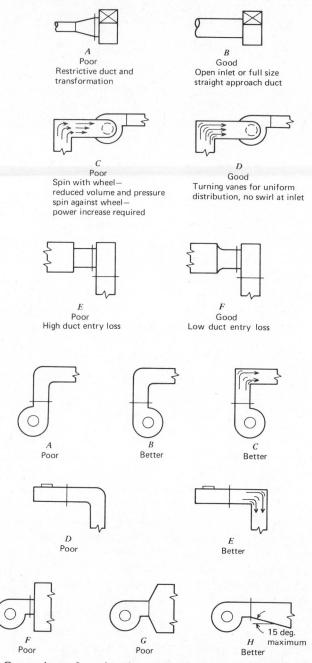

Figure 11-12. Comparison of good and poor fan connections. (Reprinted by permission from *ASHRAE Handbook and Product Directory—Equipment*, 1979.)

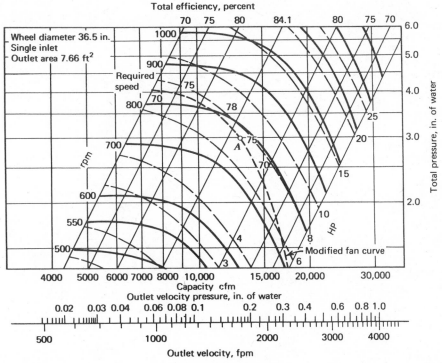

Figure 11-13. Performance curves showing modified fan curve. (Reprinted by permission from *ASHRAE Journal, 14,* Part I, 1972.)

11-5   FIELD PERFORMANCE TESTING

The design engineer is often responsible for checking the fan installation when it is put into operation. In cases of malfunction or a drastic change in performance the engineer must find and recommend corrective action. The logarithmic plot of fan performance is again quite convenient. From the original system design the specified capacity and total pressure are known and the fan model number and description establishes the fan characteristics as shown in Fig. 11-14. The system characteristic is shown as line *S-D-A-S*. The system shown was designed to operate at about 13,000 cfm and 3 in. of water total pressure. To check the system, measurements of capacity and total pressure are made in the field using a pitot tube and an inclined manometer. The use of these instruments was discussed in Chapter 9.

Several different conditions may be indicated by capacity and pressure measurements. First, if the measurements indicate operation at point *A* in Fig. 11-14, the system and fan are performing as designed. Operation at points *B* or *C* indicates that the fan is performing satisfactorily, but that the system is not operating as designed. At point *B* the system has more pressure loss than anticipated, and at point *C* the system has less pressure loss than desired. To obtain the desired capacity the

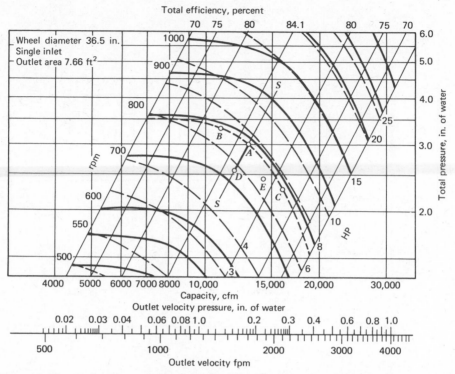

Figure 11-14. Performance curves showing field test combinations. (Reprinted by permission from *ASHRAE Journal, 14,* Part I, No. 1, 1972.)

fan speed must be increased to about 900 rpm for point *B* and reduced to about 650 rpm for point *C*. Operation at point *D* indicates that the fan is not performing as it should. This may be because an incorrect belt drive or belt slippage has caused the fan to operate at the incorrect speed. Poorly designed inlet and outlet connections to the fan may have altered fan performance as shown in Fig. 11-13. The correct capacity may be obtained by correcting the fan speed or by eliminating the undesirable inlet and outlet connections. Operation at point *E* indicates that neither the system nor the fan are operating as designed, which is the usual case found in the field. Although in this situation corrective action could be made be decreasing fan speed to about 700 rpm, any undesirable features of the fan inlet or outlet should first be eliminated to maintain a high fan efficiency. After any increase in fan speed the change in power requirements should be carefully ascertained, because fan power is proportional to the cube of the fan speed.

Example 11-5.   A duct system was designed to handle 2.5 m³/s of air with a total pressure requirement of 465 Pa. The fan of Fig. 11-10 was selected for the system. Field measurements indicate that the system is operating at 2.4 m³/s at

490 Pa total pressure. Recommend corrective action to bring the system up to the design capacity.

Solution. The system characteristics for the design condition and the actual condition may be sketched on Fig. 11-10 and are parallel to the efficiency lines. The fan is performing as specified; however, the system has more flow resistance that it was designed for. To obtain the desired volume flow rate we must reduce the system flow resistance or increase the fan speed. A check should first be made for unnecessary flow restrictions or closed dampers and, where practical, adjustments should be made to lower the flow resistance. As a last resort the fan speed must be increased, keeping in mind that the power requirements and noise level will increase.

Assuming that the duct system cannot be altered, the fan speed for this example must be increased from 900 rpm to 975 rpm to obtain 2.5 m^2/s of flow. The total pressure produced by the fan will increase from 490 to 564 Pa. The shaft power requirement at the design condition was 2100 W and will be increased to 2400 W at the higher speed. In this case the required increase in fan speed is moderate and the increase in noise level should be minimal. The motor must be checked, however, to be sure that an overload will not occur at the higher speed.

11-6 FANS AND VARIABLE AIR VOLUME SYSTEMS

The variable volume air distribution system is usually designed to supply air to a large number of spaces with the total amount of circulated air varying between some minimum and the full load design quantity. Normally the minimum is about 20 to 25 percent of the maximum. The volume flow rate of the air is controlled independent of the fan by the terminal boxes, and the fan must respond to the system. Because the fan capacity is directly proportional to fan speed and power is proportional to the cube of the speed, it seems obvious that the fan speed should be decreased as volume flow rate decreases. There are practical and economic considerations involved, however. A variable speed electric motor would be ideal, but they have very low efficiency which offsets the benefit of lowering fan speed. Fan drives which make use of magnetic couplings have been developed and are referred to as eddy current drives. These are excellent devices with almost infinite adjustment of fan speed. Their only disadvantage is high cost. Another approach is to change the fan speed by changing the diameter of the V-belt drive pulley by adjusting the pulley shives. This requires a mechanism that will operate while the drive is turning. The main disadvantage of this approach is maintenance. The eddy current and variable pulley drives appear to be the most practical.

Another approach to control of the fan is to throttle and introduce a swirling component to the air entering the fan which alters the fan characteristic in such a way that less power is required at the lower flow rates. This is done with variable inlet vanes which are a radial damper system located at the inlet to the fan. Gradual closing of the vanes reduces the volume flow rate of air and changes the fan char-

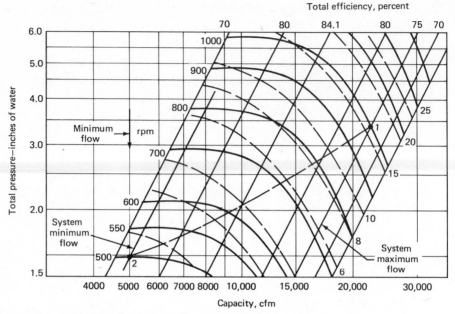

Figure 11-15. Variable speed fan in a variable volume system.

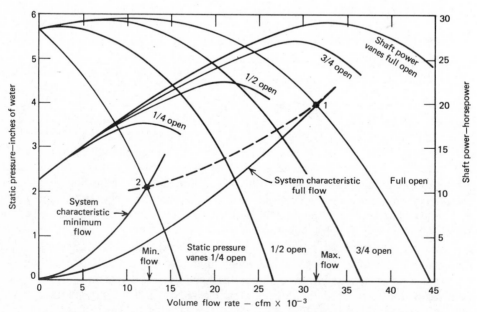

Figure 11-16. Variable inlet vane fan in a variable volume system.

acteristic as shown in Fig. 11-16. This approach is not as effective in reducing fan power as fan speed reduction, but the cost and maintenance are low.

The fan speed or inlet vane position is normally controlled to maintain a constant static pressure in the duct system. Static pressure could be sensed at the fan outlet; however, this will lead to abnormally low pressure at some terminal boxes located a large distance from the fan. Therefore, the static pressure sensor should be located about two-thirds of the total duct length from the fan and the controller set so that the most distant terminal box will have some minimum static pressure. This approach gives a more uniform static pressure throughout the duct system.

Consider the response of a fan in a *VAV* system with static pressure control as discussed above. The minimum and maximum flow rates are shown in Figs. 11-15 and 11-16 for variable speed and inlet vanes respectively. Without any fan control, the operating point must move along the constant speed and the full open vane characteristics. This results in high static pressure, low efficiency, and wasted fan power. Further it may not be possible to have stable operation of the fan at the minimum flow rate. There will also be a very high static pressure in the system which is undesirable. When the fan speed is reduced or inlet vanes are closed to maintain a fixed static pressure at some downstream location in the duct system, the fan static pressure actually decreases to that shown at point 2. This occurs because the loss in pressure between the fan and the sensing point decreases as the flow rate decreases. This is predictable because duct pressure loss is proportional to the duct velocity squared. However, the complete analysis of a variable volume system is difficult because there are infinite variations of the terminal box dampers, fan speed, and inlet vanes. It is possible to locate point 2 in Figs. 11-15 and 11-16 and system operation will then be between points 1 and 2.

11-7 AIR FLOW IN DUCTS

The general subject of fluid flow in ducts and pipes was discussed in Chapter 9. The special topic of air flow is treated in this section. Although the basic theory is the same, certain simplifications and computational procedures will be adopted to aid in the design of air ducts.

Equation (9-1b) applies to the flow of air in a duct. Neglecting the elevation head terms, assuming that the flow is adiabatic, and no fan is present, Eq. (9-1b) becomes

$$\frac{g_c}{g}\frac{P_1}{\rho} + \frac{\overline{V}_1^2}{2g} = \frac{g_c}{g}\frac{P_2}{\rho} + \frac{\overline{V}_2^2}{2g} + l_f \qquad (11\text{-}6)$$

and in terms of the total head

$$\frac{g_c}{g}\frac{P_{01}}{\rho} = \frac{g_c}{g}\frac{P_{02}}{\rho} + l_f \qquad (11\text{-}6a)$$

Equations (11-6) provide a great deal of insight to the duct flow problem. The only important terms remaining in the energy equation are the static head, the velocity head, and the lost head. The static and velocity head terms are interchangeable and

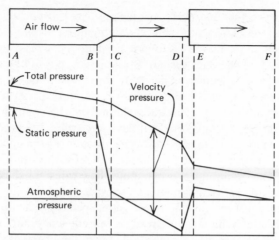

Figure 11-17. Pressure changes during flow in ducts. (Reprinted by permission from *ASH-RAE Handbook of Fundamentals, 1972.*)

may increase or decrease in the direction of flow depending on the duct cross-sectional area. Because the lost head l_f must be positive, the total pressure always decreases in the direction of flow. Figure 11-17 illustrates these principles.

For duct flow the units of each term in Eq. (11-6) are usually in. of water because of their small size. When SI units are used, each term has units of N/m^2 or Pa. Equation (11-6) then takes the following form:

$$P_1 + \frac{\rho \overline{V}_1^2}{2} = P_2 + \frac{\rho \overline{V}_2^2}{2} + \rho g l_f \tag{11-6b}$$

where l_f has the units of meters as defined in Eq. (9-6). To simplify the notation, the equations may be written

$$H_{s1} + H_{v1} = H_{s2} + H_{v2} + l_f \tag{11-6c}$$

and

$$H_{01} = H_{02} + l_f \tag{11-6d}$$

In this form each term has the units of length in any system of units. For air at standard conditions

$$H_v = \left(\frac{\overline{V}}{4005} \right)^2 \text{ in. of water} \tag{11-7}$$

where $\overline{V}$ is in ft/min.

$$P_v = \left(\frac{\overline{V}}{1.29} \right)^2 \text{ Pa} \tag{11-8}$$

where $\overline{V}$ is in m/s.

The lost head due to friction in a straight, constant area duct is given by Eq. (9-6) and the computational procedure is the same as discussed in section 9-2. Because this approach becomes tedious when designing ducts, special charts have been prepared. Figures 11-18, 11-18a, 11-18b, and 11-18c are examples of such charts for air flowing in galvanized steel ducts with approximately 40 joints per 100 ft or 30 m. The charts are based on standard air and fully developed flow. For the temperature range of 50 F or 10 C to about 100 F or 38 C there is no need to correct for viscosity and density changes. Above 100 F or 38 C, however, a correction should be made. The density correction is also small for moderate pressure changes. For elevations below about 2000 ft or 610 m the correction is small. The correction for density and viscosity will normally be less than one. For example, dry air at 100 F at an elevation of 2000 ft would exhibit a pressure loss about 10 percent less than given in Fig. 11-18. The effect of roughness is a more important consideration and difficult to assess.

A common problem to designers is determination of the roughness effect of fibrous glass duct liners and fibrous ducts. This material is manufactured in several grades with various degrees of absolute roughness. Further, the joints and fasteners necessary to install the material affect the overall pressure loss. Smooth galvanized ducts typically have a friction factor of about 0.02, whereas fibrous liners and duct materials will have friction factors varying from about 0.03 to 0.06 depending on the quality of the material and joints, and the duct diameter. The usual approach to account for this roughness effect is to use a correction factor which is applied to the pressure loss obtained for galvanized metal duct such as Fig. 11-18. Figure 11-19 shows a range of data for commercially available fibrous duct liner materials. These correction factors probably do not allow for typical joints and fasteners.

A more refined approach to the prediction of pressure loss in rough or lined ducts is to generate a chart, such as Fig. 11-18, using Eq. (9-6) and the Colebrook function (2)

$$\frac{1}{\sqrt{f}} = -2 \log_{10} \left[\frac{e}{3.7D} + \frac{2.51}{\mathrm{Re}_D \sqrt{f}} \right] \tag{11-9}$$

to express the friction factor. Eq. (11-9) is valid in the transition region where f depends on the absolute roughness e, the duct diameter D, and the Reynolds number Re_D. Equations (9-6) and (11-9) and the ideal gas property relation may be easily programmed for a small computer to calculate the lost pressure for a wide range of temperatures, pressures, and roughness. This general approach eliminates the need for corrections of any kind.

The head loss due to friction is greater for a rectangular duct than for a circular duct of the same cross-sectional area and capacity. For most practical purposes ducts of aspect ratio not exceeding 8:1 will have the same lost head for equal length and mean velocity of flow as a circular duct of the same hydraulic diameter. When the duct sizes are expressed in terms of hydraulic diameter D_h and when the equations for friction loss in round and rectangular ducts are equated for equal length and capacity, Eq. (11-10) for the circular equivalent of a rectangular duct is obtained.

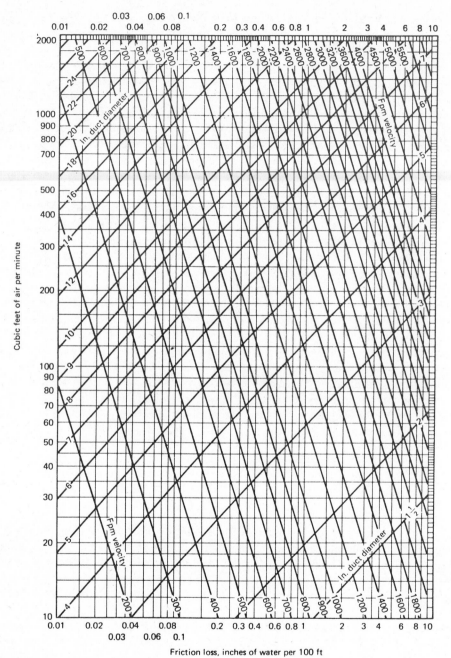

Figure 11-18.   Lost head due to friction for galvanized steel ducts. (Reprinted by permission from *ASHRAE Handbook of Fundamentals*, 1977.)

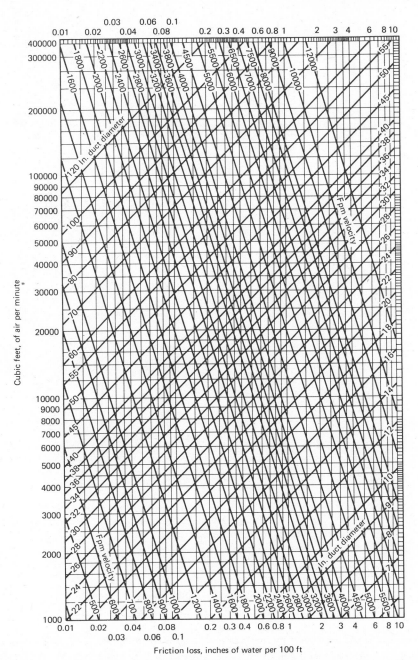

Figure 11-18a. Lost head due to friction for galvanized steel ducts. (Reprinted by permission from *ASHRAE Handbook of Fundamentals*, 1977.)

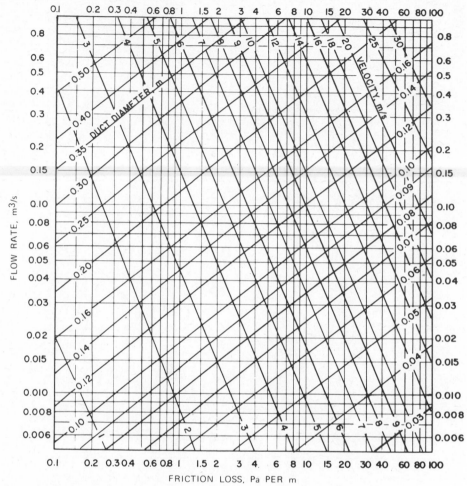

Figure 11-18b.   Lost head due to friction for galvanized steel ducts. (*Using SI units in Heating, Air Conditioning and Refrigeration* by W. F. Stoecker, © 1975, Business News Publishing Company.)

$$D_e = 1.3 \frac{(ab)^{5/8}}{(a+b)^{1/4}} \tag{11-10}$$

where a and b are the rectangular duct dimensions in any consistent units. Table 11-3 has been compiled using Eq. (11-10). A more complete table is given in the *ASHRAE Handbook* (2).

Oval ducts are sometimes used in commercial duct systems. The frictional pressure loss may be treated in the same manner as rectangular ducts using the circular equivalent of the oval duct as defined by

$$D_e = 1.55A^{0.625}/P^{0.25} \qquad (11\text{-}11)$$

with

$$A = b^2/4 + b(a - b) \qquad (11\text{-}11a)$$

and

$$P = b + 2(a - b) \qquad (11\text{-}11b)$$

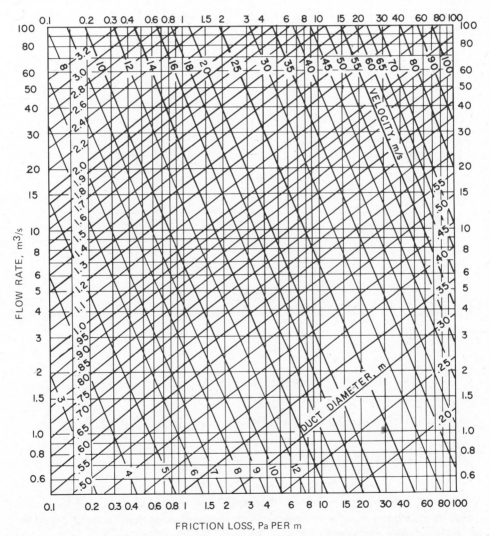

Figure 11-18c.   Lost head due to friction for galvanized steel ducts. (*Using SI units in Heating, Air Conditioning and Refrigeration* by W. F. Stoecker, © 1975, Business News Publishing Company.)

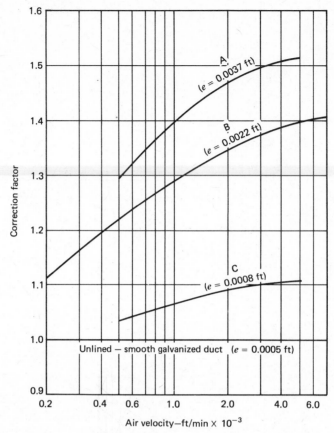

Figure 11-19. Range of roughness correction factors for commercially available duct liners.

where

      a = major diameter of oval duct, in. or m
      b = minor diameter of oval duct, in. or m

Equations (11-11) are valid for aspect ratios ranging from 2 to 4 (2).

11-8 AIR FLOW IN FITTINGS

Whenever a change in area or direction occurs in a duct or when the flow is divided and diverted into a branch, substantial losses in total pressure may occur. These losses are usually of greater magnitude than the losses in the straight pipe and are referred to as dynamic losses.

Dynamic losses vary as the square of the velocity and are conveniently represented by

$$H_0 = C(H_v) \tag{11-12}$$

Table 11-3 CIRCULAR EQUIVALENTS OF RECTANGULAR DUCTS FOR EQUAL FRICTION AND CAPACITY — dimensions in inches, feet, or meters

Side Rectangular Duct	6	7	8	9	10	11	12	13	14	15	16	17	18	19	20	22	24
6	6.6																
7	7.1	7.7															
8	7.5	8.2	8.8														
9	8.0	8.6	9.3	9.9													
10	8.4	9.1	9.8	10.4	10.9												
11	8.8	9.5	10.2	10.8	11.4	12.0											
12	9.1	9.9	10.7	11.3	11.9	12.5	13.1										
13	9.5	10.3	11.1	11.8	12.4	13.0	13.6	14.2									
14	9.8	10.7	11.5	12.2	12.9	13.5	14.2	14.7	15.3								
15	10.1	11.0	11.8	12.6	13.3	14.0	14.6	15.3	15.8	16.4							
16	10.4	11.4	12.2	13.0	13.7	14.4	15.1	15.7	16.3	16.9	17.5						
17	10.7	11.7	12.5	13.4	14.1	14.9	15.5	16.1	16.8	17.4	18.0	18.6					
18	11.0	11.9	12.9	13.7	14.5	15.3	16.0	16.6	17.3	17.9	18.5	19.1	19.7				
19	11.2	12.2	13.2	14.1	14.9	15.6	16.4	17.1	17.8	18.4	19.0	19.6	20.2	20.8			
20	11.5	12.5	13.5	14.4	15.2	15.9	16.8	17.5	18.2	18.8	19.5	20.1	20.7	21.3	21.9		
22	12.0	13.1	14.1	15.0	15.9	16.7	17.6	18.3	19.1	19.7	20.4	21.0	21.7	22.3	22.9	24.1	
24	12.4	13.6	14.6	15.6	16.6	17.5	18.3	19.1	19.8	20.6	21.3	21.9	22.6	23.2	23.9	25.1	26.2
26	12.8	14.1	15.2	16.2	17.2	18.1	19.0	19.8	20.6	21.4	22.1	22.8	23.5	24.1	24.8	26.1	27.2
28	13.2	14.5	15.6	16.7	17.7	18.7	19.6	20.5	21.3	22.1	22.9	23.6	24.4	25.0	25.7	27.1	28.2
30	13.6	14.9	16.1	17.2	18.3	19.3	20.2	21.1	22.0	22.9	23.7	24.4	25.2	25.9	26.7	28.0	29.3
32	14.0	15.3	16.5	17.7	18.8	19.8	20.8	21.8	22.7	23.6	24.4	25.2	26.0	26.7	27.5	28.9	30.1
34	14.4	15.7	17.0	18.2	19.3	20.4	21.4	22.4	23.3	24.2	25.1	25.9	26.7	27.5	28.3	29.8	31.0
36	14.7	16.1	17.4	18.6	19.8	20.9	21.9	23.0	23.9	24.8	25.8	26.6	27.4	28.3	29.0	30.7	32.0
38	15.0	16.4	17.8	19.0	20.3	21.4	22.5	23.5	24.5	25.4	26.4	27.3	28.1	29.0	29.8	31.4	32.8
40	15.3	16.8	18.2	19.4	20.7	21.9	23.0	24.0	25.1	26.0	27.0	27.9	28.8	29.7	30.5	32.1	33.6

*Reprinted by permission from *ASHRAE Handbook of Fundamentals*, 1977.

Table 11-4   TOTAL PRESSURE LOSSES DUE TO ELBOWS

Type	Illustration	Conditions	Loss Coefficient C
90 degree round 4 piece		R/D = 0.75 1.0 1.5 2.0	0.50 0.37 0.27 0.24
90 degree round		Miter	1.2
Miter with turning vanes		Plate Formed None	0.20 0.15 1.20

*Reprinted by permission from *ASHRAE Handbook of Fundamentals*, 1977.

where the loss coefficient C is a constant and Eqs. (11-7) or (11-8) express H_v. When different upstream and downstream areas are involved as in an expansion or contraction, either the upstream or downstream value of H_v may be used in Eq. (11-12) but C will be different in each case.

Fittings are classified as either constant flow, such as an elbow or transition, or as divided flow, such as a wye or tee. Tables 11-4 and 11-5 give loss coefficients for many different types of constant flow fittings. It should be kept in mind that the quality and type of construction may vary considerably for a particular type of fitting. Some manufacturers provide data for their own products.

Example 11-6.   Compute the lost pressure in a 6 in., 90 degree 4-piece elbow that has 150 cfm of air flowing through it. The ratio of turning radius to diameter is 1.5.

Solution.   The lost pressure will be computed from Eq. (11-12). From Table 11-4 the loss coefficient is read as 0.27 and the average velocity in the elbow is computed as

$$\overline{V} = \frac{\dot{Q}}{A} = \frac{\dot{Q}}{(\pi/4)D^2} = \frac{(150)4(144)}{\pi(36)} = 764 \text{ ft/min}$$

Table 11-5   LOSS COEFFICIENTS FOR AREA CHANGES*

Type	Illustration	Conditions	Loss	Coefficient
		A_1/A_2	C_1	C_2
Abrupt expansion		0.1	0.81	81
		0.2	0.64	15
		0.3	0.49	5
		0.4	0.36	2.25
		0.5	0.25	1.00
		0.6	0.16	0.45
		0.7	0.09	0.18
		0.8	0.04	0.06
		0.9	0.01	0.01
		θ, degrees	C_r	
Gradual expansion[a]		5	0.17	
		7	0.22	
		10	0.28	
		20	0.45	
		30	0.59	
		40	0.73	
		A_2/A_1	C_2	
Abrupt contraction square edge		0.0	0.34	
		0.2	0.32	
		0.4	0.25	
		0.6	0.16	
		0.8	0.06	
		θ, degrees		
Gradual contraction		30	0.02	
		45	0.04	
		60	0.07	
Equal area transformation		$A_1 = A_2$	C	
		$\theta \leq 14$	0.15	
Franged entrance			C	
		$A = \infty$	0.34	

Table 11-5   LOSS COEFFICIENTS FOR AREA CHANGES* (*Continued*)

Type	Illustration	Conditions	Loss Coefficient
Duct entrance			C
		$A = \infty$	0.85
Formed entrance			C
		$A\, \infty$	0.03
		A_0/A	C_0
Square edge orifice in duct	$A_1 = A_2$ A_0	0.0	2.50
		0.2	1.86
		0.4	1.21
		0.6	0.64
		0.8	0.20
		1.0	0.0

*Reprinted by permission from *ASHRAE Handbook of Fundamentals,* 1977.
$^a H_0 = C_1 C_r H_v$ where C_1 is for an abrupt expansion of the same area ratio.

then

$$H_0 = C\left(\frac{\overline{V}}{4005}\right)^2 = 0.27 \left(\frac{764}{4005}\right)^2 = 0.010 \text{ in. of water}$$

In SI units the elbow diameter is 15.24 cm and the flow rate is 4.25 m³/min. The average velocity is then

$$\overline{V} = \frac{\dot{Q}}{A} = \frac{4.25}{(\pi/4)(0.1524)^2(60)} = 3.88 \text{ m/s}$$

The loss coefficient C is dimensionless and is therefore unchanged. Using Eq. (11-7a), we get

$$H_0 = C\left(\frac{\overline{V}}{1.29}\right)^2 = 0.27 \left(\frac{3.88}{1.29}\right)^2 = 2.44 \text{ Pa}$$

This result compares favorably with the result obtained in English units since 0.010 in. of water may be converted directly to 2.44 Pa.

Elbows are generally efficient fittings in that their losses are small when the turn is gradual. When an abrupt turn is used without turning vanes, the lost pressure will be four or five times larger.

When considering the lost pressure in divided flow fittings, the loss in the straight-through section as well as the loss through the branch outlet must be considered. Figure 11-20 gives representative data for branch-type fittings. Note that the angle of the branch takeoff has a great influence on the loss coefficient. The loss coefficient C_b in Fig. 11-20 is based on the average velocity in the main duct just upstream of the fitting.

The straight-through flow in a divided flow fitting is treated in a slightly different way. Because several different situations can arise where the velocity may increase or decrease, experimental data have been tabulated. Table 11-6 gives representative data for the common types of divided flow fittings. A converging or diverging section is similar and therefore included on the same chart. The velocity may increase, decrease, or remain constant through the fitting. In every case there will be some loss in total pressure. The data for velocities below about 1600 ft/min or 8 m/s are not very reliable because it is difficult to measure the small pressure losses involved. At velocities below about 1000 ft/min or 5 m/s the loss can be neglected for practical purposes; however, it is good practice to allow for a small loss.

Example 11-7.   Compute the loss in total pressure for a 90 degree branch takeoff as well as for the straight-through section. The main section has a diameter of 12 in. with a flow rate of 1100 cfm. The branch flow rate is 250 cfm through a 7 in. duct. The main duct does not reduce following the branch.

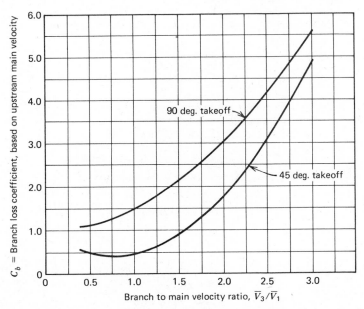

Figure 11-20.   Branch loss coefficient vs. velocity ratio. (Reprinted by permission from *ASHRAE Handbook of Fundamentals*, 1972.)

Table 11-6 LOSS IN TOTAL PRESSURE FOR TRANSITION FITTINGS
in. of water (in. × 25.4 = mm, and in. of water × 249 = Pa)

Upstream Velocity $\overline{V}_u$	Downstream Velocity $\overline{V}_d$, fpm							
	600	800	1000	1200	1400	1600	1800	2000
600	0.001	0.003	0.004					
800	0.003	0.002	0.008	0.010	0.010			
1000	0.004	0.003	0.003	0.005	0.010	0.010		
1200	0.007	0.006	0.005	0.005	0.010	0.010	0.011	0.012
1400	0.010	0.010	0.008	0.008	0.006	0.010	0.011	0.013
1600	0.017	0.015	0.014	0.010	0.008	0.012	0.011	0.013
1800	0.022	0.020	0.018	0.012	0.011	0.012	0.012	0.013
2000	0.027	0.020	0.019	0.018	0.017	0.017	0.015	0.010

Upstream Velocity $\overline{V}_u$	Downstream Velocity $\overline{V}_d$, fpm						
	2000	2500	3000	3500	4000	4500	5000
2000	0.010	0.020	0.030				
2500	0.030	0.020	0.030	0.040			
3000	0.040	0.030	0.030	0.040	0.050		
3500	0.060	0.050	0.040	0.040	0.050	0.060	0.070
4000	0.090	0.090	0.080	0.060	0.050	0.060	0.070
4500	0.120	0.110	0.100	0.090	0.070	0.060	0.070
5000	0.140	0.140	0.130	0.100	0.100	0.080	0.070

Solution. It is first necessary to compute the average velocity in each section of the fitting.

$$\overline{V}_u = \frac{\dot{Q}_u}{A_u} = \frac{1100}{(\pi/4)(12/12)^2} = 1400 \text{ ft/min}$$

$$\overline{V}_b = \frac{\dot{Q}_b}{A_b} = \frac{250}{(\pi/4)(7/12)^2} = 935 \text{ ft/min}$$

$$\overline{V}_d = \frac{\dot{Q}_d}{A_d} = \frac{850}{(\pi/4)(12/12)^2} = 1082 \text{ ft/min}$$

The ratio of the branch to upstream velocity is

$$\frac{\overline{V}_b}{\overline{V}_u} = \frac{935}{1400} = 0.67$$

The loss coefficient for the branch is then read from Fig. 11-20 as 1.2 and

$$H_{0b} = C_b \left(\frac{\overline{V}_u}{4005} \right)^2 = 1.2 \left(\frac{1400}{4005} \right)^2 = 0.417 \text{ in. of water}$$

The lost pressure for the straight-through section is determined from Table 11-6. At an upstream velocity of 1400 ft/min and a downstream velocity of 1082 ft/min the loss in total pressure is read as about 0.008 in. of water, say 0.01 in. of water.

The data discussed above for branch type fittings are for pipe of circular cross section. When rectangular duct is used, the pressure losses will vary depending on the design of the fitting. Extensive data for fittings of all kinds are given in reference 2. These data should be used in a realistic design situation.

When the average velocity of the air in the duct is below about 1000 ft/min, the loss in total pressure in fittings becomes much less dependent on velocity, and equivalent lengths may be used with reasonable accuracy to account for the lost pressure. This approach simplifies calculations and is helpful in the design of low velocity systems. Figures 11-21 and 11-22 show common fittings used in low velocity systems with their equivalent lengths. A more accurate approach is to include the effect of the pipe diameter by specifying length to diameter (L/D) ratios for the various fit-

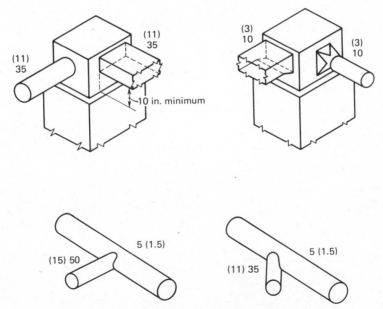

Figure 11-21.   Equivalent lengths of some plenum and branch fittings in feet and meters in parentheses. (Courtesy of National Environmental Systems Contractors Association, 1501 Wilson Blvd., Arlington, Va. 22209.)

tings. Data of this type are given in reference 2. The equivalent length of a particular fitting is then computed from

$$L_e = \left(\frac{L}{D}\right) D \qquad (11\text{-}13)$$

The equivalent length increases somewhat with diameter.

To compute the lost pressure, we add the equivalent length of the fitting to the adjacent pipe length. The following example shows the procedure.

Example 11-8.   Compute the lost pressure for each branch of the simple duct system shown in Fig. 11-23 using the equivalent length approach and SI units.

Solution.   The lost pressure will first be computed from 1 to a; then sections a to 2 and a to 3 will be handled separately.

The equivalent length of section 1 to a consists of the actual length of the 25 cm duct plus the equivalent length for the entrance from the plenum of 11 m given in Fig. 11-21. Then

$$L_{1a} = 15 + 11 = 26 \text{ m}$$

From Fig. 11-18b at 0.19 m³/s and for a pipe diameter of 25 cm, the lost pressure is 0.85 Pa/m of pipe. Then

$$\Delta P_{o1a} = (0.85)(26) = 22.1 \text{ Pa}$$

Section a to 3 has an equivalent length equal to the sum of the actual length, and the equivalent length for the 45 degree branch takeoff, one 45 degree elbow, and one 90 degree elbow.

$$L_{a3} = 12 + 11 + 1.5 + 3 = 27.5 \text{ m}$$

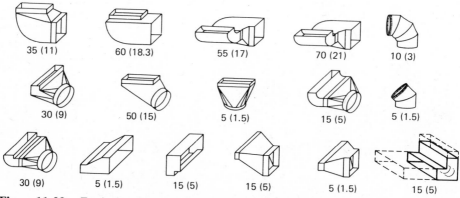

Figure 11-22.   Equivalent lengths of common duct fittings in feet and meters in parentheses. (Courtesy of National Environmental Systems Contractors Association, 1501 Wilson Blvd., Arlington, Va. 22209.)

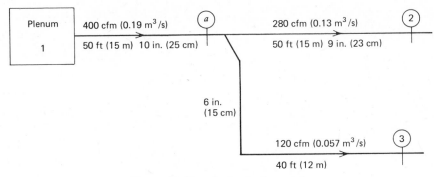

Figure 11-23.   A simple duct system.

From Fig. 11-18b, at 0.057 m^3/s and for a 15 cm diameter pipe, the lost pressure is 1.0 Pa/m of pipe. Then

$$\Delta P_{0a3} = (1.0)(27.5) = 27.5 \text{ Pa}$$

Section a to 2 has an equivalent length equal to the sum of the actual length and the equivalent length for the straight-through section of the branch fitting. A length of 2 m is assumed for this case.

$$L_{a2} = 15 + 2 = 17 \text{ m}$$

From Fig. 11-18b, the loss per meter of length is 0.6 Pa and

$$\Delta P_{0a2} = (0.6)(17) = 10.2 \text{ Pa}$$

Then the lost pressure for section 1 to 2 is

$$\Delta P_{012} = \Delta P_{01a} + \Delta P_{0a2} = 22.1 + 10.2 = 32.3 \text{ Pa}$$

and for section 1 to 3

$$\Delta P_{013} = \Delta P_{01a} + \Delta P_{0a3} = 22.1 + 27.5 = 49.6 \text{ Pa}$$

11-9   DUCT DESIGN—GENERAL CONSIDERATIONS

The purpose of the duct system is to deliver a specified amount of air to each diffuser in the conditioned space at a specified total pressure. This is to ensure that the space load will be absorbed and the proper air motion within the space will be realized. The method used to lay out and size the duct system must result in a reasonably quiet system and must not require unusual adjustments to achieve the proper distribution of air to each space. A low noise level is achieved by limiting the air velocity, by using sound-absorbing duct materials or liners, and avoiding drastic restrictions in the duct such as nearly closed dampers. Table 11-7 gives recommended and maximum duct velocities for low velocity systems. A low velocity duct

Table 11-7   RECOMMENDED AND MAXIMUM DUCT VELOCITIES
FOR LOW VELOCITY SYSTEMS*

Designation	Recommended Velocities, fpm (m/s)		
	Residences	Schools, Theaters, Public Buildings	Industrial Buildings
Outdoor air intakes[a]	500 (2.54)	500 (2.54)	500 (2.54)
Filters[a]	250 (1.27)	300 (1.52)	350 (1.78)
Heating coils[a]	450 (2.29)	500 (2.54)	600 (3.05)
Cooling coils[a]	450 (2.29)	500 (2.54)	600 (3.05)
Air washers[a]	500 (2.54)	500 (2.54)	500 (2.54)
Fan outlets	1000–1600 (5.08–8.13)	1300–2000 (6.60–10.16)	1600–2400 (8.13–12.19)
Main ducts	700–900 (3.56–4.57)	1000–1300 (5.08–6.60)	1200–1800 (6.1–9.14)
Branch ducts	600 (3.05)	600–900 (3.05–4.57)	800–1000 (4.06–5.08)
Branch risers	500 (2.54)	600–700 (3.05–3.56)	800 (4.06)
Maximum Velocities, fpm (m/s)			
Outdoor air intakes[a]	800 (4.06)	900 (4.57)	1200 (6.10)
Filters[a]	300 (1.52)	350 (1.78)	350 (1.78)
Heating coils[a]	500 (2.54)	600 (3.05)	700 (3.56)
Cooling coils[a]	450 (2.29)	500 (2.54)	600 (3.05)
Air washers[a]	500 (2.54)	500 (2.54)	500 (2.54)
Fan outlets	1700 (8.64)	1500–2200 (7.62–11.18)	1700–2800 (8.64–14.22)
Main ducts	800–1200 (4.06–6.10)	1100–1600 (5.59–8.13)	1300–2200 (6.60–11.18)
Branch ducts	700–1000 (3.56–5.08)	800–1300 (4.06–6.60)	1000–1800 (5.08–9.14)
Branch risers	650–800 (3.30–4.06)	800–1200 (4.06–6.10)	1000–1600 (5.08–8.13)

*Reprinted by permission from *ASHRAE Handbook of Fundamentals,* 1977.
[a]These velocities are for total face area, not the net free area; other velocities in table are for net free area.

Table 11-8   RECOMMENDED MAXIMUM DUCT VELOCITIES FOR
HIGH VELOCITY SYSTEMS*

cfm (m³/min) Carried by the Duct	Maximum Velocities fpm (m/s)
60,000 to 40,000 (1,700 to 1,133)	6000 (30.5)
40,000 to 25,000 (1,333 to 708)	5000 (25.4)
25,000 to 15,000 (708 to 425)	4500 (22.9)
15,000 to 10,000 (425 to 283)	4000 (20.3)
10,000 to 6,000 (283 to 170)	3500 (17.8)
6,000 to 3,000 (170 to 85)	3000 (15.2)
3,000 to 1,000 (85 to 28)	2500 (12.7)

*Reprinted by permission from *ASHRAE Handbook of Fundamentals,* 1977.

system will generally have a pressure loss of less than 0.15 in. water per 100 ft (1.23 Pa/m), whereas high velocity systems may have pressure losses up to about 0.7 in. water per 100 ft (5.7 Pa/m). Table 11-8 gives similar data for high velocity systems. The use of fibrous glass duct materials has gained wide acceptance in recent times because they are very effective for noise control. These ducts are also attractive from the fabrication point of view because the duct, insulation, and reflective vapor barrier are all the same piece of material. Metal ducts are usually lined with fibrous glass material in the vicinity of the air distribution equipment and for some distance away from the equipment. The remainder of the metal duct is then wrapped or covered with insulation and a vapor barrier. Insulation on the outside of the duct also reduces noise.

The duct system should be relatively free of leaks, especially when the ducts are outside the conditioned space. Air leaks from the system to the outdoors result in a direct loss that is proportional to the amount of leakage and the difference in enthalpy between the outdoor air and the air leaving the conditioner. At this time adequate methods of predicting duct leakage are not available; however, research is underway to gather data on duct leakage, and new standards are being established on the maximum allowable leakage. Presently 1 percent of the total volume flow of the duct system is an accepted maximum leakage rate in high velocity systems. No generally accepted level has been set up for low velocity systems. However, care should be taken to tape or otherwise seal all joints to minimize leakage in all duct systems and the sealing material should have a projected life of 20 to 30 years.

The layout of the duct system is very important to the final design of the system. Generally the location of the air diffusers and air moving equipment is first selected with some attention given to how a duct system may be installed. The ducts are then laid out with attention given to space and ease of construction. It is very important to design a duct system that can be constructed and installed in the allocated space. If this is not done, the installer may make changes in the field that lead to unsatisfactory operation.

The total pressure requirements of a duct system are an important consideration. From the standpoint of first cost, the ducts should be small; however, small
ducts tend to give high air velocities, high noise levels, and large losses in total pressure. Therefore, a reasonable compromise between first cost, operating cost, and
practice must be reached. The cost of owning and operating an air distribution system can be expressed in terms of system parameters, energy cost, life of the system,
and interest rates such that an optimum velocity or friction rate can be established
(2). A number of computer programs are available for this purpose.

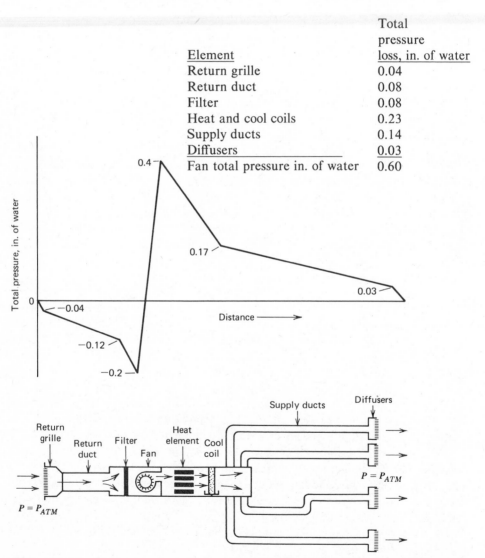

Element	Total pressure loss, in. of water
Return grille	0.04
Return duct	0.08
Filter	0.08
Heat and cool coils	0.23
Supply ducts	0.14
Diffusers	0.03
Fan total pressure in. of water	0.60

Figure 11-24.   Total pressure profile for a typical residential or light commercial system.

The total pressure requirements of a duct system are determined in two main ways. For residential and light commercial applications all of the heating, cooling, and air moving equipment is determined by the heating and/or cooling load. Therefore, the fan characteristics are known before the duct design is begun. Furthermore, the pressure losses in all other elements of the system except the supply and return ducts are known. The total pressure available for the ducts is then the difference between the total pressure characteristic of the fan and the sum of the total pressure losses of all of the other elements in the system excluding the ducts. Figure 11-24 shows a typical total pressure profile for a residential or light commercial system. In this case the fan is capable of developing 0.6 in. of water at the rated capacity. The return grille, filter, coils, and diffusers have a combined loss in total pressure of 0.38 in. of water. Therefore, the available total pressure for which the ducts must be designed is 0.22 in. of water. This is usually divided for low velocity systems so that the supply duct system has about twice the total pressure loss of the return ducts.

Large commercial and industrial duct systems are usually designed using velocity as a limiting criterion and the fan requirements are determined after the design is complete. For these larger systems the fan characteristics are specified and the correct fan is installed in the air handler at the factory or on the job.

It has been the practice of some designers to neglect velocity pressure when dealing with low velocity systems. This assumption does not simplify the duct design procedure and is unnecessary. When the air velocities are high, the velocity pressure must be considered to achieve reasonable accuracy. If static and velocity pressure are computed separately as required by some design methods, the problem becomes very complex. The trend is to use total pressure exclusively in duct design procedures because it is simpler and accounts for all of the flow energy.

11-10   DESIGN OF LOW VELOCITY DUCT SYSTEMS

The methods described in this section pertain to low velocity systems where the average velocity is less than about 1000 ft/min or 5 m/s. These methods can be used for high velocity system design, but the results will not be satisfactory in most cases.

Equal Friction Method

The principle of this method is to make the pressure loss per foot of length the same for the entire system. If the layout is symmetrical with all runs from fan to diffuser about the same length, this method will produce a good balanced design. However, most duct systems have a variety of duct runs ranging from long to short. The short runs will have to be dampered, which can cause considerable noise.

The usual procedure is to select the velocity in the main duct adjacent to the fan to provide a satisfactory noise level (Table 11-7) for the particular application. The known flow rate then establishes the duct size and the lost pressure per unit of length using Fig. 11-18. This same pressure loss per unit length is then used throughout the system. A desirable feature of this method is the gradual reduction of air velocity from fan to outlet, thereby reducing noise problems. After sizing the system the designer must compute the total pressure loss of the longest run (largest flow resistance), taking care to include all fittings and transitions. When the total pressure available for the system is known in advance, the design loss value may be established by estimating the equivalent length of the longest run and computing the lost pressure per unit length.

Example 11-9.   Select duct sizes for the simple duct system of Fig. 11-25 using the equal friction method and SI units. The system is for a residence. The total pressure available for the duct system is 0.12 in. of water or 30 Pa and the loss in total pressure for each diffuser at the specified flow rate is 0.02 in. of water or 5 Pa.

Solution.   Because the system is for a residence, the velocity in the main supply duct should not exceed 5 m/s and the branch duct velocities should not exceed 3 m/s. The total pressure available for the ducts, excluding the diffusers, is 25 Pa and the longest run is 1-2-3. The equivalent length method will be used to account for losses in the fittings. Then if we use Figs. 11-22 and 11-23,

$$L_{123} = (L_1 + L_{ent}) + (L_2 + L_{st}) + (L_3 + L_{wye} + L_{el} + L_{boot})$$
$$L_{123} = (11 + 6) + (1.5 + 4.6) + (3 + 11 + 3 + 9) = 49.1 \text{ m}$$

and

$$\Delta P'_0 = \frac{\Delta P_0}{L_{123}} = \frac{25}{49.1} = 0.509 \text{ Pa/m}$$

This value will be used to size the complete system. Table 11-9 summarizes the results showing the duct sizes, velocity in each section, and the loss in total pressure in each section. The rectangular sizes selected are rather arbitrary in this case.

It is of interest to check the actual loss in total pressure from the plenum to each outlet.

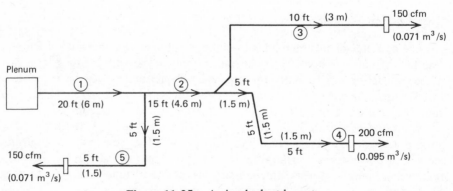

Figure 11-25.   A simple duct layout.

Table No. 11-9   SOLUTION TO EXAMPLE 11-9

Section Number	$\dot{Q}$ m^3/s	D cm	$w \times h$ cm $\times$ cm	Velocity m^3/s	$\Delta P_0'$ Pa/m	L_e m	ΔP_0 Pa
1	0.237	30	40 $\times$ 18	3.5	0.51	17	8.67
2	0.166	26.5	38 $\times$ 16	3.2	0.51	6.1	3.11
3	0.071	19	28 $\times$ 11	2.6	0.51	26	13.26
4	0.095	21.5	32 $\times$ 13	2.8	0.51	28.5	14.54
5	0.071	19	28 $\times$ 11	2.6	0.51	37	18.87

$$(\Delta P_0)_{123} = 8.67 + 3.11 + 13.26 = 25.04 \text{ Pa}$$
$$(\Delta P_0)_{124} = 8.67 + 3.11 + 14.54 = 26.32 \text{ Pa}$$
$$(\Delta P_0)_{15} = 8.67 + 18.87 = 27.54 \text{ Pa}$$

The loss in total pressure for the three different runs are unequal when it is assumed that the proper amount of air is flowing in each. However, the actual physical situation is such that the loss in total pressure from the plenum to the conditioned space is equal for all runs of duct. Therefore, the total flow rate from the plenum will divide itself among the three branches in order to satisfy the lost pressure requirement. If no adjustments are made to increase the lost pressure in sections 3 and 4, the flow rates in these sections will increase relative to section 5 and the total flow rate from the plenum will increase slightly because of the decreased system resistance. However, dampers in sections 3 and 4 could be adjusted to balance the system.

Balanced Capacity Method

This method of duct design has been referred to as the "balanced pressure loss method" (3). However, it is the flow rate or capacity of each outlet that is balanced and not the pressure. As discussed above the loss in total pressure automatically balances regardless of the duct sizes. The basic principle of this method of design is to make the loss in total pressure equal for all duct runs from fan to outlet when the required amount of air is flowing in each. In general all runs will have a different equivalent length and the pressure loss per unit length for each run will be different. It is theoretically possible to design every duct system to be balanced. This may be shown by combining Eqs. (9-6) and (11-6c) to obtain

$$H_{01} - H_{02} = l_f = f \frac{L_e}{D} H_v \qquad (11\text{-}14)$$

Then for a given duct and fluid flowing

$$H_{01} - H_{02} = l_f(L_e, D, \overline{V}) \qquad (11\text{-}15)$$

Because the volume flow rate $\dot{Q}$ is a function of the velocity $\overline{V}$ and the diameter D, Eq. (11-15) may be written as

$$H_{01} - H_{02} = l_f(\dot{Q}, L_e) \tag{11-15a}$$

For a given equivalent length the diameter can always be adjusted to obtain the necessary velocity that will produce the required loss in total pressure. There may be cases, however, when the required velocity may be too high to satisfy noise limitations and a damper or other means of increasing the equivalent length will be required.

The design procedure for the balanced capacity method is the same as the equal friction method in that the design pressure loss per unit length for the run of longest equivalent length is determined in the same way depending on whether the fan characteristics are known in advance. The procedure then changes to one of determining the required total pressure loss per unit length in the remaining sections to balance the flow as required. The method shows where dampers may be needed and provides a record of the total pressure requirements of each part of the duct system. Example 11-10 demonstrates the main features of the procedure.

Example 11-10. Design the duct system in Fig. 11-25 and Example 11-9 by using the balanced capacity method. Use SI units.

Solution. The total pressure available and the equivalent lengths will be the same as those in Example 11-9. In addition, the procedure for the design of the longest run L_{123} is exactly the same. Sections 4 and 5 must then be sized to balance the system. It is obvious from Fig. 11-25 that the lost pressure in section 4 must equal that of section 3.

$$\Delta P_{04} = \Delta P_{03} = 13.26 \text{ Pa}$$

The equivalent length of section 4 is 28.5 m; therefore,

$$\Delta P_{04}' = \frac{\Delta P_{04}}{L_4} = \frac{13.26}{28.5} = 0.465 \text{ Pa/m}$$

From Fig. 11-18b, at $\Delta P_{04}' = 0.465$ and $\dot{Q} = 0.095 \text{ m}^3/\text{s}$

$$D_4 = 22 \text{ cm} \qquad (26 \times 15 \text{ cm rectangle})$$
$$\overline{V}_4 = 2.7 \text{ m/s}$$

The loss in total pressure for section 5 is

$$\Delta P_{05} = \Delta P_{02} + \Delta P_{03} = 3.11 + 13.26 = 16.37 \text{ Pa}$$

The equivalent length of section 5 is 37 m; therefore

$$\Delta P_{05}' = \frac{\Delta P_{05}}{L_5} = \frac{16.37}{37} = 0.442 \text{ Pa/m}$$

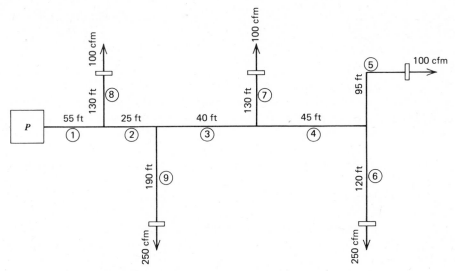

Figure 11-26.   Duct system for Example 11-11.

From Fig. 11-18*b*

$$D_5 = 19.5 \text{ cm} \qquad (22 \times 15 \text{ cm rectangle})$$
$$\overline{V}_5 = 2.4 \text{ m/s}$$

Comparison of the diameters or rectangular sizes obtained for sections 4 and 5 using the two different methods discussed thus far shows the sizes to be different for the balanced capacity method, as suggested in the discussion following Example 11-9. For these simple examples the differences were not dramatic. In real systems the equivalent lengths of the various runs vary considerably and the balanced capacity method is superior.

The only limitation of the balanced capacity method is the use of equivalent lengths for the fittings. Experience has shown this to be a minor error when duct velocities are less than 1000 ft/min or about 5 m/s (4).

Example 11-11.   Design the duct system of Fig. 11-26 by using the balanced capacity method. The velocity in the duct attached to the plenum must not exceed 900 ft/min and the overall loss in total pressure should not exceed about 0.32 in. of water. Total pressure losses for the diffusers are all equal to 0.04 in. of water. Rectangular ducts are required. The lengths shown are the *total* equivalent lengths of each section. Use English units.

Solution.   In order to hold the duct sizes to a minimum the maximum velocity criteria will be used to establish the design pressure loss for the longest run of duct. Using Fig. 11-18 the pressure per 100 ft and equivalent diameter of section 1 is

$$H'_{01} = 0.096 \text{ in. of water}/100 \text{ ft}$$
$$D_1 = 12.9 \text{ in.}$$

If this pressure loss per 100 ft is used to design sections 1-2-3-4-5, the longest run, the loss in total pressure will be

$$H_{015} = 0.096(55 + 25 + 40 + 45 + 95)/100 = 0.250 \text{ in. of water}$$

When the diffuser loss is added, the maximum lost pressure criterion is still satisfied ($H_0 = 0.29$ in. of water); therefore the longest run will be sized using $H'_{01} = 0.096$ in. of water/100 ft. Table 11-10 summarizes this procedure.

Branches 6, 7, 8, and 9 must now be sized to balance the system. Because all of the diffusers have the same lost pressure, they need not be considered in the remainder of the analysis.

$$H_{06} = H_{05} = 0.091 \text{ in. of water}$$
$$L_6 = 120 \text{ ft}$$
$$H'_{06} = 0.091(100)/120 = 0.076 \text{ in. of wat}/100 \text{ ft}$$
$$D_6 = 8.7 \text{ in.} \qquad (11 \times 6 \text{ rectangle})$$
$$\overline{V}_6 = 610 \text{ ft/min}$$
$$H_{07} = H_{04} + H_{05} = 0.044 + 0.091 = 0.135 \text{ in. of water}$$
$$L_7 = 130 \text{ ft}$$
$$H'_{07} = \frac{0.135(100)}{130} = 0.100 \text{ in. of wat}/100 \text{ ft}$$
$$D_7 = 5.8 \text{ in.} \qquad (6 \times 5 \text{ rectangle})$$
$$\overline{V}_7 = 540 \text{ ft/min}$$
$$H_{09} = H_{03} + H_{07} = 0.038 + 0.135 = 0.173 \text{ in. of water}$$
$$L_9 = 190 \text{ ft}$$
$$H'_{09} = \frac{0.173(100)}{190} = 0.091 \text{ in. of wat}/100 \text{ ft}$$
$$D_9 = 8.4 \text{ in.} \qquad (10 \times 6 \text{ rectangle})$$
$$\overline{V}_9 = 650 \text{ ft/min}$$
$$H_{08} = H_{09} + H_{02} = 0.173 + 0.024 = 0.197 \text{ in. of water}$$
$$L_8 = 130 \text{ ft}$$
$$H'_{08} = \frac{0.197(100)}{130} = 0.152 \text{ in. of wat}/100 \text{ ft}$$
$$D_8 = 5.4 \text{ in.} \qquad (6 \times 4 \text{ rectangle})$$
$$\overline{V}_8 = 630 \text{ ft/min}$$

The resulting velocities for sections 8 and 9 are slightly high but are probably acceptable.

Table 11-10 DATA FOR EXAMPLE 11-11

Section Number	$\dot{Q}$ cfm	D_e in.	$w \times h$ in. $\times$ in.	Velocity fpm	L_e ft	H_0 in. of water
1	800	12.9	18 × 8	900	55	0.053
2	700	12.1	16 × 8	880	25	0.024
3	450	10.2	15 × 6	780	40	0.038
4	350	9.4	13 × 6	740	45	0.044
5	100	5.9	6 × 5	540	95	0.091
Diffuser 5	100	—	—	—	—	0.040
				Total		0.290

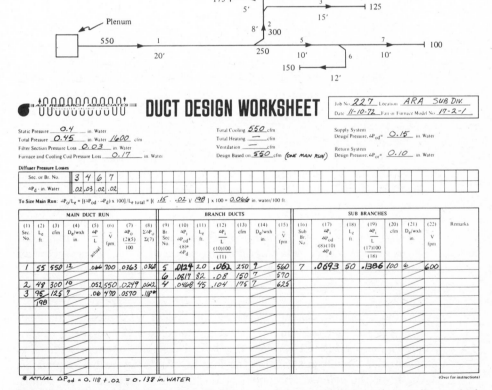

Figure 11-27. Duct design example using worksheet and balanced capacity method. (Reprinted with permission of Air Research Associates, Inc.)

The previous examples have shown the balanced capacity method to be straight-forward and to produce much detailed information about the duct system, particularly with respect to the duct velocities and the placement of dampers. However, the calculation procedure is cumbersome and time-consuming when carried out for a complex system as shown. To alleviate these undesirable features a worksheet has been devised. An example has been solved using the worksheet (Fig. 11-27). A full-sized worksheet is enclosed in the packet at the end of the text, Chart 6. Note that the worksheet has provisions for recording information concerning the various elements of the air distribution system including the fan, filters, cooling and heating coils, and the diffusers. One added feature of the worksheet is the provision for sub-branches, which sometimes occur in complex systems. The following procedure is used with the duct design worksheet.

1. Fill in the heading on the worksheet based on the equipment selected for the job.
2. Select diffusers for each outlet and record the total pressure requirements as shown.
3. Determine the main duct run (longest equivalent length). This is run 1 to 2 to 3 in the example; then section 4 is a branch, section 5 to 6 is a branch, and section 7 is a subbranch. Section 5 to 6 is selected as the branch because it has a larger equivalent length than section 5 to 7.
4. Enter the section numbers, equivalent lengths, and volume flow rate for each section as shown. Notice the arrangement of the section numbers with respect to the duct system. Reading across the first line, section 5 to 6 branches off section 1 and section 7 branches off section 5. Similarly, section 4 branches off section 2.
5. Calculate the loss in total pressure per 100 ft for the main duct run in the space provided and use the value and Fig. 11-18 to size the various sections. If rectangular ducts are to be used, record the exact duct diameter from the chart. If circular ducts are to be used, round off to the nearest standard diameter as shown for the example.
6. Record the actual loss per 100 ft in column 5, the velocity in column 6, and compute the actual loss in total pressure for each section: record in column 7. The accumulated loss in total pressure is recorded in column 8. The last number in column 8 is the actual loss in total pressure for the main duct run exclusive of the diffuser and should be less than or equal to the design value.
7. Size the branch ducts by computing the loss in total pressure for each branch as shown in column 10. Find $\Delta P/L$ in column 12 and select the duct diameter using Fig. 11-18. When a branch has two or more sections as shown, the branch is sized for the loss in total pressure and equivalent length for the complete branch. For this example, sections 5 and 6 must have a loss of 0.0817 in. of water and the total equivalent length is 102 ft. Then $\Delta P/L = 0.08$. From the friction chart $D_6 = 7$ in. and $D_5 = 8.5$ in.

D_5 must be rounded up to 9 in. and then $\Delta P/L = 0.062$. The loss in total pressure for section 5 is finally calculated as 0.0124 in. of water.

8.   The subbranches are sized in the same way as the branches. Only a few systems have more than one or two subbranches. Most systems have none.

9.   If the ducts are to be rectangular, Table 11-3 may be used to convert the diameters to rectangular dimensions.

The completed worksheet makes a good record and provides a check on the velocities in each section. A check for computational errors may be made by summing the pressure losses horizontally in columns 8, 10, and 17 plus the lost pressure for the diffuser.

Return Air Systems

The design of the return system may be carried out using the methods described above. In this case the air flows through the branches into the main duct and back to the plenum. Although the losses in constant flow fittings are the same regardless of the flow direction, divided flow fittings behave differently and different equivalent lengths or loss coefficients must be used. Figures 11-28 and 11-29 give loss coefficients for typical divided flow fittings used in return systems. Equation (11-12)

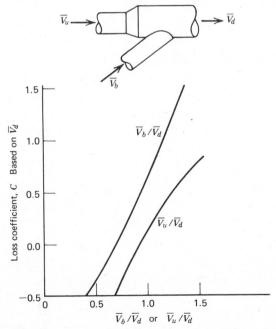

Figure 11-28.   Loss coefficients for converging flow fittings.

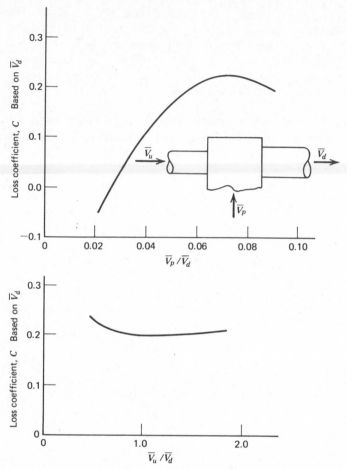

Figure 11-29.   Loss coefficients for converging flow through a plenum.

defines the lost pressure using the loss coefficient. The length to diameter ratio for the fittings shown in Figs. 11-28 and 11-29 is given approximately by

$$\frac{L}{D} \approx 50\ C \qquad\qquad (11\text{-}16)$$

Reference 2 gives considerable data for converging type fittings of both circular and rectangular cross section. It should be noted that for low velocity ratios the loss coefficient can become negative with converging flow streams. This seems contrary to natural laws; however, this behavior is a result of a high velocity stream mixing with a low velocity stream. Kinetic energy is transferred from the higher to the lower velocity air, which results in an increase in energy or total pressure of the slower stream. Low velocity return systems are usually designed using the equal friction method. The total pressure loss for the system is then estimated as discussed for supply duct systems. Dampers may be required just as with supply systems. In large commercial systems a separate fan for the return air may be required.

11-11 TURNING VANES AND DAMPERS

The two main accessory devices used in duct systems are vanes and dampers. It is the responsibility of the design engineer to specify the location and use of the devices.

Turning vanes have the purpose of preventing turbulence and consequent high loss in total pressure where turns are necessary in rectangular ducts. Although large radius turns may be used for the same purpose, this requires more space. When turning vanes are used, an abrupt 90 degree turn is made by the duct, but the air is turned smoothly by the vanes, as shown in Fig. 11-30. Turning vanes are of two basic designs (Fig. 11-31). The airfoil type is the more efficient of the two, but they are more expensive to fabricate than the single piece flat vane. The vanes shown in Figs. 11-31 *a* and 11-31 *b* are used in ducts having a maximum width of 36 in.; the vanes of Fig. 11-31 *c* are used in larger ducts.

Dampers are necessary to balance a system and to control ventilation and exhaust air. The dampers may be hand operated and locked in position after adjust-

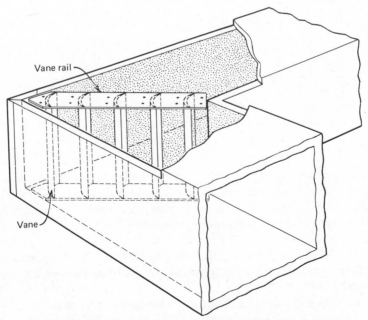

Figure 11-30.   Typical use of vanes in a rectangular duct.

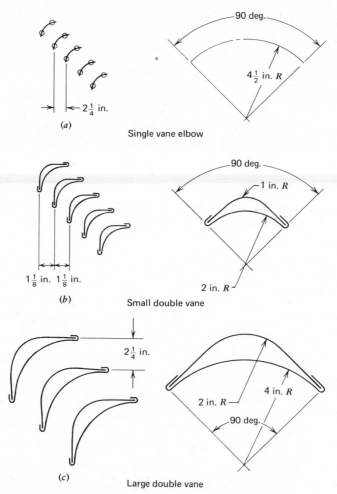

Figure 11-31. Details of single- and double-type turning vanes.

ment or may be motor operated and controlled by temperature sensors or by other remote signals. The damper may be a single blade on a shaft or a multiblade arrangement as shown in Fig. 11-32. The blades may also be connected to operate in parallel.

A combination damper and turning assembly, usually called an extractor, is shown in Fig. 11-33. The extractor is adjustable from outside the duct and may be extended into the air stream to regulate flow to a branch. Most designs completely close off flow to the branch when retracted. A four-bar linkage keeps the vanes aligned with the flow.

Pressure losses for turns equipped with turning vanes were discussed in Section 11-8. The pressure loss for an extractor will be about the same as an elbow with single plate vanes.

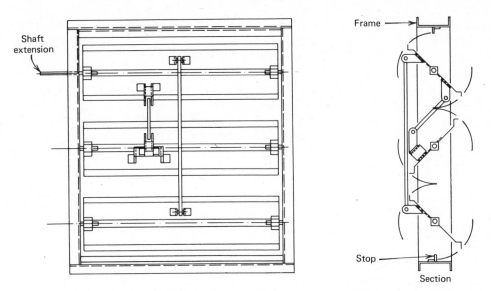

Figure 11-32.   Typical opposed blade damper assembly.

11-12   HIGH VELOCITY DUCT DESIGN

Because space allocated for ducts in large commercial structures is very limited due to the high cost of building construction, alternatives to the low velocity central system are usually sought. One approach is to use hot and cold water, which is piped to the various spaces where small central units or fan coils may be used. However, it is sometimes desirable to use rather extensive duct systems without taking up too much space. The only way this can be done is to move the air at much higher velocities. Table 11-8 indicates that high velocity systems may use velocities as high as 6000 ft/min or about 30 m/s. The use of high velocities reduces the duct sizes dramatically, but introduces some new problems.

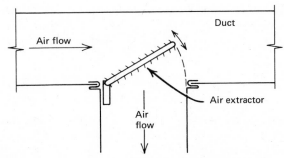

Figure 11-33.   Combination turning vane and damper assembly.

Noise is probably the most serious consequence of high velocity air movement. Special attention must be given to the design and installation of sound-attenuating equipment in the system. Generally, a sound absorber is installed just downstream of the fan and is similar to an automobile's muffler. Because the air cannot be introduced to the conditioned space at a high velocity, a device called a *terminal box* is used to throttle the air to a low velocity, control the flow rate, and attenuate the noise. The terminal box is located in the general vicinity of the space it serves and may distribute air to several outlets. These boxes take many forms.

The energy required to move the air through the duct system at high velocity is also an important consideration. The total pressure requirement is typically on the order of several inches of water and is due to the large losses associated with the high velocities. To partially offset the high fan power requirements, variable speed fans are sometimes used to take advantage of low volume flow requirements of variable volume systems at off-design conditions. Careful selection of duct fittings and the duct layout can also reduce the power requirement.

The higher static and total pressures required by high velocity systems aggravate the duct leakage problem. Ordinary duct materials such as light gauge steel and ductboard do not have suitable joints to prevent leakage and are also too weak to withstand the forces arising from the high pressure differentials. Therefore, an improved duct fabrication system has been developed for high velocity systems. The duct is generally referred to as *spiral duct* and has either a round or oval cross section. The fittings are machine formed and are especially designed to have low pressure losses and close fitting joints to prevent leakage. The pressure losses in spiral pipe and fittings are generally different from those in ordinary galvanized duct and fittings. There is also a much wider variety of fittings available for high velocity systems. For brevity, the data on pressure losses presented in Sections 11-7 and 11-8 will be used in this section. In actual practice, however, engineering data should be obtained from the manufacturer.

The criterion for designing high velocity duct systems is somewhat different from that used for low velocity systems. Emphasis is shifted from a self-balancing system to one that has minimum losses in total pressure. The high velocity system achieves the proper flow rate at each outlet through the use of the terminal box. The terminal box is designed to operate with a minimum pressure loss of about 0.5 to 1.0 in. of water or 125 to 245 Pa. Some of these devices maintain a constant volume flow rate, whereas others vary the flow rate in proportion to the space heating or cooling requirement. A double-duct type may also be used where warm and cool air are blended according to the space requirement. The run of duct which has the largest potential flow resistance should be designed to have a total pressure loss as low as possible. This run will then determine the system total pressure. The remaining runs will presumably result in a lower lost pressure but will be balanced by the terminal box.

Total pressure losses must be estimated carefully using the loss coefficient method, Eq. (11-12), because the equivalent length method is not reliable at velocities greater than about 1000 ft/min or 5 m/s. The most efficient fittings from the standpoint of lost pressure should be selected. For example, an ordinary square entrance to a duct from a plenum may be used in a low velocity system but, when used with high velocity air, the lost pressure becomes prohibitive. Table 11-5 indicates that the loss coefficient for a formed entrance is about one tenth that of a square entrance. If the duct velocity is 6000 ft/min or 30 m/s, the lost pressure for the square entrance is about 0.76 in. of water or 186 Pa, whereas the formed entrance has a loss of only about 0.067 in. of water or 20 Pa. Similar savings in total pressure may be achieved in the selection of branch fittings. Manufacturer's data should be consulted and the best fitting selected. Often the space available will not permit the use of the most efficient fitting and a less efficient fitting must be used. The designer should become acquainted with the standard practices for duct construction (7) before attempting an actual design problem.

There are a number of duct design procedures that may be used for high velocity systems. These will be described and demonstrated separately. The design procedures are relatively independent of the type of air distribution system involved. For example, a variable air volume duct system should be designed for the full load condition when air flow is at a maximum. At full load the *VAV* system operates the same as a constant volume system. However, a *VAV* system may be very extensive, serving zones that peak at much different times. In such a case, each section of the duct system must be sized to handle the maximum amount of air required.

Static Regain Method

This method systematically reduces the air velocity in the direction of flow in such a way that the increase (regain) in static pressure in each transition just balances the pressure losses in the following section. Although the method can produce a reasonably balanced system, this is not a major criterion for design of high velocity systems. The main disadvantages of the method are: (1) the very low velocities and large duct sizes that may result at the end of long runs; (2) the tedious bookkeeping and trial-and-error aspects of the method; and (3) the total pressure requirements of each part of the duct system are not readily apparent.

The general procedure for use of the static regain method is to first select a velocity for the duct attached to the fan or supply plenum. With the capacity this establishes the size of this main duct. The run of duct that appears to have the largest flow resistance is then designed first, using the most efficient fittings and layout possible. A velocity is assumed for the next section in the run and the static pressure regain is compared to the lost pressure for that section. Usually about two velocities must be checked to find a reasonable balance between the static pressure regain and the losses of a section. It must also be kept in mind that standard duct sizes must be used, which usually prevents an exact balance between regain and losses. The spiral duct is available in diameters of 3 to 24 in. using increments of one inch and from 24 to 50 in. using increments of two inches. Standard metric pipe ranges from about 8 to 60 cm in 1 cm increments and from 60 to 120 cm in 2 cm increments. The following example demonstrates the procedure for a simple duct system.

Example 11-12. Design the duct system shown in Fig. 11-34 using the static regain method. Each outlet has a terminal box that requires a minimum of 0.5 in. of water total pressure. Other pertinent data are shown on the sketch.

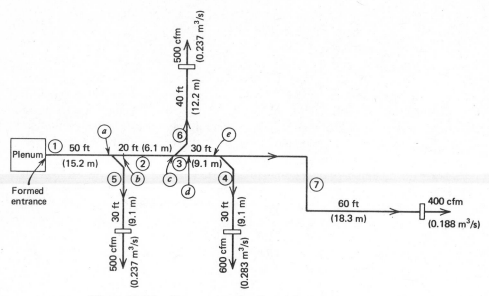

Figure 11-34. Duct system for Examples 11-13 and 11-15.

Solution.   The first step in the solution is to select a velocity for the section of duct connected to the plenum. Table 11-8 indicates that a maximum velocity of 2500 fpm should be used when the duct carries less than 3000 cfm. The section may then be sized using Fig. 11-18. A 12 in. duct will result in a velocity of about 2550 fpm, which is acceptable. Then

$$D_1 = 12 \text{ in.}; \qquad \overline{V}_1 = 2550 \text{ fpm}; \qquad H_{v1} = 0.405 \text{ in. of water}$$

From Table 11-5 the loss coefficient for the formed entrance is 0.03 and from Fig. 11-18 the pressure loss per 100 ft of duct is $H'_{01} = 0.75$ in. of water/100 ft. Then

$$(H_0)_p = 0.03 H_{v1} = 0.03(0.405) = 0.012 \text{ in. of water}$$
$$(H_0)_{pa} = \frac{H'_{01}L}{100} = \frac{0.75(50)}{100} = 0.375 \text{ in. of water}$$
$$H_{01} = 0.387 \text{ in. of water}$$

The total pressure required at the plenum to meet the needs of section 1 is then given by

$$H_{0p} = H_{01} + H_{v1} = 0.387 + 0.405 = 0.792 \text{ in. of water}$$

If the static regain method could be followed exactly, this would be the system total pressure, exclusive of the terminal box requirement, because velocity pressure is to be converted to static pressure to offset the lost pressure.

The run of duct with the largest flow resistance appears to be sections 1-2-3-4, but this must be checked at the conclusion of the deisgn. To size section 2 it is required that the increase in static pressure (decrease in velocity pressure) from point a to point b must equal the lost pressure from point b to point c. This is equivalent to

$$H_a - H_c = H_{vc} - H_{va} + (H_0)_{ab} + (H_0)_{bc} = 0 \qquad (11\text{-}17)$$

or

$$H_a - H_c = H_{v2} - H_{v1} + \Sigma(H_0)_{ac} = 0 \qquad (11\text{-}17\text{a})$$

and this may be generalized as

$$H_u - H_d = H_{vd} - H_{vu} + \Sigma(H_0) = 0 \qquad (11\text{-}17\text{b})$$

where the subscripts u and d refer to upstream and downstream, respectively. To size section 2 a diameter of 11 in. will be assumed and calculations made to check against Eq. (11-17b).

$$D_2 = 11 \text{ in.}; \qquad \overline{V}_2 = 2000 \text{ fpm}; \qquad H'_{02} = 0.53 \text{ in. of water/100 ft}$$
$$H_{v2} = 0.25 \text{ in. of water}; \qquad \dot{Q}_2 = 1500 \text{ cfm}$$
$$(H_0)_{ab} = 0.03 \text{ in. of water} \qquad (\text{Table 11-6})$$
$$(H_0)_{bc} = 0.53(20)/100 = 0.106 \text{ in. of water}$$
$$\Sigma(H_0)_{ac} = 0.136 \text{ in. of water}$$

Then using Eq. (11-17b)

$$H_a - H_c = 0.25 - 0.405 + 0.136 = -0.019 \text{ in. of water}$$

This indicates that a larger increase in static pressure than required was achieved in the transition. Although the next smaller pipe size could be tried, the 11 in. duct would prove to be the best solution.

The same general procedure is used to size section 3 assuming a 10 in. diameter duct.

$$D_3 = 10 \text{ in.}; \qquad \overline{V}_3 = 1850 \text{ fpm}; \qquad H'_{03} = 0.5 \text{ in. of water/100 ft}$$
$$H_{v3} = 0.213 \text{ in. of water}; \qquad \dot{Q}_3 = 1000 \text{ cfm}$$
$$(H_0)_{cd} = 0.015 \text{ in. of water} \qquad (\text{Table 11-6})$$
$$(H_0)_{de} = \frac{0.5(30)}{100} = 0.15 \text{ in. of water}$$
$$\Sigma(H_0)_{ce} = 0.165 \text{ in. of water}$$
$$H_c - H_e = 0.2313 - 0.25 + 0.165 = 0.128 \text{ in. of water}$$

Assume another diameter for section 3.

$$D_3 = 11 \text{ in.}; \qquad \overline{V}_3 = 1520 \text{ cfm}; \qquad H'_{03} = 0.31 \text{ in. of water/100 ft}$$
$$H_{v3} = 0.144 \text{ in. of water}; \qquad \dot{Q}_3 = 1000 \text{ cfm}$$
$$(H_0)_{cd} = 0.017 \text{ in. of water} \qquad (\text{Table 11-6})$$
$$(H_0)_{de} = 0.31(30)/100 = 0.093 \text{ in. of water}$$
$$\Sigma(H_0)_{ce} = 0.11 \text{ in. of water}$$
$$H_c - H_e = 0.004 \text{ in. of water}$$

This is a very good solution for section 3.

It is obvious that the design of a system using the above procedure can become quite tedious. A tabular form reduces the work considerably and will be used to conclude this design. Table 11-11 contains all of the design calculations for this example.

The total pressure requirement for the system is determined by the run with the maximum cumulative static pressure loss plus the velocity pressure in section 1.

$$H_0 = (H_0)_{1237} + H_{v1} = 0.392 + 0.405 = 0.797 \text{ in. of water}$$

where $H_{1237} = \Sigma(H_u - H_d)$ from Table 11-11. However, this does not include the pressure loss for the terminal box. In allowing for this pressure, remember that the total pressure just upstream of the box is equal to the velocity pressure. Section 7 has a velocity pressure of 0.052 in. of water. Because the terminal box requires at least 0.5 in. of water, the system total pressure requirement should be increased to

$$H_0 = 0.797 + (0.5 - 0.052) = 1.245 \text{ in. of water}$$

The required total pressure for this example is relatively low because of its small size. As high velocity systems become larger, the maximum velocity may be increased according to Table 11-8 with a corresponding increase in required total pressure.

Example 11-13. Suppose the duct system of Example 11-12 handles a total volume flow of 23 m^3/s. Design section 1 for the maximum velocity allowable from Table 11-8 and estimate the total pressure required for the system assuming static regain design. Assume a minimum total pressure of 245 Pa for the terminal boxes.

Solution. Table 11-8 shows a maximum velocity of 30.5 m/s for this case. Then using Fig. 11-18c the duct (section 1) is sized as

$$D_1 = 1 \text{ m}; \quad \overline{V}_1 = 30 \text{ m/s}; \quad \Delta P'_{01} = 7.6 \text{ Pa/m}$$

The loss coefficient for the formed entrance is 0.03 and the velocity pressure in section 1 is

$$P_{v1} = (30/1.29)^2 = 540.8 \text{ Pa}$$

Then

$$(\Delta P_0)_p = 0.03(540.8) = 16.2 \text{ Pa}$$
$$(\Delta P_0)_{pa} = 7.6(15.2) = 115.5 \text{ Pa}$$

The total pressure loss for section 1 is then 131.7 Pa. A conservative estimate of the system total pressure is given by

$$P_0 = P_{01} + P_{v1} + (P_0)_{box}$$
$$P_0 = 131.7 + 540.8 + 245 = 917.5 \text{ Pa}$$

Table 11-11 STATIC REGAIN DESIGN METHOD

Section Number	Item	D in.	Q cfm	$\bar{V}$ fpm	H_v in. of water	L ft	H_0' or C in. of water per 100 ft	H_0 in. of water	ΣH_0 in. of water	$(H_u - H_d)^a$ in. of water	$\Sigma(H_u - H_d)$ in. of water
1	Duct	12	2000	2550	0.405	50	0.75	0.375	0.387	0.387	0.387
	Entrance						0.03	0.012			
2	Duct	11	1500	2000	0.250	20	0.53	0.106	0.136	−0.019	0.368
	Fitting							0.030			
3	Duct	10	1000	1850	0.213	30	0.50	0.150	0.165	0.128	
	Fitting							0.015			
	Duct	11	1000	1520	0.144	30	0.31	0.093	0.110	0.004	0.372
	Fitting							0.017			
4	Duct	9	600	1350	0.114	30	0.32	0.096	0.147	0.117	
	Fitting						0.45	0.051			
	Duct	10	600	1100	0.075	30	0.19	0.057	0.087	0.018	0.390
	Fitting						0.40	0.030			
	Branch 7 off Section 3										
7	Duct	9	400	910	0.052	60	0.15	0.090			
	Fitting							0.012			
	Fitting						0.200	0.010	0.112	0.020	0.392
	Branch 6 off Section 2										
6	Duct	8	500	1450	0.131	40	0.41	0.164	0.216	0.097	
	Fitting						0.40	0.052			
	Duct	9	500	1140	0.081	40	0.23	0.092	0.133	0.036	0.332
	Fitting						0.50	0.041			
	Branch 5 off Section 1										
5	Duct	8	500	1450	0.131	30	0.41	0.123	0.189	−0.085	0.302
	Fitting						0.50	0.066			

^{a}Eq. (11-17b).

The total pressure method is an adaptation of the static regain method. The method provides insight into the actual energy requirements of each part of the duct system. The objective is to size each section so that the total pressure available at the beginning of a section is just enough to balance the loss in total pressure in the section.

Assumed Velocity Method

This method evolved following general acceptance of the total pressure concept in duct design. Appreciation of the total energy requirements of the duct system has placed emphasis on control of the local velocities and total pressure losses during the design of the system. At the same time almost no effort is made to design a self-balancing system. The method is based on the selection of acceptable air velocities at the inlet and in the branches of the system and then on a gradual reduction of the velocity in between. The total pressure losses are then calculated for each section proceeding from either end. The various air velocities are selected such that standard duct sizes will result in order to avoid double effort on the part of the designer.

This design method is recommended for the *VAV* duct system with zones that peak at widely varying times. Because the air flow rates in the various parts of the system change continually, there is no advantage to using a sophisticated design method. Careful attention to the maximum velocities attained in each section is of greatest importance.

A solution to the duct system of Fig. 11-34 using the assumed velocity method is given in Table 11-12 in SI units. The known characteristics, flow rate and duct length, were first entered in the table. The maximum velocities for sections 1, 4, 5, 6, and 7 were then selected along with the duct diameters for the same sections and entered in the table. The velocities and duct diameters were then selected for the intermediate sections. The reductions in velocity were divided as evenly as possible. The various losses were then computed and summed as shown for each run of duct. After completion of the table the results can be checked for any abnormally high losses; in such a case the duct diameter or a fitting may be changed to lower the overall total pressure requirement of the system. The system as designed in Table 11-12 requires about 503 Pa total pressure.

Accessories

The main accessories used with high velocity duct systems are sound absorbers and dampers. Sound absorbers were mentioned earlier and may take many forms. Although the design of these elements is beyond the scope of this book, manufacturer's data or other references on acoustics and noise should be consulted.

Balancing dampers may be used in high velocity systems; however, this function is usually handled by the terminal box. Fire dampers are of great importance in this system. The function of the fire damper is to stop the flow of air to a particular area if a fire should occur. Large commercial structures have partitions, floors, and ceilings that confine a fire to a given area for some specified time. When an air duct

Table 11-12   ASSUMED VELOCITY DESIGN METHOD

Section Number	Item	D cm	$\dot{Q}$ m³/s	$\overline{V}$ m/s	L m	$\Delta P_0'$ Pa/m	C	H_0 Pa	ΣH_0 Pa
1	Duct	30	0.945	14	15.2	7.4		112.5	
	Entrance						0.03	3.9	116.4
2	Duct	28	0.70	11.2	6.1	5.4		32.9	
	Fitting							8.7	158.0
3	Duct	23	0.47	11.0	9.1	6.2		56.4	
	Fitting							3.5	217.9
4	Duct	19	0.283	9.4	9.1	6.0		54.6	
	Fitting						0.45	23.9	
	Box							124.5	420.9
	Branch 7 off Section 3								
7	Duct	16	0.188	9.8	18.3	8.0		146.4	
	Fitting							2.5	
	Fitting						0.20	11.5	
	Box							124.5	502.8
	Branch 6 off Section 2								
6	Duct	18	0.237	9.0	12.2	6.0		73.2	
	Fitting						0.40	19.5	
	Box							124.5	375.2
	Branch 5 off Section 1								
5	Duct	18	0.237	9.0	9.1	6.0		54.6	
	Fitting						0.45	21.9	
	Box							124.5	317.4

passes through one of these fire barriers, a fire damper is generally required. The fire damper is normally open but automatically closes in the event of a fire. Some of these dampers are held open by fusible links while others are controlled by smoke detectors or other such devices.

The location and control of the fire dampers are the responsibility of the design engineer who must become acquainted with the governing codes for the particular application and city and state in which he or she is practicing.

REFERENCES

1. *ASHRAE Handbook and Product Directory,* American Society of Heating, Refrigerating and Air-Conditioning Engineers, New York, 1979.

2. *ASHRAE Handbook of Fundamentals,* American Society of Heating, Refrigerating and Air-Conditioning Engineers, New York, 1977.

3. Burgess H. Jennings, *Environmental Engineering Analysis and Practice,* International Textbook Company, Scranton, Pennsylvania, 1970.

4. F. C. McQuiston, "Duct Design for Balanced Air Distribution in Low Velocity Systems," *Proceedings,* Conference on Improved Efficiency in HVAC Components, Purdue University, Lafayette, Indiana, 1974.

5. Low Velocity Duct Construction Standards, 4th Ed., Sheet Metal and Air-Conditioning Contractors National Association, Washington, D.C., 1969.

6. J. B. Graham, "Methods of Selecting and Rating Fans," *ASHRAE Journal,* 1972.

7. High Pressure Duct Construction Standards, 3rd Ed., Sheet Metal and Air-Conditioning Contractors National Association, Washington D.C., 1975.

8. "AMCA Fan Application Manual," Parts 1, 2, and 3, Air Movement and Control Association, Inc., 30 West University Drive, Arlington Heights, IL.

PROBLEMS

11-1 A centrifugal fan is delivering 1700 cfm of air at a static pressure differential (across the fan) of 1.0 in. of water. The fan has an outlet area of 0.71 ft² and requires 0.7 horsepower shaft input. Compute (a) the static power, (b) the static efficiency, (c) the total efficiency, and (d) the fan total pressure.

11-2 A fan is delivering 2.1 m³/s of air at a total pressure of 280 Pa. The fan has an outlet area of 0.19 m² and requires 1100 W of shaft power. Compute (a) the total efficiency, (b) the static efficiency, and (c) the static pressure differential.

11-3 The fan of Problem 11-1 is operating at 1350 rpm. The fan speed is reduced to 1200 rpm. Compute the capacity, static and total pressure, and the shaft power at the lower speed.

11-4 The fan of Problem 11-2 is operating at 700 rpm. The speed is increased to 800 rpm. Compute the capacity, static and total pressure, and the shaft power at the higher speed.

11-5 The fan of Problem 11-1 discharges into a duct as shown in Fig. 11-35. Estimate the static and total pressure change from (i) to (o). Also find the average velocity at point (o).

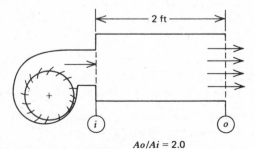

$Ao/Ai = 2.0$

Figure 11-35. Schematic of fan for Problem 11-5.

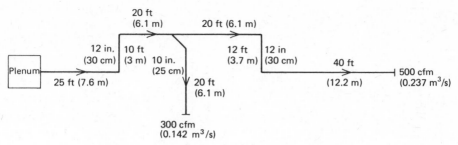

Figure 11-36. Schematic for Problems 11-6 and 11-7.

11-6 Compute the loss in total pressure for each run of the duct system shown in Fig. 11-36. The ducts are of round cross section. Turns and fittings are as shown. Use the loss coefficient and the equivalent length approaches and compare the answers. Use English units.

11-7 Compute the loss of total pressure for each run of the duct system shown in Fig. 11-36. Use SI units and the loss coefficient method.

11-8 The duct system shown in Fig. 11-37 is one branch of the complete air distribution system for a home. The system is a perimeter type located below the floor. The diffuser boots are shown in Fig. 11-22. Size the various sections of the system using the equal friction method and round pipe. A total pressure of 0.13 in. of water is available at the plenum. Compute the actual loss in total pressure for each run assuming the proper amount of air is flowing.

11-9 Consider the duct layout shown in Fig. 11-38. The system is supplied air by a rooftop unit that develops 0.25 in. of water total pressure external to the unit. The return air system requires 0.10 in. of water. The ducts are to be of round cross section and the maximum velocity in the main run is 850 ft/min, whereas the branch velocities must not exceed 650 ft/min. (a) Size the ducts using the equal friction method. Show the location of any required dampers. Compute the total pressure loss for the system. (b) Size the ducts using the balanced capacity method and the duct design

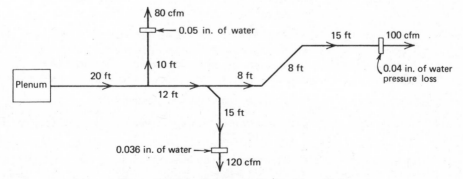

Figure 11-37. Schematic for Problem 11-8.

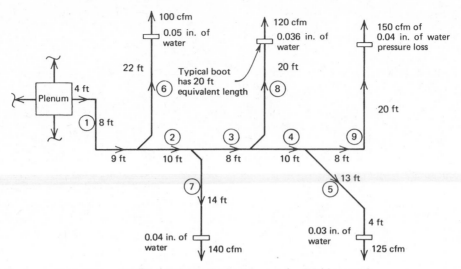

Figure 11-38.   Schematic duct layout for Problem 11-9.

worksheet. Show the location of any required dampers. Compute the loss in total pressure for the system.

11-10   Design the duct system shown in Fig. 11-39 for circular ducts. The fan produces a total pressure of 0.70 in. of water at 1000 cfm. The lost pressure in the filter, furnace, and evaporator is 0.35 in. of water. The remaining total pressure should be divided between the supply and return with about 60 to 65 percent used for the supply system. Diffuser and grille losses are shown on the diagram. Use equivalent lengths to account for fitting losses. (a) Use the equal friction method to size the ducts. (b) Use the balanced capacity method and the duct design worksheet to size the ducts.

11-11   Design the duct system shown in Fig. 11-40 using the balanced capacity method and the duct design worksheet. Circular ducts are to be used and installed below a concrete slab. The total pressure available at the plenum is 0.18 in. of water, and each diffuser has a loss in total pressure of 0.025 in. of water. Use the equivalent length method of accounting for losses in the fittings.

11-12   A high velocity duct layout is shown in Fig. 11-41. (a) Size the system using the static regain design method. Use maximum allowable velocities from Table 11-8. (b) Specify the required total pressure in the plenum. (c) Specify the fan characteristics using Fig. 11-7, shaft horsepower, efficiency, speed, and total pressure.

11-13   Repeat Problem 11-12 using the assumed velocity method.

11-14   Solve Problem 11-12 using the assumed velocity method and SI units. Specify fan characteristics.

11-15   Comment on the desirability of using the fan described in Fig. 11-6 to circulate (a) 10,000 cfm at about 1.8 in. of water total pressure, (b) 15,000 cfm at about 0.9 in. of water total pressure, (c) 2500 cfm at 2.0 in. of water total pressure.

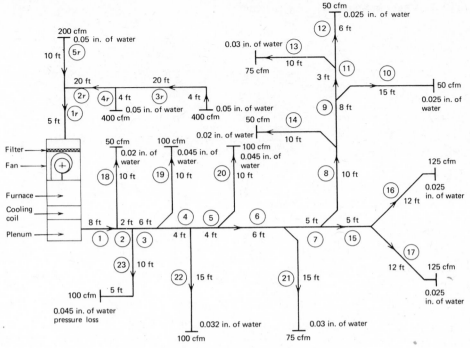

Figure 11-39. Layout for Problem 11-10.

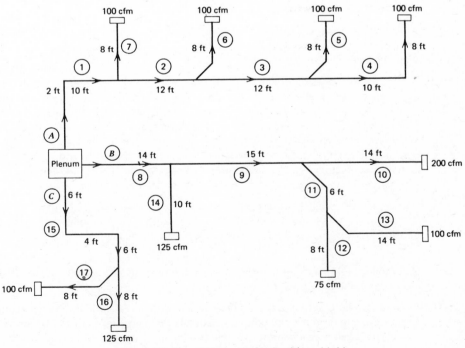

Figure 11-40. Schematic for Problem 11-11.

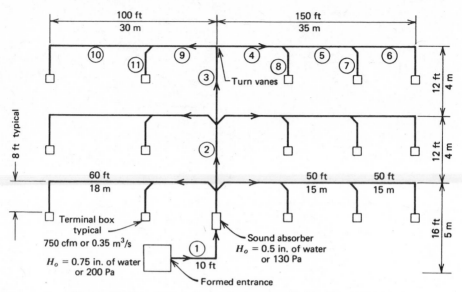

Figure 11-41.   High velocity duct system for Problem 11-12.

11-16 Would the fan shown in Fig. 11-7 be suitable for use in a system requiring (a) 5,000 cfm at 1.5 in. of water total pressure? (b) 30,000 cfm at 5.0 in. of water total pressure? (c) 15,000 cfm at 4.0 in. of water total pressure? Explain.

11-17 A duct system has been designed for 150 m³/min at 600 Pa total pressure. Would the fan shown in Fig. 11-10 be suitable for this application? Explain and estimate the total efficiency, fan speed, and shaft power.

11-18 A duct system has been designed to have 2.5 in. of water total pressure loss. It is necessary to use an elbow at the fan outlet which turns the air 90 degrees to the side. Reference 8 gives a system effect factor of 0.2 in. of water for this situation. The fan inlet also has an elbow with a system effect factor of 0.15 in. of water. (a) What total pressure should the fan for this system produce to circulate the desired amount of air? (b) Sketch the system and fan characteristics with and without the system effect factors. (c) The design volume flow rate is 15,000 cfm. Using Fig. 11-7 estimate the resulting flow rate when the system effect factors are not used.

11-19 The fan shown in Fig. 11-10 is operating in a system at 900 rpm with a flow rate of 150 m³/s. The system was designed to circulate 170 m³/min. Assuming that the duct pressure losses were accurately calculated, estimate the system effect factor for the fan.

11-20 Refer to Fig. 11-15 and compute the percent difference in shaft power between flow conditions 1 and 2.

11-21 A variable volume system using a fan as shown in Fig. 11-15 will operate at about 15,000 cfm a majority of the time. (a) Estimate the power saved in kW-hr for one day as compared with a constant volume system operating at point 1. (b) Estimate the power saving as compared with a VAV system with no fan speed control.

11-22 Consider the fan of Fig. 11-16 operating as shown in a *VAV* system. Compute the percent decrease in shaft power between flow conditions 1 and 2.

11-23 Assume the fan and *VAV* system shown in Fig. 11-16 operate at an average capacity of 25,000 cfm over a given 24-hour period. (a) Estimate the power savings as compared with a constant volume system operating at point 1. (b) Estimate the power saving as compared with a *VAV* system with no fan control. That is, point 1 will move along the full open characteristic

11-24 Estimate the lost pressure in 50 ft of 22 in. × 20 in. metal duct with an air flow rate of 3000 cfm. There are two mitered elbows with formed, 2-piece turning vanes and the duct is lined with one inch of the type B liner shown in Fig. 11-19.

11-25 A circular metal duct 20 ft in length has an abrupt contraction at the inlet and an abrupt expansion at the exit. Both have an area ratio of 0.6. The duct has a diameter of 10 in. with a flow rate of 600 cfm. Estimate the loss in total pressure for the duct including the contraction and expansion.

11-26 Consider a 6 in. circular branch duct which is tapped into the side of a large rectangular duct. The branch is a flexible fiber glass material with roughness similar to the type C liner of Fig. 11-19. The branch is 20 ft in length with a flow rate of 150 cfm. Estimate the lost pressure for the takeoff and duct assuming the takeoff and duct are similar to an abrupt contraction with an area ratio of 0.0.

11-27 The schematic of a large office complex is shown in Fig. 11-42. Halls, closets, and so on, have been omitted for clarity. The air quantities shown are the maximum that each space will require but the spaces will not all peak at the same time. (a) Lay out a duct system starting from the equipment room running due west and turning into two trunk ducts running north and south between the three rows of spaces. Each space requires a terminal box connected to the trunk duct with circular duct. (b) Estimate the maximum flow for the system and each section of the duct using the

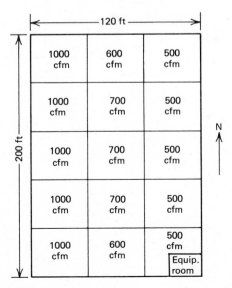

Figure 11-42. Schematic for Problem 11-27.

following guidelines: The east rooms will peak about 10:00 A.M. and remain constant until about 3:00 P.M. The interior rooms peak at 9:00 P.M., and remain constant until 4:00 P.M. The north and south rooms in the center peak at noon and demand gradually decreases after that. The west rooms reach one-half capacity at noon and peak at 5:00 P.M. The minimum overall load occurs at 5:00 A.M. and is about 20 percent of the maximum. (c) Size the ducts using a maximum velocity of about 2500 ft/min and the assumed velocity method. Use a velocity of about 1000 ft/min for the branch connections to the terminal boxes (d) Estimate the total pressure requirements of the system at maximum flow. (e) Locate the static pressure sensor for the fan speed control. (f) Refer to Fig. 11-15 and locate operating points 1 and 2. Scale the size of the fan up or down as required by changing the capacity values along the abscissa. (g) How much power does the system require at the minimum and maximum operating points?

CHAPTER TWELVE

Mass Transfer and the Measurement of Humidity

Although the psychrometric chart is very useful in solving moist air problems, to use the chart we must know two independent properties plus the barometric pressure. The dry bulb temperature and barometric pressure may be determined easily but the measurement of the one remaining property is a problem.

The wet bulb thermometer is the most practical device available to measure a property related to the moisture content of the air and is a reliable instrument when properly applied. The main objective of this chapter is to discuss its application in detail. However, some discussion of convective mass transfer is at first necessary.

12-1   COMBINED HEAT AND MASS TRANSFER

Many problems in engineering are concerned with the simultaneous transfer of mass and heat. In this book these problems deal mainly with heated and cooled air–water vapor mixtures that result in the evaporation or condensation of the water.

It is well known that a link exists between the transport of momentum, heat, and mass (1). Consider the two-dimensional boundary layer equations for an incompressible, constant property fluid with a zero pressure gradient.

$$u \frac{\partial C}{\partial x} + v \frac{\partial C}{\partial y} = D \frac{\partial^2 C}{\partial y^2} \tag{12-1}$$

$$u \frac{\partial u}{\partial x} + v \frac{\partial u}{\partial y} = \nu \frac{\partial^2 u}{\partial y^2} \tag{12-2}$$

$$u \frac{\partial t}{\partial x} + v \frac{\partial t}{\partial y} = \alpha \frac{\partial^2 t}{\partial y^2} \tag{12-3}$$

The boundary conditions are:

$$y = 0; \quad u = 0, \quad v = v_w, \quad t = t_w, \quad C = C_w \tag{12-4}$$
$$y = \infty; \quad u = u_\infty \quad \quad \quad t = t_\infty, \quad C = C_\infty \tag{12-5}$$

The quantities D, ν, and α are the diffusivities for mass, momentum, and heat, respectively. When these quantities are equal and the wall velocity v_w is zero, the solutions to Eqs. (12-1) through (12-3) are identical and the concentration C, velocity u, and temperature profiles t are similar. In this case the Prandtl number Pr and Schmidt number Sc are both equal to one, where

$$\text{Pr} = \frac{\nu}{\alpha} \tag{12-6}$$

and

$$\text{Sc} = \frac{\nu}{D} \tag{12-7}$$

The ratio of the Schmidt number to the Prandtl number is the Lewis number

$$\text{Le} = \frac{\text{Sc}}{\text{Pr}} = \frac{\alpha}{D} \tag{12-8}$$

which is also equal to one in this case.

To clearly show the connection between heat, mass, and momentum transfer consider an energy balance on a surface

$$h(t_\infty - t_w) = k\left(\frac{\partial t}{\partial y}\right)_{y=0} \tag{12-9}$$

where h and k are the convective heat transfer coefficient and the fluid thermal conductivity, respectively. When the solution to Eq. (12-3) for t and the distance y are nondimensionalized

$$t' = \frac{t - t_w}{t_\infty - t_w} \qquad y' = \frac{y}{L} \tag{12-10}$$

and inserted in Eq. (12-9), the following result is obtained:

$$\text{Nu} = \frac{hL}{k} = \left(\frac{\partial t'}{\partial y'}\right)_{y=0} \tag{12-11}$$

where Nu is the Nusselt number, a dimensionless heat transfer coefficient.

Fick's law may be written as

$$\dot{m}_w = -DA\left(\frac{\partial C}{\partial y}\right)_{y=0} \tag{12-12}$$

where $\dot{m}_w$ is the mass transfer rate, mass per unit time. A mass transfer coefficient is introduced:

$$\dot{m}_w = h_m A(C_w - C_\infty) \tag{12-12a}$$

where h_m has dimensions of length per unit time and C is the concentration, mass per unit volume. Combination of Eqs. (12-12) and (12-12a) and the use of the dimensionless concentration and length

$$C' = \frac{(C_w - C)}{(C_w - C_\infty)} \qquad y' = \frac{y}{L} \qquad (12\text{-}13)$$

yields

$$\text{Sh} = \frac{h_m L}{D} = \left(\frac{\partial C'}{\partial y'}\right)_{y=0} \qquad (12\text{-}14)$$

where Sh is the Sherwood number, a dimensionless mass transfer coefficient.

When the temperature and concentration profiles are similar, the dimensionless profiles C' and t' are identical. As a result the dimensionless heat and mass transfer coefficients given by Eqs. (12-11) and (12-14) are also identical. The link to momentum transfer or friction can be shown in a similar way. A similar development may be carried out to show the same general relationship between heat, mass, and momentum transfer in turbulent flow.

It is well known that expressions for Nusselt and Sherwood numbers have the form

$$\text{Nu} = f(\text{Re}, \text{Pr}) \qquad (12\text{-}15)$$
$$\text{Sh} = f(\text{Re}, \text{Sc})$$

where the functional relations are given by

$$\text{Nu} = C_1 \text{Re}^a \text{Pr}^b \qquad (12\text{-}15a)$$
$$\text{Sh} = C_1 \text{Re}^a \text{Sc}^b$$

Reynolds analogy was first used to show the connection between heat and momentum transfer and appears as

$$\frac{h}{\rho c_p \overline{V}} = \frac{f}{2} \qquad (12\text{-}16)$$

where f is the Fanning friction factor. The analogy was later extended to mass transfer:

$$\frac{h_m}{\overline{V}} = \frac{h}{\rho c_p \overline{V}} = \frac{f}{2} \qquad (12\text{-}16a)$$

The Reynolds analogy has long been recognized as giving reasonable results when $\text{Pr} \approx \text{Sc} \approx 1$ and the temperature potential is moderate. Many other analogies have been proposed to account for the effect of Prandtl number. Chilton and Colburn (2) have proposed one of the most widely accepted analogies called the j-factor analogy:

$$j = j_m = \frac{f}{2} \qquad (12\text{-}17)$$

where

$$j = \frac{h}{\rho c_p \overline{V}} \text{Pr}^{2/3} = \frac{h}{G c_p} \text{Pr}^{2/3} \qquad (12\text{-}18)$$

and

$$j_m = \frac{h_m}{\overline{V}} Sc^{2/3} \tag{12-19}$$

From Eqs. (12-18) and (12-19)

$$\frac{h}{\rho c_p h_m} = \left(\frac{Sc}{Pr}\right)^{2/3} = Le^{2/3} \tag{12-20}$$

where $Le^{2/3}$ is approximately 1.0 for moist air at usual conditions. In air-conditioning calculations it is generally more convenient to use the concentration in the form of the humidity ratio W rather than C in mass of water per unit volume. The relation between the two is

$$C = W\rho \tag{12-21}$$

where ρ is the mass density of the dry air, mass per unit volume. Equation (12-12a) then becomes

$$\dot{m}_w = h_m A\rho(W_w - W_\infty) \tag{12-13a}$$

or

$$\dot{m}_w = h_d A(W_w - W_\infty) \tag{12-13b}$$

where

$$h_d = \rho h_m \tag{12-22}$$

The dimension of h_d is mass of air per unit area and time. The analogy of Eq. (12-20) then becomes

$$\frac{h}{c_p h_d} = Le^{2/3} \tag{12-20a}$$

and the j factor of Eq. (12-19) becomes

$$j_m = \frac{h_d}{\rho \overline{V}} Sc^{2/3} \tag{12-19a}$$

The use of the above analogy in moist air problems requires caution. Over the range of temperatures of 50 to 140 F or 10 to 60 C and from completely dry to saturated air the Lewis number ranges from about 0.81 to 0.86 (3), whereas the Schmidt and Prandtl numbers have values of about 0.6 and 0.7. Therefore, the theoretical basis for the analogy is not entirely satisfied. The analogy is also based on ideal surface and flow field conditions. For example, there is undisputed evidence that the water deposited on a surface during dehumidification roughens the surface and upsets the analogy because h is for a smooth dry surface (4). There is also evidence that disturbances in the flow field may influence the transfer phenomena and sometimes render the analogy invalid (5). The analogy seems to be most valid when there is direct contact between the air and water. In situations such as a dehumidifying heat exchanger, the condensate that collects on the surface upsets the fundamental basis for the analogy. Research recently completed related to this phenomena will be discussed in Chapter 14. We will use the *j*-factor analogy most extensively for direct contact processes.

12-2 THE PSYCHROMETER

The wet bulb thermometer or a variation of it is the most convenient and simplest means of measuring a state property related to the moisture content of an air–water vapor mixture, and practically all computational methods and charts are based on its use. Figures 2-2 and 2-3 show typical arrangements of the wet bulb and dry bulb thermometers that together make up the psychrometer.

Figure 12-1 shows a schematic arrangement of a psychrometer. Although mercury thermometers are shown, thermocouples or other temperature-sensing devices can be used. To carry out an analysis of the psychrometer the following assumptions are made.

1. Heat conduction along the stem of the thermometer is negligible.
2. The air velocity is low enough for the static and stagnation air temperatures to be essentially equal.

Notice that each thermometer may exchange heat with the air by convection and with the surroundings by radiation. In many cases radiation effects are very small for the dry bulb thermometer. Because the wet bulb thermometer is generally at a

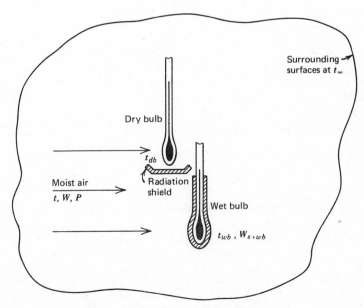

Figure 12-1. Schematic psychrometer.

different temperature than the dry bulb thermometer and the surroundings, radiation effects must usually be considered. It should always be kept in mind that a temperature-sensing device indicates its own temperature, which may not be the temperature of the surrounding medium.

If we assume that the psychrometer has reached a steady state condition, an energy balance on the dry bulb thermometer yields

$$h(t_{db} - t) = h_r(t_\infty - t_{db}) \tag{12-23}$$

where

t = actual air dry bulb temperature, F or C
t_{db} = dry bulb thermometer temperature, F or C
t_∞ = temperature of surrounding enclosure, F or C
h = convective film coefficient for the thermometer, Btu/(hr-ft²-F) or W/(m²-C)
h_r = radiation heat transfer coefficient, Btu/(hr-ft²-F) or W/(m²-C)

Solving for the air dry bulb temperature gives

$$t = t_{db} - \left(\frac{h_r}{h}\right)_{db} (t_\infty - t_{db}) \tag{12-24}$$

A method of handling the (h_r/h) term will be described later.

The energy balance on the wet bulb thermometer must account for the evaporation of water from the wick.

$$h_d(W_{s,wb} - W)i_{fg,wb} = h(t - t_{wb}) + h_r(t_\infty - t_{wb}) \tag{12-25}$$

where

$W_{s,wb}$ = humidity ratio for saturated air at the temperature indicated by the wet bulb thermometer, lbmw/lbma or kgw/kga
h_d = mass transfer coefficient, lbma/(ft²-hr) or kga/(m²-s)
$i_{fg,wb}$ = enthalpy of vaporization of water at the indicated wet bulb temperature, Btu/lbmw or J/kgw
t_{wb} = indicated wet bulb temperature, F or C

One may solve for the humidity ratio of the air

$$W = W_{s,wb} - B(t - t_{wb}) \tag{12-26}$$

where B is the *wet bulb coefficient* defined by

$$B = \frac{Le^{2/3}c_p}{i_{fg,wb}} \left| 1 + \left(\frac{h_r}{h}\right)\frac{(t_\infty - t_{wb})}{(t - t_{wb})} \right| \tag{12-27}$$

where the analogy of Eq. (12-20a) has been used to obtain the mass transfer coefficient h_d. The specific heat c_p may be accurately represented by

$$c_p = c_{pa} + c_{pv}\frac{(W + W_{s,wb})}{2} \tag{12-28}$$

and the combination of Eqs. (12-26) through (12-28) becomes

$$B = \cfrac{c_{pa} + c_{pv}W_{s,wb}}{\cfrac{i_{fg,wb}}{Le^{2/3}\left|1 + \cfrac{h_r}{h}\cfrac{(t_\infty - t_{wb})}{(t - t_{wb})}\right|} + \cfrac{c_{pv}}{2}(t - t_{wb})} \qquad (12\text{-}29)$$

where c_{pa} is 0.24 Btu/(lbma-F) or 1004 J/(kg-C) and c_{pv} is 0.45 Btu/(lbmv-F) or 1860 J/(kg-C). Equations (12-26) and (12-29) permit calculation of the air-humidity ratio W for the general case. The humidity ratio of saturated air at the wet bulb temperature may be obtained from Table A-2.

For the usual situation, where the surrounding surfaces are at the dry bulb temperature ($t = t_\infty$), Eq. (12-29) reduces to

$$B = \cfrac{c_{pa} + c_{pv}W_{s,wb}}{\cfrac{i_{fg,wb}}{Le^{2/3}\left(1 + \cfrac{h_r}{h}\right)} + \cfrac{c_{pv}}{2}(t - t_{wb})} \qquad (12\text{-}30)$$

When the wet bulb is perfectly shielded from radiation effects, Eq. (12-29) reduces to

$$B = \cfrac{c_{pa} + c_{pv}W_{s,wb}}{\cfrac{i_{fg,wb}}{Le^{2/3}} + \cfrac{c_{pv}}{2}(t - t_{wb})} \qquad (12\text{-}31)$$

The determination of the wet bulb coefficient B is the main problem in finding the humidity ratio W. Although the Lewis number is often assumed to be unity, for dry bulb temperatures of 50 to 100 F or 10 to 38 C and for completely dry to saturated air a value of 0.85 is more accurate. The radiation heat transfer coefficient h_r may be estimated from the relation

$$h_r = \sigma\varepsilon \frac{[(T_\infty)^4 - (T_{wb})^4]}{(t_\infty - t_{wb})} \qquad (12\text{-}32)$$

where ε is the emittance of the wick covering the wet bulb and the Stefan-Boltzman constant σ is 0.1712×10^{-8} Btu/(hr-ft^2-R^4) or 5.673×10^{-8} W/(m^2-K^4) depending on the system of units utilized. Equation (12-32) may be expressed in a more convenient form by multiplying and dividing by 10^8. Thus

$$h_r = \sigma'\varepsilon \frac{\left|\left(\dfrac{T_\infty}{100}\right)^4 - \left(\dfrac{T_{wb}}{100}\right)^4\right|}{(t_\infty - t_{wb})} \qquad (12\text{-}32a)$$

where $\sigma' = \sigma \times 10^8$.

The convection heat transfer coefficient h may be computed from the relation

$$Nu_d = \frac{hd}{k} = C\,Re_d^n \qquad (12\text{-}33)$$

Table 12-1   COEFFICIENTS FOR
EQUATION 12-33

$\text{Re}_d = \bar{V}d/\nu$	C	n
4–40	0.821	0.385
40–4,000	0.615	0.466
4,000–40,000	0.174	0.618

where the Prandtl number has been assumed equal to one and the coefficients C and n are given in Table 12-1. The fluid properties k and ν should be evaluated at the *mean film temperature*, t_f.

$$t_f = \frac{t + t_{wb}}{2} \tag{12-34}$$

Table B-4 gives the properties of air. For convenience in computing the wet bulb coefficient B, Figures 12-2 and 12-3 have been constructed using Eqs. (12-32) and (12-33), *for the special case of the surrounding temperature t_∞ equal to the air dry bulb temperature t.* A value of $\varepsilon_{wb} = 0.9$ was used and a wet bulb thermometer diameter of 0.3 in. or 7.5 mm applies to Fig. 12-2, whereas a diameter of 0.1 in. or 2.5 mm was used in Fig. 12-3. Figures 12-2 and 12-3 apply only to Eq. (12-30) where $t = t_\infty$. For the general case a new expression must be derived for the wet bulb coefficient B if Figs. 12-2 and 12-3 are to be used or Eq. (12-29) may be used with h_r and h as computed from Eqs. (12-32) and (12-33).

Example 12-1.   A wet bulb psychrometer is inserted in a 10 in. diameter, insulated air duct. The volume flow rate of the air is 450 cfm and it is thoroughly mixed. The dry bulb thermometer indicates 60 F, whereas the wet bulb reading is 50 F. The thermometers are unshielded and have a diameter of about 0.3 in. Determine the humidity ratio W.

Solution.   Because the duct is insulated the inside surface will approach the air dry bulb temperature very closely

$$t_\infty = t_{db} = t$$

Equation (12-26) will be used to compute W, whereas $W_{s,wb}$ may be read from Table A-2 at $t_{wb} = 50$ F, as 0.007658 lbmw/lbma. The wet bulb coefficient B is given by Eq. (12-30) for this case with h_r/h read from Fig. 12-2. The enthalpy of vaporization $i_{fg,wb}$ is obtained from Table A-1 as 1065 Btu/lbmw and the Lewis number is taken as 0.85.

The average air velocity in the duct is given by

$$\bar{V} = \frac{\dot{Q}}{A} = \frac{450}{(\pi/4)(10/12)^2} = 825 \text{ ft/min}$$

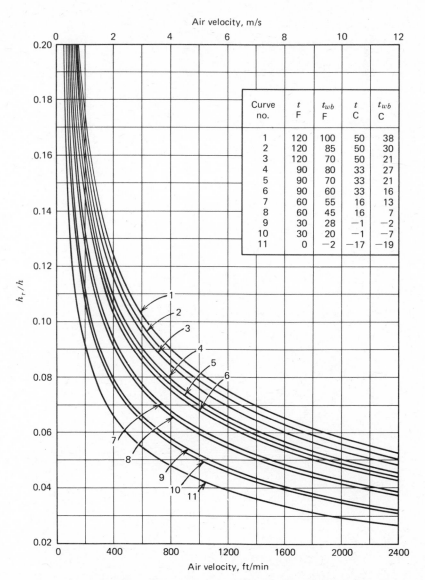

Air velocity, m/s

Curve no.	t F	t_{wb} F	t C	t_{wb} C
1	120	100	50	38
2	120	85	50	30
3	120	70	50	21
4	90	80	33	27
5	90	70	33	21
6	90	60	33	16
7	60	55	16	13
8	60	45	16	7
9	30	28	−1	−2
10	30	20	−1	−7
11	0	−2	−17	−19

Air velocity, ft/min

Figure 12-2. Ratio h_r/h for a wet bulb diameter of 0.3 in. (7.5 mm). (James L. Threlkeld, *Thermal Environmental Engineering,* 2nd edition, © 1970, p. 202. Reprinted by permission of Prentice-Hall, Inc., Englewood Cliffs, N.J.)

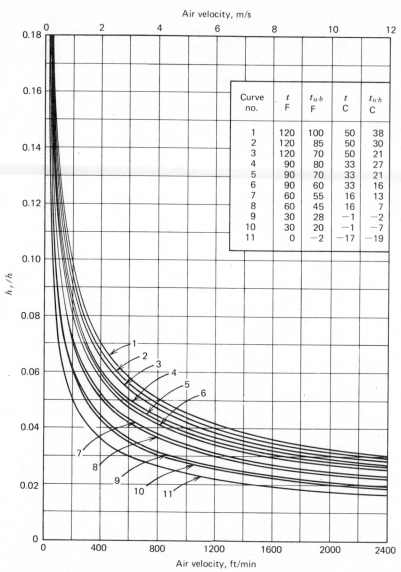

Figure 12-3. Ratio h_r/h for a wet bulb diameter of 0.1 in. (2.5 mm). (James L. Threlkeld, *Thermal Environmental Engineering*, 2nd edition, © 1970, p. 203. Reprinted by permission of Prentice-Hall, Inc., Englewood Cliffs, N.J.)

By interpolating between curves 7 and 8 in Fig. 12-2, $h_r/h = 0.067$. By using Eq. (12-30) we get

$$B = \frac{0.240 + 0.45(0.007658)}{\dfrac{1065}{(0.85)^{2/3}(1 + 0.067)} + 0.225(60 - 50)} = 2.184 \times 10^{-4}$$

Then by Eq. (12-26)

$$W = 0.007658 - (2.184)10^{-4}(60 - 50) = 0.005474 \text{ lbmw/lbma}$$

Example 12-2.   An uninsulated metal duct has air flowing at 600 ft/min. An inserted thermometer indicates an air dry bulb temperature of 120 F, whereas the pipe wall temperature is estimated to be 90 F. The thermometer is not shielded and its diameter is about 0.3 in. Find the actual air dry bulb temperature.

Solution.   Equation (12-24) applies here. The ratio h_r/h must be determined analytically using Eqs. (12-32) and (12-33). The Reynolds number will first be computed as

$$\text{Re}_d = \frac{\overline{V}d}{\nu} = \frac{600}{60} \frac{0.3}{12} \frac{10^4}{1.8} = 1389$$

where the viscosity ν was obtained from Table B-4 at 105 F. Then using Eq. (12-33) with coefficients from Table 12-1, we find that

$$h = \frac{k}{d}(0.615)\text{Re}_d^{0.466} = \frac{(15.6)10^{-3}}{(0.3/12)}(0.615)(1389)^{0.466}$$
$$h = 11.2 \text{ Btu/(hr-ft}^2\text{-F)}$$

where k was obtained from Table B-4 at 105 F. Eq. (12-32) with $\varepsilon = \varepsilon_{db} = 0.8$, $t_\infty = 90$ F, and t_{wb} replaced with $t_{db} = 120$ F yields

$$h_r = 0.1713(0.8)\frac{\left|\left(\dfrac{550}{100}\right)^4 - \left(\dfrac{580}{100}\right)^4\right|}{(90 - 120)} = 1.0 \text{ Btu/(hr-ft}^2\text{-F)}$$

Then

$$\frac{h_r}{h} = \frac{1.0}{11.2} = 0.089$$

Substitution in Eq. (12-24) then gives

$$t = 120 - (0.089)(90 - 120) = 122.7 \text{ F}$$

It is evident that radiation can introduce significant errors in the dry bulb temperature. The error is quite dependent on the emittance of the thermometer bulb. If the bulb of Example 12-2 had an emittance of about 0.2, the error in the measurement is less than one degree F.

Example 12-3. An unshielded wet bulb thermometer is placed in the duct used in Example 12-2 and indicates a temperature of 100 F. Find the humidity ratio W.

Solution. It will be necessary to estimate the wet bulb coefficient using Eq. (12-29) and Eq. (12-26) will give W. In this case

$$h_r = 0.1713(0.9) \frac{\left|\left(\dfrac{550}{100}\right)^4 - \left(\dfrac{560}{100}\right)^4\right|}{(90 - 100)} = 1.05 \text{ Btu/(hr-ft}^2\text{-F)}$$

and h will be assumed equal to that calculated in Example 12-2, $h = 11.2$ Btu/(hr-ft^2-F). Other data required are: $W_{s,wb} = 0.0432$ lbmw/lbma; $i_{fg,wb} = 1036.7$ Btu/lbmw; Le = 0.85; and $t = 122.7$ F from Example 12-2. Then

$$B = \frac{0.240 + 0.45(0.0432)}{(0.85)^{2/3} \left|1 + \dfrac{1.05}{11.2}\dfrac{(90 - 100)}{(122.7 - 100)}\right|} + 0.225(122.7 - 100)$$

$$B = 2.144 \times 10^{-4} \text{ lbmw/(lbma-F)}$$

and

$$W = 0.0432 - (2.144)10^{-4}(122.7 - 100) = 0.0383 \text{ lbmw/lbma}$$

In summary, the most reliable results are obtained with the wet bulb psychrometer when the air velocity is relatively high and radiation effects are small. There will always be some uncertainty in computing the wet bulb coefficient B; this effect will be greatest when W is small. Almost all humidity measuring devices have the same limitation.

12-3 ADIABATIC SATURATION TEMPERATURE VERSUS WET BULB TEMPERATURE

In the previous section the humidity ratio W was determined by measurements of wet and dry bulb temperatures with heat and mass transfer theory. The wet bulb temperature was not considered as a thermodynamic state property and indeed it should not be, because it depends on air velocity and the bulb geometry, in addition to the state of the moist air. It should be emphasized that the wet bulb temperature and the adiabatic saturation temperature are distinctly different. The adiabatic saturation temperature is hypothetical and can only be approached in practice. When the wet bulb temperature appears in psychrometric equations and charts, it is really the adiabatic saturation temperature that is being considered.

Threlkeld (3) has analyzed the problem and correlated wet bulb temperature with the adiabatic saturation temperature (sometimes called thermodynamic wet bulb temperature). The results are shown in Figs. 12-4 and 12-5 for thermometers with diameters of 0.3 and 0.1 in. or 7.5 and 2.5 mm. The data are shown as percent deviation ($t_{wb} - t^*$) of the wet bulb depression ($t - t_{wb}$) for shielded and unshielded wet bulbs. Three different temperature combinations are given and several general conclusions may be drawn. It appears that the unshielded wet bulb will generally more closely approximate the adiabatic saturation temperature. Temperatures taken with air velocities greater than about 200 ft/min or 1 m/s generally have the least deviation. Lower air velocities may be used with smaller wet bulbs. Threlkeld (3) draws the following general conclusion: For atmospheric temperatures above freezing, where the wet bulb depression does not exceed about 20 F or 11 C, and where no unusual radiation circumstances exist, t_{wb} should differ from t^* by less than about

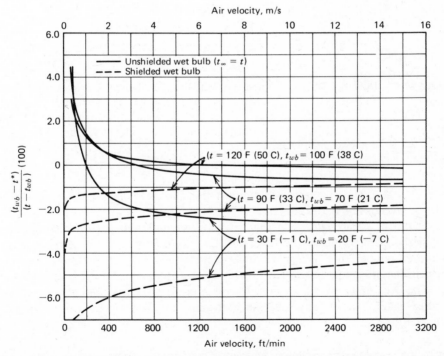

Figure 12-4. Deviation ($t_{wb} - t^*$) in percent of the wet bulb depression ($t - t_{wb}$) for a wet bulb diameter of 0.3 in. (7.5 mm) and a barometric pressure of 14.696 psia (101.325 kPa). (James L. Threlkeld, *Thermal Environmental Engineering*, 2nd edition, © 1970, p. 208. Reprinted by permission of Prentice-Hall, Inc., Englewood Cliffs, N.J.)

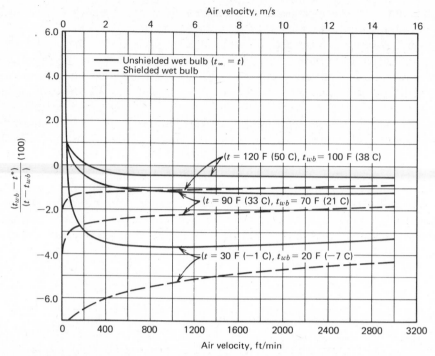

Figure 12-5. Deviation $(t_{wb} - t^*)$ in percent of the wet bulb depression $(t - t_{wb})$ for a wet bulb diameter of 0.1 in. (2.5 mm) and a barometric pressure of 14.696 psia (101.325 kPa). (James L. Threlkeld, *Thermal Environmental Engineering,* 2nd edition, © 1970, p. 209. Reprinted by permission of Prentice-Hall, Inc., Englewood Cliffs, N.J.)

0.5 F or 0.27 C for an unshielded mercury-in-glass thermometer as long as the air velocity exceeds about 100 ft/min or 0.5 m/s.

Thus, for most engineering problems the wet bulb temperature obtained from a properly operated unshielded psychrometer may be used directly as the adiabatic saturation temperature. When wet bulb depression is large or radiation effects are unusual, the procedure in Section 12-2 should be used.

The wick on the wet bulb should be kept clean and thoroughly moistened with distilled water. Air velocity over the bulb must be maintained to obtain reliable results.

12-4 OTHER METHODS OF MEASURING HUMIDITY

In addition to the wet bulb psychrometer there are other devices that measure humidity. Most of these devices are more complicated and less reliable than the psychrometer. One device that gives the relative humidity as a direct reading is a hygrometer. Several general types of hygrometers exist.

Dewpoint Hygrometer

In the most common form of this hygrometer means are provided for cooling and observing the temperature of a surface exposed to moist air. The highest temperature at which detectable condensation occurs on the surface is the dewpoint temperature. Charts and tables are used with the dewpoint temperature to find the relative humidity. A bright metallic surface cooled by ether or some other refrigerant is usually employed. There are many variations of the dewpoint hygrometer that are more fully discussed in reference 6.

Dimensional Change Hygrometer

Changes in humidity cause many organic materials to shrink or expand; this action may be tied to a suitable linkage to actuate a dial or switch. Materials commonly used are human hair, nylon, dacron, wood, and paper. Unfortunately these materials do not consistently have a reproducible behavior, which causes problems in calibration. The most common application of this type of hygrometer is in domestic humidistats used to control humidifiers.

Electrical Impedance Hygrometer

This type of hygrometer correlates the changing moisture content of a substance with a change in electrical impedance. An electrical or electronic signal processing system is therefore required. The advantage of the impedance hygrometer is that it provides an electrical signal for recording that may be used in telemetering data from remote locations. Reference 6 provides additional details.

Electrolytic Hygrometer

In this device a highly effective desiccant absorbs moisture from a regulated stream of the moist air. The absorbed moisture is electrolyzed, and the required current is related to the humidity level. The desiccant is usually phosphorous pentoxide and the apparatus is used at moisture-air ratios of 1 to 1000 parts per million.

Gravimetric Hygrometer

This hygrometer removes water vapor from a known quantity of moist air and weighs it. Obviously precise laboratory equipment is required. Powerful desiccants or freezing processes are used to remove the water vapor from the air.

All of the hygrometers discussed above must be calibrated. This may be done by maintaining an atmosphere at a known temperature and humidity and exposing the hygrometer to it. Complete calibration requires comparison at several different conditions. The National Bureau of Standards maintains a primary reference standard. Additional information and reference material on calibration and standards can be found in reference 6.

REFERENCES

1. J. D. Parker, J. H. Boggs, and E. F. Blick, *Introduction to Fluid Mechanics and Heat Transfer,* Addison-Wesley, Reading, Massachusetts. 1969.

2. T. H. Chilton and A. P. Colburn, "Mass Transfer (Absorption) Coefficients— Prediction from Data on Heat Transfer and Fluid Friction," *Industrial and Engineering Chemistry,* November 1934.

3. James L. Threlkeld, *Thermal Environmental Engineering,* Second Edition, Prentice-Hall, Englewood Cliffs, N.J., 1970.

4. J. L. Guillory and F. C. McQuiston, "An Experimental Investigation of Air Dehumidification in a Parallel Plate Exchanger," *ASHRAE Trans., 29,* June 1973.

5. Wayne A. Helmer, "Condensing Water Vapor—Air Flow in a Parallel Plate Heat Exchanger," Ph.D. Thesis, Purdue University, May 1974.

6. *ASHRAE Handbook of Fundamentals,* American Society of Heating, Refrigerating and Air-Conditioning Engineers, New York, 1977.

PROBLEMS

12-1   Compute the heat transfer coefficient where air is flowing over a surface at a velocity of 1200 ft/min and has a mean temperature of 100 F. The *j* factor is known to be 0.008.

12-2   Compute the heat transfer coefficient for air flowing at a temperature of 40 C and a velocity of 6 m/s. The *j* factor is 0.01.

12-3   Estimate the rate at which water is evaporated from a 1000 acre lake on an August day when the dry bulb and wet bulb temperatures are 100 and 75 F, respectively (43,560 ft^2 = 1 acre). Assume a heat transfer coefficient *h* of 5 Btu/(hr-ft^2-F) between the moist air and the lake surface and a water surface temperature of 80 F.

12-4   A psychrometer is mounted in a 6-in. diameter insulated duct. Forty cfm of air is flowing in the duct. The thermometers indicate 90 and 70 F and are unshielded. The bulb diameters are approximately $\frac{1}{4}$ in. Determine (a) the humidity ratio using Chart 1 and (b) the humidity ratio by the procedures of Section 12-2. (c) Compare the results of (a) and (b). (d) Using Fig. 12-4, select an optimum air velocity for this situation. How could this be achieved without changing the flow rate?

12-5   Consider a wet bulb psychrometer installed in an insulated air duct, 100 mm in diameter. The thermometers are indicating temperatures of 16 and 13 C and are not shielded. The bulb diameters are about 8 mm. The air flow rate is 0.02 m^3/s. (a) Compute the humidity ratio. (b) What minimum air velocity should be used to obtain a constant error in the adiabatic saturation temperature?

12-6   The sensible heat transfer coefficient for a dry surface has been determined to be 9 Btu/(hr-ft²-F) at a certain Reynolds number. Estimate the total heat transfer to the surface per square foot at a location where the wall temperature is 50 F and the state of the moist air flowing over the surface is given by 75 F dry bulb and 65 F wet bulb. Assume that the Reynolds number does not change and the Lewis number is 0.82. The condensate present on the surface will increase the transfer coefficients by about 15 percent.

12-7   Estimate the total heat transfer rate per square meter to a surface where the surface temperature is 9 C and the moist air flowing over the surface is at 25 C dry bulb and 19 C wet bulb. The sensible heat transfer coefficient for the same surface under dry conditions and at the same Reynolds number is known to be 55 W/(m²-C). Assume that the effect of the wet surface is to increase the heat transfer coefficient by 25 percent.

12-8   A psychrometer is installed in a duct where the air velocity is 700 ft/min. The thermometers are unshielded. The dry bulb thermometer indicates a temperature of 110 F and the duct wall temperature is estimated to be 85 F. The thermometer has a diameter of 0.15 in. Determine the actual air dry bulb temperature.

12-9   The wet bulb thermometer of Problem 12-8 indicates a temperature of 75 F. Determine the humidity ratio W.

12-10 An unshielded psychrometer installed in an insulated duct indicates temperatures of 60 and 45 F. The bulb diameters are about 0.10 in. and the air velocity is about 800 ft/min. Estimate the adiabatic saturation temperature of the air.

12-11 An unshielded psychrometer installed in an insulated duct indicates temperatures of 33 and 21 C. Estimate the adiabatic saturation temperature if the air velocity is 6 m/s and the diameters of the temperature probes are about 2 mm.

12-12 An unshielded psychrometer is to be used where the temperatures are about 105 and 85 F. Bulb diameters are about 0.1 in. The difference between the wet bulb and adiabatic saturation temperature must be relatively constant. (a) What minimum air velocity should be used? (b) Give the approximate difference.

12-13 Moist air flows over a wet bulb thermometer which is cylindrical in shape with a diameter of 0.25 in. (including the wick). If the air velocity is 1000 ft/min, the dry bulb air temperature is 80 F, and the wet bulb temperature is 70 F, compute the (a) sensible heat transfer coefficient h, (b) the mass transfer coefficient h_d.

12-14 A housekeeper hangs a wet blanket out to dry in a high wind. The blanket weighs 4 lb dry and 16 lb wet and has dimensions of 7 ft by 8 ft. Assume outdoor conditions of 90 F db and 50 percent relative humidity and that the blanket is at a temperature of 90 F. Estimate the time required for the blanket to become dry if the average heat transfer coefficient on both sides of the blanket is 4 Btu/(hr-ft²-F).

12-15 The air temperature in an uninsulated metal duct is to be measured with an unshielded thermometer about 0.20 inches in diameter ($\varepsilon = 0.8$) located on the duct centerline where the air velocity is 1000 ft/min. The air pressure is near standard and the thermometer reads 150 F. The air surrounding the duct is at 50 F and the ratio of the inside to outside heat transfer coefficients for the duct is 5. Estimate the actual air temperature in the duct.

12-16 A psychrometer using two thermocouples is suspended in a room where the mean air velocity is 0.25 m/s. The dry bulb thermocouple is 1 mm in diameter, whereas the wet bulb temperature is measured with a thermocouple with wick 2 mm in diameter. The thermocouples are unshielded and the surrounding surfaces are at a temperature about 2 C greater than the dry bulb. If the indicated dry bulb and wet bulb temperatures are 25 C and 19 C, estimate (a) the true dry bulb temperature, (b) the humidity ratio.

12-17 Suppose the psychrometer of Problem 12-16 is used in an uninsulated duct where moist air is flowing at an indicated dry bulb and wet bulb of 50 C and 35 C. The heat transfer coefficients on the inside and outside of the duct have a ratio of 7 and the ambient temperature is 28 C. The mean air velocity in the duct is 1.0 m/s. Estimate (a) the dry bulb temperature of the air, (b) the humidity ratio of the air.

CHAPTER THIRTEEN

Direct Contact Heat and Mass Transfer

The process of air humidification was discussed in Chapter 2 on the basis of thermodynamics. In that case the condition line on the psychrometric chart was found to be solely a function of the water enthalpy, and all of the water mixed with the air stream was assumed to evaporate and become a part of the air-water vapor mixture.

In this chapter we will consider problems in which the quantity of water in contact with the air is much larger than the quantity added or withdrawn from the air stream. Depending on the moist air state and the water temperature a variety of results are possible. The air may be cooled and humidified or dehumidified, or heated and humidified by direct contact with water.

Only heat and mass transfer will be considered. As discussed in Chapter 3 the air may be cleansed of dust and water soluble vapors by contact with water.

13-1 BASIC RELATIONS FOR THE AIR WASHER

The basic relations for all types of direct contact equipment are quite similar. The air washer is considered first; we shall introduce modifications later for spray dehumidifiers and cooling towers.

The primary reason for treating direct contact equipment as a separate group arises from the difficulty in evaluating the heat and mass transfer areas. For the air washer or any spray type device that does not have packing materials, the heat and mass transfer areas are approximately equal. For a cooling tower the difference may be considerable.

The basic relations may be written as follows; see Fig. 13-1. The mass transfer is given by

$$-dG_l = G_a\, dW = h_d a_m (W_i - W)\, dL \qquad (13\text{-}1)$$

where W_i is the humidity ratio at the interface between the water and moist air. The quantity a_m is the mass transfer area per unit volume of the chamber. G_l and G_a are the water and air mass velocities $\dot{m}_l/A_c$ and $\dot{m}_a/A_c$ in mass per unit time and area

451

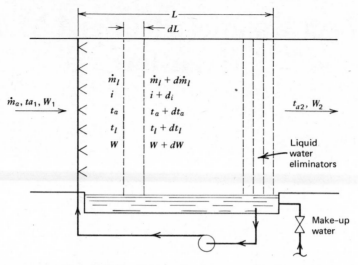

Figure 13-1.   Schematic of a spray chamber.

where A_c is the cross-sectional area of the chamber. The water evaporated equals the increase in moisture of the air, which must equal the mass transfer rate. The sensible heat transfer to the air is given by

$$G_a c_{pa}\, dt_a = h_a a_h(t_i - t_a)\, dL \tag{13-2}$$

where t_i is the interface temperature and a_h is the heat transfer area per unit volume. The total energy transfer to the air is

$$G_a(c_{pa}\, dt_a + i_{fg}\, dW) = [h_d a_m(W_i - W)i_{fg} + h_a a_h(t_i - t_a)]\, dL \tag{13-3}$$

The concept of enthalpy potential, which is discussed in detail in Chapter 14, may be used to simplify Eq. (13-3). The term in brackets on the left-hand side of Eq. (13-3) is di where i is the enthalpy of the moist air in Btu/lbma. If it is assumed that $a_h = a_m$ and the analogy of Eq. (12-20a) is used to relate the heat and mass transfer coefficients h and h_d with $L_e = 1$, Eq. (13-3) becomes

$$G_a\, di = h_d a_m(i_i - i)\, dL \tag{13-3a}$$

where $(i_i - i)$ is the enthalpy potential. An energy balance yields

$$G_a\, di = \pm G_l c_l\, dt_l + c_l t_l dG_l \tag{13-4}$$

The negative sign refers to parallel flow of air and water, whereas the positive sign refers to counterflow. The last term in Eq. (13-4) is very small and will be neglected in the following development. The heat transfer to the water may be expressed as

$$G_l c_l\, dt_l = h_l a_h(t_l - t_i)\, dL \tag{13-5}$$

Equations (13-1) through (13-5) are the basic relations for direct contact equipment with the possible exception of chambers that contain packing, such as cooling towers.

Various relations useful in equipment design may be derived from the basic relations. Combining Eqs. (13-3a), (13-4), and (13-5) gives

$$\frac{i - i_i}{t_l - t_i} = -\frac{h_l a_h}{h_d a_m} = -\frac{h_l}{h_d} \qquad (13\text{-}6)$$

This shows that the ratio of driving potentials for total heat transfer through the air and liquid films is equal to the ratio of film resistances for the gas and liquid film. Combining Eq. (13-2) and (13-3a) yields

$$\frac{di}{dt_a} = \frac{i - i_i}{t_a - t_i} \qquad (13\text{-}7)$$

while use of Eqs. (13-1) and (13-2) gives

$$\frac{dW}{dt_a} = \frac{W - W_i}{t_a - t_i} \qquad (13\text{-}8)$$

Equation (13-8) indicates that at any cross section in the spray chamber the instantaneous slope of the process path on the psychrometric chart is a straight line. This is illustrated in Fig. 13-2, where state 1 represents the state of the moist air entering the chamber and point $1i$ represents the interface saturation state. The initial path of the process is then in the direction of the line connecting points 1 and $1i$. As the

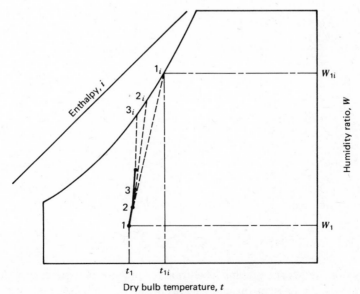

Figure 13-2.   Air washer humidification process.

air is heated and humidified, the water is cooled and the interface state changes. In the case shown the interface state gradually moves downward along the saturation curve. Points 2 and 3 represent other states along the process curve. Note that the path is directed toward the interface in each case. The interface states are defined by Eqs. (13-4) and (13-6). Equation (13-4) relates the air enthalpy change to the water temperature change, whereas Eq. (13-6) describes the way in which the interface state changes to accommodate the transport coefficients and the air state. Solution of Eqs. (13-4) and (13-6) for the interface state is rather complex but can be done by trial and error or by the use of a complex graphical procedure. A simpler method utilizes a psychrometric chart with enthalpy and temperature as coordinates (1). The use of these coordinates makes it possible to plot Eqs. (13-4) and (13-6) for easy graphical solution. The following example illustrates the procedure.

Example 13-1. A parallel flow air washer is to be designed as shown in Fig. 13-1. The design conditions are as follows.

Water temperature at the inlet, $\quad t_{l1} = 90\ F$

Water temperature at the outlet, $\quad t_{l2} = 75\ F$

Air dry bulb temperature at the inlet, $\quad t_{a1} = 60\ F$

Air wet bulb temperature at the inlet, $\quad t_{wb1} = 42\ F$

Air mass flow rate per unit area, $\quad G_a = 1250\ lbm/(hr\text{-}ft^2)$

Spray ratio $\quad G_l/G_a = 0.75$

Air heat transfer coefficient per unit volume, $\quad h_a a_h = 75\ Btu/(hr\text{-}F\text{-}ft^3)$

Liquid heat transfer coefficient per unit volume, $\quad h_l a_h = 1000\ Btu/(hr\text{-}F\text{-}ft^3)$

Air volume flow rate, $\quad \dot{Q} = 7000\ cfm$

Solution. The mass flow rate of the dry air is given by

$$\dot{m}_a = \frac{\dot{Q}}{v_1} = \frac{7000}{13.1} = 534\ lbm/min$$

Then the spray chamber must have a cross-sectional area of

$$A_c = \frac{\dot{m}_a}{G_a} = \frac{534(60)}{1250} = 25.6\ ft^2$$

The Colburn analogy of Eq. (12-20a) with Le = 1 will be used to obtain the mass transfer coefficient (assuming $a_m = a_h$).

$$h_d a_m = \frac{h_a a_h}{c_{pa}} = \frac{75}{0.24} = 313\ lbm/(hr\text{-}ft^3)$$

Figure 13-3 shows the graphical solution for the interface states and the process path for the air passing through the air washer. The solution is carried out as follows.

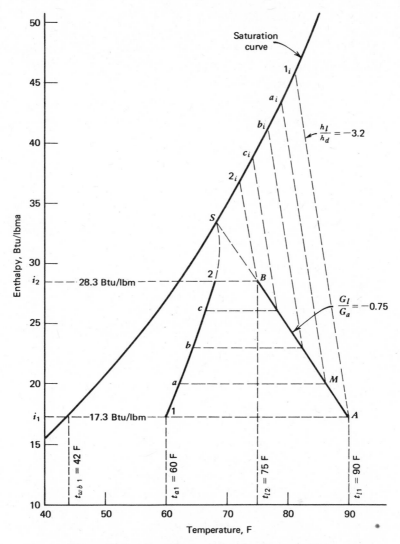

Figure 13-3.   Graphical solution for Example 13-1.

1. Locate state 1 as shown at the intersection of t_{a1} and i_1. Point A is a construc-
    tion point defined by the entering water temperature t_{l1} and i_1. Note that
    the temperature scale is used for both the air and water.

2. The energy balance line is constructed from point A to point B and is
    defined by Eq. (13-4):

$$\frac{di}{dt_l} = -\frac{G_l c_l}{G_a} = -\frac{G_l}{G_a} = -0.75$$

since $c_l = 1$ Btu/(lbmw-F). Point B is determined by the temperature of leaving water $t_{l2} = 75$ F. The negative slope is a consequence of parallel flow. The line AB has no physical significance.

3. The line $A1_i$ is called a tie line and is defined by Eq. (13-6):

$$\frac{i - i_i}{t_l - t_i} = \frac{-h_l a_h}{h_d a_m} = \frac{-1000}{313} = -3.2$$

The intersection of the tie line having a slope of -3.2 with the saturation curve defines the interface state 1_i. The combination of the line AB and line $A1_i$ represents graphical solutions of Eqs. (13-4) and (13-6).

4. The initial slope of the air process path is then given by a line from state 1 to state 1_i. The length of the line $1a$ depends on the required accuracy of the solution and the rate at which the curvature of the path is changing.

5. The procedure is repeated by constructing the line aM and the tie line Ma_i, which has the same slope as $A1_i$. The path segment ab is on a line from a to a_i. Continue in the same manner until the final state of the air point 2 is reached. State point 2 is on a horizontal line passing through point B. The final state of the air is defined by $t_{a2} = 68$ F, $i_2 = 28.3$ Btu/lbma, and $t_{wb2} = 62.3$ F.

6. To complete the solution, it is necessary to determine the length of the air washer. Equation (13-3a) gives

$$dL = \frac{G_a}{h_d a_m} \frac{di}{(i_i - i)} \tag{13-3 b}$$

or

$$L = \frac{G_a}{h_d a_m} \int_1^2 \frac{di}{(i_i - i)} \tag{13-3c}$$

Equation (13-3c) can be evaluated graphically or numerically. A plot of $1/(i_i - i)$ versus i is shown in Fig. 13-4 and the area under the curve represents the value of the integral. Using Simpson's rule with four equal increments yields

$$y = \int_1^2 \frac{di}{(i_i - i)} \approx \frac{\Delta i}{3} (y_1 + 4y_2 + 2y_3 + 4y_4 + y_5)$$

With $\Delta i = 2.75$

$$y = \left(\frac{2.75}{3}\right)(0.036 + 4(0.044) + 2(0.057) + 4(0.076) + 0.12) = 0.688$$

The design length is then

$$L = \frac{1250}{313}(0.688) = 2.75 \text{ ft}$$

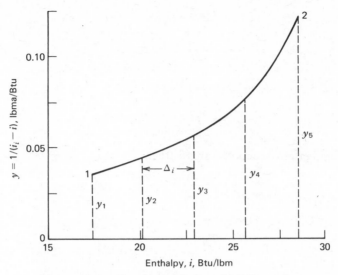

Figure 13-4. Graphical solution of $\int di/(i_i - i)$.

The method used in the above example can be used to predict the performance of existing direct contact equipment. Furthermore, test data from a unit may be used in a trial and error procedure to determine the transport coefficients.

The graphical procedure outlined above can be programmed rather easily for a small digital computer. If more than occasional calculations are needed, this would be worthwhile. Use of a computer would also permit refinements and variation of design parameters.

13-2 THE SPRAY DEHUMIDIFIER

The spray dehumidifier can often be used to advantage when a source of cold water, such as a well or spring, is available. To be effective the spray dehumidifier must be used in counterflow, which can best be achieved as shown in Fig. 13-5. The following example illustrates the design procedure for this case.

Example 13-2. A counterflow spray dehumidifier is to be designed as shown schematically in Fig. 13-5. These are the design conditions:

Water temperature at the inlet, $\quad t_{l1} = 7$ C
Water temperature at the outlet, $\quad t_{l2} = 13$ C
Air dry bulb temperature at the inlet, $\quad t_{a1} = 29$ C
Air wet bulb temperature at the inlet, $\quad t_{wb1} = 23$ C
Air mass flow rate per unit area, $\quad G_a = 1.36$ kg/(s-m^2)

Spray ratio, $\quad \dfrac{G_l}{G_a} = 0.70$

Air heat transfer coefficient per unit volume, $\quad h_a a_h = 121$ W/(C-m^3)

Liquid heat transfer coefficient per unit volume, $h_l a_h = 1470 \text{ W/(C-m}^3)$

Air volume flow rate, $\dot{Q} = 2.83 \text{ m}^3/\text{s}$

Find the cross-sectional area and final state of the air.

Solution.   The mass flow rate of the dry air is

$$\dot{m}_a = \frac{\dot{Q}}{v_1} = \frac{2.83}{0.875} = 3.24 \text{ kg/s}$$

The chamber cross-sectional area is then

$$A_c = \frac{\dot{m}_a}{G_a} = \frac{3.24}{1.36} = 2.38 \text{ m}^2$$

Using the Colburn analogy and assuming that $a_m = a_h$

$$h_d a_m = \frac{h_a a_h}{c_{pa}} = \frac{121}{1000} = 0.121 \text{ kg/(s-m}^3)$$

The graphical solution for the interface states and the process path is shown in

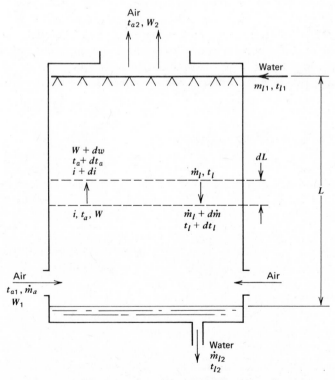

Figure 13-5.   Schematic of counterflow spray chamber.

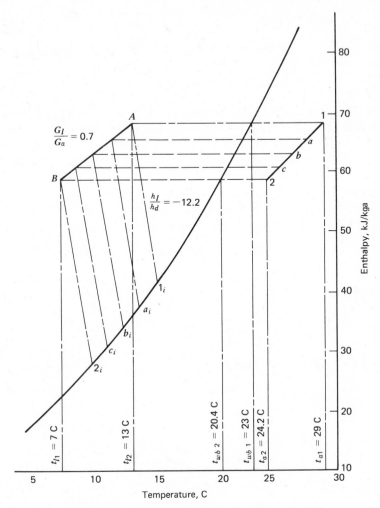

Figure 13-6.   Graphical solution for Example 13-2.

Fig. 13-6. The procedure is quite similar to that given in Example 13-1. Note, however, that the energy balance line *AB* has a positive slope because of the counterflow arrangement. Notice also that because of counterflow point *A* corresponds to t_{l2} and point *B* corresponds to t_{l1}. The final state of the air is defined by $t_{a2} = 24$ C and $t_{wb2} = 20$ C.

The height of the spray chamber is determined by using Eq. (13-3a) with the procedure given in Example 13-1.

The use of spray dehumidifiers is less frequent than extended surface heat exchangers, which are discussed in Chapter 14.

13-3 COOLING TOWERS

A typical cooling tower used in HVAC applications is shown in Fig. 8-6. The particular model shown is a packaged mechanical draft unit.

The function of the cooling tower is to reject heat to the atmosphere by reducing the temperature of water circulated through condensers or other heat rejection equipment. For this reason the state of the air and its path on the psychometric chart is of little interest. The cooling range for the water (decrease in temperature) and the approach of the leaving water temperature to the ambient wet bulb temperature are most important in determining the size and cost of a cooling tower. This will be evident in the analysis to follow.

The general equations derived in Section 13-2 are still valid. To facilitate calculations, however, the procedure is somewhat different in this case. Equation (13-3a) for the total energy transfer to the air is recalled.

$$G_a \, di = G_l c_l \, dt_l = h_d a_m (i_i - i) \, dL \qquad (13\text{-}3a)$$

To avoid consideration of the interfacial conditions, an overall coefficient U_i is adopted that relates the driving potential to the enthalpy i_l at the bulk water temperature t_l. Equation (13-3a) then becomes

$$G_a \, di = G_l c_l \, dt_l = U_i a_m (i_l - i) \, dL \qquad (13\text{-}9)$$

and, when Eq. (13-9) is integrated,

$$\frac{U_i a_m L}{G_l c_l} = \int \frac{dt_l}{(i_l - i)} \qquad (13\text{-}9a)$$

Now

$$\dot{m}_l = G_l A_c \qquad \text{and} \qquad V = A_c L$$

then

$$\frac{U_i a_m V}{\dot{m}_l c_l} = \int \frac{dt_l}{(i_l - i)} \qquad (13\text{-}10)$$

To review:

U_i = overall mass transfer coefficient between the water and air, lbm/(hr-ft^2) or kg/(s-m^2)

a_m = mass transfer surface area per unit volume associated with U_i, ft^2/ft^3 or m^2/m^3

V = total cooling tower volume, ft^3 or m^3

$\dot{m}_l$ = mass flow rate of water through the tower, lbm/hr or kg/s

c_l = specific heat of the water, Btu/(lbm-F) or kJ/(kg-C)

t_l = water temperature at a particular location in the tower, F or C

i_l = enthalpy of saturated moist air at t_l, Btu/lbm or kJ/kg

i = enthalpy of the moist air at temperature t, Btu/lbm or kJ/kg

The left-hand side of Eq. (13-10) is a measure of the cooling tower size and has the familiar form of the NTU parameter used in heat exchanger design.

Equation (13-10) cannot be integrated in a straightforward mathematical way; however, a step-by-step finite element approach can be used. The following example illustrates the procedure. The remainder of the design procedure is considered after the example.

Example 13-3.   Water is to be cooled from 100 to 85 F in a counterflow cooling tower when the outside air has a 75 F wet bulb temperature. The water-to-air flow ratio ($\dot{m}_l/\dot{m}_a$) is 1.0. Calculate the transfer units as defined by Eq. (13-10).

Solution.   Figure 13-7 is the cooling diagram for the given conditions. As the water is cooled from t_{l1} to t_{l2}, the enthalpy of the saturated air at the interface i_l follows the saturation curve from A to B. The air entering at wet bulb temperature t_{wb1} has enthalpy i_1. (This assumes the air enthalpy is only a function

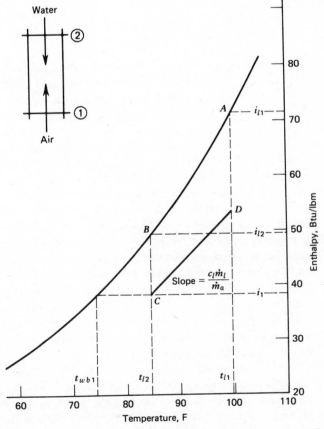

Figure 13-7.   Counterflow cooling diagram for Example 13-3.

of wet bulb temperature.) The leaving water temperature t_{l2} and the enthalpy i_1 define point C and the initial driving potential is represented by the distance BC. The enthalpy increase of the air is a straight line function with respect to the water temperature as defined by Eq. (13-9). The slope of the air operating line CD is therefore $(c_l \dot{m}_l / \dot{m}_a)$.

Point C represents the air conditions at the inlet and point D represents the air conditions leaving the tower. Note that the driving potential gradually increases from the bottom to the top of the tower. Counterflow integration calculations start at the bottom of the tower where the air conditions are known. Evaluation of the integral of Eq. (13-10) may be carried out in a manner similar to that described in Example 13-1 by plotting t_l versus $1/(i_l - i)$; however, another method will be used here (2). The step-by-step procedure is shown in Table 13-1. Water temperatures are listed in column 1 in increments of one to two degrees. Although smaller increments will give greater accuracy, this must be balanced against an increase in calculation time. The film enthalpies shown in column 2 are obtained from Table A-2 as the enthalpy of saturated air at the water temperatures. Column 3 shows the air enthalpy, which is determined from

$$\Delta i = \frac{c_l \dot{m}_l}{\dot{m}_a} \Delta t_l \qquad (13\text{-}9b)$$

where the initial air enthalpy i_1 is 38.5 Btu/lbma, $c_l = 1.0$ Btu/lbmw-F, $\dot{m}_l / \dot{m}_a = 1.0$, and Δt is read in column 1 of Table 13-1. The data of columns 4 and 5 are obtained from columns 2 and 3. Column 6 is the average of two steps from column 5 multiplied by the water temperature increment (column 1) for the same step. The number of transfer units are then given in column 7 as the summation of column 6. Column 8 gives the temperature range over which the water has been cooled. The last entry in column 7 is the number of transfer units required for this problem. It is evident from Table 13-1 that either an increase in the cooling range or a decrease in the leaving water temperature will increase the number of transfer units. As mentioned earlier, these two factors are quite important in cooling tower design. The heat exchangers with which the cooling tower is connected should be designed with the cooling tower in mind. It may be more economical to enlarge the heat exchangers and/or increase the flow rate of the water than increase the size of the cooling tower. Obviously, this becomes an optimization problem where the digital computer would be useful.

To continue the problem of tower design, we need information on the overall mass transfer coefficient per unit volume, $U_i a_m$. There is little theory to predict this coefficient; therefore, we must rely on experiments. The tower characteristic may generally be represented as a log-log plot of $U_i a_m V / \dot{m}_l$ versus $\dot{m}_l / \dot{m}_a$ for a constant air mass flow rate and water inlet temperature t_{l1} (Fig. 13-8). Although different air flow rates and water temperatures cause the characteristic to move up or down, variations in air flow rate of about ± 20 percent and water temperature variations of

Table 13-1   COUNTERFLOW COOLING TOWER INTEGRATION CALCULATIONS

1	2	3	4	5	6	7	8
Water Temperature t_l degrees F	Enthalpy of Film i_l Btu/lbma	Enthalpy of Air i Btu/lbma	Enthalpy Difference $(i_l - i)$ Btu/lbma	Reciprocal of Enthalpy Difference $\dfrac{1}{(i_l - i)}$ lbma/Btu	Average $\dfrac{\Delta t_l}{(i_l - i)}$ F-lbma/Btu	Summation $\dfrac{\Delta t_l}{(i_l - i)}$ F-lbma/Btu	Cooling Range degrees F
85	49.4	38.5	10.9	0.0917	0.0905	0.0905	1
86	50.7	39.5	11.2	0.0893	0.1748	0.2653	3
88	53.2	41.5	11.7	0.0855	0.1661	0.4314	5
90	55.9	43.5	12.4	0.0806	0.1558	0.5872	7
92	58.8	45.5	13.3	0.0752	0.1451	0.7323	9
94	61.8	47.5	14.3	0.0699	0.1348	0.8671	11
96	64.9	49.5	15.4	0.0649	0.1248	0.9919	13
98	68.2	51.5	16.7	0.0599	0.1148	1.1067	15
100	71.7	53.5	18.2	0.0549			

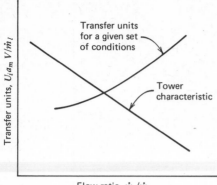

Flow ratio, $\dot{m}_l/\dot{m}_a$ **Figure 13-8.**   Tower characteristic.

±10 F do not move the characteristic significantly. The slope of the characteristic varies from about −0.4 to −1.0. The required transfer units for a given set of water and air conditions are also shown in Fig. 13-8. The intersection of the two curves defines the operating point for the tower.

Assuming that the tower characteristics are known, we may determine the mass transfer coefficient per unit volume $U_i a_m$. After many tests have been made on towers of a similar type, it is possible to predict $U_i a_m$ with reasonable accuracy. Then the volume of the tower required for a given set of conditions is given by

$$V = \frac{N \dot{m}_l c_l}{U_i a_m} \tag{13-11}$$

where N is the number of transfer units given by Eq. (13-10). The cross-sectional area of the tower is defined by

$$A_c = \frac{\dot{m}_a}{G_a} = \frac{\dot{m}_l}{G_l} \tag{13-12}$$

and the height of the tower is given by

$$L = \frac{V}{A_c} \tag{13-13}$$

Example 13-4.   Suppose the cooling tower of Example 13-3 must handle 1000 gpm of water. It has been determined that an air mass velocity of 1500 lbma/(hr-ft²) is acceptable without excessive water carry-over (drift). The overall mass transfer coefficient per unit volume $U_i a_m$ is estimated to be 120 lbm/(hr-ft³) for the type of tower to be used. Estimate the tower dimensions for the required duty.

Solution.   The transfer units N required for the tower were found to be 1.107 in Example 13-3. Then the total volume of the tower is given by Eq. (13-11) as

$$V = \frac{1.107(1000)8.33(60)(1.0)}{120} = 4611 \text{ ft}^3$$

The cross-sectional area of the tower may be determined from Eq. (13-12) using the mass velocity of the air and the water-to-air ratio.

$$A_c = \frac{\dot{m}_a}{G_a} = \left(\frac{\dot{m}_l}{\dot{m}_a}\right)\frac{\dot{m}_a}{G_a} = \frac{\dot{m}_l}{G_a}$$

$$A_c = \frac{1000(8.33)(60)}{1500} = 333 \text{ ft}^2$$

which is equivalent to an 18×18 ft cross section. Then from Eq. (13-13)

$$L = \frac{V}{A_c} = \frac{4611}{333} = 13.8 \text{ ft}$$

Caution must be exercised in using mass transfer data from the literature for tower design. There are many variations in construction that affect the transport coefficients dramatically. The scale of the tower also seems to be important because of the ratio of wall surface area to total volume.

For most HVAC applications, relatively small factory assembled cooling towers are used. Performance data are usually presented in a form such that a certain standard size may be selected. Figure 13-9 and Table 13-2 are typical of what might be furnished by a manufacturer for a line of towers. The entering water temperature, air wet bulb temperature, and the water flow rate determine the model to be selected for a fixed leaving water temperature of 85 F. The example shown as a dotted line illustrates use of the chart. This procedure usually causes the tower to be slightly

Table 13-2 PERFORMANCE DATA FOR SOME FACTORY ASSEMBLED COOLING TOWERS

Model	Nominal Rating		No. of Cells	No. of Fans	CFM	Motor HP
	Tons	GPM				
A	50	120	1	1	10,500	5
B	100	240	1	2	21,000	10
C	100	240	1	2	21,000	2–5
D	150	360	1	3	31,500	15
E	200	480	1	4	42,000	20
F	200	480	1	4	42,000	2–10
G	250	600	1	5	52,500	25
H	300	720	1	6	63,000	30
I	300	720	1	6	63,000	2–15
J	350	840	1	8	84,840	2–20
K	400	960	1	8	84,000	2–20
L	500	1200	1	10	105,000	2–25
M	600	1440	1	12	126,000	2–30

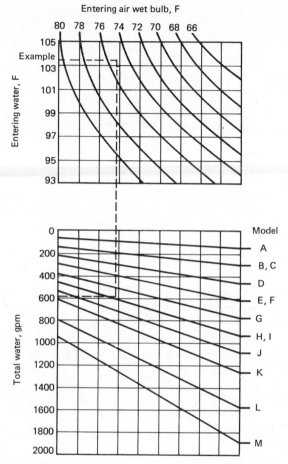

Figure 13-9. Selection chart for some factory assembled cooling towers with a fixed cold water temperature of 85 F.

oversized. The cooling range may be computed from the entering water temperature and flow rate, and the tower capacity.

Example 13-5. Select a cooling tower using Fig. 13-9 and Table 13-2 for the conditions of Examples 13-3 and 13-4. Compute the cooling range, approach, and heat transfer rate.

Solution. The entering water temperature and air wet bulb temperature are 100 F and 75 F, respectively, with a water flow rate of 1000 gpm. Referring to Fig. 13-9, a model L would be the obvious choice; however, it is somewhat oversized. Suppose that the heat exchangers (condensers) in the circuit could be changed so that the water would enter the tower at 103 F with a flow rate of

835 gpm. This is still the same duty of about 625 tons. Referring back to Fig. 13-9, a model K tower fits this situation. The cooling range is

$$t_{l1} - t_{l2} = 103 - 85 = 18 \text{ F}$$

and the approach is

$$t_{l2} - t_{wb2} = 85 - 75 = 10 \text{ F}$$

the heat transfer rate from the water is

$$\dot{q} = 835(60)8.33(103 - 85) = 7.5 \times 10^6 \text{ Btu/hr}$$

or

$$\dot{q} = 626 \text{ tons}$$

The reader is referred to the equipment volume of the *ASHRAE Handbook and Product Directory* (2), which has a great deal of information on cooling tower performance and selection.

REFERENCES

1. *ASHRAE Handbook of Fundamentals,* American Society of Heating, Refrigerating and Air-Conditioning Engineers, New York. 1977.

2. *ASHRAE Handbook and Product Directory—Equipment,* American Society of Heating, Refrigerating and Air-Conditioning Engineers, New York, 1979.

PROBLEMS

13-1 Determine the final state of the air, the cross-sectional area, and the height of a counterflow spray dehumidifier that operates as follows:

$t_{l1} = 50 \text{ F}$	$t_{a1} = 95 \text{ F}$
$t_{l2} = 60 \text{ F}$	$t_{wb1} = 82 \text{ F}$
$G_a = 1200 \text{ lbm/(hr-ft}^2)$	$\dot{Q}_a = 5000 \text{ cfm}$
$\dot{Q}_w = 55 \text{ gpm}$	$h_a a_h = 60 \text{ Btu/(hr-F-ft}^3)$
$h_t a_h = 800 \text{ Btu/(hr-F-ft}^3)$	

13-2 A counterflow spray dehumidifier operates as stated below. Determine (a) the final state of the air, (b) the cross-sectional area, and (c) the height of the chamber.

$t_{l1} = 10 \text{ C}$	$t_{a1} = 35 \text{ C}$
$t_{l2} = 16 \text{ C}$	$t_{wb1} = 28 \text{ C}$
$G_a = 1.63 \text{ kg/(s-m}^2)$	$\dot{Q}_a = 2.36 \text{ m}^3/\text{s}$
$\dot{Q}_w = 3.5 \times 10^{-3} \text{ m}^3/\text{s}$	$h_a a_h = 104 \text{ W/(C-m}^3)$
$h_t a_h = 1385 \text{ W/(C-m}^3)$	

13-3 Redesign the air washer in Example 13-1 assuming counterflow of the air and water.

13-4 Solve Problem 13-1 assuming parallel flow of the air and water spray.

13-5 Solve Problem 13-2 assuming parallel flow of the air and water.

13-6   A counterflow cooling tower cools water from 104 to 85 F when the outside air has a wet bulb temperature of 76 F. The water flow rate is 2000 gpm and the air flow rate is 210,000 cfm. Calculate the transfer units for the tower.

13-7   A counterflow cooling tower cools water from 44 to 30 C. The outdoor air has a wet bulb temperature of 22 C. Water flows at the rate of 0.32 m³/s and the water-to-air mass flow ratio is 1.0. Estimate the transfer units for the tower.

13-8   Estimate the tower dimensions for Problem 13-6. The air mass velocity may be assumed to be 1800 lbma/(hr-ft²) and the overall mass transfer coefficient per unit volume $U_i a_m$ is about 125 lbm/(hr-ft³).

13-9   Estimate the tower dimensions for Problem 13-7. Assume a mass velocity for the air of 2.7 kg/(s-m²) and an overall mass transfer coefficient per unit volume $U_i a_m$ of 0.56 kg/(s-m³).

13-10   Use Fig. 13-9 and Table 13-2 to select a suitable tower(s) for the conditions of Problem 13-6.

13-11   Complete Example 13-2 to find the height of the chamber.

13-12   Repeat Example 13-3 changing the entering air wet bulb temperature to 79 F. Compare the transfer units with those of Table 13-1. How will this affect the tower dimensions?

13-13   Repeat Example 13-3, changing the entering water temperature to 105 F. Compare the transfer units with those of Table 13-1. How will this affect the tower dimensions?

13-14   The condensers for a centrifugal chiller plant require 200 gpm with water entering at 85 F and leaving at 100 F, the outdoor ambient air wet bulb temperature is 76 F. (a) Select a suitable cooling tower using Fig. 13-9 and Table 13-2, (b) compute the cooling range, approach, and the tower capacity.

CHAPTER FOURTEEN

Extended Surface Heat Exchangers

The term *heat exchanger* is usually applied to a device in which two fluid streams separated by a solid surface exchange heat energy. These devices may take many forms. However, ordinary metal tubes are the main components of many types. The heat exchanger is the most widely used device in HVAC applications. Although it is only a part of the overall system, a heat exchanger is required in every heating and cooling design problem that the engineer solves.

The major applications in the HVAC field are as follows: refrigerant-to-water in the case of chillers and water-cooled condensers where shell and tube configurations are often used; water-to-air and refrigerant-to-air where finned tubes are often used; and air-to-air in the case of heat recovery applications where plate-fin type surfaces or rotating heat exchangers may be used.

The heat transfer surface may operate under conditions where only sensible heat transfer occurs or where latent and sensible heat transfer occur simultaneously as with dehumidifying coils. Both of these cases are treated in this chapter. Pressure loss is an important consideration in the design of heat exchangers; this will be considered in conjunction with the heat transfer in the discussions that follow.

In general, complications arise when one attempts to describe the rate of heat transfer from one fluid to the other in a heat exchanger. For practical purposes

$$\dot{q} = UA \, \Delta t_m \tag{14-1}$$

where U is the familiar overall heat transfer coefficient and A is the surface area associated with U. The mean temperature difference between the streams Δt_m must be used because Δt varies continuously throughout the heat exchanger. The overall heat transfer coefficient U is also a variable resulting from changing physical properties and hydrodynamic characteristics from one part of the exchanger to another. In addition, the calculation of U is difficult and subject to large errors in some cases.

The use of Eq. (14-1) requires the assumption of an average value for U and the determination of a suitable mean temperature difference. An alternate approach

469

SCRNB

to heat exchanger problems involves the use of an average value of U and the concept of an effectiveness.

14-1   THE *LMTD* METHOD

With suitable assumptions it is possible to derive an expression for the mean temperature difference required in Eq. (14-1) for parallel and counterflow. The assuumptions are as follows:

1. The overall heat tranfer coefficient U, the mass flow rates $\dot{m}_c$ and $\dot{m}_h$, and the specific heats c_c and c_h are all constants where the subscripts c and h refer to the cold and hot streams.
2. There is no heat loss or gain external to the heat exchanger and there is no axial conduction in the heat exchanger.
3. A single bulk temperature applies to each stream at a given cross section.

Figures 14-1 and 14-2 show counterflow and parallel flow heat exchangers that represent the simplest types. The subscripts i and o refer to inlet and outlet, respectively. For both counterflow and parallel flow the appropriate mean temperature is given by

$$\Delta t_m = \frac{\Delta t_2 - \Delta t_1}{\ln(\Delta t_2/\Delta t_1)} \tag{14-2}$$

where Δt_2 and Δt_1 are defined in Figs 14-1 and 14-2. This particular mean temperature difference is the *log mean temperature difference* and is designated *LMTD*.

Complex Flow Patterns

In many cases the flow paths in the heat exchanger are quite complex and in some cases expressions may be developed for Δt_m; however, they are generally so complicated that charts have been developed to replace the equations. The concept of a correction factor F is used where

$$\dot{q} = UA(F)(\Delta t_m) \tag{14-3}$$

where Δt_m is computed in the same manner as the *LMTD* for an equivalent counterflow exchanger. Figures 14-3 and 14-4 show charts for two common flow configurations encountered in HVAC applications. The shell and tube configuration shown in Fig. 14-3 is often used for chillers and water-cooled condensers while the crossflow configuration occurs with many heating and cooling applications where air flows normal to a bank of finned tubes. The parameters P and R are defined as

$$P = \frac{t_{co} - t_{ci}}{t_{hi} - t_{ci}} \quad \text{and} \quad R = \frac{t_{hi} - t_{ho}}{t_{co} - t_{ci}} \tag{14-4}$$

P may be thought of as an effectiveness of the heat exchanger when $(\dot{m}c_p)_c < (\dot{m}c_p)_h$ where $(\dot{m}c_p)$ is the *fluid capacity rate C*. If the numerator and denominator of P are

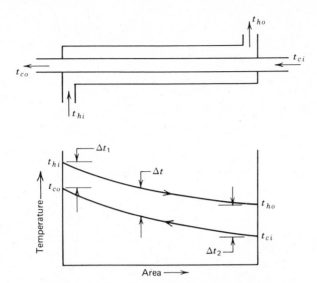

Figure 14-1.   Counterflow heat exchanger.

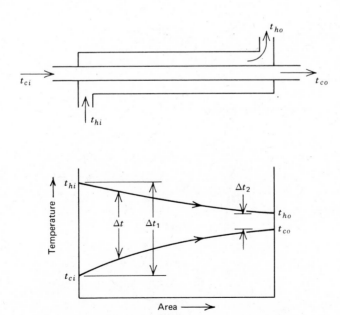

Figure 14-2.   Parallel flow heat exchanger.

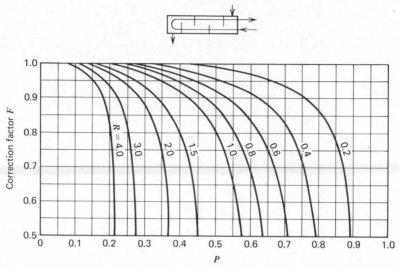

Figure 14-3. Correction factor plot for exchanger with one shell pass and two, four, or any multiple of tube passes.

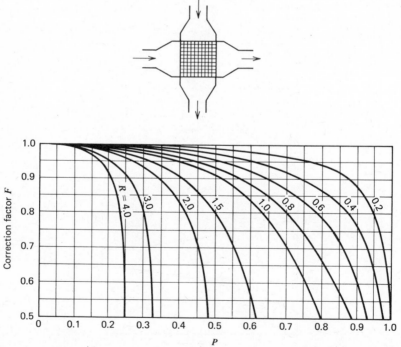

Figure 14-4. Correction factor plot for single pass cross flow exchanger, both fluids unmixed.

multipled by C_c, P becomes the ratio of the actual heat transfer to the cold fluid to the heat transfer if the cold fluid is heated to the temperature of the entering hot fluid. P is thus the actual heat transfer divided by the theoretical maximum heat transfer to the cold fluid. When $C_h < C_c$, P is the effectiveness divided by R. It may also be shown that $R = C_c/C_h$.

Calculations involving the *LMTD* are straightforward when the fluid inlet and outlet temperatures are known. If three of the temperatures are known, together with the fluid capacity rates of the streams, the fourth temperature may be calculated by a simple energy balance. When two of the temperatures are unknown, however, trial and error procedures are required because of the form of the equation for the *LMTD*, Eq. (14-2). In the design of heat exchangers this is often true.

14-2   THE *NTU* METHOD

The *NTU* method has the advantage of eliminating the trial and error procedure of the *LMTD* method for many practical problems when only the inlet fluid temperatures are known. *NTU* stands for the "Number of Transfer Units."

Heat exchanger effectiveness was mentioned in the previous article as

$$\varepsilon = \frac{\text{actual heat transfer rate}}{\text{maximum possible heat transfer rate}}$$

The actual heat transfer rate is given by

$$\dot{q} = C_h(t_{hi} - t_{ho}) = C_c(t_{co} - t_{ci}) \tag{14-5}$$

The maximum possible heat transfer rate is expressed by

$$\dot{q}_{max} = C_{min}(t_{hi} - t_{ci}) \tag{14-6}$$

This is true because the maximum heat transfer will occur when one of the fluids undergoes a temperature change equal to the maximum in the heat exchanger, which is $(t_{hi} - t_{ci})$. The fluid experiencing the maximum temperature change must be the one with the minimum value of C to satisfy the energy balance.

The fluid with the minimum value of C may be the hot or the cold fluid. For $C_h = C_{min}$, using Eq. (14-5) and (14-6),

$$\varepsilon = \frac{\dot{q}}{\dot{q}_{max}} = \frac{C_h(t_{hi} - t_{ho})}{C_{min}(t_{hi} - t_{ci})} = \frac{(t_{hi} - t_{ho})}{(t_{hi} - t_{ci})} \tag{14-7}$$

For $C_c = C_{min}$

$$\varepsilon = \frac{\dot{q}}{\dot{q}_{max}} = \frac{C_c(t_{co} - t_{ci})}{C_{min}(t_{hi} - t_{ci})} = \frac{(t_{co} - t_{ci})}{(t_{hi} - t_{ci})} \tag{14-8}$$

It is therefore necessary to have two expressions for the effectiveness, Eqs. (14-7) and (14-8). When effectiveness is known, the outlet temperature may be easily computed. For example, when $C_h < C_c$,

$$t_{ho} = \varepsilon(t_{ci} - t_{hi}) + t_{hi} \tag{14-7a}$$

Also

$$t_{co} = \frac{q}{C_c} + t_{ci} = \frac{C_h}{C_c}(t_{hi} - t_{ho}) + t_{ci} \qquad (14\text{-}9)$$

or

$$t_{co} = \frac{C_h}{C_c}\varepsilon(t_{hi} - t_{ci}) + t_{ci} \qquad (14\text{-}9a)$$

Expressions for ε for several flow configurations are shown in Table 14-1. Figures 14-5, 14-6, and 14-7 show graphical representations of the effectiveness for three common cases. The NTU parameter is defined as UA/C_{min} and may be thought of as a heat transfer size factor. It may also be observed that flow configuration is unimportant when $C_{min}/C_{max} = 0$. This corresponds to the situation of one fluid undergoing a phase change where c_p may be thought of as being infinite. Evaporating or condensing refrigerants as well as condensing water vapor are examples where $C_{min}/C_{max} = 0$.

    The NTU method has gained greatest acceptance in connection with design of *compact heat exchangers* where a large surface area per unit volume exists. An

Table 14-1   THERMAL EFFECTIVENESS OF HEAT EXCHANGERS WITH VARIOUS FLOW ARRANGEMENTS

Parallel flow:	$\varepsilon = \dfrac{1 - \exp[-NTU(1 + C)]}{1 + C}$
Counterflow:	$\varepsilon = \dfrac{1 - \exp[-NTU(1 - C)]}{1 - C\exp[-NTU(1 - C)]}$
Cross flow (both streams are unmixed):[a]	$\varepsilon = 1 - \exp\{(\dfrac{1}{C\eta})[\exp((-NTU)(C)(\eta)) - 1]\}$ where $\eta = NTU^{-0.22}$
Cross flow (both streams are mixed):	$\varepsilon = NTU\left\{\dfrac{NTU}{1 - \exp(-NTU)} + \dfrac{(NTU)(C)}{1 - \exp[-(NTU)(C)]} - 1\right\}^{-3}$
Cross flow (stream C_{min} is unmixed):	$\varepsilon = \dfrac{1}{C}\{1 - \exp[-C(1 - \exp(-NTU))]\}$
Cross flow (stream C_{max} is unmixed):	$\varepsilon = 1 - \exp\{-\dfrac{1}{C}[1 - \exp(-(NTU)(C))]\}$
1-2 Parallel counterflow:	$\varepsilon = 2\left\{1 + C + \dfrac{1 + \exp[-NTU(1 + C^2)^{1/2}]}{1 - \exp[-NTU(1 + C^2)^{1/2}]}(1 + C^2)^{1/2}\right\}^{-1}$ where $NTU = (UA/C_{min})$, and $C = C_{min}/C_{max}$

[a]This is an approximate expression.

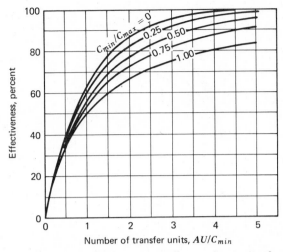

Figure 14-5.   Effectiveness for counterflow exchanger performance.

arbitrary definition has been proposed which states that the ratio of surface area to volume in a compact exchanger is greater than 200 ft²/ft³ or 656 m²/m³. The common finned tube exchangers used extensively in HVAC systems generally fall into this category. Compact heat exchangers are used in applications where at least one fluid is a gas. Because the thermal resistance associated with the gas film is high compared with a fluid such as water, fins are used to increase the heat transfer area and decrease the thermal resistance.

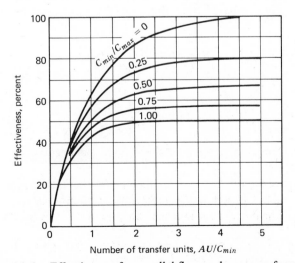

Figure 14-6.   Effectiveness for parallel flow exchanger performance.

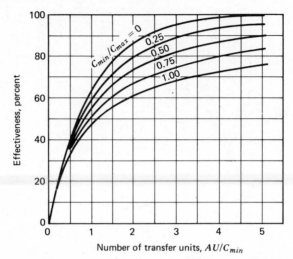

Figure 14-7.   Effectiveness for cross flow exchanger with fluids unmixed.

14-3   HEAT TRANSFER—SINGLE COMPONENT FLUIDS

The heat transfer rate from one fluid to the other in a heat exchanger is expressed by Eq. (14-3):

$$\dot{q} = UA(F)\,\Delta t_m \tag{14-3}$$

when the *LMTD* method is used. When the *NTU* method is utilized, the heat transfer rate is generally computed from the temperature change for either fluid. For example

$$\dot{q} = (\dot{m}c_p)_h(t_{hi} - t_{ho}) \tag{14-10}$$

since

$$NTU = \frac{UA}{(\dot{m}c_p)_{min}} \tag{14-11}$$

It is evident that an average value of the overall coefficient U must be known for both design methods.

The concept of overall thermal resistance and the overall heat transfer coefficient was discussed in Chapter 4. The general procedure is the same for heat exchangers. For a simple heat exchanger without fins the overall coefficient U is given by

$$\frac{1}{UA} = \frac{1}{h_o A_o} + \frac{\Delta x}{kA_m} + \frac{1}{h_i A_i} + \frac{R_{fi}}{A_i} + \frac{R_{fo}}{A_o} \tag{14-12}$$

where

h_o = heat transfer coefficient on the outside, Btu/(hr-ft²-F) or W/(m²-C)

h_i = heat transfer coefficient on the inside, Btu/(hr-ft²-F) or W/(m²-C)

Δx = thickness of the separating wall, ft or m

k = thermal conductivity of the separating wall, (Btu-ft)/(ft²-hr-F) or (W-m)/(m²-C)

A = area, ft² or m² where o, m, and i refer to outside, mean, and inside, respectively

R_f = fouling factor, (hr-ft²-F)/Btu or (m²-C)/W

In general the areas A_o, A_m, A_i are not equal and U may be referenced to any one of the three. Let $A = A_o$, then

$$\frac{1}{U_o} = \frac{1}{h_o} + \frac{\Delta x}{k(A_m/A_o)} + \frac{1}{h_i(A_i/A_o)} + \frac{R_{fi}}{(A_i/A_o)} + R_{fo} \qquad (14\text{-}12a)$$

Many of the heat exchangers used in HVAC systems have fins on one or both sides. Because the fins do not have a uniform temperature, the fin efficiency η is used to describe the heat transfer rate.

$$\eta = \frac{\text{actual heat transfer}}{\text{(heat transfer with fin all at the base temperature } t_b)}$$

Figure 14-8 shows a simple fin. Since the base on which the fin is mounted also transfers heat, and another parameter similar to fin efficiency is defined, called the surface effectiveness η_s.

$$\eta_s = \frac{\text{actual heat transfer for fin and base}}{\text{(heat transfer for fin and base when the fin is at the base temperature } t_b)}$$

If we assume that h is uniform over the fin and base surface, the actual heat transfer rate is then given by

$$\dot{q} = hA\eta_s(t_b - t_\infty) \qquad (14\text{-}13)$$

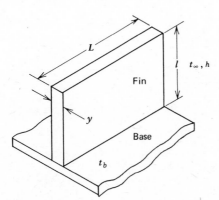

L

t_∞, h

Fin

y

Base

t_b

Figure 14-8.   A simple fin of uniform cross-section.

where A is the total surface area of the fin and base. The thermal resistance is given by

$$R' = \frac{1}{hA\eta_s} \tag{14-14}$$

For a case where both sides of the heat exchanger have fins, the overall coefficient U is, assuming no fouling,

$$\frac{1}{UA} = \frac{1}{h_o A_o \eta_{so}} + \frac{\Delta x}{kA_m} + \frac{1}{h_i A_i \eta_{si}} \tag{14-15}$$

If $A = A_o$,

$$\frac{1}{U_o} = \frac{1}{h_o \eta_{so}} + \frac{\Delta x}{k(A_m/A_o)} + \frac{1}{h_i \eta_{si}(A_i/A_o)} \tag{14-15a}$$

The second term on the right-hand side of Eqs. (14-15) and (14-15a) represents the thermal resistance of the base and is often negligible.

Fin Efficiency

Extended or finned surfaces may take on many forms ranging from the simple plate of uniform cross section shown in Fig. 14-8 to complex patterns attached to tubes. Several common configurations are considered in this section starting with the fin of uniform cross section. The heat transfer rate for the fin may be shown to be

$$\dot{q} = (t_b - t_\infty)mkA_c \tanh(ml) \tag{14-16}$$

when heat transfer from the tip is zero (1). The parameter m is given by

$$m = \left[\frac{hP}{kA_c} \right]^{1/2} \tag{14-17}$$

where

$k =$ thermal conductivity of the fin material, (Btu-ft)/(ft²-hr-F) or (W-m)/(m²-C)
$P =$ perimeter of the fin 2(L + y), ft or m
$A_c =$ cross-sectional area of the fin (Ly), ft² or m²
$l =$ fin height, ft or m

If we use the definition of fin efficiency given above,

$$\eta = \frac{(t_b - t_\infty)mkA_c \tanh(ml)}{hA(t_b - t_\infty)} \tag{14-18}$$

or

$$\eta = \frac{\tanh(ml)}{(ml)} \tag{14-19}$$

The heat transfer coefficient used to define m is assumed to be a constant. In a practical heat exchanger h will vary over the surface of the fins and will probably change between the inlet and outlet of the exchanger. The only practical solution is to use an average value of h for the complete surface. Equation (14-19) may be applied to a surface such as that shown in Fig. 14-25 on page 497.

Most fins are very thin and $L \gg y$. In this case the parameter m defined by Eq. (14-17) may be simplified by setting $P = 2L$. Then

$$m = \left[\frac{2hL}{kLy} \right]^{1/2} = \left[\frac{2h}{ky} \right]^{1/2} \tag{14-20}$$

This approximation is often applied without explanation.

Finned tube heat exchangers are very popular for water-to-air or refrigerant-to-air applications. Figure 14-9 shows a sketch of a tube with circular fins. The diagram is somewhat idealized, since in practice the fin is usually wound on the tube in a helix from one continuous strip of material. A typical circular finned tube water coil is shown in Fig. 14-10. Typically the fin will be quite thin. In the case of the circular fin the solution for the fin efficiency is very complex and is not generally used for practical problems; however, Fig. 14-11 shows a plot of the solution. An approximate but quite accurate method of predicting η for a circular fin has been developed by Schmidt (2). The method is largely empirical but has many advantages when an analytical expression is required. The method is summarized as follows:

$$\eta = \frac{\tanh(mr\phi)}{(mr\phi)} \tag{14-21}$$

where m is defined by Eq. (14-20) and

$$\phi = \left(\frac{R}{r} - 1 \right)[1 + 0.35 \ln(R/r)] \tag{14-22}$$

When R/r is between 1.0 and 8 and η falls between 0.5 and 1.0, the error is less than one percent of the value of the fin efficiency taken from Fig. 14-11.

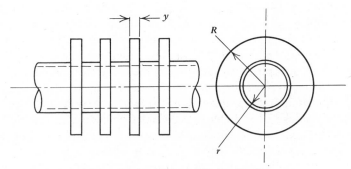

Figure 14-9.   Tube with circular fins.

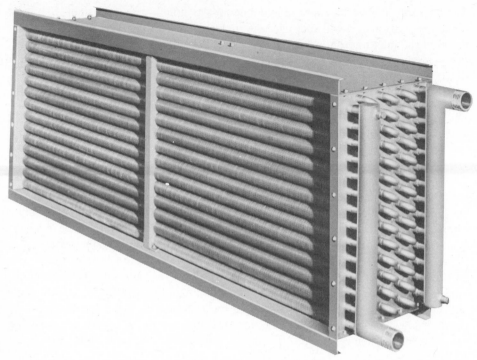

Figure 14-10. Circular finned tube water coil. (Courtesy Thermal Corporation, Houston, Texas.)

Continuous plate fins are also used extensively in finned tube heat exchangers. In this case each fin extends from tube to tube. Figure 14-12 shows such an arrangement. It is not possible to obtain a closed analytical solution for this type of fin and approximate methods are necessary. Consider the rectangular tube array of Fig. 14-13 with continuous plate fins. When it is assumed that the heat transfer coefficient is constant over the fin surface, an imaginary rectangular fin may be defined as shown. The outline of the fin is an equipotential line where the temperature gradient is zero. The problem is then to find η for a rectangular fin. Zabronsky (3) has suggested that a circular fin of equal area be substituted for purposes of calculating η; however, Carrier and Anderson (4) have shown that the efficiency of a circular fin of equal area is not accurate; they recommend the sector method. Rich (6) developed charts shown in reference 5 to facilitate use of the sector method. Schmidt (2) describes an approach to this problem that is nearly as accurate as the sector method and has the advantage of simplicity. Again the procedure is empirical; however, Schmidt tested the method statistically using maximum and minimum values of η that must bracket the actual fin efficiency. The method is based on the selection of a circular fin with a radius R_e that has the same fin efficiency as the rectangular fin.

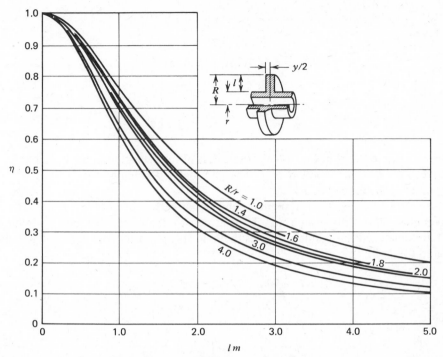

Figure 14-11.   Performance of circumferential fins of rectangular cross section. (Reprinted from *ASME Trans., 67,* 1945.)

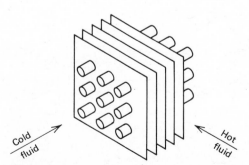

Figure 14-12.   Continuous plate-fin-tube heat exchanger.

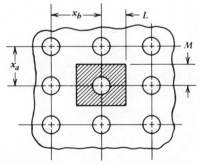

Figure 14-13.   Rectangular tube array.

After R_e is determined, Eq. (14-21) is used for the calculation of η. For the rectangular fin

$$\frac{R_e}{r} = 1.28\psi(\beta - 0.2)^{1/2} \tag{14-23}$$

where

$$\psi = \frac{M}{r} \quad \text{and} \quad \beta = \frac{L}{M}$$

M and L are defined in Fig. 14-13 where L is always selected to be greater than or equal to M. In other words $\beta \geq 1$. The parameter ϕ given by Eq. (14-22) is computed using R_e instead of R.

Figure 14-14 shows a triangular tube layout with continuous plate fins. Here a hexangular fin results, which may be analyzed by the sector method (5). Schmidt (2) also analyzed this result and gives the following empirical relation, which is similar to Eq. (14-23):

$$\frac{R_e}{r} = 1.27\psi(\beta - 0.3)^{1/2} \tag{14-24}$$

where

$$\psi = \frac{M}{r} \quad \text{and} \quad \beta = \frac{L}{M}$$

M and L are defined in Fig. 14-14 where $L \geq M$. Equations (14-21) and (14-22) are used to compute η.

Special types of fins are sometimes used such as spines or fins of nonuniform cross section. Reference 5 contains data pertaining to these surfaces.

In the discussion above we have assumed that the fins are rigidly attached to the base material so that zero thermal contact resistance exists. This may not always be true, particularly for plate-fin-tube surfaces. Eckels (24) has developed an empir-

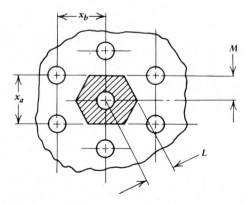

Figure 14-14.   Hexangular tube array.

ical relation to predict the unit contact resistance for plate-fin-tube surfaces as follows:

$$R_{ct} = C \left[\frac{D_t \left(\frac{s}{y} - 1 \right)^2}{y} \right]^{0.6422} \tag{14-25}$$

where

R_{ct} = Unit contact resistance (hr-ft^2-F)/Btu or (m^2-C)/W
C = A constant, 2.222 × 10^{-6} for English Units and 3.913 × 10^{-7} for SI units
D_t = Outside tube diameter, in. or m
s = Fin spacing, in. or m
y = Fin thickness, in. or m

This unit contact resistance is associated with the outside tube area and added to Eq. (14-15). Because this contact resistance is undesirable as well as difficult to predict, every effort should be made to eliminate it in the manufacture of the heat exchanger. If tests are made for a surface, the contact resistance is usually reflected in the heat transfer coefficients obtained.

Surface Effectiveness

Surface effectiveness was defined earlier as the actual heat transfer to the fin and base divided by the heat transfer to the fin and base when the fin is all at the base temperature t_b. This may be written

$$\eta_s = \frac{\dot{q}}{hA(t_b - t_\infty)} = \frac{hA_b(t_b - t_\infty) + hA_f\eta(t_b - t_\infty)}{hA(t_b - t_\infty)} \tag{14-26}$$

where $A = A_b + A_f$. Assuming h is constant over the fin and base, we see that

$$\eta_s = \frac{A_b + \eta A_f}{A} = 1 - \frac{A_f}{A}(1 - \eta) \tag{14-27}$$

and

$$\dot{q} = hA\eta_s(t_b - t_\infty) \tag{14-28}$$

14-4   TRANSPORT COEFFICIENTS INSIDE TUBES

Most HVAC heat exchanger applications of flow inside tubes and passages involve water, water vapor, and boiling or condensing refrigerants. The smooth copper tube is by far the most common geometry with these fluids. Forced convection turbulent flow is the most important mode; however, laminar flow sometimes occurs.

Turbulent Flow of Liquids Inside Tubes

Probably the most widely used heat transfer correlation for this common case is the Dittus-Boelter Equation (1)

$$\frac{\bar{h}D}{k} = 0.023\mathrm{Re}_D^{0.8}\mathrm{Pr}^n \qquad (14\text{-}29)$$

where

$$n = 0.4 \qquad t_{wall} > t_{bulk}$$
$$n = 0.3 \qquad t_{wall} < t_{bulk}$$

Equation (14-29) applies under conditions of $\mathrm{Re}_D > 10{,}000$, $0.7 < \mathrm{Pr} < 100$, and $L/D > 60$. All fluid properties should be evaluated at the arithmetic mean bulk temperature of the fluid. Appendix B gives the thermophysical properties required in Eq. (14-29) for some common liquids and gases. Reference 5 gives other, similar correlations for special conditions. Equation (14-29) may be used for annular or non-circular cross sections for approximate calculations. In this case the tube diameter D is replaced by the hydraulic diameter D_h.

$$D_h = \frac{4(\text{cross-sectional area})}{(\text{wetted perimeter})} \qquad (14\text{-}30)$$

Reference 7 gives extensive data for noncircular flow channels when more accurate values are required.

Pressure drop for flow of liquids inside pipes and tubes was discussed in Chapter 9. The same procedure applies to heat exchanger tubes; we must still take into account the considerable increase in equivalent length caused by the many U-turns, tube inlets and exits, and the headers required in most heat exchangers.

Laminar Flow of Liquids Inside Tubes

The recommended correlation for predicting the average film coefficient in laminar flow in tubes is

$$\frac{\bar{h}D}{k} = 1.86\left[\mathrm{Re}_D\mathrm{Pr}\frac{D}{L}\right]^{1/3}\left(\frac{\mu}{\mu_s}\right)^{0.14} \qquad (14\text{-}31)$$

When the term in brackets is less than about 20, Eq. (14-31) becomes invalid; however this will not occur for most heat exchanger applications. Properties should be evaluated at the arithmetic mean bulk temperature except for μ_s, which is evaluated at the wall temperature.

A word of caution is appropriate concerning the transition from laminar to turbulent flow. This region is defined approximately by $2000 < \mathrm{Re}_D < 10{,}000$. Prediction of heat transfer and friction coefficients is uncertain during transition. The usual practice is to avoid the region by proper selection of tube size and flow rate.

Pressure drop is computed as described above for turbulent flow in tubes and in Chapter 9. For laminar flow the friction factor (Moody) is given by

$$f = \frac{64}{\text{Re}_D} \qquad (14\text{-}32)$$

Ethylene Glycol-Water Solutions

In many systems it is necessary to add ethylene glycol to the water to prevent freezing and consequent damage to the heat exchangers and other components. The effect of the glycol on flow friction was discussed in Chapter 9, and it was shown that the lost head is generally increased when a glycol-water solution is used. The heat transfer is also adversely affected. Figures 14-15 and 14-16 give the specific heat and thermal conductivity of ethylene glycol solutions as a function of temperature and concentration. Similar data for specific gravity and viscosity are given in Chapter 9. It is very important to anticipate the use of glycol solutions during the design phase of a project because the heat transfer coefficient using a 30 percent glycol solution may be as much as 40 percent less than the coefficient using pure water. This is mainly because of the lower thermal conductivity and specific heat of the glycol solution.

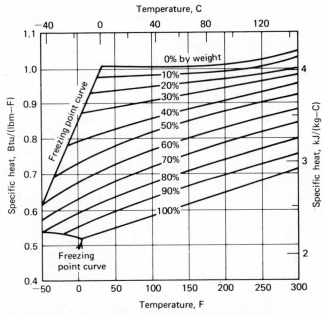

Figure 14-15. Specific heat of aqueous solutions of ethylene glycol. (Reprinted by permission from *ASHRAE Handbook of Fundamentals*, 1977.)

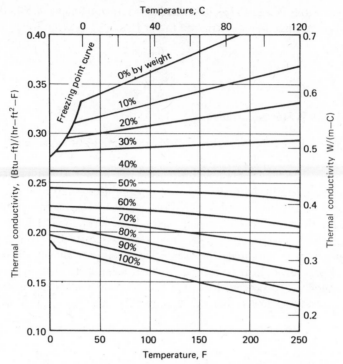

Figure 14-16. Thermal conductivity of aqueous solutions of ethylene glycol. (Reprinted by permission from *ASHRAE Handbook of Fundamentals*, 1977.)

Condensation and Evaporation Inside Horizontal Tubes

The prediction of heat transfer and pressure drop in two-phase flow is much more uncertain than with single phase flow. The mixture of vapor and liquid can vary considerably in composition and hydrodynamic behavior, and it is generally not possible to describe all conditions with one relation. Two-phase flow inside horizontal tubes is the most common situation in HVAC systems, and one or two correlations are presented for this case.

The following relations from reference 5 apply to film condensation, the dominant mode

$$\frac{hD}{k_l} = 13.8(\text{Pr})_l^{1/3} \left(\frac{i_{fg}}{c_p \, \Delta t} \right)^{1/6} \left[\frac{DG_v}{\mu_l} \left(\frac{\rho_l}{\rho_v} \right)^{1/2} \right]^{0.2} \tag{14-33}$$

where

$$\frac{DG_l}{\mu_l} < 5000 \quad \text{and} \quad 1000 < \frac{DG_v}{\mu_l} \left(\frac{\rho_l}{\rho_v} \right)^{1/2} < 20,000$$

The subscripts l and v refer to liquid and vapor, respectively, and Δt is the difference between the fluid saturation temperature and the wall surface temperature. When

$$20,000 < \frac{DG_v}{\mu_l}\left(\frac{\rho_l}{\rho_v}\right)^{1/2} < 100,000 \tag{14-34}$$

$$\frac{hD}{k_l} = 0.1(\text{Pr})_l^{1/3}\left(\frac{i_{fg}}{c_p\,\Delta t}\right)^{1/6}\left[\frac{DG_v}{\mu_l}\left(\frac{\rho_l}{\rho_v}\right)^{1/2}\right]^{2/3}$$

Equations (14-33) and (14-34) are for condensing saturated vapor; however, little error is introduced for superheated vapor when the wall temperature is below the saturation temperature and h is calculated for saturated vapor. Appendixes A and B give the required properties.

The average heat transfer coefficients for evaporating R-12 and R-22 may be estimated from the following relation from reference 5:

$$\frac{hD}{k_l} = C_1\left[\left(\frac{GD}{\mu_l}\right)^2\left(\frac{J\,\Delta x\,i_{fg}g_c}{Lg}\right)\right]^n \tag{14-35}$$

where

J = Joules equivalent, 778 (ft-lbf)/Btu or 1 for SI units

Δx = change in quality of the refrigerant, mass of vapor per unit mass of the mixture

i_{fg} = enthalpy of vaporization, Btu/lbm or J/kg

L = length of the tube, ft or m

C_1 = constant = 9×10^{-4} when $x_e < 0.9$, and 8.2×10^{-3} when $x_e \geq 1.0$
(x_e is the quality of the refrigerant leaving the tube)

n = constant = 0.5 when $x_e < 0.9$, and 0.4 when $x_e \geq 1.0$

The correlation was obtained from tests made using copper tubes having diameters of 0.47 and 0.71 in. and lengths from 13 to 31 ft. Evaporating temperatures varied from -4 to 32 F. Equation (14-35) is sufficient for most HVAC applications where Appendixes A and B give the required properties.

The pressure drop which occurs with a gas-liquid flow is of interest. Experience has shown that pressure drops in two-phase flow are usually much higher than would occur for either phase flowing along at the same mass rate. As in any flow, the total pressure drop along a tube depends upon three factors: (1) friction, due to viscosity, (2) change of elevation, and (3) acceleration of the fluid.

Friction is present in any flow situation, although in some cases it may contribute less than the other two factors. In horizontal flow the change in elevation is zero, and there would be no pressure drop due to this factor. Where there is a small change in gas density or little evaporation occurring, the pressure drop due to acceleration would usually be small. In flow with large changes of density or where evaporation is present, however, the acceleration pressure drop may be very significant.

There has been a variety of schemes proposed for predicting the pressure drop in two-phase flow. Some investigators have attempted to develop theories and equations valid only for a particular flow pattern. These have the disadvantage of requiring that the type of flow pattern be known, and they lead to complexities in those situations where flow patterns change along the flow. Other investigators have attempted to develop single theories valid over several flow regimes. These attempts have not been completely successful.

A widely used method of predicting pressure drop in two-phase flow was developed by Lockhart and Martinelli [27]. Their method, which is essentially empirical, is based upon measurements of isothermal flow of air and several liquids under essentially incompressible conditions. *The pressure drop predicted is that due to friction only.*

Extensive work has been devoted to the two-phase pressure loss problem, but available methods remain very complex and impractical for general use. The general topic is discussed in reference 5. Pressure loss across tube banks is discussed in reference 28.

14-5   TRANSPORT COEFFICIENTS OUTSIDE TUBES AND COMPACT SURFACES

Air is the most common flow medium in this case except for shell and tube evaporators and condensers where heat is transferred between a refrigerant inside the tubes to water outside the tubes. Water-to-water applications may also occur. In the case of bare tube bundles the flow may be across the tubes, parallel to the tubes, or a combination of the two when baffles are used. Compact surfaces such as finned tubes or plate fins will usually have air flowing parallel to the fins and normal to the tubes.

Bare Tubes with Parallel Flow

Kays (8) discusses both laminar and turbulent heat transfer with flow parallel to the tubes and presents results for constant heat flux per unit length of tube for fully developed velocity and temperature profiles. References 9 and 10 also pertain to this problem as well as pressure loss with flow parallel to tubes.

Bare Tubes with a Combination of Parallel and Cross Flow

This flow pattern occurs when baffles are used in a tube and shell exchanger (Fig. 14-17). Empirical data are usually employed because of the complex nature of the flow pattern. The most successful of these methods developed from the work of Tinker and was clarified by Devore (11); the shell side flow was split into the basic components of pure cross flow, baffle-to-baffle leakage flow, tube-to-baffle leakage flow, and bundle-to-shell bypass flow. Tinker made gross simplifications to describe the individual flow streams and developed a set of equations that could be solved simultaneously. Tinker's method was utilized in part to develop the Delaware method reported by Bell (12). This method avoided some of the complexities of Tinker's

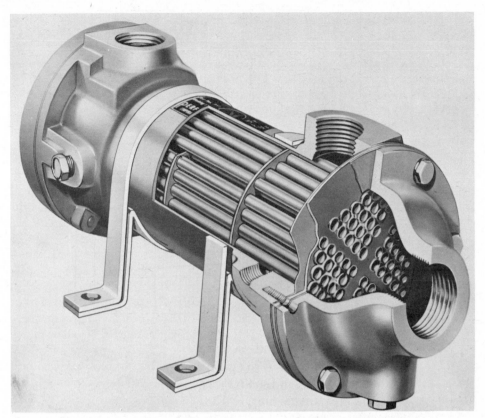

Figure 14-17.   A one shell pass, one tube pass heat exchanger. (Courtesy of McQuay-Perfex, Inc., Milwaukee, Wis.)

method by assuming independent correction factors for the individual streams. A study by Taborek and Palen (13) using extensive data banks and the digital computer has led to a method that improves upon the older ones.

Bare Tubes in Cross Flow

The most common application of bare tubes in pure cross flow involves air. Although this application is rapidly going out of style in favor of finned tubes, considerable data are available for tubes in cross flow as shown in Fig. 14-18 and reference 7. The manner of presentation is quite typical of that used for all types of compact heat exchanger surfaces where the j-factor

$$j = \frac{h}{Gc_p} \Pr^{2/3} \tag{14-36}$$

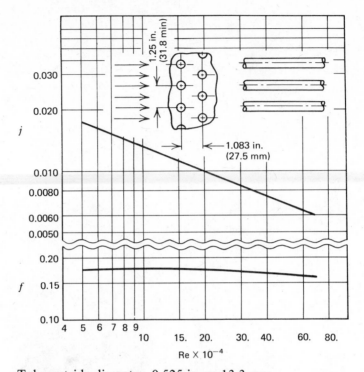

Tube outside diameter, 0.525 in. or 13.3 mm
Hydraulic diameter, 0.1604 ft or 48.9 mm
Free flow area/frontal area, σ, 0.581
Heat transfer area/total volume, α, 14.5 ft²/ft³ or 47.6 m²/m³

Figure 14-18.   Heat transfer and flow friction data for a staggered tube bank, four rows of
tubes (23). (Reprinted by permission from *ASHRAE Trans.*, 79, Part II,
1973.)

and the Fanning friction factor f are plotted versus the Reynolds number

$$Re = \frac{GD_h}{\mu} \tag{14-37}$$

The number of rows of tubes in the flow direction has an effect on the j-factor and
the heat transfer coefficient h. The data of Fig. 14-18 are applicable to an exchanger
with four rows of tubes (23). For bare tubes in cross flow, the relation between the
heat transfer coefficient for a finite number of tube rows N_r and that for an infinite
number of tube rows is given approximately by

$$\frac{h}{h_\infty} = 1 - 0.32e^{-0.15N_r} \tag{14-38}$$

when $2 < N_r < 10$. One might expect the friction to also depend on the number of tube rows; however, this does not hold true. The assumption is that since a contraction and expansion occur for each row, the friction factor is the same for each row.

The mechanical energy equation, Eq. (9-1c), with the elevation and work terms zero, expresses the lost head for a bank of tubes oriented as shown in Fig. 14-19.

$$\frac{(P_{01} - P_{02})g_c}{\rho_m g} = l_h \tag{14-39}$$

where l_h is made up losses resulting from a change in momentum, friction, and entrance and exit contraction and expansion losses. Integration of the momentum equation through the heat exchanger core, Fig. 14-19 yields (7)

$$\frac{\Delta P_0 g_c}{\rho_m g} = \frac{G_c^2}{2 g \rho_m \rho_1} \left[(K_i + 1 - \sigma^2) \right.$$
$$\left. + 2 \left(\frac{\rho_1}{\rho_2} - 1 \right) + f \frac{A}{A_c} \frac{\rho_1}{\rho_m} - (1 - \sigma^2 - K_e) \frac{\rho_1}{\rho_2} \right] \tag{14-40}$$

where K_i and K_e are entrance and exit loss coefficients that will be discussed in the next section and σ is the ratio of minimum flow to frontal area of the exchanger. It may be shown that

$$\frac{A}{A_c} = \frac{4L}{D_h} \tag{14-41}$$

which is a result of the hydraulic diameter concept.

A = total heat transfer area, ft^2 or m^2
A_c = flow cross-sectional area, ft^2 or m^2

Referring to Eq. (14-40) the first and last terms in the brackets account for entrance and exit losses, whereas the second and third terms account for flow acceleration and friction, respectively. In the case of tube bundles the entrance and exit effects are included in the friction term, that is, $K_i = K_e = 0$. Equation (14-40) then becomes

$$l_h = \frac{G_c^2}{2 g \rho_m \rho_1} \left[(1 + \sigma^2) \left(\frac{\rho_1}{\rho_2} - 1 \right) + f \frac{A}{A_c} \frac{\rho_1}{\rho_m} \right] \tag{14-42}$$

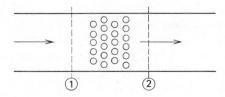

Figure 14-19.   Flow normal to a tube bank.

G_c is based on the minimum flow area and ρ_m is the mean density between inlet and outlet given by

$$\rho_m = \frac{1}{A} \int_A \rho \, dA \tag{14-43}$$

Equation (14-43) is difficult to evaluate. An arithmetic average is usually a good approximation except for parallel flow.

$$\rho_m \approx \frac{\rho_1 + \rho_2}{2} \tag{14-44}$$

A useful nondimensional form of Eq. (14-42) is given by

$$\frac{\Delta P_0}{P_{01}} = \frac{G_c^2}{2g_c\rho_1 P_{01}} \left[(1 + \sigma^2)\left(\frac{\rho_1}{\rho_2} - 1\right) + f\frac{A}{A_c}\frac{\rho_1}{\rho_m} \right] \tag{14-45}$$

where ΔP_0 and P_{01} have units of lbf/ft^2 or Pa. Equations (14-42) and (14-45) are also valid for finned tubes or any other surface that does not have abrupt contractions or expansions.

Finned Tube Heat Transfer Surfaces

Heat transfer surfaces may take many forms; however, finned tubes are the most popular in HVAC applications. The manner in which data are presented is the same as that shown for bare tubes in Fig. 14-18 and the lost head may be computed using Eq. (14-45). Rich (14, 23) has studied the effect of both fin spacing and tube rows for the plate-fin-tube geometry. Both the j factor and friction factor decrease as the fin spacing is decreased. The decrease in j factor was about 50 percent and the decrease in friction factor was about 75 percent as fin pitch was increased from 3 to 20 over the Reynolds number (G_cD_h/μ) range of 500 to 1500. Figure 14-20 shows how the data correlated when the Reynolds number was based on the tube row spacing χ_b. For a given fin pitch it was found that the j factors decreased as the number of tube rows was increased from 1 to 6 in the useful Reynolds number range. This is contrary to the behavior of bare tubes and results from the difference in the flow fields in each case. The one row coil had significantly higher j factors than the others, as was known from previous work done by Shepherd (15). Figure 14-21 shows the j factor data for the coils with various numbers of tube rows. Note that the Reynolds number is based on the tube row spacing. The combination of Figs. 14-20 and 14-21 therefore give performance data for all heat exchangers of this one tube diameter and tube pattern with variable fin pitch and number of tube rows. It is to be expected that other surfaces with tube diameters and patterns in the same range will behave similarly. The study of tube row effect (14), also showed that all rows in a plate-fin-tube coil do not have the same heat transfer rate (Fig. 14-22). The j factors are less for each successive row in the useful (low) Reynolds number range.

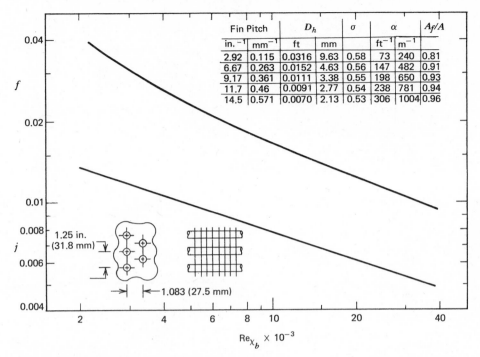

Fin Pitch		D_h		σ	α		A_f/A
in.$^{-1}$	mm^{-1}	ft	mm		ft^{-1}	m^{-1}	
2.92	0.115	0.0316	9.63	0.58	73	240	0.81
6.67	0.263	0.0152	4.63	0.56	147	482	0.91
9.17	0.361	0.0111	3.38	0.55	198	650	0.93
11.7	0.46	0.0091	2.77	0.54	238	781	0.94
14.5	0.571	0.0070	2.13	0.53	306	1004	0.96

Tube outside diameter, 0.525 in. or 13.3 mm
Fin thickness, 0.006 in. or 0.152 mm

Figure 14-20.   Heat transfer and friction data for a plate-fin-tube coil with various fin spacings and 5 rows of tubes. (Reprinted by permission from *ASHRAE Trans., 79*, Part II, 1973.)

The friction factors behave in a manner similar to that discussed above for bare tube banks; therefore, it is assumed there is no tube row effect.

Caution should be exercised in using published data for plate-fin-tube heat exchangers, especially if the number of rows is not given. Recent research by McQuiston (25, 26) has resulted in the correlation of plate-fin-tube transport data which account for geometric variables as well as hydrodynamic effects. Figure 14-23 shows j factors plotted versus the parameter JP which is defined as

$$JP = Re_D^{-0.4}(A/A_t)^{-0.15} \tag{14-46}$$

where

$$Re_D = G_c D/\mu \tag{14-47}$$

and

$$A/A_t = \frac{4}{\pi} \frac{\chi_b}{D_h} \frac{\chi_a}{D} \tag{14-48}$$

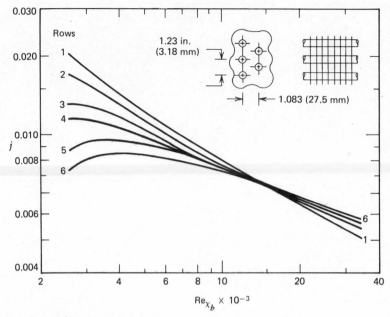

Tube outside diameter, 0.521 in. or 13.2 mm
Fin pitch, 14.5 fins/in. or 0.571 fins/mm
Fin thickness, 0.006 in. or 0.152 mm
Free flow area/frontal area, 0.536

Figure 14-21.   Heat transfer data for plate-fin-tube coils with various number of tube rows. (Reprinted with permission from *ASHRAE Trans., 81,* Part I, 1975.)

In this case the Reynolds number is based on the outside tube diameter and A/A_t is the ratio of the total heat transfer area to the area of the bare tubes without fins. Note that A/A_t becomes 1.0 for a bare tube bank and the correlation takes a familiar form. The tube row effect is not accounted for in Fig. 14-23 and must be done separately using Fig. 14-21 which is well described by

$$j_n/j_1 = 1 - 1280 \ N_r \text{Re}_{x_b}^{-1.2} \tag{14-49}$$

where the subscripts n and 1 pertain to the number of tube rows. Because Fig. 14-23 is for 4 rows of tubes it is more convenient to write

$$j_n/j_4 = \frac{1 - 1280 \ N_r \text{Re}_{x_b}^{-1.2}}{1 - 5120 \ \text{Re}_{x_b}^{-1.2}} \tag{14-50}$$

where j_4 is read from Fig. 14-23.

Generalized correlation of friction data is more involved than that for heat transfer data. Figure 14-24 shows such a correlation using a parameter FP defined as

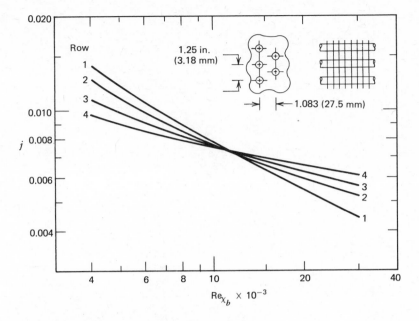

Tube outside diameter, 0.521 in. or 13.2 mm
Fin pitch, 14.5 fins/in. or 0.571 fins/mm
Fin thickness, 0.006 in. or 0.152 mm
Free flow area/frontal area, 0.536

Figure 14-22.   Heat transfer data for each row of a 4 row plate-fin-tube coil. (Reprinted with permission from *ASHRAE Trans., 81,* Part I, 1975.)

$$FP = \mathrm{Re}_D^{-0.25} \left(\frac{D}{D^*} \right)^{0.25} \left[\frac{\chi_a - D}{4(s - y)} \right]^{-0.4} \left[\frac{\chi_a}{D^*} - 1 \right]^{-0.5} \tag{14-51}$$

where D^* is a hydraulic diameter defined by

$$D^* = \frac{D(A/A_t)}{1 + (\chi_a - D)/s} \tag{14-52}$$

The correlating parameters of Eqs. (14-51) and (14-52) have evolved over a long period of time from observations of experimental data. The friction data scatter more than the heat transfer data of Fig. 14-23, which are typical. Note that the data from reference 25 are much more consistent that some of the other data which date back more than 20 years.

The data presentations of Figures 14-23 and 14-24 have the advantage of generality and are also adaptable to the situation where moisture is condensing on the surface. This will be discussed later in this chapter. These same type correlations may be used for other types of finned surfaces such as circular and wavy fins.

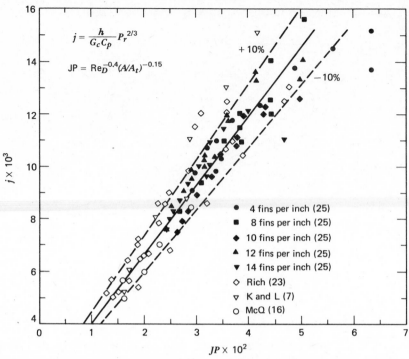

Figure 14-23. Heat transfer correlation for smooth plate-fin-tube coils with 4 rows of tubes. (Reprinted with permission from *ASHRAE Trans., 84,* Part 1, 1978.)

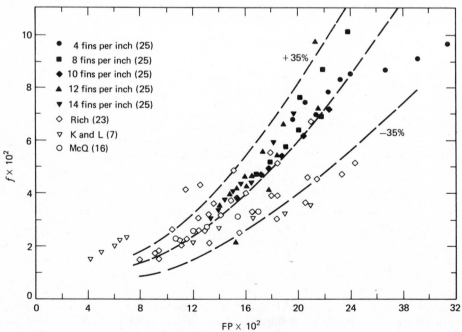

Figure 14-24. Correlation of friction data for smooth plate-fin-tube coils. (Reprinted with permission from *ASHRAE Trans., 84,* Part 1, 1978.)

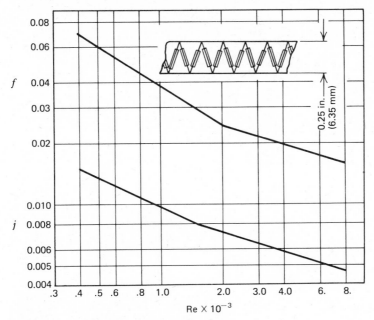

Fin pitch, 11 fins/in. or 0.433 fins/mm
Plate spacing, 0.25 in. or 6.35 mm
Louver spacing, 0.5 in. or 12.7 mm
Hydraulic diameter, 0.0101 ft or 3.08 mm
Fin thickness, 0.006 in. or 0.15 mm
Heat transfer area/volume between the plates, 367 ft^2/ft^3 or 1204 m^2/m^3
Fin area/total area, 0.656

Figure 14-25.   Heat transfer and friction data for a louvered plate-fin surface.

Plate Fin Heat Transfer Surfaces

Figure 14-25 illustrates the plate fin heat transfer surface. The fins may have several variations such as louvers, strips, or waves. Plain smooth fins are generally not used because of the low heat transfer coefficients that arise when the flow length becomes long. The types mentioned above disturb the boundary layer so that the length does not influence the heat transfer or flow friction coefficients. Figure 14-25 is an example of data for a louvered plate fin surface.

In computing the lost head for these surfaces, we must consider the entrance and exit losses resulting from abrupt contraction and expansion. The entrance and exit losses are expressed in terms of a loss coefficient K and the velocity head inside the heat exchanger core. Thus for the entrance

$$\Delta P_{0i} = K_i \frac{G_c^2}{2\rho_i g_c} \qquad (14\text{-}53)$$

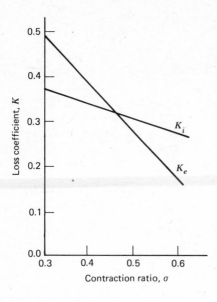

Figure 14-26. Entrance and exit pressure loss coefficients for a plate-fin heat exchanger with flow interruptions.

and for the exit

$$\Delta P_{0e} = K_e \frac{G_c^2}{2\rho_e g_c} \tag{14-54}$$

Equations (14-53) and (14-54) are included in Eq. (14-45)

$$\frac{\Delta P_0}{P_{01}} = \frac{G_c^2}{2g_c P_{01}\rho_1} \left[(K_i + 1 - \sigma^2) \right.$$
$$\left. + 2\left(\frac{\rho_1}{\rho_2} - 1\right) + f\frac{A}{A_c}\frac{\rho_1}{\rho_m} - (1 - \sigma^2 - K_e)\frac{\rho_1}{\rho_2} \right] \tag{14-55}$$

The entrance and exit loss coefficients depend on the type of surface, the contraction ratio, and the Reynolds number $G_c D_h/\mu$. The degree to which the velocity profile has developed is also important. Reference 7 gives entrance and exit loss coefficients that apply to surfaces such as that shown in Fig. 14-25. The loss coefficients are a function of Reynolds number. However, most plate fin surfaces have flow interruptions that cause continual redevelopment of the boundary layer, which is equivalent to a very high Reynolds number condition. Figure 14-26 gives loss coefficients applicable to plate fin-type surfaces with flow interruptions such as that of Fig. 14-25.

14-6 DESIGN PROCEDURES FOR SENSIBLE HEAT TRANSFER

It is difficult to devise one procedure for designing all heat exchangers because the given parameters vary from situation to situation. All of the terminal temperatures may be known or only the inlet temperatures may be given. The mass flow rates may be fixed in some cases and a variable in others. Usually the surface area is not given.

Earlier in the chapter the *LMTD* and *NTU* methods were described as the two general heat exchanger design procedures. Either method may be used, but the *NTU* method has certain advantages. Consider the design problem where t_{hi}, t_{ho}, t_{ci}, $\dot{m}_c$, and $\dot{m}_h$ are known and the surface area A is to be determined. With either approach the heat transfer coefficients must be determined as previously discussed so that the overall coefficient U can be computed. The *NTU* approach then proceeds as follows:

1. Compute the effectiveness ε and C_{min}/C_{max} from the given data.
2. Determine the *NTU* for the particular flow arrangement from the ε-NTU curve, such as Fig. 14-5 or Table 14-1.
3. Compute A from $A = NTU(C_{min}/U)$.

The LMTD approach is as follows:

1. Compute P and R from the given terminal temperatures.
2. Determine the correction factor F from the appropriate curve such as Fig. 14-3.
3. Calculate the *LMTD* for an equivalent counterflow exchanger using Eq. (14-2).
4. Calculate A from $A = \dot{q}/U(F)(LMTD)$ where $\dot{q} = C_c(t_{co} - t_{ci}) = C_h(t_{hi} - t_{ho})$.

The *NTU* approach requires somewhat less effort in this case.

Consider the design problem where A, U, $\dot{m}_c$, $\dot{m}_h$, t_{hi}, and t_{ci} are given, and it is necessary to find the outlet temperatures t_{ho}, t_{co}. The *NTU* approach is as follows:

1. Calculate the $NTU = UA/C_{min}$ from given data.
2. Find ε from the appropriate curve for the flow arrangement using *NTU* and C_{min}/C_{max}.
3. Compute one outlet temperature from Eq. (14-7) or (14-8).
4. Compute the other outlet temperature from

$$\dot{q} = C_c(t_{co} - t_{ci}) = C_h(t_{hi} - t_{ho})$$

The LMTD approach requires iteration as follows:

1. Calculate R from $R = C_c/C_h$.
2. Assume one outlet temperature in order to compute P (first approximation).
3. Find F from the appropriate curve (first approximation).
4. Evaluate *LMTD* (first approximation).
5. Determine $\dot{q} = UAF(LMTD)$ (first approximation).
6. Calculate outlet temperature to compare with the assumption of step 2.
7. Repeat steps 2 through 6 until satisfactory agreement is obtained.

It is obvious that the *NTU* method is much more straightforward in this later problem. A complete design problem is illustrated in the following example.

Example 14-1. Design a water-to-air heating coil of the continuous plate-fin-tube type. The required duty for the coil is as follows:

Heat outdoor air from 50 to about 100 F
Air flow rate of 2000 cfm
Entering water temperature, 150 F
Leaving water temperature, 140 F
Air face velocity should not exceed 1000 fpm
Water-side head loss should not exceed 10 ft of water
Water connections must be on the same end of the coil
Air side pressure drop should not exceed about 1.2 in. of water

Solution. Figure 14-27 is a schematic of a typical water-to-air heating coil that has multiple rows of tubes. Although the water may be routed through the tubes in many different ways, the circuiting is usually such that counterflow will be approached as shown. Counterflow can usually be assumed when three or more rows are used. Because the water inlet and outlet connections must be on the same end of the coil in this case, a multiple of two rows is used if possible.

The first step in the solution will be to compute the overall heat transfer coefficient U, which will be based on the air-side area. Equation (14-15a) applies where η_{si} is equal to one and the wall thermal resistance is negligible.

$$\frac{1}{U_o} = \frac{1}{h_o \eta_{so}} + \frac{1}{h_i(A_i/A_o)}$$

The subscript o refers to the air side and the i refers to the water side. Equation (14-29) will be used to find the coefficient h_i assuming a water velocity of 4 ft/sec. Experience has shown that velocities greater than 5 ft/sec or 1.5 m/s result in very high lost head. Since at this point the tube diameter must be established, a surface geometry must be selected.

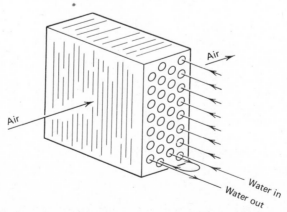

Figure 14-27. A typical heating coil circuited to approach counterflow.

One standard plate-fin-tube surface uses $\frac{1}{2}$ in. tubes in a triangular layout as shown in Fig. 14-14 with χ_a of 1.25 in. and χ_b of 1.083 in. The fin pitch is 8 fins/in. and the fin thickness is 0.006 in. As a result of fabrication of the coil, the final tube outside diameter is 0.525 in. with a wall thickness of 0.015 in. Other geometric data will be given as required and the j-factor and friction factor will be obtained from Figs. 14-23 and 14-24. The Reynolds number based on the tube inside diameter is then

$$\text{Re}_D = \frac{\rho \overline{V} D}{\mu} = \frac{61.5(4)(0.4831/12)}{(1.04/3600)} = 34{,}275$$

where ρ and μ are evaluated at 145 F. The Prandtl number is

$$\text{Pr} = \frac{\mu c_p}{k} = \frac{(1.04)(1.0)}{(0.38)} = 2.74$$

Then using Eq. (14-29)

$$h_i = 0.023 \frac{k}{D} \text{Re}_D^{0.8} \text{Pr}^{0.3}$$
$$h_i = 0.023 \frac{0.38}{(0.483)/12} (34{,}275)^{0.8}(2.74)^{0.3}$$
$$h_i = 1250 \text{ Btu/(hr-ft}^2\text{-F)}$$

where the exponent on the Prandtl number is for $t_{wall} < t_{bulk}$ and L/D has been assumed to be larger than 60.

To compute the air-side heat transfer coefficient it is necessary to know the air velocity or air mass velocity inside the core. Because the coil face velocity cannot exceed 1000 ft/min a face velocity of 900 ft/min will be assumed. Then

$$\dot{m}_a = G_{fr} A_{fr} = G_c A_c$$

and

$$G_c = G_{fr} \frac{A_{fr}}{A_c} = \frac{G_{fr}}{\sigma}$$

where the subscript fr refers to the face of the coil and c refers to the minimum flow area inside the coil. The ratio of minimum flow to frontal area for this case has been calculated to be 0.555.

$$G_{fr} = \rho_{fr}\overline{V}_{fr} = \frac{14.7(144)(900)60}{53.35(510)} = 4200 \text{ lbm/(hr-ft}^2\text{)}$$

and

$$G_c = \frac{4200}{(0.555)} = 7569 \text{ lbm/(hr-ft}^2\text{)}$$

The j-factor correlation of Fig. 14-23 is based on the parameter JP which is defined by Eq. (14-46). The Reynolds number is then

$$\text{Re}_D = \frac{G_c D}{\mu} = \frac{7569(0.525/12)}{0.044} = 7578$$

and the parameter A/A_t defined by Eq. (14-48) is

$$A/A_t = \frac{4(1.083)1.25(0.555)}{\pi(0.01312)12(0.525)} = 11.6$$

where the hydraulic diameter is another known dimension of the coil. The parameter JP is

$$JP = (7578)^{-0.4}(11.6)^{-0.15} = 0.0194$$

The j-factor is now read from Fig. 14-23 as 0.0066. Then

$$\text{StPr}^{2/3} = \left(\frac{h_o}{G_c c_p}\right)\left(\frac{\mu c_p}{k}\right)^{2/3} = 0.0066$$

or

$$h_o = 0.0066(7569)(0.24)(0.71)^{-2/3} = 15.1 \text{ Btu/(hr-ft}^2\text{-F)}$$

The next step is to compute the fin efficiency and the surface effectiveness. Equations (14-20), (14-21), (14-22), and (14-24) will be used. The equivalent fin radius R_e is first computed from Eq. (14-24). The dimensions L and M are found as follows by referring to Fig. 14-14.

$$\text{Dim}_1 = \frac{\chi_a}{2} = \frac{1.25}{2} = 0.625 \text{ in.}$$

$$\text{Dim}_2 = \frac{[(\chi_a/2)^2 + \chi_b^2]^{1/2}}{2}$$

$$= \frac{[(0.625)^2 + (1.083)^2]^{1/2}}{2} = 0.625 \text{ in.}$$

Because Dim_1 is equal to Dim_2 in this case

$$L = M = 0.625 \text{ in.}$$

Then

$$\psi = \frac{M}{r} = \frac{0.625}{(0.525/2)} = 2.38$$

$$\beta = \frac{L}{M} = \frac{0.625}{0.625} = 1.0$$

and

$$\frac{R_e}{r} = 1.27(2.38)(1.0 - 0.3)^{1/2} = 2.53$$

From Eq. (14.22)

$$\phi = (2.53 - 1)(1 + 0.35 \ln 2.53) = 2.03$$

and using Eq. (14-20)

$$m = \left[\frac{2(15.1)}{100(0.006/12)} \right]^{1/2} = 24.6 \text{ ft}^{-1}$$

where the thermal conductivity k of the fin material has been assumed equal to 100 (Btu-ft)/(ft²-hr-F), which is typical of aluminum fins. Then from Eq. (14-21),

$$\eta = \frac{\tanh \left[(24.6)(0.525/24)(2.03) \right]}{(0.525/24)(2.03)} = 0.73$$

The surface effectiveness η_{so} is then computed using Eq. (14-26) where A_f/A is 0.919.

$$\eta_{so} = 1 - 0.919(1 - 0.73) = 0.75$$

The ratio of the water-side to air-side heat transfer areas must finally be determined. The ratio of the total air side heat transfer area to the total volume (A_o/V) or α is given as 170 ft⁻¹. The ratio of the water-side heat transfer area to the total volume (A_i/V) is closely approximated by

$$\frac{A_i}{V} = \frac{D_i \pi}{\chi_a \chi_b}$$

$$\frac{A_i}{A_o} = \left(\frac{A_i}{V} \right) \Big/ \left(\frac{A_o}{V} \right) = \frac{\pi D_i}{\chi_a \chi_b \alpha} \qquad (14\text{-}56)$$

$$\frac{A_i}{A_o} = \frac{\pi(0.483/12)}{(1.25/12)(1.083/12)(170)} = 0.079$$

The overall coefficient U is then given by

$$\frac{1}{U_o} = \frac{1}{15.1(0.75)} + \frac{1}{1248(0.079)} = 0.098$$

and

$$U_o = 10.2 \text{ Btu}/(\text{hr-ft}^2\text{-F})$$

The fluid capacity rates will now be computed. For the air

$$\dot{m} = \rho \dot{Q} = \frac{14.7(144)}{53.35(510)} (2000)(60) = 9{,}336 \text{ lbm/hr}$$

and

$$C_{air} = C_c = 0.24(9336) = 2{,}241 \text{ Btu}/(\text{hr-F})$$

For the water

$$\dot{q} = C_w(t_{wi} - t_{wo}) = C_{air}(t_{ao} - t_{ai})$$

and

$$C_w = C_h = C_{air} \frac{(t_{ao} - t_{ai})}{(t_{wi} - t_{wo})}$$

$$C_w = 2241 \frac{(100 - 50)}{(150 - 140)} = 11,205 \text{ Btu/(hr-F)}$$

Since $C_w > C_{air}$, $C_{air} = C_{min} = C_c$, $C_w = C_h = C_{max}$, and

$$\frac{C_{min}}{C_{max}} = \frac{2241}{11,205} = 0.20$$

The effectiveness ε is given by Eq. (14-8)

$$\varepsilon = \frac{t_{co} - t_{ci}}{t_{hi} - t_{ci}} = \frac{100 - 50}{150 - 50} = 0.50$$

Now assuming that the flow arrangement is counterflow, the NTU is read from Fig. 14-5 at $\varepsilon = 0.5$ and $C_{min}/C_{max} = 0.2$ as 0.74. Then

$$NTU = \frac{U_o A_o}{C_{min}}$$

$$A_o = \frac{0.74(2241)}{10.2} = 163 \text{ ft}^2$$

The total volume of the heat exchanger is given by

$$V = \frac{A_o}{\alpha} = \frac{163}{170} = 0.96 \text{ ft}^3$$

Since a face velocity of 900 ft/min was assumed, the face area is

$$A_{fr} = \frac{\dot{Q}}{V_{fr}} = \frac{2000}{900} = 2.22 \text{ ft}^2$$

and the depth is

$$L = \frac{V}{A_{fr}} = 0.96/2.22 = 0.43 \text{ ft} = 5.18 \text{ in.}$$

The number of rows of tubes N_r will then be

$$N_r = \frac{L}{X_b} = 5.18/1.083 = 4.78$$

Since N_r must be an integer and a multiple of two for the flow arrangement of Fig. 14-27, six rows must be used. This will overdesign the heat exchanger. Another possibility is to use five rows with a different circuiting arrangement so that the water connections are on the same end of the coil. This will be considered later when the lost head on the water side is computed.

The lost head on the air side of the exchanger is given by Eq. (14-45) where the ratio A/A_c is given by

$$\frac{A}{A_c} = \frac{(\alpha V)}{(\sigma A_{fr})} = \frac{170(0.96)}{0.480(2.22)} = 153$$

The mass velocity G_c was previously computed as 7569 lbm/(hr-ft²) and the mean density ρ_m is approximately

$$\rho_m = \frac{P}{2R}\left(\frac{1}{T_{ci}} + \frac{1}{T_{co}}\right)$$

$$\rho_m = \frac{14.7(144)}{2(53.35)}\left[\frac{1}{510} + \frac{1}{560}\right] = 0.074 \ \text{lbm/ft}^3$$

The friction factor is read from Fig. 14-24 with FP computed from Eq. (14-51). Using Eq. (14-52)

$$\frac{D^*}{D} = \frac{11.6}{1 + (1.25 - 0.525)/0.125} = 1.71$$

and

$$FP = (7578)^{-0.25}(1.71)^{-0.25}\left[\frac{1.25 - 0.525}{4(0.125 - 0.006)}\right]^{-0.4}\left[\frac{1.25}{0.898} - 1\right]^{-0.5}$$

$$= 0.126$$

then from Fig. 14-24, $f = 0.041$.

$$\Delta P_0 = \frac{(7569)^2}{2(0.078)(32.2)(3600)^2}\left\{[1 + (0.555)^2]\left(\frac{0.078}{0.071} - 1\right)\right.$$
$$\left. + \ 0.041(153)\left(\frac{0.078}{0.074}\right)\right\}$$

$$\Delta P_0 = 5.93 \ \text{lbf/ft}^2$$

and

$$l_h = \frac{\Delta P_o}{\rho_w} = \frac{5.93}{62.4} = 0.095 \ \text{ft of water} = 1.14 \ \text{in. wat.}$$

The lost head on the tube side of the exchanger must now be computed. Recall that a velocity of 4 ft/sec was assumed to compute the heat transfer coefficient h_i. It has also been determined that at least five rows of tubes are required and the water connections must be on the same end of the exchanger. Therefore, consider the arrangement shown in Fig. 14-28. If we use two passes per row of tubes, the water enters and leaves the same end of the coil. For the coil shown there are five separate water circuits.

The flow cross-sectional area for the water may be determined from the fluid capacity rate for the water and the continuity equation.

$$m_w = \overline{V}A\rho = \frac{C_h}{c_p}$$

and

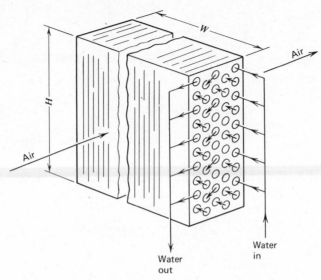

Figure 14-28.   Five row coil with two passes per row.

$$A = \frac{C_h}{\overline{V}\rho c_p} = \left(\frac{11{,}205}{3600}\right)\frac{1}{(4)(61.5)(1.0)}$$

$$A = 0.01265 \text{ ft}^2$$

For N number of tubes

$$A = N\frac{\pi}{4}D_i^2$$

and

$$N = \frac{4A}{\pi D_i^2} = \frac{4(0.01265)(144)}{\pi(0.483)^2} = 9.94$$

Since N must be an integer, 10 tubes are required and the water velocity is reduced somewhat. This reduction in velocity will not significantly reduce the heat transfer coefficient h_i. To adapt to the flow arrangement of Fig. 14-28 a coil that is 20 tubes high must be used. Then the height H becomes

$$H = 20\chi_a = 20(1.25) = 25 \text{ in.}$$

The frontal area A_{fr} was previously found to be 2.22 ft². Then the width W is

$$W = \frac{A_{fr}}{H} = \frac{2.22}{(25/144)} = 12.8 \text{ in.}$$

This arrangement will meet all of the design requirements; however, the shape of the coil, height of 25 in. and width of 12.8 in., may be unacceptable. If so, another alternative must be sought such as using six rows of tubes, placing the headers on opposite ends, or some other modification.

The lost head l_{fw} will be computed using Eq. (9-6). Lost head in the return bends will be accounted for by assuming a loss coefficient of 2 for each bend. The flow length L_w is

$$L_w = 2(5)(12.8/12) = 10.7 \text{ ft}$$

and the Moody friction factor is 0.023 from Fig. 9-1 at a Reynolds number of 34,275, which takes into account the lower water velocity. There are nine return bends in each circuit. Then

$$l_{fw} = 0.023 \frac{(10.7)}{(0.483/12)} \frac{(4)^2}{(64.4)} + 2(9) \frac{(4)^2}{(64.4)}$$
$$l_{fw} = 6 \text{ ft of water}$$

This value of the lost head does not include the losses in the inlet and outlet headers. Header losses may be substantial, depending on their design and fabrication, and may be equal to the losses in the tubes and return bends.

The given design criteria have been satisfied with the possible exception of the shape of the coil, mentioned above.

There are many different ways the heat exchanger design problem may be posed. About the same amount of work is involved in every case, however. The previous example shows that the process is laborious and time-consuming. Therefore, almost all manufacturers have devised computer programs that carry out the design process quickly and accurately. Because of the speed of a computer, a simulation rather than a design approach may be used where the performance of a given configuration is determined.

14-7 COMBINED HEAT AND MASS TRANSFER

When the heat exchanger surface in contact with moist air is at a temperature below the dew point temperature for the air, condensation of vapor will occur. Typically the air dry bulb temperature and the humidity ratio both decrease as the air flows through the exchanger. Therefore, sensible and latent heat transfer occur simultaneously. This process is similar to that occurring in the spray dehumidifier discussed in Chapter 13 and can be analyzed using the same procedure; however, this is not generally done.

The problem of cooling coil analysis and design is complicated by the uncertainty in determining the transport coefficients h, h_d, and f. It would be very convenient if heat transfer and friction data for dry heating coils, such as that shown in Figs. 14-23 and 14-24 could be used with the Colburn analogy of Eq. (12-20a) to obtain the mass transfer coefficients. But it has been known for some time that this approach is not reliable, and recent work (17, 18, 25, 26) has shown that the analogy does not always hold true. Figure 14-29 shows j factors for a simple parallel plate

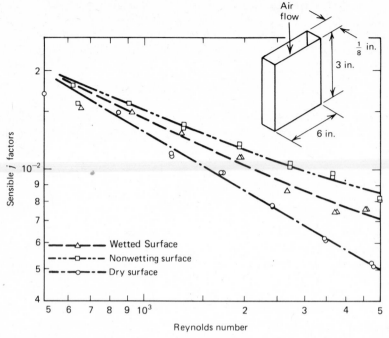

Figure 14-29. Sensible heat transfer j factors for a parallel plate exchanger (18). (Reprinted by permission from *ASHRAE Trans., 82,* Part II, 1976.)

exchanger that were obtained for different surface conditions. Although these particular j factors are for the sensible heat transfer, the mass transfer j factors and the friction factors exhibited the same behavior. Note that the dry surface j factors fall below those obtained under dehumidifying conditions with the surface wet. The converging-diverging nature of the curves of Fig. 14-29 can be explained by the roughness introduced by the water on the surface and the nature of the boundary layers at different Reynolds numbers. The velocity, temperature, and concentration boundary layer thicknesses can all be approximated by

$$\frac{\delta}{x} = \frac{5}{\sqrt{Re_x}} \tag{14-57}$$

where

δ = boundary layer thickness
x = distance from inlet, measured in the same units as δ
Re_x = Reynolds number based on x

Equation (14-57) shows that at low Reynolds numbers the boundary layer grows quickly; the droplets are soon covered and have little effect on the flow field. As the Reynolds number is increased, the boundary layer becomes thin and more of the total flow field is exposed to the droplets. The roughness caused by the droplets induces mixing and larger j factors. Note that the data of Fig. 14-29 cannot be applied to all surfaces because the length of the flow channel is also an important variable. It seems certain, however, that the water collecting on the surface is responsible for breakdown of the j factor analogy. The j factor analogy is approximately true when the surface conditions are identical (18). That is, when the surface is wetted, the sensible and mass transfer j factors are in close agreement. This is of little use, however, because a wet test must be made to obtain this information. Under some conditions it is possible to obtain a film of condensate on the surface instead of droplets. For example, aluminum when thoroughly degreased and cleaned with a harsh detergent in hot water experiences filmwise condensation (25). Figure 14-30 shows j factor and friction data for a plate-fin-tube surface under dry conditions and with filmwise and dropwise condensation. Note that the trends are the same as those shown in Fig. 14-29. The friction factors are influenced by the water on the surface over the complete Reynolds number range, whereas the j factors are only affected at the higher Reynolds numbers. The data shown correspond to face velocities of 200 to 800 ft/min with air at standard conditions. Although not shown, the mass transfer j factors show the same trends and are in reasonable agreement with the wet surface j factors shown in Fig. 14-30.

Recent research involving plate-fin-tube surfaces (26) has resulted in correlations which relate dry sensible j and f factors to those for wetted dehumidifying surfaces. Expressions were developed that modify the parameters JP and FP of Figs. 14-23 and 14-24 for wet surface conditions. In developing these functions, it was found that a Reynolds number based on fin spacing and the ratio of fin spacing to space between the fins were useful. For film-type condensation, the modifying functions are:

Sensible j factor

$$J(s) = 0.84 + 4 \times 10^{-5}\mathrm{Re}_s^{1.25} \tag{14-58}$$

Total j factor

$$J_i(s) = (0.95 + 4 \times 10^{-5}\mathrm{Re}_s^{1.25})\left(\frac{s}{s-y}\right)^2 \tag{14-59}$$

Friction factor

$$F(s) = (1 + \mathrm{Re}_s^{-0.4})\left(\frac{s}{s-y}\right)^{1.5} \tag{14-60}$$

For dehumidifying conditions, the abscissa of Fig. 14-23 is changed to $J(s)JP$ and $J_i(s)JP$. The abscissa of Fig. 14-24 is changed to $F(s)FP$.

Enthalpy Potential

The enthalpy potential was mentioned in Chapter 13 and will be more fully justified here. The heat transfer from moist air to a surface at a temperature below the air dew point may be expressed as

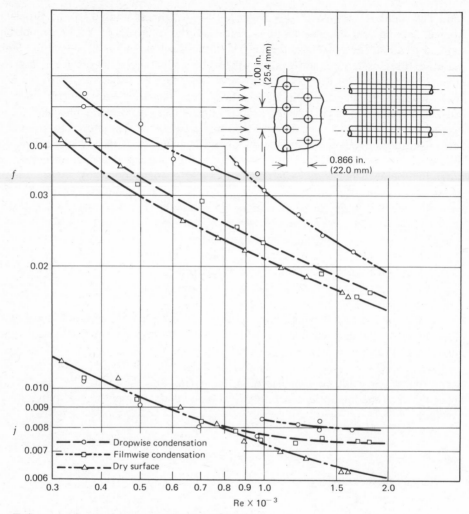

Tube outside diameter, 0.390 in. or 9.96 mm
Fin pitch, 8 fins/in. or 0.315 fins/mm
Hydraulic diameter, 0.0123 ft or 3.74 mm
Fin thickness, 0.006 in. or 0.152 mm
Free flow area/frontal area, σ, 0.56
Heat transfer area/total volume, α, 182 ft^{-1} or 600 m^{-1}
Fin area/total area, 0.92

Figure 14-30.   Heat transfer and friction data with mass transfer for a plate-fin-tube surface, four rows of tubes.

$$\frac{\dot{q}}{A} = h(t_w - t_\infty) + h_d(W_w - W_\infty)i_{fg} \tag{14-61}$$

Using the analogy of Eq. (12-20a) with Le = 1 we see that

$$\frac{\dot{q}}{A} = h_d[c_p(t_w - t_\infty) + (W_w - W_\infty)i_{fg}] \tag{14-62}$$

The enthalpy of vaporization i_{fg} is evaluated at the wall temperature. Even though the Colburn analogy does not appear to be precise, there is a proportionality between h and h_d, which is all that is required here. The enthalpy of the saturated moist air at the wall is given by

$$i_w = c_{pa}t_w + W_w(i_f + i_{fg}) \tag{14-63}$$

where i_f and i_{fg} are evaluated at the wall temperature. For the moist air in the free stream

$$i_\infty = c_{pa}t + W[i_f + i_{fg} + c_{ps}(t - t_w)] \tag{14-64}$$

The temperature t_w should be the dew point temperature and i_f and i_{fg} should be evaluated at the dew point temperature. However, the errors tend to compensate and Eq. (14-64) is a very good approximation. The difference in enthalpy between the surface i_w and the free stream i_∞ is then

$$i_w - i_\infty = c_p(t_w - t_\infty) + i_{fg}(W_w - W_\infty) + i_f(W_w - W_\infty) \tag{14-65}$$

Comparison of Eqs. (14-62) and (14-65) then yields

$$\frac{\dot{q}}{A} = h_d[(i_w - i_\infty) - i_f(W_w - W_\infty)] \tag{14-66}$$

The last term is typically about 0.5 percent of $(i_w - i_\infty)$ and can be neglected. Thus, the driving potential for simultaneous transfer of heat and mass is enthalpy to a close approximation, whereas temperature and concentration are the driving potentials for sensible heat and mass, respectively.

Equation (14-66) expresses the total heat transfer at a particular location in the heat exchanger; however, the moist air enthalpy at the surface i_w and in the free stream i_∞ vary throughout the exchanger, as shown in Fig. 14-31 for counterflow. In addition, most coils will have fins that must be accounted for. Then

$$\dot{q} = h_d A \eta_{ms} \Delta i_m \tag{14-67}$$

where η_{ms} is the surface effectiveness with combined heat and mass transfer and Δi_m is some mean enthalpy difference. With suitable assumptions it can be shown that Δi_m has the same form as the *LMTD* for counterflow:

$$\Delta i_m = \frac{\Delta i_1 - \Delta i_2}{\ln \dfrac{\Delta i_1}{\Delta i_2}} \tag{14-68}$$

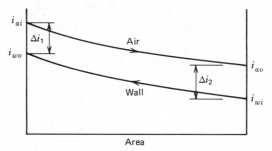

Figure 14-31. Enthalpy difference in a counterflow dehumidifying coil.

This is true because i_w is directly proportional to t_c, the refrigerant temperature. It should be noted that Eq. (14-67) expresses the total heat transfer rate from the wall to the air stream where the wall temperature is now known explicitly. However, the heat transfer rate from the refrigerant to the wall is given by

$$\dot{q} = h_i A_i (\Delta t_m)_i \tag{14-69}$$

where $(\Delta t_m)_i$ expresses the mean temperature difference between the refrigerant and the wall and the thermal resistance of the thin wall has been neglected. A simple iterative procedure is then necessary to solve Eqs. (14-67) and (14-69) for the total heat transfer rate.

It was mentioned earlier in this chapter that the heat transfer coefficient decreases from the inlet to the exit of the coil as shown in Fig. 14-22. This has a direct effect on total heat transfer calculation because the coil surface temperature is higher than expected at the inlet due to a higher heat transfer rate there. This should be taken into account because a portion of the coil near the air inlet may be at a temperature greater than the dewpoint with no mass transfer occurring.

The sensible heat transfer from the moist air to the refrigerant is computed for counterflow by Eq. (14-3):

$$\dot{q}_s = UA(LMTD) \tag{14-3}$$

where $LMTD$ is given by Eq. (14-2) and U is given by Eq. (14-15a) with η_s equal to η_{ms}.

The latent heat transfer is then easily computed from

$$\dot{q}_l = \dot{q} - \dot{q}_s \tag{14-70}$$

It is also true that

$$\dot{q}_l = \dot{m}_a (W_i - W_o) i_{fg} \tag{14-71}$$

Fin Efficiency with Mass Transfer

The fin efficiency with combined heat and mass transfer is lower than the value obtained with only sensible heat transfer. Although the basic definition is unchanged from that given in section 14-3, the analysis is more complex and not exact. Threlkeld (20) has solved the problem; his solution, however, depends on knowing the water film thickness on the fin. This is a very nebulous quantity and impractical to use. An accepted method is outlined in reference 19 and is an adaptation of the work of Ware and Hacha (21) and others. Although this method appears to give good results for performance and rating purposes, it has some undesirable features since the coil surface temperature is assumed to be the only parameter affecting the fin efficiency regardless of the moist air conditions. Another disturbing feature is failure of the solution to reduce to the dry coil case when the surface and moist air conditions warrant this. These inconsistencies are troublesome when making general coil studies.

A fin of uniform cross section as shown in Fig. 14-8 has been analyzed by McQuiston (22). The method is approximate but reduces to the case of zero mass transfer and is adaptable to circular and plate-fin-tube surfaces. The analysis is outlined as follows: An energy balance on an elemental volume yields the following differential equation, assuming one dimensional heat transfer and constant properties:

$$\frac{d^2t}{dx^2} = \frac{P}{kA_c}[h(t - t_\infty) + h_d i_{fg}(W - W_\infty)] \tag{14-72}$$

where

t = temperature of the element, F or C
x = distance measured from base of fin, ft or m
P = circumference of the fin, ft or m
k = thermal conductivity of the fin material, (Btu-ft)/(hr-ft²-F) or (W-m)/(m²-C)
A_c = cross-sectional area of the fin, ft² or m²
h = convective heat transfer coefficient, Btu/(hr-ft²-F) or W/(m²-C)
t_∞ = temperature of the air-vapor mixture flowing around the fin, F or C
h_d = convective mass transfer coefficient, lbm/(ft²-hr) or kg/(m²-s)
i_{fg} = latent heat of vaporization of water, Btu/lbm or J/kg
W = specific humidity of saturated air at temperature t, lbmw/lbma or kgw/kga
W_∞ = specific humidity of the air-vapor mixture, lbmw/lbma or kgw/kga

The analogy of Eq. (4-20a) will be used to obtain the mass transfer coefficient h_d with Le = 1

$$h_d = \frac{h}{c_p} \tag{14-73}$$

where

c_p = specific heat capacity of the dry air, Btu/(lbm-F) or W/(kg-C)

As suggested, the coefficient h should be for a wet surface. Other correlations may also be used. Combining Eqs. (14-72) and (14-73) gives

$$\frac{d^2t}{dx^2} = \frac{hP}{kA_c}\left[(t - t_\infty) + \frac{i_{fg}}{c_p}(W - W_\infty)\right] \qquad (14\text{-}72a)$$

Now, if $(W - W_\infty)$ is simply related to $(t - t_\infty)$, Eq. (14-72a) can be easily solved for the temperature distribution in the fin. To justify such a simplification consider the physical aspects of a typical cooling and dehumidifying coil.

Let the air-vapor mixture enter an exchanger at a fixed condition designated by point 1 on the psychrometric chart of Fig. 14-32. Consider an evaporator with a constant temperature refrigerant operating such that the moist air very near the wall is at a temperature designated by point w2. The humidity ratio of the leaving air W_2 will approach W_{w2} as shown in Fig. 14-33. The process line 1-2 in Fig. 14-32 can be approximated by a straight line and a simple relationship between $(W - W_\infty)$ and $(t - t_\infty)$ exists. In fact,

$$(W_{w2} - W_1) = C(t_{w2} - t_1) \qquad (14\text{-}74)$$

where C is a constant. An examination of data for many coils and various operating conditions shows that C will typically vary less than 10 percent from inlet to exit. It then seems reasonable to use an average value such as

$$C_{avg} = \frac{(C_1 + C_2)}{2} \qquad (14\text{-}75)$$

Due to the shape of the saturation curve, precise location of the point w2 on Fig. 14-32 does not greatly affect the value of C for a particular coil condition. On the other

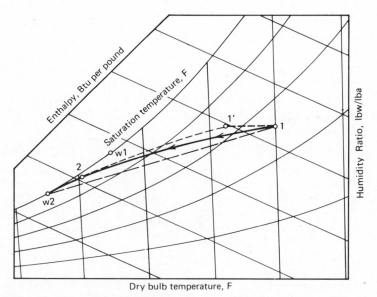

Figure 14-32. Cooling and dehumidifying processes. (Reprinted by permission from *ASHRAE Trans., 81*, Part I, 1975.)

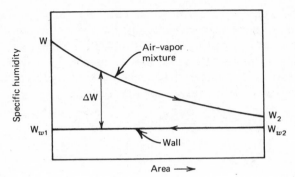

Figure 14-33.   Specific humidity for a constant temperature refrigerant.

hand, C is very sensitive to the location of point 1. For example, for $t_w = 45$ F and $t_1 = 80$ F, C varies from 1.4×10^{-4} to 0.0 as ϕ_1 varies from 50 to 29 percent. For the last condition there will be no condensation on the surface and only sensible heat transfer will occur.

When chilled water is used as a cooling medium in a counterflow arrangement, Fig. 14-34 applies. In this case the wall temperature is somewhat higher where the air enters the exchanger. Typical conditions for the moist air very near the wall are shown on Fig. 14-32 as points $w1$ and $w2$. Here the surface is completely wetted.

At the inlet to the exchanger the value of C is given by

$$C_1 = \frac{W_{w1} - W_1}{t_{w1} - t_1} \tag{14-76}$$

whereas at the exit

$$C_2 = \frac{W_{w2} - W_2}{t_{w2} - t_2} \tag{14-76a}$$

Again C_1 is less than C_2 and an average value should be used. In most cases C will change less than 10 percent from inlet to outlet.

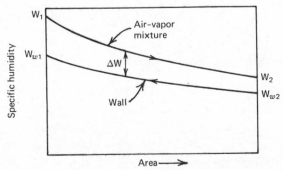

Figure 14-34.   Specific humidity for chilled water as the refrigerant.

Process 1-1'-2 approximates a situation where the coil is partially dry. In this case C_1 is zero until the air reaches the location in the coil where the surface temperature is below the air dew point temperature. C then increases to the value of C_2 at the exit. Again an average value of C may be used or C_2 may be used to obtain a conservative solution. The differential equation describing the temperature distribution in a thin fin of uniform cross section thus becomes

$$\frac{d^2(t - t_\infty)}{dx^2} = M^2(t - t_\infty) \tag{14-77}$$

where

$$M^2 = \frac{hP}{kA_c}\left[1 + \frac{Ci_{fg}}{c_p}\right] \tag{14-78}$$

When there is no condensation $C = 0$ and

$$M^2 = \frac{hP}{kA_c} = m^2 \tag{14-78a}$$

The fin efficiency derived from the well-known solution of Eq. (14-77) where t_∞ and M are constants and the following boundary conditions are used:

$$x = 0 \qquad t = t_w \tag{14-79}$$
$$x = l \qquad \frac{dt}{dx} = 0$$

becomes

$$\eta_m = \frac{\tanh(Ml)}{(Ml)} \tag{14-80}$$

The approximation of Eq. (14-20) is also used here

$$\frac{hP}{kA_c} \approx \frac{2h}{ky} \tag{14-81}$$

Equation (14-80) is identical in form to the equation for the fin efficiency with no mass transfer, Eq. (14-19).

The solution may be applied to circular fins on a tube, Eq. (14-21), or to the case of plate-fin-tube heat transfer surfaces, Eqs. (14-23) and (14-24). Figure 14-11 may also be used with m replaced by M.

The surface effectiveness has the same form as Eq. (14-27)

$$\eta_{ms} = 1 - \frac{A_f}{A}(1 - \eta_m) \tag{14-82}$$

The method presented is thought to be the most accurate available and is simple and straightforward to use. The method is readily adapted to the computer and is easy to use with hand calculations.

Transport Coefficients

The heat tranfer and friction coefficients on the refrigerant side of the exchanger are determined by the methods discussed in Section 14-4. Chilled water and evaporating refrigerants are the usual cases.

The heat, mass, and friction coefficients on the air side of the exchanger should be obtained from correlations based on test data, if at all possible, since the analogy method is unreliable. The correlations of Figs. 14-23 and 14-24 as modified using $J(s)$, $J_i(s)$, and $F(s)$ (discussed at the beginning of this section), are recommended for plate-fin-tube coils. Other finned tube surfaces have similar behavior. For example, the dry surface heat transfer coefficients, for circular finned tubes in a staggered tube pattern, are well correlated by

$$j = 0.38 \ JP \tag{14-83}$$

and the friction factors are given by

$$f = 1.53 \ FP^2 \tag{14-84}$$

JP and FP may then be modified for a wet surface using Eqs. (14-58), (14-59), and (14-60). No information is available on the effect of tube rows on mass transfer coefficients; however, it should be similar to that for sensible heat transfer for a dry surface, Fig. 14-21.

Example 14-2.   Estimate the heat, mass, and friction coefficients for a 4-row cooling coil that has the geometry of Fig. 14-20 with 12 fins per in. The face velocity of the air is 600 ft/min and has an entering temperature of 80 F. The air leaves the coil at a temperature of 60 F. The air is at standard barometric pressure.

Solution.   The correlations of Figs. 14-23 and 14-24 will be used with the parameters JP, FP, $J(s)$, and $F(s)$ computed from Eqs. (14-46) through (14-60). The mass velocity is

$$G_c = \frac{60(600)14.7(144)}{0.48(53.35)540} = 5{,}511 \ \frac{\text{lbm}}{\text{hr-ft}^2}$$

The Reynolds number based on tube diameter is then

$$\frac{G_c D}{\mu} = \frac{5511(0.525/12)}{0.044} = 5{,}480$$

where data from Fig. 14-20 and Table B-4 are used. To compute the parameter A/A_t, assume a 4-row coil, one foot in length, with 10 tubes in the face. This coil has a volume of

$$V = \frac{10(1.25)(12)4(1.083)}{1728} = 0.376 \ \text{ft}^3$$

and a total of 40 tubes. The total outside surface area of the tubes is

$$A_t = 40 \pi D L = 40 \pi (0.525/12)(1) = 5.498 \text{ ft}^2$$

From Fig. 14-20, $A/V = \alpha = 238 \text{ ft}^2/\text{ft}^3$, then

$$A/A_t = 238 \times 0.376/5.498 = 16.28$$

Using Eq. (14-46)

$$JP = (5480)^{-0.4}(16.28)^{-0.15} = 0.021$$

The Reynolds number based on the fin spacing is

$$\text{Re}_s = \text{Re}_D \left(\frac{s}{D} \right) = 5480 \left(\frac{0.0855}{0.525} \right) = 892$$

then

$$J(s) = 0.84 + 4 \times 10^{-5}(892)^{1.25} = 1.035$$

and

$$J_i(s) = [0.95 + 4 \times 10^{-5}(892)^{1.25}] \left[\frac{0.0855}{0.0855 - 0.006} \right]^2 = 1.324$$

then

$$JP \, J(s) = 0.0210(1.035) = 0.022$$

and

$$JP \, J_i(s) \, 0.0210(1.324) = 0.028$$

Using Fig. 14-23, $j = 0.0071$ and $j_i = 0.0088$
From Eq. (14-36)

$$h = j \, G_c \, c_p/\text{Pr}^{2/3} = 0.0071(5511)0.24/(0.7)^{2/3} = 11.9 \text{ Btu/(hr-ft}^2 \text{ F)}$$

and from Eq. (12-19a)

$$h_d = j_i \, G_c/\text{Sc}^{2/3} = 0.0088(5511)/(0.6)^{2/3} = 68.2 \text{ lbma/(ft}^2\text{-hr)}$$

The friction factor will be determined from Fig. 14-24.
Using Eqs. (14-51) and (14-52)

$$D^* = \frac{0.525(16.28)}{1 + (1.25 - 0.525)/0.0855} = 0.902$$

and

$$FP = (5480)^{-0.25} \left(\frac{0.525}{0.902} \right)^{0.25} \left[\frac{1.25 - 0.525}{4(0.0855 - 0.006)} \right]^{-0.4}$$
$$\left[\frac{1.25}{0.902} - 1 \right]^{-0.5} = 0.118$$

Using Eq. (14-60)

$$F(s) = (1 + 892^{-0.4}) \left(\frac{0.0855}{0.0855 - 0.006} \right)^{1.5} = 1.19$$

Then

$$FP\ J(s) = 0.118(1.19) = 0.14$$

and from Fig. 14-24, $f = 0.031$.

Note that the presence of moisture on the surface increases h, h_d, and f.

Design Calculations

The design of a heat exchanger for simultaneous transfer of heat and mass is more complicated than for sensible heat transfer because the total energy transfer from the air has two components.

$$\dot{q} = \dot{m}_c c_{pc}(t_{co} - t_{ci}) = \dot{m}_a c_p(t_{ai} - t_{ao}) + \dot{m}_a i_{fg}(W_{ai} - W_{ao}) \quad (14\text{-}85)$$

The simplest case occurs when the terminal conditions are all known because

$$\dot{q} = \dot{m}_a(i_{ai} - i_{ao}) = \dot{m}_c c_{pc}(t_{co} - t_{ci}) \quad (14\text{-}86)$$

and

$$\dot{q}_s = \dot{m}_a c_p(t_{ai} - t_{ao}) = UA(LMTD) \quad (14\text{-}87)$$

When only inlet conditions are known, a double iteration loop is required and hand calculations become impractical. The digital computer may be used to great advantage in this case. Figure 14-35 shows a flow diagram for a successful computer program.

The method discussed in Chapter 13 for the spray dehumidifier may be used to predict performance of a dehumidifying coil when hand calculations are necessary. One modification is necessary in Eq. (13-6) where the liquid film coefficient h_l must be replaced by an overall coefficient defined by

$$\frac{1}{UA} = \frac{1}{h_c A_c} + \frac{\Delta x}{kA_m} \quad (14\text{-}88)$$

The first term on the right is the refrigerant side film resistance and the second term is the wall thermal resistance. Otherwise the procedure is identical; however, the results will be approximate.

The American Refrigeration Institute has a certification program for heating and cooling coils (19) that embodies a calculation procedure for dehumidifying coils. The standard is intended to standardize the rating of coils by setting up detailed procedures to calculate performance. Hand calculations are feasible through the use of charts and graphs. The method has many disadvantages when used for coil design, however.

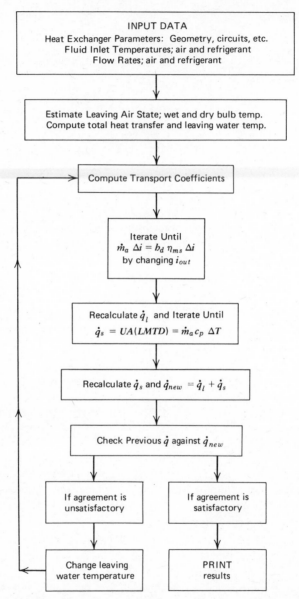

Figure 14-35.   Flow diagram for heat exchanger computer program.

REFERENCES

1. J. D. Parker, J. H. Boggs, and E. F. Blick, *Introduction to Fluid Mechanics and Heat Transfer,* Addison-Wesley, Reading, Massachusetts, 1969.

2. T. E. Schmidt, "La Production Calorifique des Surfaces Munies D'ailettes," *Annexe Du Bulletin De L'Institut International Du Froid,* Annexe G-5, 1945–46.

3. H. Zabronsky, "Temperature Distribution and Efficiency of a Heat Exchanger Using Square Fins on Round Tubes," *ASME Transactions, Journal of Applied Mechanics, 77,* December 1955.

4. W. H. Carrier and S. W. Anderson, "The Resistance to Heat Flow Through Finned Tubing," *Heating, Piping, and Air Conditioning,* May 1944.

5. *ASHRAE Handbook of Fundamentals,* American Society of Heating, Refrigerating and Air-Conditioning Engineers, New York, 1977.

6. D. G. Rich, "The Efficiency and Thermal Resistance of Annular and Rectangular Fins," Presented at the 1966 International Heat Transfer Conference.

7. W. M. Kays and A. L. London, *Compact Heat Exchangers,* 2nd Ed., McGraw-Hill, San Francisco, California, 1964.

8. W. M. Kays, *Convective Heat and Mass Transfer,* McGraw-Hill, San Francisco, California, 1966.

9. E. M. Sparrow, A. L. Loeffler, Jr., and H. A. Hubbard, *Trans. ASME, J. Heat Transfer,* November 1961.

10. R. G. Diessler and M. F. Taylor, *NACA TN 2351,* Washington, D.C., 1955.

11. A. Devore, *Petroleum Refiner, 40,* No. 5, May 1961 and *41,* No. 12, December 1962.

12. K. J. Bell, "Bulletin No. 5," University of Delaware Engineering Experimental Station, 1963.

13. J. W. Palen and Jerry Taborek, "Solution of Shell Side Flow Pressure Drop and Heat Transfer by Stream Analysis Method," Chemical Engineering Progress Symposium Series, *65,* No. 92, 1969.

14. Donald G. Rich, "The Effect of the Number of Tube Rows on Heat Transfer Performance of Smooth Plate Fin-and-Tube Heat Exchangers," *ASHRAE Trans., 81,* Part I, 1975.

15. D. G. Shepherd, "Performance of One-Row Tube Coils with Thin-Plate Fins, Low Velocity Forced Connection," *Heating, Piping, and Air Conditioning,* April 1956.

16. F. C. McQuiston, "Heat Transfer and Flow Friction Data for Two Fin-Tube Surfaces," *ASME Journal of Heat Transfer, 93,* May 1971.

17. J. L. Guillory and F. C. McQuiston, "An Experimental Investigation of Air Dehumidification in a Parallel Plate Exchanger," *ASHRAE Transactions, 29,* June 1973.

18. F. C. McQuiston, "Heat, Mass, and Momentum Transfer in a Parallel Plate Dehumidifying Exchanger," *ASHRAE Trans., 82,* Part II, 1976.

19. *ARI Standard 410-72,* Air Conditioning and Refrigeration Institute, Arlington, Virginia, 1972.

20. J. L. Threlkeld, *Thermal Environmental Engineering,* 2nd Edition, Prentice-Hall, Englewood Cliffs, N.J., 1970.

21. C. D. Ware and T. H. Hacha, "Heat Transfer From Humid Air to Fin and Tube Extended Surface Cooling Coils," ASME Paper, No. 60-HT-17, August 1960.

22. F. C. McQuiston, "Fin Efficiency with Combined Heat and Mass Transfer," *ASHRAE Trans., 81,* Part I, 1975.

23. D. G. Rich, "The Effect of Fin Spacing on the Heat Transfer and Friction Performance of Multirow Smooth Plate Fin-and-Tube Heat Exchangers," *ASHRAE Transactions, 79,* June 1973.

24. R. W. Eckels, "Contact Conductance of Mechanically Expanded Plate Finned Tube Heat Exchangers," Westinghouse Research Laboratories, Scientific Paper No. 77-1E9-SURCO-P1, March 1977.

25. F. C. McQuiston, "Heat, Mass, and Momemtum Transfer Data for Five Plate-Fin-Tube Heat Transfer Surfaces," *ASHRAE Trans., 84,* Part 1, 1978.

26. F. C. McQuiston, "Correlation of Heat, Mass, and Momemtum Transport Coefficients for Plate-Fin-Tube Heat Transfer Surfaces," *ASHRAE Trans., 84,* Part 1, 1978.

27. R. W. Lockhart, and Martinelli, R. D., "Proposed Correlation of Data for Isothermal, Two-Phase, Two-Component Flow in Pipes," *Chem. Eng. Prog., 45,* 39, 1949.

28. K. Ishihara, J. W. Palen, and J. Taborek, "Critical Review of Correlations for Predicting Two-Phase Flow Pressure Drop Across Tube Banks," *Heat Transfer Engineering,* Vol. 1, No. 3, Jan–Mar. 1980.

PROBLEMS

14-1 Five thousand cfm of standard air (60 F, 29.92 in. mercury) are heated to 120 F in a crossflow heat exchanger. The heating medium is water at 200 F, which is cooled to 180 F. Determine (a) the *LMTD* and correction factor F, (b) the fluid capacity rates C_c and C_h, (c) the flow rate of the water in gpm, (d) the overall conductance UA, (e) the *NTU* based on C_{min}, and (f) the effectiveness ε.

14-2 One m³/s of standard air is heated to 45 C in a cross flow exchanger. Water is the hot fluid and is cooled from 90 to 85 C. Find (a) the $LMTD$ and correction factor F, (b) the fluid capacity rates C_c and C_h, (c) the flow rate of the water in m³/s, (d) the overall conductance UA, (e) the NTU based on C_{min}, and (f) the effectiveness ε.

14-3 A tube has circular fins as shown in Fig. 14-10, R is 1.0 in., whereas r is 0.5 in. The fin thickness y is 0.008 in. and the fin pitch is 10 fins/in. The average heat transfer coefficient for the fin and tube is 10 Btu/(hr-ft²-F), whereas the thermal conductivity k of the fin and tube material is 90 (Btu-ft)/(ft²-hr-F). Determine (a) the fin efficiency η using Fig. 14-11, (b) the fin efficiency using Eq. (14-21), and (c) compare (a) and (b).

14-4 Find the surface effectiveness for the finned tube in Problem 14-3 assuming the ratio of fin area to total area is 0.9.

14-5 Water flows through the finned tube in Problem 14-4. Find the overall heat transfer coefficient assuming a water heat transfer coefficient of 1100 Btu/(hr-ft²-F).

14-6 A plate fin heat exchanger has fins that are 6 mm high, 0.16 m thick, with a fin pitch of 0.47 fins/mm. The average heat transfer coefficient is 57 W/(m²-C) and the thermal conductivity of the fin material is 173 W/(m-C). Find the fin efficiency.

14-7 Assume the ratio of fin area to total area in Problem 14-6 is 0.85. Find the surface effectiveness.

14-8 Refrigerant 12 flows on the opposite side of the surface in Problem 14-6. Determine the overall heat transfer coefficient if the refrigerant heat transfer coefficient is 1.4 kW/(m²-C). Neglect the thermal resistance of the plate.

14-9 Find the fin efficiency for a plate-fin-tube surface like that shown in Fig. 14-14 using the data shown in Table 14-2.

14-10 Determine the average heat transfer coefficient for water flowing in a $\frac{1}{2}$ in. O.D. copper tube that has a wall 0.016 in. thick. The water flows at a rate of $2\frac{1}{2}$ gpm and undergoes a temperature change from 180 to 150 F. The tube is over 4 ft long.

14-11 Repeat Problem 14-10 for a 30 percent ethylene glycol solution.

14-12 Estimate the heat transfer coefficient for water flowing with an average velocity of 4 ft/sec in a long rectangular channel 3 × 1 in. The bulk temperature of the water is 45 F. Assume (a) cooling and (b) heating.

14-13 Repeat Problem 14-12 for 30 percent ethylene glycol instead of water.

14-14 Coolant flows through a long 12 mm I.D. tube with an average velocity of 1.5 m/s. The coolant undergoes a temperature change from 40 to 50 C. Compute the average heat transfer coefficient for (a) water and (b) 30 percent ethylene glycol solution.

Table 14-2   DATA FOR PROBLEM 14-9

	a	b	y	D	h	k
(a)	1.12 in.	1.35 in.	0.010 in.	0.64 in.	10 Btu/(hr-ft²-F)	90 (Btu-ft)/(hr-ft²-F)
(b)	25 mm	22 mm	0.18 mm	10 mm	68 W/(m²-C)	170 W/(m-C)

14-15 Water flows through a 0.34 in. I.D. tube with an average velocity of $\frac{1}{2}$ ft/sec. The mean bulk temperature of the water is 45 F and the length of the tube is 10 ft. Estimate the average heat transfer coefficient for (a) water and (b) ethylene glycol (30 percent solution).

14-16 Estimate the average heat transfer coefficient for water flowing in 10 mm I.D. tubes at a velocity of 0.15 m/s. The length of the tubes is 3 m and the mean bulk temperature of the water is 40 C.

14-17 Estimate the average heat transfer coefficient for condensing water vapor in $\frac{5}{8}$ in. O.D. tubes that have walls 0.018 in. thick. The vapor is saturated at 5 psia and flows at the rate of 1 lbm/hr at the tube inlet. The quality of the vapor is 10 percent when it exits from the tube.

14-18 Water vapor condenses in 15 mm I.D. tubes. The flow rate is 0.126×10^{-3} kg/s per tube at an absolute pressure of 35 kPa. The vapor leaving the tubes has a quality of 12 percent. Estimate the average heat transfer coefficient.

14-19 Refrigerant 22 enters the $\frac{3}{8}$ in. O.D. tubes of an evaporator at the rate of 80 lbm/hr per tube. The refrigerant enters at 70 psia with a quality of 20 percent and leaves with 10 F of superheat. The effective tube length is 5 ft. Estimate the average heat transfer coefficient.

14-20 Estimate the average heat transfer coefficient for evaporating R-22 in 8.5 mm I.D. tubes that are 2 m in length. The mass velocity of the refrigerant is 200 kg/(s-m²) and the pressure and quality at the inlet are 210 kPa and 30 percent, respectively. Saturated vapor exits from the tubes.

14-21 For the surface shown in Fig. 14-20 with 6.7 fins per in., the air mass velocity G_{fr} is 1800 lbm/(hr-ft²) and the air is heated from 70 to 120 F. Compute the heat transfer coefficient and friction factor using (a) Fig. 14-20, (b) Figs. 14-23 and 14-24.

14-22 For the surface of Fig. 14-20, with 0.263 fins per mm, the air mass velocity is 4.5 kg/(m²-s) and the mean bulk temperature is 20 C. Estimate the heat transfer coefficient and friction factor using (a) Fig. 14-20, (b) Figs. 14-23 and 14-24.

14-23 Compute the lost pressure for Problem 14-21 assuming standard atmospheric pressure at the inlet.

14-24 Compute the lost pressure for Problem 14-22 assuming standard atmospheric pressure at the inlet.

14-25 Find the heat transfer coefficient and friction factor for the surface of Fig. 14-25 assuming a mass velocity of 2700 lbm/(hr-ft²). The air enters at 75 F and leaves at 55 F.

14-26 Compute the lost pressure for Problem 14-25 assuming a pressure of 14.6 lbf/in.² at the inlet and flow length of 4 in.

14-27 Design a condenser for the following given conditions:
Tube side: R-22, 2 phase, $t = 120$ F, $h_i = 200$ Btu/(hr-ft²-F);
Finned side: $t_{in} = 100$ F, Heat transfer rate = 40,000 Btu/hr,
Maximum head loss = 0.3 in. water; *Construction:* All aluminum, $k = 100$ Btu/(hr-ft²-F). Assume a five-row coil and use the surface of Fig. 14-20 with 8 fins per in.

14-28 Design a hot water heating coil for a capacity of 200,000 Btu/hr. Use the surface of Fig. 14-20 with two rows of tubes and 6.7 fins per in. Air enters the coil at 70 F and water enters at 150 F. Sketch the circuiting arrangement for the tube side. Compute the head loss for both the water and air. Make the width of the exchanger about twice the height.

14-29 Air enters a fin tube heat exchanger described by Fig. 14-20 (6.7 fins per in.) at the rate of 2000 cfm at 14.696 psia and with a face velocity of 1000 ft/min. The entering and leaving air temperatures are 40 and 110 F, respectively. Hot water flows through the tubes with an average velocity of 3.5 ft/sec, entering at 200 F and leaving at 190 F. Assume counterflow and use the *NTU* method to determine (a) the heat transfer area required and (b) the heat exchanger dimensions and the number of tube rows in the air flow direction. (c) Compute the lost head on the air side in inches of water, and (d) select circuiting for the tubes and compute the head loss for the water side.

14-30 Compute the surface effectiveness for the surface of Fig. 14-25 under dehumidifying conditions. Moist air enters at 78 F dry bulb and 67 F wet bulb. Assume a base plate temperature of 47 F. The face velocity of the entering air is 600 ft/min at 14.7 psia.

14-31 The surface of Fig. 14-20 is to be sized for a condenser in an air-conditioning unit. The condenser must reject 4.4 kW. Assume Refrigerant 22 enters the condenser at 51 C as saturated vapor; it leaves as saturated liquid at the same temperature. Air enters the condenser at 35 C. Use a 5-row coil with aluminum fins with 0.571 fins per mm. Compute the air-side head loss in Pa.

14-32 A heat exchanger surface like Fig. 14-20 has a face area of 1 ft², 5 rows of tubes and aluminum fins (11.7 fins per in.). Moist air enters the exchanger at 80 F dry bulb and 68 F wet bulb, at standard atmospheric pressure and with a face velocity of 650 ft/min. Compute the (a) leaving air conditions, (b) total heat transfer rate, (c) sensible heat transfer rate, and (d) head loss for the air. The refrigerant is water that undergoes a temperature change from 45 to 50 F in a counterflow arrangement.

14-33 A heat exchanger made like the surface of Fig. 14-20 has a face area of 1 m² and 4 rows of tubes and aluminum fins (0.461 fins per mm). Air enters the coil at 26 C dry bulb and 20 C wet bulb with a face velocity of 3.5 m/s and a pressure of 101 kPa absolute pressure. Determine the (a) leaving air conditions, (b) total heat transfer rate, (c) sensible heat transfer rate, (d) head loss for the air. The refrigerant is water that undergoes a change in temperature from 7 to 10 C in counterflow. Determine the average heat transfer coefficient in the tubes.

14-34 The surface of Problem 14-21 has 8 rows of tubes. Compute the heat transfer coefficient.

14-35 The surface of Problem 14-22 has 6 rows of tubes. Compute the heat transfer coefficient.

14-36 Air enters a finned coil at 80 F db, 67 F wb and standard atmospheric pressure. The chilled water has a temperature of 50 F at this position in the coil. Will moisture condense from the air for the following assumed conditions? Copper tubes with wall thickness of 0.018 in. and thermal conductivity of 190 Btu/(ft-hr-F). The heat transfer coefficient on the inside of the tubes is 1000 Btu/(hr-ft²-F). The product of total heat transfer coefficient and surface effectiveness ($h_d\eta_o$) is 60 lbma/(ft^2-hr) and the ratio of total surface area to inside tube surface area is 12.

14-37 Consider a cooling coil with air entering at 27 C db, 19 C wb, and standard atmospheric pressure. The coil is all aluminum construction with tube wall thickness of 0.50 mm and thermal conductivity of 58 W/(m-C). The heat transfer coefficient on the inside of the tubes is 53 W/(m^2-C). The product of total heat transfer coefficient and surface effectiveness ($h_d\eta_o$) is 2.5 kg/(m^2-hr). The ratio of total surface area to inside tube surface area is 14. Will condensation occur on the surface at this location if the chilled water has a temperature of 14.3 C at this location in the coil?

14-38 Suppose the coil of Example 14-2 had 6 rows of tubes instead of 4 and estimate the heat, mass, and friction coefficients.

CHAPTER FIFTEEN

Refrigeration

The refrigerator or heat pump as it is sometimes called is basic to the heating, ventilating, and air-conditioning industry. Applications vary from food processing, packaging, storing, and transportation to environmental control and comfort. There are basically three refrigeration methods:

1. Vapor compression
2. Absorption
3. Thermoelectric

The first two are by far the most common. Although these systems have many complex variations, only the basic cycles will be discussed in this chapter. References 2, 3, 4, and 5 should be consulted for a more in-depth treatment.

The performance of a refrigerating system is expressed in terms of the *coefficient of performance* or *COP*.

$$COP = \frac{\text{useful refrigerating effect}}{\text{net energy input}} \tag{15-1}$$

The *refrigerating efficiency* η_R is the ratio of the coefficient of performance of a cycle or system to that of an ideal cycle or system:

$$\eta_R = \frac{COP}{(COP)_{ideal}} \tag{15-2}$$

The capacity of a refrigeration cycle can be measured in any one of several units such as Btu/hr or kW. A common unit of capacity in the United States is the *ton of refrigeration*, equal to 200 Btu/min or 12,000 Btu/hr of useful heat removal. The term is a carryover from the days when ice was used to cool. The latent enthalpy of fusion of ice is 144 Btu/lbm. One ton of ice melting per day requires the transfer of about 200 Btu/min of heat from the cold space.

The power input required in a gas or vapor compression system for a given refrigerating effect depends on the coefficient of performance. In terms of horsepower per ton the expression is

$$\text{hp/ton} = \frac{(12{,}000) \text{ Btu/ton-hour}}{(2545) \text{ Btu/hp-hour }(COP)} = \frac{4.72}{COP} \qquad (15\text{-}3)$$

15-1   THE CARNOT REFRIGERATION CYCLE

The reversible cycle is a useful thermodynamic concept. Two important characteristics of a reversible cycle are:

1.   No refrigeration cycle can have a higher coefficient of performance than that of a reversible cycle operating between the same source and sink temperatures.
2.   All reversible refrigeration cycles operated between the same source and sink temperatures have identical coefficients of performance.

The most convenient reversible cycle to use as an ideal refrigerating cycle is the *Carnot refrigeration cycle* consisting of two isothermal processes and two adiabatic processes. Such a cycle is shown in Fig. 15-1 on temperature-entropy coordinates. Because of characteristic 2 above, no particular working medium needs to be specified.

During the isothermal process 2–3, heat is received by the working fluid (refrigerant) at the evaporator temperature T_e. At the end of the isothermal process the refrigerant is then compressed adiabatically. Because a reversible adiabatic process is also isentropic, the process 3–4 appears as a vertical line on the $T - s$ coordinates. The temperature of the refrigerant increases as the adiabatic process proceeds. By this work process the refrigerant temperature is increased from T_3 to T_4, enabling the refrigerant to give up heat to the environment or to some coolant. This cooling of the refrigerant occurs at the constant condenser temperature T_c shown as process 4–1. To complete the cycle the refrigerant is brought back to its initial condition (state 2) by an adiabatic expansion from state 1 to state 2.

The heat absorbed by the refrigerant during the isothermal process 2–3 is

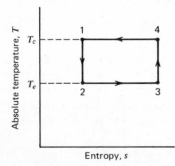

Figure 15-1.   The Carnot refrigeration cycle.

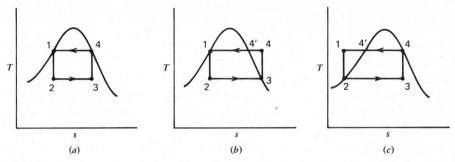

Figure 15-2. Possible Carnot cycles using a phase change to accomplish part or all of the isothermal processes.

$$q_{23} = T_e(s_3 - s_2)$$

The usual thermodynamic convention is followed in which heat added to the system (the refrigerant) is positive.

The heat given up by the refrigerant during the isothermal process 4-1 is

$$q_{41} = T_c(s_1 - s_4) = T_c(s_2 - s_3) = -T_c(s_3 - s_2)$$

The net work done on the cycle is given by the first law of thermodynamics, using the convention that work done by the system is positive.

$$w_{net} = q_{net} = q_{23} + q_{41} = +T_e(s_3 - s_2) - T_c(s_3 - s_2)$$

From Eq. (15-1) the *COP* is

$$COP = \frac{T_e(s_3 - s_2)}{-[T_e(s_3 - s_2) - T_c(s_3 - s_2)]} = \frac{T_e}{T_c - T_e} \qquad (15\text{-}4)$$

If we attempt to actually build a working Carnot cycle, we would immediately be confronted with some serious problems. The isothermal processes, for example, might be accomplished by having a change of phase occur in the refrigerant. Three possibilities are shown in Fig. 15-2.

In case (a) the evaporation of the refrigerant occurs between points 2 and 3 and condensation occurs between points 4 and 1. In attempting to stop the evaporation at point 3 there would be difficulty in sensing the proper condition at exit since no easily measurable property (such as temperature or pressure) is changing along process 2–3. In addition, the adiabatic compression 3–4 would be complicated by the presence of moisture at the inlet. The two-phase expansion that occurs in process 1–2 is also very difficult to accomplish in a real expander.

In case (b) the point 3 is easily sensed because the temperature begins to rise in a constant pressure process as soon as the saturated vapor condition is reached. The adiabatic compression 3–4 is also easily accomplished (at least approximately), and no liquid will enter the compressor. The isothermal compression 4–4 however, requires an extra compressor and is difficult to approach in real equipment with significant capacity. The process 1–2 is also impractical, as was discussed in the previous case.

In case (c) the point 1 is at a very high pressure relative to the pressure at 4′ and would be impractical for that reason.

No practical heat engine or refrigeration device operating on the Carnot cycle has been developed, and it appears best simply to use the concept as a standard of perfection and to design the real cycle to approach this ideal as close as practical.

15-2   THE THEORETICAL SINGLE STAGE CYCLE

To meet practical considerations in the adiabatic expansion and compression processes, we may replace the expander by a simple throttling valve and carry out the adiabatic compression all the way to the condenser pressure. Such a modified cycle is described in Fig. 15-3.

Refrigerant entering the compressor is assumed to be dry saturated vapor at the evaporator pressure. This is a convenient place to begin an analysis because we can easily determine all fluid properties here. The compression process 3–4 is assumed to be reversible and adiabatic and, therefore, isentropic, and is continued until the condenser pressure is reached. Point 4 is obviously in the superheated vapor region. The process 4–1 is carried out at constant pressure with the temperature of the vapor decreasing until the saturated vapor condition is reached at 4′; then the process is both at constant temperature and constant pressure during the condensation from 4′

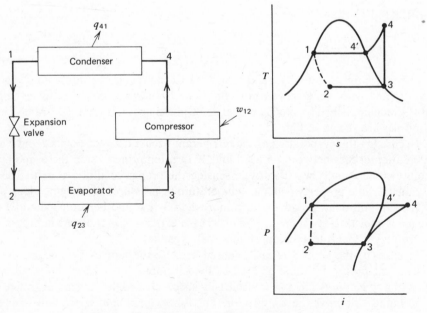

Figure 15-3.   The theoretical single stage cycle.

to 1. At point 1 the refrigerant leaves the condenser as a saturated liquid. It is then expanded through a throttling valve where partial evaporation occurs as the pressure drops across the valve. The throttling process 1–2 is irreversible with an increase in entropy occurring. For this reason the process is shown as a dashed line in Fig. 15-3. For a throttling process the enthalpy at exit and at inlet are equal. To determine the coefficient of performance of this cycle the useful refrigerating effect and the net energy input must be determined. For steady flow of one pound of refrigerant:

$$i_3 - i_2 = i_3 - i_1$$

Useful refrigerating effect $= q_{23} = i_3 - i_2 = i_3 - i_1$

Net energy input $= w_{34} = i_4 - i_3$

$$COP = \frac{i_3 - i_1}{i_4 - i_3}$$

Example 15-1   For a theoretical single stage cycle operating with a condenser pressure of 250 psia and an evaporator pressure of 40 psia and using refrigerant 22 as the working fluid determine the (a) *COP*, (b) temperature at exit to compressor, (c) temperature in condenser, and (d) temperature in evaporator.

Solution.   Use the pressure-enthalpy diagram for refrigerant 22, Chart 5.

(a) $COP = \dfrac{i_3 - i_1}{i_4 - i_3} = \dfrac{104.6 - 43.1}{125.4 - 104.6} = \dfrac{61.5}{20.8} = 2.96$

(b) $t_4 = 168$ F

(c) $t_1 = 116$ F

(d) $t_2 = 2$ F

The coefficient of performance is closely related to the evaporating and condensing temperatures. For the highest *COP* the cycle should be operated at the lowest possible condensing temperature and the highest possible evaporating temperature.

In the theoretical single stage cycle no work is produced in the expansion process 1–2, in contrast to the adiabatic expansion in the Carnot cycle. This is lost work that cannot be put back into the system to drive the compressor; therefore, additional work is needed for the compression process in the actual cycle. In addition, extra work is required in the adiabatic compression because the enthalpy at exit from the compressor at point 4 is higher than in the Carnot cycle. On a $T - s$ diagram the area under a reversible constant pressure process represents the change of enthalpy for that process. In Fig. 15-4, for example, the area under the constant pressure process 4–4'–1–1' is equal to the change of enthalpy $(i_4 - i_{1'})$. The process 1–1' can be assumed to be along the saturated liquid line if subcooling of point 1' is not extremely large. The change of enthalpy $(i_3 - i_{1'})$ is equal to the area under the line 1'–3. Thus, because

$$w_{34} = i_4 - i_3 = (i_4 - i_{1'}) - (i_3 - i_{1'})$$

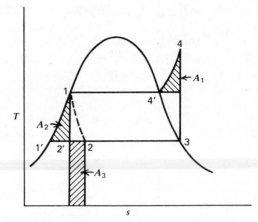

Figure 15-4. Comparison of Carnot and theoretical cycles.

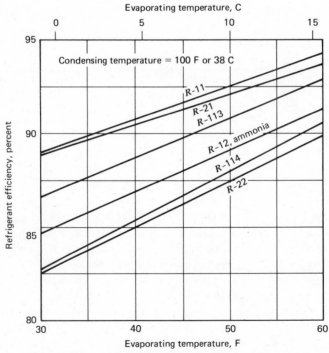

Figure 15-5. Refrigerating efficiency of theoretical single stage cycle for various refrigerants. (James L. Threlkeld, *Thermal Environmental Engineering*, 2nd edition,
© 1970, p. 53, reprinted by permission of Prentice-Hall, Inc., Englewood
Cliffs, N.J.)

then the enclosed area 3–4–4′–1–1′–2′–2–3 is equal to the compressor work required per unit mass of refrigerant flow in the theoretical single stage cycle. The additional net work above that required for the equivalent Carnot cycle w_c is equal to the sum of the two areas A_1 and A_2.

$$w_{34} = w_c + A_1 + A_2$$

The throttling process 1–2 reduces the amount of refrigerating effect possible since point 2 is shifted to the right by the irreversible process. This loss of refrigerating capability is given by area A_3. The refrigerating effect of the theoretical cycle q_{41} is given by

$$q_{2'3} - A_3 = q_c - A_3$$

where q_c is the heat flow into the evaporator of the Carnot cycle.

The refrigerating efficiency of the theoretical single stage cycle can be expressed in terms of the three areas described above and the evaporator heat flow q_c and net work w_c of an equivalent Carnot cycle.

$$\eta_R = \frac{COP}{COP_c} = \frac{(q_c - A_3)/(w_c + A_1 + A_2)}{q_c/w_c} = \frac{1 - A_3/q_c}{1 + (A_1 + A_2)/w_c}$$

The magnitude of the areas A_1 and A_2, and A_3 depends on the evaporating and condensing temperatures and on the particular refrigerant used. The variation of η_R among the more common refrigerants is not large. Figure 15-5 shows the variation of refrigerating efficiency with evaporating temperature for several common refrigerants at a fixed condensing temperature.

Example 15-2. Compute the *COP* and refrigerating efficiency for a theoretical single stage cycle operating with a condenser pressure of 1219 kPa and an evaporator pressure of 204 kPa using R-12 as the refrigerant. The work for the cycle is 31 kJ/kg of refrigerant.

Solution. From Fig. 15-3, the *COP* may be expressed as

$$COP = \frac{i_3 - i_1}{i_4 - i_3} = \frac{i_3 - i_1}{w_{34}}$$

And from Table A-3a

$$i_3 = 248.9 \text{ kJ/kg}; \qquad i_1 = 346.3 \text{ kJ/kg}$$

Then

$$COP = \frac{346.3 - 248.9}{31} = 3.14$$

A Carnot cycle operating between the saturation temperatures corresponding to the given condenser and evaporator pressures has a *COP* given by Eq. (15-4) where from Table A-3a

$$T_e = -12 + 273 = 261 \text{ K}$$
$$T_c = 50 + 273 = 323 \text{ K}$$
$$COP_c = \frac{261}{323 - 261} = 4.2$$

The refrigerating efficiency is then

$$\eta_R = \frac{COP}{COP_c} = \frac{3.14}{4.2} = 0.75 \text{ or } 75 \text{ percent}$$

15-3 REFRIGERANTS

The working medium used in refrigeration systems is called the *refrigerant*. A refrigerant's suitability for a given application depends on many factors including its cost, and its physical, thermodynamic, and chemical properties. In this book we shall discuss three classes of compounds used as refrigerants. They are the halocarbons, the hydrocarbons, and the inorganic compounds. Table 15-1 shows some of the more common refrigerants in each of these classes. The ASHRAE designation system has been used in assigning a number to each refrigerant. In this system the halocarbons and hydrocarbons are designated as follows:

1. The first digit on the right is the number of fluorine (F) atoms in a molecule.
2. The second digit from the right is one more than the number of hydrogen (H) atoms in a molecule.
3. The third digit from the right is one less than the number of carbon (C) atoms in a molecule. If this is zero, then the number is omitted.

Inorganic refrigerants are designated by adding 700 to the molecular weight of the compound. Thus water has a molecular weight of 18 and its ASHRAE designation as a refrigerant is 718.

Example 15-3. Determine the ASHRAE number designation for dichloro-tetra-fluoroethane $CClF_2CClF_2$.

Solution. There are 4 fluorine atoms, no hydrogen atoms, and 2 carbon atoms per molecule. Thus the ASHRAE designation is

$$(2 - 1)(0 + 1)(4) = 114$$

This is the designation listed in Table 15-1.

It is desirable that a refrigerant have the characteristics listed below. The significance or relative importance of each characteristic varies from one application to the next, and there is no such thing as an ideal refrigerant for all applications.

Table 15-1 PROPERTIES OF COMMON REFRIGERANTS*

Number	Chemical Name	Chemical Formula	Molecular Weight	Boiling Point at 1 atm		Freezing Temperature		Critical Temperature		Critical Pressure		
				F	K	F	K	F	K	psia	Pa × 10⁻⁵	ATM
	Halocarbons											
11	Trichloromono-fluoromethane	CCl_3	137.4	74.9	297.0	−168.0	162.0	388.4	471.2	639.5	4.409	43.52
12	Dichlorodifluoro-methane	CCl_2F_2	120.9	−21.6	243.4	−252.0	115.4	233.6	385.2	596.9	4.116	40.62
13	Monochlorotri-fluoromethane	$CClF_3$	104.5	−114.6	191.7	−294.0	92.0	83.9	302.0	561.0	3.868	38.17
14	Carbontetra-fluoride	CF_4	88.0	−198.3	145.2	−299.0	89.3	−50.2	227.5	543.0	3.737	36.88
21	Dichloromono-fluoromethane	$CHCl_2F$	102.9	47.8	282.0	−211.0	138.2	353.3	451.7	750.0	5.171	51.03
22	Monochlorodi-fluoromethane	$CHClF_2$	86.5	−41.4	232.4	−256.0	113.2	204.8	369.2	721.9	4.978	49.12
113	Trichlorotri-fluoroethane	CCl_2FCClF_2	187.4	117.6	320.7	−31.0	238.2	417.4	487.3	498.9	3.413	33.68
114	Dichlorotetra-fluorethane	$CClF_2CClF_2$	170.9	38.8	277.0	−137.0	179.3	294.3	418.9	473.0	3.268	32.25
	Hydrocarbons											
50	Methane	CH_4	16.0	−258.7	111.7	−296.0	90.9	−116.5	190.7	673.1	4.641	45.80
170	Ethane	CH_3CH_3	30.0	−127.9	184.3	−297.0	90.4	90.0	305.4	709.8	4.894	48.30
290	Propane	$CH_3CH_2CH_3$	44.1	−43.7	231.1	−305.8	85.5	206.3	370.0	617.4	4.257	42.01
	Inorganic Compounds											
717	Ammonia	NH_3	17.0	−28.0	239.8	−107.9	195.4	271.4	406.2	1657.0	—	112.8
718	Water	H_2O	18.0	212.0	373.2	32.0	273.2	705.6	647.4	3208.0	—	219.5
729	Air	—	29.0	−317.8	78.8	—	—	−221.3	132.4	547.4	—	37.22
744	Carbon dioxide	CO_2	44.0	−109.2	194.7	−69.9	216.5	87.9	304.2	1070.0	—	72.88
764	Sulfur dioxide	SO_2	64.0	14.0	263.2	−103.9	197.7	315.5	430.7	1143.0	—	77.67

*(Abridged by permission from *ASHRAE Handbook of Fundamentals,* 1977.

535

Thermodynamic Characteristics

1. *High latent enthalpy of vaporization.* This means a large refrigerating effect per unit mass of the refrigerant circulated. In small capacity systems too low a flow rate may actually lead to problems.
2. *Low freezing temperature.* The refrigerant must not solidify during normal operating conditions.
3. *Relatively high critical temperatures.* Large amounts of power would otherwise be required for compression.
4. *Positive evaporating pressure.* Pressure in the evaporator should be above atmospheric to prevent air from leaking into the system.
5. *Relatively low condensing pressure.* Otherwise expensive piping and equipment will be required.

Physical and Chemical Characteristics

1. *High dielectric strength of vapor.* This permits the use in hermetically sealed compressors where vapor may come in contact with motor windings.
2. *Good heat transfer characteristics.* Thermophysical properties should be such that high heat transfer coefficients can be obtained. This covers properties such as density, specific heat, thermal conductivity, and viscosity.
3. *Satisfactory oil solubility.* Oil may be dissolved in the refrigerant or refrigerant in the oil. This can affect lubrication and heat transfer characteristics and lead to oil logging in the evaporator. A system must be designed with the oil solubility characteristics in mind.
4. *Low water solubility.* Water in a refrigerant can lead either to freezeup in the expansion devices or corrosion.
5. *Inertness and stability.* The refrigerant must not react with materials that will contact and its own chemical makeup must not change with time.

Safety

1. *Nonflammability.* The refrigerant should not burn or support combustion when mixed with air.
2. *Nontoxicity.* The refrigerant should not be harmful to humans directly or indirectly through foodstuffs.
3. *Nonirritability.* The refrigerant should not irritate humans (eyes, nose, lungs, or skin).

In addition to the above characteristics the refrigerant should be of relative low cost and be easy to detect in case of leaks.

15-4 COMPRESSORS

The compressor is one of the four essential parts of the vapor compression refrigeration system and may be of the reciprocating, rotary, helical rotary (screw), or centrifugal type.

Reciprocating Compressors

Most reciprocating compressors are single-acting, using pistons driven directly through a pin and connecting rod from the crankshaft. Double-acting compressors, which utilize piston rods, crossheads, stuffing boxes, and oil injection, are not extensively used.

This section covers two classifications that include most reciprocating refrigeration compressors, the halocarbon compressor, and the ammonia compressor. The halocarbon compressor is the most widely used; it is manufactured in three types of design: (1) open, (2) semi- or bolted hermetic, and (3) the welded shell hermetic. Ammonia compressors are manufactured only in the open-type design. In the open-type compressor the shaft extends through a seal in the crankcase to an external drive.

In hermetic compressors the motor and compressor are contained within the same pressure vessel with the motor shaft as part of the compressor crankshaft, and with the motor in contact with the refrigerant. Figure 15-6a illustrates a typical hermetic compressor. A semi-hermetic, accessible, or serviceable hermetic compressor is of bolted construction capable of field repair, Fig. 15-6b. A welded shell (sealed) hermetic compressor is one in which the motor-compressor is mounted inside a steel shell that in turn is sealed by welding.

The idealized reciprocating compressor is assumed to operate in a reversible adiabatic manner; pressure losses in the valves, intake, and exhaust manifolds are neglected. We will now consider the effect of some of the unavoidable realities of the compressor. The development follows that in reference 2.

Figure 15-7 shows a schematic indicator diagram for a reciprocating compressor. The diagram is idealized because the pressure in the cylinder is assumed constant during the exhaust process $P_c = P_d$ and during the intake process P_a and P_b. However, the pressure loss in the valves is to be considered. The gas remaining in the clearance volume V_d expands in a polytropic process from state d to state a. State b is generally different from that at a because of the mixing of the expanded clearance volume vapor and the intake vapor. The vapor is compressed from state b to

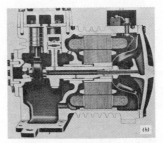

Figure 15-6.   Reciprocating compressors. (Courtesy of Copeland Corporation, Sidney, Ohio.)

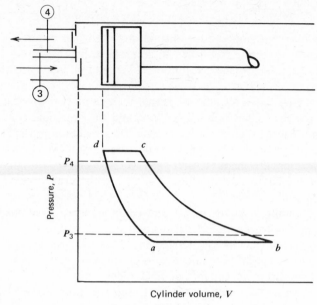

Figure 15-7.   Schematic indicator diagram for a reciprocating compressor.

state c in a polytropic process. Heat transfer may occur during the exhaust process c–d. Figure 15-8 is a pressure-specific volume diagram for the compressor of Fig. 15-7. In the state diagram of Fig. 15-8 it is assumed that:

1. The same polytropic exponent n applies to the compression process b–c and the expansion process d–a.
2. Heat transfer during the exhaust process is negligible; therefore states c and d are identical.
3. The state of the mixture of reexpanded clearance volume gas a and the intake gas 3 is the same as state b and is designated as a'. This is actually a result of assumptions 1 and 2 above. Note that the work required to compress the clearance volume vapor is just balanced by the work done in the expansion of the clearance volume vapor.

The *volumetric efficiency* η_v is defined as the ratio of the actual mass of the vapor compressed to the mass of vapor that could be compressed if the intake volume equaled the piston displacement and the state of the vapor at the beginning of the compression was equal to state 3 of Fig. 15-8. The total mass in the cylinder at the end of the intake stroke is V_b/v_b. The mass of clearance vapor is V_a/v_a. Then

$$\text{Actual mass of intake vapor} = \frac{V_b}{v_b} - \frac{V_a}{v_{a'}} = \frac{V_b - V_a}{v_b} \qquad (15\text{-}5)$$

Mass of intake vapor with volume equal to piston displacement

$$= \frac{V_b - V_d}{v_3} \quad (15\text{-}6)$$

The volumetric efficiency thus becomes

$$\eta_v = \frac{(V_b - V_a)v_3}{(V_b - V_d)v_b} \quad (15\text{-}7)$$

The clearance factor C is defined as the ratio of the clearance volume to the piston displacement or

$$C = \frac{V_d}{V_b - V_d} \quad (15\text{-}8)$$

By using Eqs. (15-7) and (15-8) and the polytropic relation

$$V_a = V_d \left(\frac{P_d}{P_a}\right)^{1/n} = V_d \left(\frac{P_c}{P_b}\right)^{1/n} \quad (15\text{-}9)$$

It may be shown that

$$\eta_v = \left[1 + C - C\left(\frac{P_c}{P_b}\right)^{1/n}\right]\frac{v_3}{v_b} \quad (15\text{-}10)$$

The three main factors that influence the volumetric efficiency are accounted for in Eq. (15-10). The term within the brackets describes the effect of the clearance volume and reexpansion of the clearance volume vapor, whereas the term v_3/v_b accounts for the pressure drop and heating of the intake vapor.

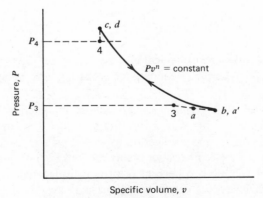

Figure 15-8. Pressure-specific volume diagram for compressor of Fig. 15-7.

From Eq. (15-7) is follows that volumetric efficiency may also be expressed as

$$\eta_v = \frac{\dot{m} v_3}{PD} \tag{15-11}$$

where PD is the piston displacement in volume per unit of time. Then combining Eqs. (15-10) and (15-11) the mass flow rate of the refrigerant is given by

$$\dot{m} = \left[1 + C - C \left(\frac{P_c}{P_b} \right)^{1/n} \right] \frac{PD}{v_b} \tag{15-12}$$

Although the polytropic exponent n must be determined experimentally, it may be approximated by the isentropic exponent k when other data are not available. Table 15-2 gives some representative values of k.

An expression for the compressor work may be derived subject to the same assumptions used in the analysis for volumetric efficiency. Referring to Fig. 15-7 we represent the work per cycle by the enclosed area of the indicator diagram.

$$W = \oint V \, dP = \int_b^c V \, dP - \int_a^d V \, dP \tag{15-13}$$

Using the realtion, $PV^n = $ constant, to describe the process paths

$$W = \frac{n}{(n-1)} P_b (V_b - V_a) \left[\left(\frac{P_c}{P_b} \right)^{(n-1)/n} - 1 \right] \tag{15-14}$$

The work per unit mass of vapor compressed may be obtained by dividing Eq. (15-14) by the mass of intake vapor, Eq. (15-5).

$$w = \frac{n}{(n-1)} P_b v_b \left[\left(\frac{P_c}{P_b} \right)^{(n-1)/n} - 1 \right] \tag{15-15}$$

The power requirement for the compressor is given by

$$\dot{W} = \frac{\dot{m} w}{\eta_m} \tag{15-16}$$

where η_m is the compressor mechanical efficiency. Figure 15-9 shows representative performance data for a hermetic compressor operating with refrigerant 22.

Example 15-4. Refrigerant 12 vapor enters the suction header of a single stage reciprocating compressor at 45 psia and 40 F. The discharge pressure is 200 psia. Pressure drop in the suction valve is 2 psi and the pressure loss in the discharge valve is 4 psi. The vapor is superheated 12 F during the intake stroke. The clearance volume is 5 percent of piston displacement. Determine the (a) volumetric efficiency, (b) compressor pumping capacity if the piston displacement is 10 in.^{3} and the crankshaft rotates at 1725 rpm, and (c) shaft horsepower required for a mechanical efficiency of 70 percent.

Table 15-2   ISENTROPIC EXPONENT k FOR SOME REFRIGERANT VAPORS

	Refrigerant		
	12	**22**	**Ammonia**
Vapor temperature, F/C	50/10	86/30	70/21
$k = c_p/c_v$	1.13	1.16	1.31

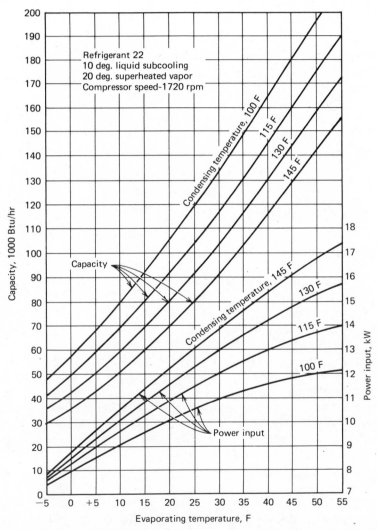

Figure 15-9.   Typical capacity and power input curves for a hermetic reciprocating compressor. (Reprinted by permission from *ASHRAE Handbook and Product Directory—Equipment*, 1975.)

Solution. The $P - v$ diagram of Fig. 15-8 will be used to aid in the problem solution. From the given data

$$P_4 = 200 \text{ psia}; \quad P_c = P_d = P_4 + 4 = 204 \text{ psia}$$
$$P_3 = 45 \text{ psia}; \quad P_b = P_3 - 2 = 43 \text{ psia}$$

From Chart 4, $v_3 = 0.92$ ft³/lbm and at point b, $t_b = 52$ F. $P_b = 43$ psia and $v_b = 0.98$ ft³/lbm. The clearance factor is given as 0.05, and it will be assumed that $n = k = 1.13$. Then using Eq. (15-10)

$$\eta_v = \left[1 + 0.05 - (0.05)\left(\frac{204}{43}\right)^{1/1.13} \right] \frac{0.92}{0.98} = 0.8$$

Now from Eq. (15-11)

$$\dot{m} = \frac{\eta_v(PD)}{v_3} = \frac{(0.8)}{(0.92)}\frac{(10)(1725)}{(1728)} = 8.68 \text{ lbm/min}$$

The shaft power required will be computed using Eqs. (15-15) and (15-16)

$$\dot{W} = \frac{\dot{m}w}{\eta_m} = \left(\frac{\dot{m}}{\eta_m}\right)\frac{n}{(n-1)}P_b v_b \left[\left(\frac{P_c}{P_b}\right)^{(n-1)/n} - 1\right]$$
$$\dot{W} = \left(\frac{8.68}{0.7}\right)\frac{1.13}{(1.13-1)}(43)(144)0.98 \left[\left(\frac{204}{43}\right)^{(1.13-1)/1.13} - 1\right]$$
$$\dot{W} = 128{,}298 \text{ (ft-lbf)/min} = 3.9 \text{ hp}$$

Rotary Compressors

Rotary compressors are characterized by their circular or rotary motion as opposed to reciprocating motion. Their positive displacement compression process is nonreversing and either continuous or cyclical depending on the mechanism employed. Most are direct drive machines.

Figures 15-10 and 15-11 show two common types of rotary compressors: the rolling piston type and the rotating vane type. These two machines are very similar in respect to size, weight, thermodynamic performance, field of greatest applications, range of sizes, durability, and sound level.

The rotary compressor performance is characterized by high volumetric efficiency because of its small clearance volume, and correspondingly low reexpansion loss. Figure 15-12 shows performance typical of compressors now in production.

Helical Rotary Compressors

The helical rotary or screw compressor also belongs to the broad class of positive displacement compressors. Although it was developed in the early 1930s and has been used in the gas and process industries, it was first introduced to the refrigeration industry in the late 1950s.

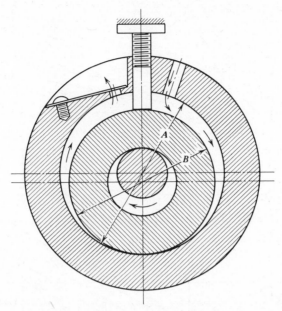

Figure 15-10.   Rolling piston-type rotary compressor. (Reprinted by permission from *ASH-RAE Handbook and Product Directory—Equipment*, 1975.)

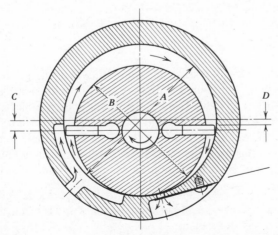

Figure 15-11.   Rotating vane-type rotary compressor. (Reprinted by permission from *ASHRAE Handbook and Product Directory—Equipment*, 1975.)

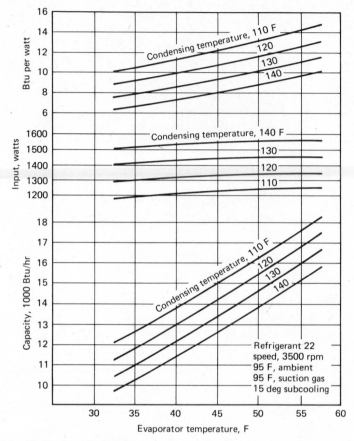

Figure 15-12. Typical performance curves for a rotary compressor. (Reprinted by permission from *ASHRAE Handbook and Product Directory—Equipment*, 1975.)

The machine essentially consists of two mating helically grooved rotors, a male (lobes) and a female (gullies), in a stationary housing with suitable inlet and outlet gas ports, Fig. 15-13. The flow of gas in the rotors is both radial and axial. With a four-lobe male rotor rotating at 3600 rpm the six-lobe female rotor follows at $\frac{4}{6}$ (3600 rpm) or 2400 rpm. The female rotor can be driven by the male rotor.

Compression in this machine is obtained by direct volume reduction with pure rotary motion. The four basic continuous phases of the working cycle are:

1. *Suction.* A pair of lobes unmesh on the inlet port side and gas flows in the increasing volume formed between the lobes and the housing until the lobes are completely unmeshed.

2. *Transfer.* The trapped pocket isolated from the inlet and outlet is moved circumferentially at constant suction pressure.

3. *Compression.* When remeshing starts at the inlet end, the trapped volume is reduced and the charge is gradually moved helically and compressed simultaneously toward the discharge end as the lobes' mesh point moves along axially.
4. *Discharge* starts when the compressed volume has been moved to the axial ports on the discharge end of the machine and continues until all the trapped gas is completely squeezed out.

During the remeshing period of compression and discharge a fresh charge is drawn through the inlet on the opposite side of the meshing point. With four male lobes rotating at 3600 rpm, four interlobe volumes are filled and discharged per revolution, providing 14,400 discharges per minute. Since the intake and discharge cycles overlap effectively, a smooth continuous flow of gas results.

Typical performance levels for oil-flooded screw compressors are given in Fig. 15-14.

Centrifugal Compressors

Centrifugal compressors or turbocompressors are members of a family of turbomachines that includes fans, propellors, and turbines. Such machines are characterized by a continuous exchange of angular momentum between a rotating mechanical element and a steadily flowing fluid. Because their flows are continuous, turbomachines have greater volumetric capacities, size-for-size, than do positive displacement devices. For effective momentum exchange their rotative speeds must be higher, but little vibration or wear results because of the steadiness of the motion and the absence of contacting parts.

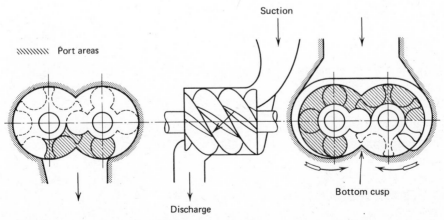

Figure 15-13. Helical rotary (screw) compressor. (Reprinted by permission from *ASHRAE Handbook and Product Directory—Equipment*, 1975.)

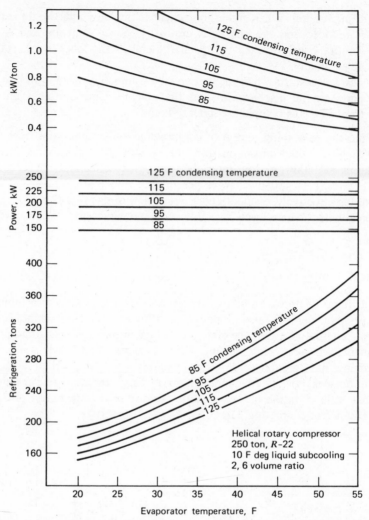

Figure 15-14. Performance levels for oil-flooded compressors. (Reprinted by permission from *ASHRAE Handbook and Product Directory—Equipment*, 1975.)

Centrifugal compressors are used in a wide variety of refrigeration and air-conditioning applications. As many as eight or nine stages can be installed in a single casing.

15-5   REGRIGERATION CONTROL VALVES

Early in the development of refrigeration cycles it became evident that the system should respond to changing load and environmental conditions to better serve the system objectives and to prevent damage to the system components. The most important component in the vapor compression cycle from the standpoint of system control is the expansion valve. Proper manipulation of the expansion valve either manually or automatically can go a long way toward optimizing system performance; other devices are sometimes required, however, to control evaporator temperature or compressor load. The most common control valves are described in the following paragraphs.

Hand Expansion Valves

The earliest refrigeration systems had only hand-operated needle valves for expansion valves. The operator monitored the system and adjusted the valve according to the refrigeration load. Although hand valves are seldom used today, they are sometimes used to bypass various automatic valves so that system operation may be maintained while the automatic valve is being serviced.

Constant Pressure Expansion Valve

The constant pressure or automatic expansion valve was the earliest type of automatic valve used in refrigeration systems. The valve acts to maintain a fixed evaporator pressure. Therefore, when evaporator pressure drops, the valve opens to allow more refrigerant flow. The opposite action results from a rise in evaporator pressure. This valve action is opposite to that desired in variable load applications. When the refrigerating load on an evaporator increases, the pressure will increase and more refrigerant is needed to handle the increased heat transfer rate. Conversely, when evaporator load decreases, pressure decreases, and less refrigerant is required. The automatic expansion valve is inadequate when the load is variable; it is therefore in very limited use at the present time.

Thermostatic Expansion Valve

The thermostatic expansion valve does not have the limitations of the constant pressure valve because of a separate mechanism that positions the valve spool to admit the refrigerant as required by evaporator load, Fig. 15-15. A bulb containing a small refrigerant charge is connected by a small tube to the chamber above the valve diaphragm. The bulb is clamped to the refrigerant line where refrigerant leaves the evaporator. Spring pressure tends to close the valve, whereas bulb pressure tends to open it. The bulb is essentially a temperature-sensing element, and several degrees superheat of the refrigerant leaving the evaporator are required before the expansion valve will open. As the load on the evaporator increases, superheat will increase and the valve opens to allow a larger refrigerant flow rate. As the load decreases, the valve will close maintaining a superheated vapor at the evaporator exit. This action protects the compressor from liquid slugging and decreases the compressor load as evaporator load decreases. Most thermostatic expansion valves are adjustable. About 5 to 10 degrees superheat is usually maintained.

It is possible for the thermostatic expansion valve to overload the compressor as

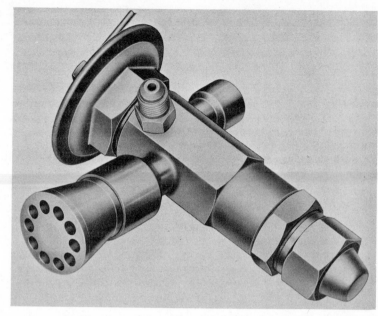

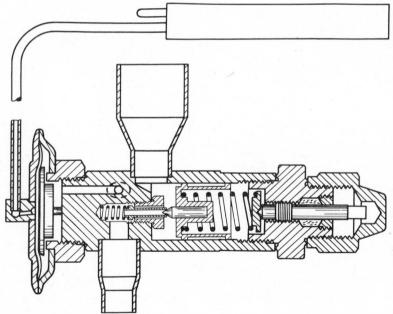

Figure 15-15. Exterior and interior views of thermostatic expansion valve. (Courtesy of Sporlan Valve Company, St. Louis, Missouri.)

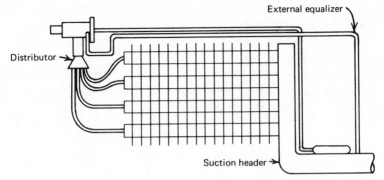

Figure 15-16.   External equalizer with refrigerant distributor feeding parallel circuits.

evaporator load continues to increase. To help prevent this condition a liquid-vapor charge may be used in the bulb. The charge is such that all the liquid is evaporated at some predetermined temperature. Very little bulb pressure rise will result with a temperature rise above this point, and the valve will not open significantly with further increase in evaporator load. Thus the load on the compressor will be limited.

When an evaporator is subjected to a large variation in load, the pressure drop of the refrigerant as it flows through the coil will vary. The refrigerant saturation temperature will therefore vary at the coil exit, and different amounts of superheat will result at light and heavy loads. This condition is controlled by the use of an external equalizer (Fig. 15-16). The valve then senses pressure at about the same point as it senses temperature, and the superheat will be correct for whatever pressure exists in the suction line. The external equalizer must be used when one expansion valve feeds parallel circuits.

Any number of evaporators may be operated in parallel on the same compressor where a thermostatic expansion valve controls the feed to each one. Each valve controls the flow as required by that evaporator.

The thermostatic expansion valve must be selected to match the refrigeration load, the refrigerant type, the pressure difference across the valve, and the refrigerant temperature. The valves may be adjusted but do not have sufficient range to work well with different refrigerants.

Capillary Tube

The capillary tube is not a valve but effectively replaces the expansion valve in many applications. It is a long thin tube placed between the condenser and evaporator. The small diameter and relatively large length of the tube produces a large pressure drop. Effective control of the system results because the tube allows the flow of liquid more readily than vapor. Although the capillary tube operates most efficiently at one particular set of conditions, there is only a slight loss in efficiency at off-design conditions in small systems.

The main advantage of the capillary tube is its simplicity and low cost. Being a passive device it is not subject to wear. However, the very small bore of the tube is subject to plugging if precautions are not taken to maintain a clean system. Moisture can also cause plugging due to ice formation.

Almost all domestic refrigerators and room air conditioners use capillary tubes as well as domestic split system air-conditioning equipment. Extra care must be taken to prevent tube plugging with the latter type of equipment.

Subcooled liquid refrigerant enters a properly sized capillary tube as shown in Fig. 15-17. Pressure loss because of friction causes the refrigerant to reach saturation at some point in the tube. At that point pressure loss becomes much more pronounced as vapor forms and the refrigerant temperature decreases rapidly to the tube exit. Critical flow conditions may occur at the tube exit with a pronounced pressure loss at that point. The amount of refrigerant flowing through the tube depends on the overall pressure difference between the condenser and the evaporator and the saturation pressure of the refrigerant entering the tube. Therefore, a change in evaporator load or environmental conditions that increases the pressure difference or decreases the saturation pressure will cause an increase in refrigerant flow rate. Fortunately an increase in load increases pressure difference and a decrease in condensing medium temperature decreases the pressure difference. This partially explains the favorable control characteristics of the capillary tube.

Although there are calculation procedures for the sizing of capillary tubes, the final selection must be made by trial and error. The equipment volume of the *ASHRAE Product Directory* (4) contains charts for preliminary selection of capillary tubes.

The charging of capillary tube systems is rather critical. This is because part of the passive control of the system results from the flooding of the condenser at part-load conditions, which acts to decrease the condenser size, decrease subcooling, and increase the saturation pressure at the capillary inlet. If other than the proper charge is used, this important effect is lost. Too much refrigerant can also cause compressor flooding during the off cycle and result in valve damage upon startup.

Evaporator Pressure Regulator

The purpose of an evaporator pressure regulator is to maintain a relatively constant minimum pressure in the evaporator. Because most of the evaporator surface is subjected to two-phase refrigerant, a constant minimum temperature will also be maintained. Figure 15-18 shows a typical evaporator pressure regulator. Evaporator pressure is sensed internally and balanced by a spring-loaded diaphragm. When evaporator pressure falls below a set value, the valve will close, restricting the flow of refrigerant so that evaporator pressure will rise.

The main applications of evaporator pressure regulators are to set a minimum evaporator temperature and to permit the use of different evaporators at different pressures on the same compressor.

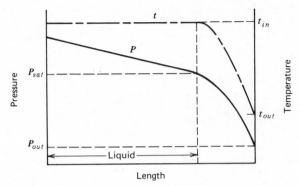

Figure 15-17.   Pressure and temperature of the refrigerant in a capillary tube.

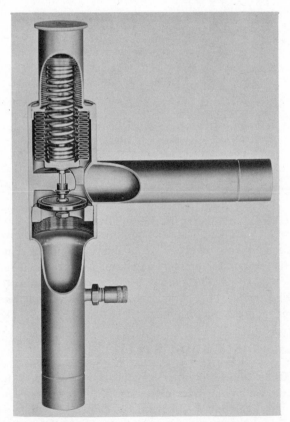

Figure 15-18.   Evaporator pressure regulating valve. (Courtesy of Sporlan Valve Company, St. Louis, Missouri.)

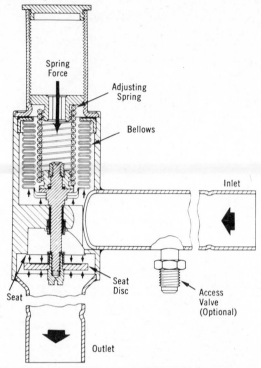

Figure 15-19.   Exterior and interior views of suction pressure regulating valve. (Courtesy of Sporlan Valve Company, St. Louis, Missouri.)

Suction Pressure Regulator

The suction pressure regulator has the function of limiting the maximum pressure at the compressor suction. Since the compressor load is determined by suction pressure, the suction pressure regulator is a load-limiting device. This valve functions very much like the evaporator pressure regulator except that it senses compressor suction pressure, Fig. 15-19. The suction pressure regulator also reduces the compressor load during the start-up period because the valve will remain closed until suction pressure is reduced to a set pressure.

15-6   THE REAL SINGLE STAGE CYCLE

Figure 15-20 illustrates a practical single stage cycle schematically. The practical aspects of the compressor were discussed in a previous section and are shown again as state points 3 through 4. Other factors cause the complete cycle to deviate from the theoretical. These are the pressure losses in all connecting tubing that increase the power requirement for the cycle and the heat transfer to and from the various components. Heat transfer will generally be away from the system on the high pressure side of the cycle and will improve cycle performance. Exposed surfaces on the low pressure side of the cycle will generally be at a lower temperature than the environment and any heat transfer will generally degrade cycle performance. As shown in Fig. 15-20, $\dot{q}_3$ will increase compressor load and q_4 decreases compressor load.

It should be mentioned that the condenser and evaporator heat transfer surfaces have pressure losses that also contribute to the overall compressor power requirement. The following example illustrates a complete cycle analysis.

Example 15-5.   A 10 ton refrigeration unit operates on R-22 and has a single stage reciprocating compressor with a clearance volume of 3 percent and operates at 1725 rpm. The refrigerant leaves the evaporator at 70 psia and 40 F. Pressure loss in the evaporator is 5 psi. Pressure loss in the suction valve is 2 psi, and there is 20 F superheat at the beginning of the compression stroke. The

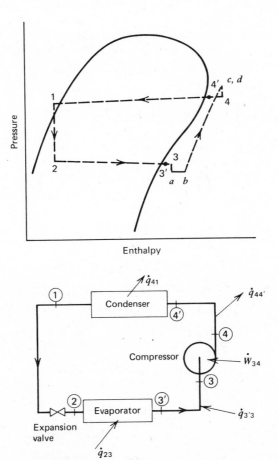

Figure 15-20.   A practical single stage cycle.

compressor discharge pressure is 300 psia, a pressure loss of 5 psi exists in the discharge valve, and the vapor is desuperheated 10 F between the compressor discharge and the condenser inlet. Neglect pressure loss in tubing between the compressor and condenser but the condenser pressure loss is 5 psi. The liquid has a temperature of 110 F at the expansion valve. Determine the following: (a) compressor volumetric efficiency, (b) compressor piston displacement, (c) shaft power input with mechanical efficiency of 75 percent, (d) heat rejected in the condenser, and (e) power input to compressor per ton of cooling effect.

Solution. Most of the state points as shown in Fig. 15-20 may be located from the given data.

$$P_1 = 290 \text{ psia}; \qquad T_1 = 110 \text{ F}; \qquad i_2 = i_1$$
$$P_{3'} = 70 \text{ psia}; \qquad t_{3'} = 40 \text{ F}; \qquad P_2 = P_{3'} + 5 = 75 \text{ psia}$$
$$P_a = P_b = P_{3'} - 2 = 68 \text{ psia}; \qquad t_b = 50 \text{ F}$$
$$P_c = P_d = 300 \text{ psia}; \qquad P_4 = P_{4'} = P_c - 5 = 295 \text{ psia}$$

States c, d, and 4 may be completely determined following the compressor and evaporator analysis. From Chart 5, $v_3 = 0.78 \text{ ft}^3/\text{lbm}$ and $v_b = 0.81 \text{ ft}^3/\text{lbm}$. Assuming that $n = k = 1.16$, the volumetric efficiency may be computed from Eq. (15-10) as

$$\eta_v = \left[1 + 0.03 - (0.03)\left(\frac{300}{68}\right)^{1/1.16}\right]\frac{0.78}{0.81} = 0.89$$

For the evaporator

$$\dot{q}_e = \dot{m}(i_{3'} - i_2) = \dot{m}(i_{3'} - i_1)$$

or

$$\dot{m} = \frac{\dot{q}_e}{(i_{3'} - i_1)}$$

From Chart 5, $i_{3'} = 109 \text{ Btu/lbm}$, $i_1 = 41 \text{ Btu/lbm}$

$$\dot{m} = 10(12{,}000)/(109 - 41) = 1765 \text{ lbm/hr}$$

From Eq. (15-11)

$$\eta_v = \frac{\dot{m}v_3}{PD} \qquad \text{or} \qquad PD = \frac{\dot{m}v_3}{\eta_v}$$
$$PD = 1765(0.78)/0.89 = 1547 \text{ ft}^3/\text{hr}$$

or

$$\frac{PD}{\text{cycle}} = 25.8 \text{ in.}^3$$

The work per pound of refrigerant may be computed from Eq. (15-15):

$$w = \frac{n}{n-1} P_b v_b \left[\left(\frac{P_c}{P_b} \right)^{(n-1)/n} - 1 \right]$$

$$w = \frac{1.16}{(1.16-1)} 68(144)(0.81) \left[\left(\frac{300}{68} \right)^{(1.16-1)/1.16} - 1 \right]$$

$$w = 13,064 \ (\text{ft-lbf})/\text{lbm}$$

$$\dot{W} = \dot{m}w = \frac{13,064(1765)}{778} = 29,638 \ \text{Btu/hr}$$

$$\dot{W}_{sh} = \frac{\dot{W}}{\eta_m} = 39,517 \ \text{Btu/hr} = 15.5 \ \text{hp}$$

The heat rejected in the condenser is given by

$$\dot{q}_c = \dot{m}(i_1 - i_{4'})$$

However, state 4′ has not been completely determined. Since the polytropic exponent was assumed equal to k, the isentropic compression process follows a line of constant entropy on Chart 5. Then $t_c = t_d = 188$ F, $t_4 \approx 185$ F, and $t_{4'} = t_4 - 10 = 175$ F. Then $i_{4'} = 125$ Btu/lbm.

$$\dot{q}_c = 1765(41 - 125) = -148,260 \ \text{Btu/hr}$$

The total power input to the compressor per ton of cooling effect is

$$\frac{\text{hp}}{\text{ton}} = \frac{15.5}{10} = 1.55$$

There are many practical problems that involve variations of the vapor compression cycle. One of the most interesting involves the system behavior when the load or the environment temperature changes. A new operating condition must be attained such that equilibrium is established and in general the cycle as shown on the $P - i$ diagram will be relocated. The automatic control valves previously discussed aid in the control of the system under changing conditions and prevent damage to the compressor. In some cases the system may be cycled off. Analysis of the vapor compression cycle becomes complex under nonsteady state conditions. However, some computer routines are available for this purpose. It is sometimes possible through hand calculations and the $P - i$ diagrams to find a new equilibrium-operating condition after some change is introduced.

15-7 ABSORPTION REFRIGERATION

The mechanical vapor compression refrigeration system described in the previous articles is an efficient and practical method. However, the required energy input is shaft work or power, which is high grade energy and expensive. The relatively large amount of work required is because of the compression of the vapor that undergoes a large decrease in volume.

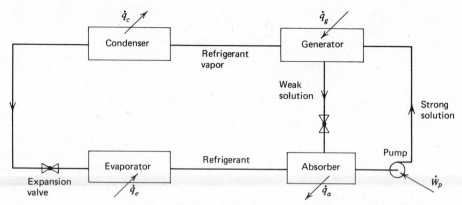

Figure 15-21. Basic absorption cycle.

It is possible to replace the vapor compression process with a series of processes where the refrigerant vapor is absorbed by a liquid and then the liquid solution is pumped to the higher pressure. Although the shaft work required is much less, large quantities of heat must be transferred to the system. Figure 15-21 illustrates the essential features of the absorption cycle. Note that the compressor of the vapor compression cycle has been replaced by the generator, absorber, and pump, and that the solution of refrigerant and absorbent circulates through these elements. The refrigerant alone flows through the condenser, expansion valve, and evaporator. Heat is transferred to the solution in the generator, and the refrigerant vapor is separated from the solution, whereas the opposite process occurs in the absorber. The ease with which these processes may be carried out in practice depends on the refrigerant–absorbent pair and the many refinements to the cycle required in actual practice. These factors will be discussed later. The overall performance of the absorption cycle in terms of refrigerating effect per unit of energy input is generally poor; however, waste heat such as that rejected from a power plant can often be used to achieve better overall energy conservation.

The use of solar energy is also possible although the required temperature of the heating medium presents some problems in flat plate solar collector performance and in energy storage.

15-8 PROPERTIES AND PROCESSES OF HOMOGENEOUS BINARY MIXTURES

A homogeneous mixture is uniform in composition and cannot be separated into its constituents by pure mechanical means. The properties such as pressure, density, and temperature are uniform throughout the mixture. The thermodynamics of binary mixtures is extensively covered by Bosnjakovic (1).

The thermodynamic state of a mixture cannot be specified by two independent properties as may be done with a pure substance. The composition of the mixture as described by the *concentration x,* the ratio of the mass of one constituent to the mass of the mixture, is required in addition to two other independent properties such as pressure and temperature.

The miscibility of a mixture is an especially important characteristic for an absorption system. A mixture is miscible if it does not separate after mixing. A miscible mixture is homogeneous. Some mixtures may not be miscible under all conditions with temperature being the main property influencing miscibility. Oil-Refrigerant 22 is miscible at higher temperatures but nonmiscible at low temperatures. Binary mixtures for an absorption refrigeration cycle must be completely miscible in both the liquid and vapor phases with no miscibility gaps.

The behavior of a binary mixture in the vicinity of the saturation region is important to absorption refrigeration. To understand this behavior the use of imaginary experiments and a temperature versus concentration diagram is helpful. Figure 15-22 a shows a piston-cycinder arrangement containing a binary mixture in the liquid phase with a concentration of material B, x_1. The piston has a fixed mass and is frictionless, so that the pressure of the mixture is always constant. Figure 15-22 c is the $t - x$ diagram for the mixture with the initial state shown as point 1. As heat is slowly transferred to the mixture, the temperature will rise and at point 2 vapor will begin to form and collect under the piston as shown in Fig. 15-22 b. If the experiment is stopped at some point above this temperature and the concentrations of the vapor and liquid determined, a rather surprising result is observed. For example, the state of the liquid might be at point 3 while the state of the vapor is at point 4 and the concentration of material B in the vapor x_4 is larger than in the liquid x_3. We shall see later that this phenomenon is quite helpful in the absorption cycle.

Continued heating of the mixture will gradually vaporize all of the liquid (state point 5) with concentration $x_5 = x_1$. Further heating will superheat the vapor to point 6. When the superheated vapor is cooled at constant pressure, the entire process will be reversed as shown in Fig. 15-22 d. Additional experiments carried out at different concentrations but at the same pressure establish the boiling and condensing lines shown in Figs. 15-22 c and 15-22 d. If the pressure is changed and the experiments repeated at various concentrations, the boiling and condensing lines will be displaced as shown in Fig. 15-22 e. Unlike a pure substance, binary mixtures do not have a single saturation temperature for each pressure, because the saturation temperatures depend on the concentration.

A more useful representation of the properties of a binary mixture is the enthalpy-concentration diagram. Figure 15-23 is a schematic $i - x$ diagram including the liquid and vapor regions for a homogeneous binary mixture. The condensing and boiling lines for a given pressure do not converge at $x = 0$ and $x = 1$ but are separated by a distance proportional to the enthalpy of vaporization of each constituent. Lines of constant temperature are shown in the liquid and vapor regions but not in the saturation region. When the boiling and condensing lines for more than one pressure are shown on one chart, it becomes difficult to distinguish the proper isotherms for each pressure in the saturation region. However, these isotherms may be located as needed.

Chart 2 is an enthalpy-concentration diagram for an ammonia–water mixture. Chart 2 covers the subcooled and saturation region but does not have temperature

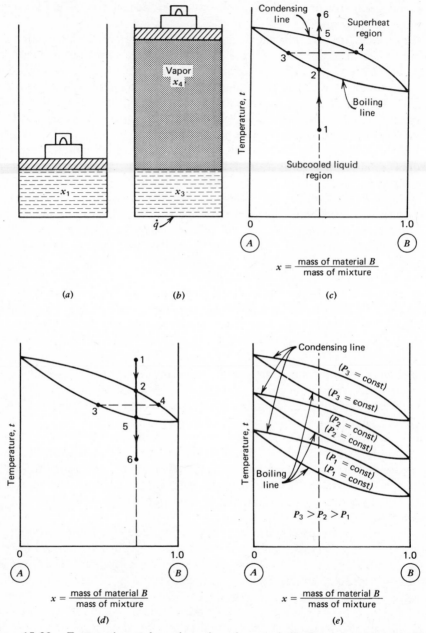

Figure 15-22. Evaporation and condensation characteristics for a homogeneous binary mixture.

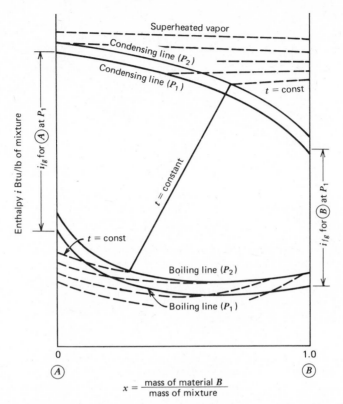

Figure 15-23. Schematic $i - x$ diagram for liquid and vapor regions for a homogeneous binary mixture.

lines in the vapor region. Lines of constant temperature in the saturation region are located using the equilibrium construction lines. A vertical line is drawn upward from the saturated liquid state to the equilibrium construction line for the same pressure. From that point a horizontal line is drawn to the saturated vapor line for the same pressure. This intersection gives the saturated vapor state point. The isotherm for the given temperature is the straight line connecting the saturated liquid and vapor states.

The most common thermodynamic processes involved in industrial systems and absorption refrigeration are described. These are the adiabatic and nonadiabatic mixing of two streams, heating and cooling including vaporization and condensation, and throttling. An understanding of these processes will help in understanding the absorption cycle. The $i - x$ diagram will be quite useful in the solution of these problems.

Adiabatic Mixing of Two Streams

Figure 15-24 shows a mixing chamber where two binary streams of different concentration and enthalpy mix in a steady flow process. Determination of the state

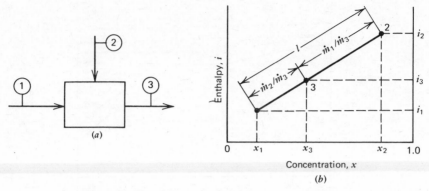

Figure 15-24.   Steady flow adiabatic mixing process.

of the stream leaving the mixing chamber involves energy and mass balances on the control volume defined by the mixing chamber. The energy balance is

$$i_1\dot{m}_1 + i_2\dot{m}_2 = i_3\dot{m}_3 \tag{15-17}$$

while the overall mass balance is given by

$$\dot{m}_1 + \dot{m}_2 = \dot{m}_3 \tag{15-18}$$

and the mass balance for one constituent is

$$\dot{m}_1 x_1 + \dot{m}_2 x_2 = \dot{m}_3 x_3 \tag{15-19}$$

Elimination of $\dot{m}_3$ from Eqs. (15-17), (15-18), and (15-19) gives

$$\frac{\dot{m}_1}{\dot{m}_2} = \frac{i_2 - i_3}{i_3 - i_1} = \frac{x_2 - x_3}{x_3 - x_1} \tag{15-20}$$

Equation (15-20) defines a straight line on the $i - x$ diagram as shown in Fig. 15-24 and state 3 must lie on this line. It may be shown that

$$x_3 = x_1 + \frac{\dot{m}_2}{\dot{m}_3}(x_2 - x_1) \tag{15-21}$$

$$i_3 = i_1 + \frac{\dot{m}_2}{\dot{m}_3}(i_2 - i_1) \tag{15-22}$$

The $i - x$ diagram may be used to advantage to solve mixing problems, but the procedure is somewhat involved when the final state is in the mixture region.

Example 15-6.   Twenty lbm per minute of liquid water–ammonia solution at 150 psia, 220 F, and concentration of 0.25 lbm ammonia per lbm of solution is mixed in a steady flow adiabatic process with 10 lbm per minute of saturated water–ammonia solution at 150 psia and 100 F. Determine the enthalpy, concentration, and temperature of the mixture.

Solution.   Chart 2 will be used to carry out the solution and Fig. 15-25 shows the procedure. State 1 is a subcooled condition located on the diagram at $t = 220$ F and $x = 0.25$. State 2 is a saturated liquid and is located at the intersection of the 150 psia boiling line and the 100 F temperature line. States 1 and 2 are joined by a straight line. State 3 is located on the connecting line and is determined by the concentration x_3 or enthalpy i_3. Using Eq. (15-21), we get

$$x_3 = x_1 + \frac{\dot{m}_2}{\dot{m}_3}(x_2 - x_1) = 0.25 + \frac{10}{30}(0.72 - 0.25)$$

$$x_3 = 0.41 \text{ lbm ammonia/lbm mixture}$$

Because the resulting mixture state lies within the saturation region, it is a mixture of vapor and liquid. To determine the temperature t_3 a graphical trial-and-error procedure shown in Fig. 15-25 is used. The fractional proportions of liquid and vapor in the mixture may also be determined from Chart 2 in a graphical manner.

$$\frac{\dot{m}_v}{\dot{m}_3} = \frac{\overline{l3}}{\overline{lv}} = 0.038 \tag{15-23}$$

Therefore, the mixture is 3.8 percent vapor and 96.2 percent liquid.

Mixing of Two Streams with Heat Exchange

This type of process is quite common and occurs in the absorber of the absorption refrigeration device. In this case shown in Fig. 15-26 the energy balance becomes

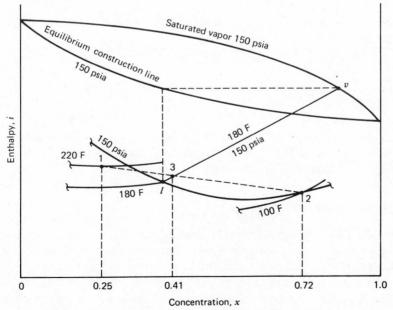

Figure 15-25.   Schematic $i - x$ diagram for Example 15-6.

$$\dot{m}_1 i_1 + \dot{m}_2 i_2 = \dot{m}_3 i_3 + \dot{q} \tag{15-24}$$

The mass balance equations are identical to those for adiabatic mixing

$$\dot{m}_1 + \dot{m}_2 = \dot{m}_3 \tag{15-18}$$

$$\dot{m}_1 x_1 + \dot{m}_2 x_2 = \dot{m}_3 x_3 \tag{15-19}$$

The equation for the concentration x_3 is the same as Eq. (15-21), however, the enthalpy i_3 is given by

$$i_3 = i_1 + \frac{\dot{m}_2}{\dot{m}_3}(i_2 - i_1) - \frac{\dot{q}}{\dot{m}_3} \tag{15-25}$$

Equation (15-25) differs from Eq. (15-19) only in the last term. The significance of this is shown in the $i - x$ diagram of Fig. 15-26. Point 3′ represents a state that would occur with adiabatic mixing. Point 3 is located a distance $\dot{q}/\dot{m}_3$ directly below point 3′ because $x_{3'} = x_3$ and heat is removed. If heat were added, then point 3 would be above point 3′.

> **Example 15-7.**   Five lbm per minute of saturated liquid aqua-ammonia at 100 psia and 220 F are mixed with 10 lbm per min. of saturated vapor aqua-ammonia at 100 psia and 220 F. Heat transfer from the mixture is 4200 Btu/min. Find the enthalpy, temperature, and concentration of the final mixture.
>
> *Solution.*   The concentration of the mixture x_3 is given by Eq. (15-21). Using Chart 2, we get
>
> $$x_3 = 0.214 + \frac{10}{15}(0.845 - 0.214) = 0.635 \frac{\text{lbm ammonia}}{\text{lbm of mixture}}$$
>
> The total mass flow rate $\dot{m}_3$ is the sum of the two inlet flow rates and
>
> $$\frac{\dot{q}}{\dot{m}_3} = \frac{4200}{15} = 280 \text{ Btu/lbm}$$
>
> Then using Chart 2, point 3 is located at $x_3 = x_{3'} = 0.635$ and
>
> $$i_3 = i_{3'} - \frac{\dot{q}}{\dot{m}_3} = 578 - 280 = 298 \text{ Btu/lbm}$$
>
> Utilizing a straightedge and the equilibrium construction line, we find that the final temperature t_3 is about 148 F.

Heating and Cooling Processes

The absorption refrigeration cycle requires vaporization and condensation processes. For ammonia-water systems rectification is necessary to produce high purity ammonia vapor. This is done by alternate heating and cooling processes. Figure 15-27 shows a simplified arrangement to accomplish this. To better visualize the processes the liquid and vapor phases are separated following each heat exchanger.

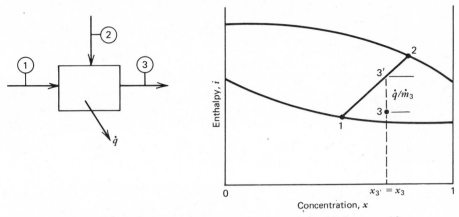

Figure 15-26.   Steady flow mixing of two streams with heat transfer.

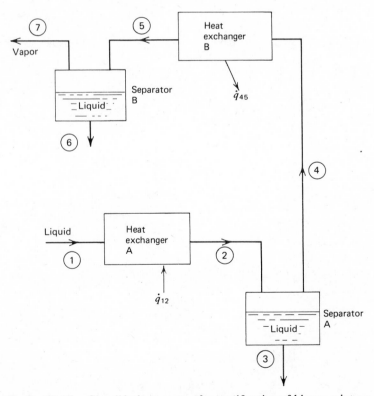

Figure 15-27.   Simplified apparatus for rectification of binary mixture.

For heat exchanger A

$$\dot{q}_{12} = \dot{m}_1(i_2 - i_1)$$
$$\dot{m}_1 = \dot{m}_2$$
$$x_1 = x_2$$

For separator A,

$$\dot{m}_2 i_2 = \dot{m}_3 i_3 + \dot{m}_4 i_4$$
$$\dot{m}_2 = \dot{m}_3 + \dot{m}_4$$
$$\dot{m}_2 x_2 = \dot{m}_3 x_3 + \dot{m}_4 x_4$$

Combination of the above energy and mass balance equations yields

$$\frac{\dot{m}_3}{\dot{m}_2} = \frac{x_4 - x_2}{x_4 - x_3} = \frac{i_4 - i_2}{i_4 - i_3} \tag{15-26}$$

$$\frac{\dot{m}_4}{\dot{m}_2} = \frac{x_2 - x_3}{x_4 - x_3} = \frac{i_2 - i_3}{i_4 - i_3} \tag{15-27}$$

Figure 15-28 shows the state points 1, 2, 3, and 4 on the $i - x$ diagram. The heat

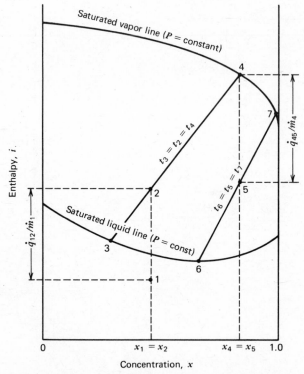

Figure 15-28. Schematic $i - x$ diagram for Fig. 15-27.

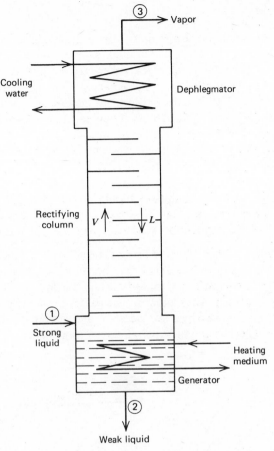

Figure 15-29. The rectifying column.

transfer $\dot{q}_{12}/\dot{m}_1$ may be determined graphically as shown, and the fractional components for the separator may be determined directly from the diagram.

$$\frac{\dot{m}_3}{\dot{m}_2} = \frac{\overline{24}}{\overline{34}} \quad \text{and} \quad \frac{\dot{m}_4}{\dot{m}_2} = \frac{\overline{32}}{\overline{34}}$$

The symbol $\overline{24}$ refers to the length of the line connecting points 2 and 4. Heat exchanger B and separator B may be analyzed in exactly the same way. The state points are shown in Fig. 15-28. Notice that the vapor at state 7 is almost 100 percent ammonia. This is necessary for the aqua-ammonia absorption cycle to operate efficiently. In actual practice the simple arrangement of Fig. 15-27 is inadequate for the separation of a binary mixture, and a rectifying column must be introduced between the two heat exchangers as shown in Fig. 15-29. Heat exchanger A is called the generator and heat exchanger B is referred to as the *dephlegmator*. Analysis of the dephlegmator is somewhat involved and the reader is referred to reference 2.

Throttling Process

The throttling process occurs in most refrigeration cycles. A throttling valve is shown schematically in Fig. 15-30a. Although evaporation occurs in throttling, and the temperature of the mixture changes, an energy balance gives $i_2 = i_1$ and the concentration remains constant $x_2 = x_1$. The state points 1 and 2 are identical on the $i - x$ diagram of Fig. 15-30b; however, it should be noted that state 1 is at pressure P_1 and state 2 is at a pressure P_2. The line $\overline{f2g}$ is located by trial and error using a straightedge and the equilibrium construction line. The temperature t_2 will generally be less than t_1. The fractional components of liquid and vapor may be determined from the line segment ratios of $\overline{f2g}$.

Example 15-8.   Saturated aqua-ammonia at 100 psia, 240 F, with a concentration of 0.3 lbm ammonia/lbm of mixture is throttled in a steady flow process to 30 psia. Find the temperature and fractions of liquid and vapor at state 2.

Solution.   State 1 is first located using Chart 2 and is shown schematically in Fig. 15-31. The isotherm for 240 F has been located using a straightedge. In this case state 1 is a two-phase mixture. State 2 is located at the same point as state 1 on the diagram. The isotherm that passes through point 2 for a pressure of 30 psia is then located by trial and error and is found to be 180 F as shown in Fig. 15-31. The fraction of vapor at state 2 is given by

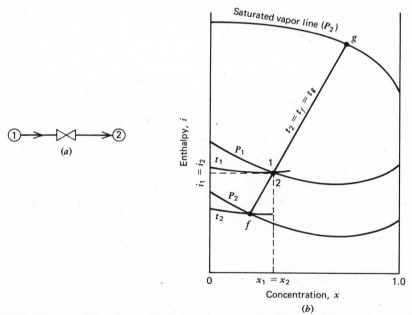

Figure 15-30.   Throttling of a binary liquid mixture under steady-flow conditions.

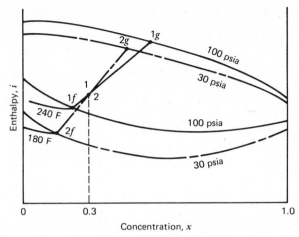

Figure 15-31.   Schematic of throttling process of Example 15-8.

$$\frac{\dot{m}_v}{\dot{m}} = \frac{\overline{22f}}{\overline{2f2g}} = \frac{2.2 \text{ in.}}{7.5 \text{ in.}} = 0.289 \, \frac{\text{lbm vapor}}{\text{lbm mixture}}$$

$$\frac{\dot{m}_f}{\dot{m}} = 1 - 0.289 = 0.711 \, \frac{\text{lbm liquid}}{\text{lbm mixture}}$$

15-9   THE THEORETICAL ABSORPTION REFRIGERATION SYSTEM

Figure 15-32 shows a schematic arrangement of components for a simple theoretical absorption cycle. For simplicity we shall assume that the absorbent does not vaporize in the generator; thus only the refrigerant flows through the condenser, expansion valve, and evaporator. The vapor leaving the evaporator is absorbed by the weak liquid in the absorber as heat is transferred from the mixture. The refrigerant-enriched solution is then pumped to the pressure level in the generator where refrigerant vapor is driven off by heat transfer to the solution while the weak solution returns to the absorber by way of the intercooler.

An ideal system would be completely reversible; however, this is not theoretically possible with the system of Fig. 15-32. The Carnot cycle meets this requirement for a vapor-compression system. The maximum attainable coefficient of performance for the absorption cycle may be determined following a method developed by Bosnjakovic (1).

By the first law of thermodynamics

$$\dot{q}_a + \dot{q}_c = \dot{q}_e + \dot{q}_g + \dot{W}_p \tag{15-28}$$

where the waste heat is

$$\dot{q}_o = \dot{q}_a + \dot{q}_c \tag{15-29}$$

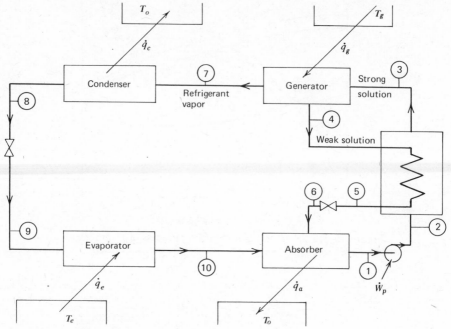

Figure 15-32. Simple absorption refrigeration system.

It will be assumed that the environment temperature T_o, the generator heating medium temperature T_g, and the refrigerated substance temperature T_e are all constant, absolute temperatures. The second law of thermodynamics requires that the net change in entropy for the system plus the surroundings must be greater than or equal to zero. Since the working fluid undergoes a cycle, its change in entropy is zero. Therefore,

$$\Delta S_{total} = \Delta S_g + \Delta S_e + \Delta S_o \geq 0 \qquad (15\text{-}30)$$

Because the reservoirs are internally reversible, their entropy changes may be computed as

$$\Delta S_g = -\frac{\dot{q}_g}{T_g}; \qquad \Delta S_e = -\frac{\dot{q}_e}{T_e}; \qquad \Delta S_o = \frac{\dot{q}_o}{T_o} \qquad (15\text{-}31)$$

and

$$\Delta S_{total} = -\frac{\dot{q}_g}{T_g} - \frac{\dot{q}_e}{T_e} + \frac{\dot{q}_o}{T_o} \geq 0 \qquad (15\text{-}32)$$

By combining Eqs. (15-31) and (15-32) it may be shown that

$$\dot{q}_g \frac{T_g - T_o}{T_g} \geq \dot{q}_e \frac{T_o - T_e}{T_e} - \dot{W}_p \qquad (15\text{-}33)$$

If the pump power $\dot{W}_p$ is neglected, Eq. (15-33) may be rearranged to give

$$COP = \frac{\dot{q}_e}{\dot{q}_g} \le \frac{T_e(T_g - T_o)}{T_g(T_o - T_e)} \qquad (15\text{-}34)$$

and when all the processes are reversible

$$(COP)_{max} = \frac{T_e(T_g - T_o)}{T_g(T_o - T_e)} \qquad (15\text{-}35)$$

Equation (15-35) shows that the maximum coefficient of performance for an absorption cycle is equal to the COP for a Carnot cycle operating between temperatures T_e and T_o multiplied by the efficiency of a Carnot engine operating between temperatures T_g and T_o. It is also shown that for a given environment temperature T_o the COP will increase with an increase in T_g or T_e. Unfortunately, practical absorption cycles have COPs much less than that given by Eq. (15-35).

15-10   THE AQUA-AMMONIA ABSORPTION SYSTEM

The aqua-ammonia system is one of the oldest absorption refrigeration cycles. The ammonia is the refrigerant while the water is the absorbent. Because both the water and ammonia are volatile, the generator of the simple cycle must be replaced by the combination of generator, rectifying column, and dephlegmator as shown in Fig. 15-33. This is necessary to separate almost all of the water vapor from the ammonia vapor. Note that one additional heat exchanger has also been added. A complete cycle will now be analyzed making use of the basic processes previously described.

Example 15-9.   Consider the cycle of Fig. 15-33 and the following given data: condensing pressure, 200 psia; evaporating pressure, 30 psia; generator temperature, 240 F; temperature of vapor leaving dephlegmator, 130 F; and temperature of the strong solution entering the rectifying column, 200 F. The heat exchanger lowers the temperature of the liquid leaving the condenser 10 F.

States 1, 3, 4, 7, 8, and 12 are saturated. Pressure drop in components and connecting lines is negligible. The system produces 100 tons of refrigeration. Determine (a) properties P, t, x, and i for all state points of the system, (b) mass flow rate for all parts of the system, (c) horsepower required for the pump assuming 75 percent mechanical efficiency, (d) system coefficient of performance, and (e) system refrigerating efficiency.

Solution.   Table 15-3 is a tabulation of thermodynamic properties and flow rates for the problem. Figure 15-34 is a schematic $i - x$ diagram showing the state points.

The given data establish all of the pressures, and the temperatures t_3, t_4, and t_7 are given. States 3, 4, and 7 are saturated states and can be located on Chart 2 and the concentration and enthalpy values read. States 7 through 12 have the same concentration because no mixing occurs. It is also true that states

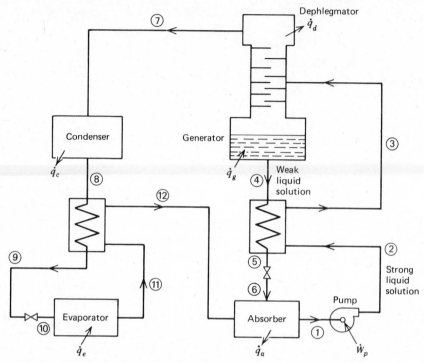

Figure 15-33. Industrial aqua-ammonia absorption system.

Table 15-3   PROPERTIES AND FLOW RATES FOR EXAMPLE 15-9

State	P psia	t F	x lbm NH$_3$/lbm	i Btu/lbm	ṁ lbm/min
1	30	79	0.408	−25	262.7
2	200	79	0.408	−24.4	262.7
3	200	200	0.408	109	262.7
4	200	240	0.298	159	221.3
5	200	97	0.298	0	221.3
6	30	97	0.298	0	221.3
7	200	130	0.996	650	41.4
8	200	97	0.996	148	41.4
9	200	86	0.996	137	41.4
10	30	0	0.996	137	41.4
11	30	40	0.996	620	41.4
12	30	57	0.996	631	41.4

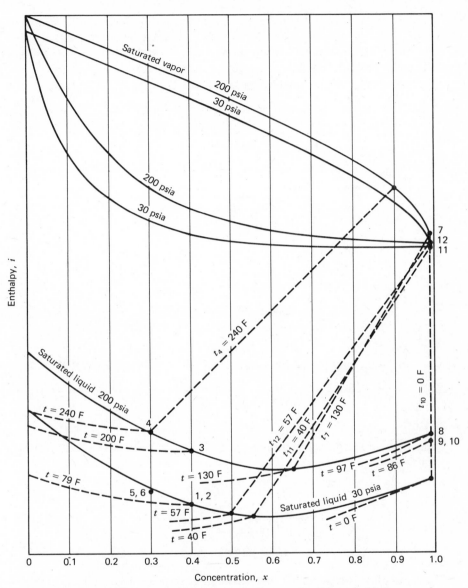

Figure 15-34.   Schematic $i - x$ diagram for Example 15-9.

1, 2, and 3 have the same concentration and $x_4 = x_5 = x_6$ as well. Now states 1, 8, and 12 are saturation states and may be located on Chart 2 making use of the known pressures and concentrations and i_1, i_8, and i_{12} read. Since $t_9 = t_8 - 10$, state 9 is determined from t_9 and x_9 and i_9 is read from the chart.

An energy balance between points 8 and 12 in Fig. 15-34 yields

$$\dot{m}_8 i_8 + \dot{q}_e = \dot{m}_{12} i_{12} = \dot{m}_8 i_{12}$$

$$\dot{m}_8 = \frac{\dot{q}_e}{i_{12} - i_8} = \frac{(100)(200)}{(631 - 148)} = 41.4 \text{ lbm/min}$$

From Fig. 15-33 it is obvious that $\dot{m}_7 = \dot{m}_8 = \dot{m}_9 = \dot{m}_{10} = \dot{m}_{11} = \dot{m}_{12}$. Mass balance on the absorber gives

$$\dot{m}_{12} + \dot{m}_6 = \dot{m}_1$$
$$\dot{m}_{12} x_{12} + \dot{m}_6 x_6 = \dot{m}_1 x_1$$

and

$$\dot{m}_6 = \dot{m}_{12} \frac{(x_{12} - x_1)}{(x_1 - x_6)} = 41.4 \frac{(0.996 - 0.408)}{(0.408 - 0.298)}$$

$$\dot{m}_6 = 221.3 \text{ lbm/min}$$

$$\dot{m}_1 = \dot{m}_{12} + \dot{m}_6 = 221.3 + 41.4 = 262.7 \text{ lbm/min}$$

From Fig. 15-34, $\dot{m}_1 = \dot{m}_2 = \dot{m}_3$ and $\dot{m}_4 = \dot{m}_5 = \dot{m}_6$.

The pump work may be expressed as

$$w_{12} = i_1 - i_2 = \frac{(P_1 - P_2) v_1}{J}$$

where $J = 778$ (ft-lbf)/Btu. The specific volume v_1 is an empirically determined quantity and is given in Table 45, Chapter 31 of reference 3. For $x_1 = 0.408$ and $t_1 = 79$ F, $v_1 = 0.0187$ ft^3/lbm. Then

$$w_{12} = \frac{144(30 - 200)(0.0187)}{778}$$

$$w_{12} = -0.588 \text{ Btu/lbm}$$

$$\dot{W}_p = w_{12} \dot{m}_1 = (-0.588)(262.7) = -154.6 \text{ Btu/min}$$

$$i_2 = i_1 - w_{12} = -25 - (-0.588) = -24.4 \text{ Btu/lbm}$$

The enthalpy, pressure, and concentration at state 2 then establish t_2.

The energy balance on the solution heat exchanger yields

$$\dot{m}_2 i_2 + \dot{m}_4 i_4 = \dot{m}_3 i_3 + \dot{m}_5 i_5$$
$$\dot{m}_4 = \dot{m}_5 \quad \text{and} \quad \dot{m}_2 = \dot{m}_3$$

Then

$$i_5 = i_4 - \frac{\dot{m}_2}{\dot{m}_4}(i_3 - i_2) = 159 - \frac{262.7}{221.3}[109 - (-24.4)]$$

$$i_5 = 0.64 \approx 0 \text{ Btu/lbm}$$

The temperature t_5 is then read from Chart 2 as a function of P_5, x_5, and i_5. The enthalpy $i_6 = i_5$ are coincident on Chart 2; however state 5 is at 200 psia while state 6 is at 30 psia. Since state 6 is subcooled, $t_6 = t_5$.

States 9 and 10 are coincident on Chart 2 since the process is one of throttling. Temperature t_{10} may be found using the method described in Example 15-7.

The energy balance on the evaporator gives

$$\dot{m}_{10}i_{10} + \dot{q}_e = \dot{m}_{11}i_{11}$$

$$\dot{m}_{10} = \dot{m}_{11}$$

$$i_{11} = \frac{\dot{q}_e}{\dot{m}_{10}} + i_{10}$$

$$i_{11} = \frac{(200)(100)}{41.4} + 137 = 620 \text{ Btu/lbm}$$

State 11 is a mixture of liquid and vapor and its temperature is found from Chart 2 to be 40 F.

The pump horsepower requirement may be computed from $\dot{W}_p$ above.

$$\text{Power} = \frac{\dot{W}_p}{\eta_m} = \frac{154.6(778)}{0.75(33,000)} = 4.86 \text{ hp}$$

To compute the coefficient of performance it is necessary to establish the generator heat transfer rate. However, this cannot be done without an analysis of the complete generator–rectifier column dephlegmator unit. To make a reasonable analysis, details of the column design and experimental data are required. For purposes of simplicity it will be assumed that the generator heat transfer rate is 115 percent of the net heat transfer for the complete rectifying unit. An energy balance gives

$$\dot{m}_3 i_3 + \dot{q}_g = \dot{m}_4 i_4 + \dot{m}_7 i_7 + \dot{q}_d$$
$$\dot{q}_{net} = (\dot{q}_g - \dot{q}_d) = \dot{m}_4 i_4 + \dot{m}_7 i_7 - \dot{m}_3 i_3$$
$$= 221.3(159) + 41.4(650) - 262.7(109)$$
$$\dot{q}_{net} = 33,462.4 \text{ Btu/min}$$

Then

$$\dot{q}_g = 1.15 \, \dot{q}_{net} = 38,480 \text{ Btu/min}$$
$$\dot{q}_d = 38,480 - 33,462 = 5020 \text{ Btu/min}$$

Neglecting the pump work

$$COP = \frac{\dot{q}_e}{\dot{q}_g} = \frac{100(200)}{38,480} = 0.52$$

The maximum COP may be calculated from Eq. (15-35). It will be assumed that $t_o = 80$ F and $t_g = 240$ F and $t_e = 0$ F were given.

$$(COP)_{max} = \frac{460(700 - 540)}{700(540 - 460)} = 1.31$$

and

$$\eta_R = \frac{COP}{(COP)_{max}} = \frac{0.52}{1.31} = 0.40$$

15-11   THE LITHIUM BROMIDE–WATER ABSORPTION SYSTEM

In recent times the lithium bromide–water absorption system has become popular for air-conditioning applications. Water is the refrigerant, whereas the lithium bromide is the absorbent. Atlhough lithium bromide is normally solid, when mixed with water a liquid solution may be formed.

The greatest advantage of the system is the nonvolatility of the lithium bromide. Only water vapor is driven off in the generator. The lithium bromide–water system is simpler and operates with a higher *COP* than the water–ammonia system. The main disadvantage is the relatively high evaporating temperatures and very low system pressures. Here is an example of a typical system analysis.

Example 15-10. The following data are given for a lithium bromide–water system of the type shown in Fig. 15-32: condensing temperature, 100 F; evaporating temperature, 40 F; temperature of strong solution leaving absorber, 100 F; temperature of strong solution entering generator, 180 F; and generator temperature, 200 F.

Saturated conditions exist for states 3, 4, 8, and 10. Pressure drop in components and lines is negligible. The system cooling capacity is 10 tons. Determine (a) the thermodynamic properties of *P, t, x,* and *i* for all necessary state points, (b) the mass flow rate for each part of the system, (c) the system coefficient of performance, (d) the system refrigerating efficiency, and (e) the steam rate for the generator if saturated steam at 220 F is used.

Solution. Parts (a) and (b) will be solved concurrently. Table 15-4 shows a tabulation of data. Properties of pure water were obtained from Table A-1, whereas solution properties were obtained from Chart 3. Note that Chart 3 expresses concentration in terms of the absorbent lithium bromide.

The high and low side pressures are found from the temperatures t_8 and t_{10}, and i_8 and i_{10} are found in Table A-1. The enthalpy of the pure water vapor at state 7 may be computed by Eq. (2-18) because it behaves as an ideal gas at the low pressure. States 3 and 4 are saturated conditions and may be located on Chart 3 as shown in Fig. 15-35. Enthalpy and concentration values may then be read for states 3 and 4.

For the evaporator

$$\dot{m}_9 = \frac{\dot{q}_e}{(i_{10} - i_9)} = \frac{10(200)}{1011} = 1.98 \text{ lb/min}$$

Table 15-4   PROPERTIES AND FLOW RATES FOR
EXAMPLE 15-10

State	P mm/Hg	t F	x lb LiBr/lbm	i Btu/lbm	$\dot{m}$ lbm/min
1	6.3	100	0.6	—	25.8
2	49.1	—	0.6	—	25.8
3	49.1	180	0.6	−35	25.8
4	49.1	200	0.65	−27	23.8
5	49.1	—	0.65	—	23.8
6	6.3	—	0.65	—	23.8
7	49.1	200	0	1151	1.98
8	49.1	100	0	68	1.98
9	6.3	40	0	68	1.98
10	6.3	40	0	1079	1.98

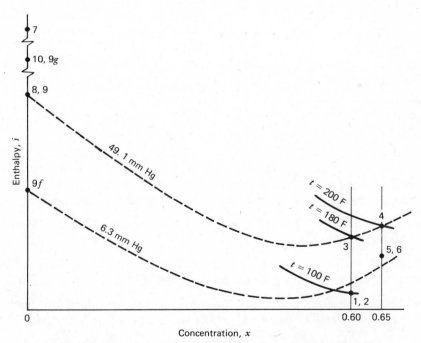

Figure 15-35.   Schematic $i - x$ diagram for Example 15-10.

For the absorber

$$\dot{m}_6 = \dot{m}_{10} \frac{(x_1 - x_{10})}{(x_6 - x_1)} = \frac{1.98(0.60)}{0.05} = 23.8 \text{ lb/min}$$

$$\dot{m}_1 = \dot{m}_6 + \dot{m}_{10} = 25.8 \text{ lb/min}$$

For the generator

$$\dot{q}_g = \dot{m}_4 i_4 + \dot{m}_7 i_7 - \dot{m}_3 i_3$$
$$= 23.8(-27) + 1.98(1151) - 25.8(-35)$$
$$\dot{q}_g = 2538 \text{ Btu/min}$$

Neglecting the pump work

$$COP = \frac{\dot{q}_e}{\dot{q}_g} = \frac{200(10)}{2538} = 0.79$$

For a generator source temperature of 220 F, refrigerated medium temperature of 45 F, and assuming an environment temperature of 100 F the maximum COP becomes

$$(COP)_{max} = \frac{505(680 - 560)}{680(560 - 505)} = 1.62$$

and

$$\eta_R = \frac{0.79}{1.62} = 0.49$$

Assuming saturated water at 220 F leaves the steam coil

$$\dot{m}_s = \frac{60\dot{q}_g}{i_{fg}} = \frac{60(2538)}{965} = 158 \text{ lb/hr}$$

REFERENCES

1. Fran Bosnjakovic, *Technical Thermodynamics,* translated by Perry L. Blackshear, Jr., Holt, Rinehart, and Winston, New York, 1965.

2. James L. Threlkeld, *Thermal Environmental Engineering,* Second Edition, Prentice-Hall, Englewood Cliffs, N.J., 1970.

3. *ASHRAE Handbook of Fundamentals,* American Society of Heating, Refrigerating and Air-Conditioning Engineers, New York, 1977.

4. *ASHRAE Handbook and Product Directory Equipment,* American Society of Heating, Refrigerating and Air-Conditioning Engineers, New York, 1979.

5. W. F. Stoecker, *Refrigeration and Air Conditioning,* McGraw-Hill, New York, 1958.

PROBLEMS

15-1 A vapor compression refrigeration cycle operates on R-22 and follows the theoretical single-stage cycle. The condensing temperature is 110 F and the evaporating temperature is 40 F. The system produces 10 tons of cooling effect. Determine the (a) coefficient of performance, (b) refrigerating efficiency, (c) horsepower per ton of refrigeration, (d) mass flow rate of the refrigerant, (e) theoretical horsepower input to the compressor, and (f) theoretical piston displacement of the compressor in ft^3/min.

15-2 A vapor compression refrigeration cycle uses R-22 and follows the theoretical single stage cycle. The condensing temperature is 48 C and the evaporating temperature is −18 C. Power input to the cycle is 2.5 kW and the mass flow rate of refrigerant is 0.05 kg/s. Determine (a) heat rejected from the condenser, (b) the coefficient of performance, (c) the enthalpy at the compressor exit, and (d) the refrigerating efficiency.

15-3 An R-12 system is arranged as shown in Fig. 15-36. Compression is isentropic. Assume frictionless flow. Find the system horsepower/ton.

15-4 Consider a 4 cylinder, 3 in. bore by 4 in. stroke, 800 rpm, single-acting compressor for use with R-12. Proposed operating conditions for the compressor are 100 F condensing temperature and 40 F evaporating temperature. It is estimated that the refrigerant will enter the expansion valve as a saturated liquid, that vapor will leave the evaporator at a temperature of 45 F, and that vapor will enter the compressor at a temperature of 55 F. Assume a compressor volumetric efficiency of 70 percent and frictionless flow. Calculate the refrigeration capacity in tons.

15-5 Saturated R-22 at 45 F enters the compressor of a single stage system. Discharge pressure is 275 psia. Suction valve pressure drop is 2 psi. Discharge valve pressure drop is 4 psi. Assume the vapor is superheated 10 F in the cylinder during the intake

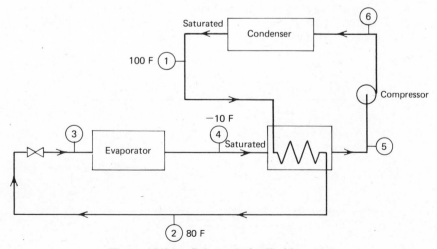

Figure 15-36.   Schematic for Problem 15-3.

stroke. Piston clearance is 5 percent. Determine the (a) volumetric efficiency, (b) pumping capacity in lbm/min for 20 ft³/min piston displacement, and (c) horsepower requirement if mechanical efficiency is 80 percent.

15-6 Consider the single stage vapor compression cycle shown in Fig. 15-37. Design conditions using R-12 are:

$$\dot{q}_L = 30,000 \text{ Btu/hr} \qquad P_3 = 200 \text{ psia}$$
$$P_1 = 60 \text{ psia saturated} \qquad P_3 - P_4 = 2 \text{ psi}$$
$$P_2 = 55 \text{ psia} \qquad C = 0.04$$
$$T_2 = 60 \text{ F} \qquad \eta_m = 0.90$$
$$PD = 9.4 \text{ cfm}$$

(a) Determine $\dot{W}$, $\dot{q}_H$, $\dot{m}_{12}$, and sketch the cycle on a P vs. i diagram. If the load $\dot{q}_L$ decreases to 24,000 Btu/hr and the system comes to equilibrium with $P_2 = 50$ psia and $T_2 = 50$ F, (b) determine $\dot{W}$, $\dot{q}_H$, $\dot{m}$ and locate the cycle on a P vs. i diagram.

15-7 Consider an ordinary single stage vapor compression air-conditioning system. Because of clogged filters the air flow over the evaporator is gradually reduced to a very low level. Explain how the evaporator and compressor will be affected if the system continues to operate.

15-8 A vapor compression cycle is subject to short periods of very light load; it is not practical to shut the system down. During these periods of light load moisture condenses from the air flowing over the evaporator and freezes. Suggest a modification to the system to prevent this condition.

15-9 A vapor compression cycle is subject to occasional overload that leads to the tripping of circuit breakers. Explain how the system can be modified to prevent compressor overload without shutting the system off.

15-10 Saturated water vapor at 50 F is mixed in a steady flow process with a saturated lithium bromide–water solution having a concentration of 0.60 lbm Li Br per lbm of mixture. The mass of the liquid solution is five times the mass of the water vapor mixed. The mixing process occurs at constant pressure. Find (a) the concentration of the resulting mixture, and (b) the heat that must be removed in Btu/lbm of the final mixture if a saturated liquid solution is produced.

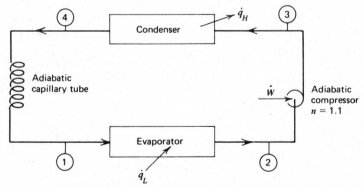

Figure 15-37. Schematic for Problem 15-6.

15-11 A saturated liquid aqua-ammonia solution at 220 F and 200 psia is throttled to a pressure of 10 psia. Find (a) the temperature after the throttling process, and (b) the relative portions of liquid and vapor in the mixture after throttling.

15-12 A solution of ammonia and water at 180 F, 100 psia, and with a concentration of 0.25 lbm ammonia per lbm of solution is heated at constant pressure to a temperature of 280 F. The vapor is then separated from the liquid and cooled to a saturated liquid at 100 psia. What are the temperature and concentration of the saturated liquid?

15-13 It is proposed to use hot water at 180 F or 82 C from a solar collector system to operate a simple absorption cycle. Compute the maximum possible coefficient of performance assuming an environment temperature of 100 F or 38 C and an air-cooled evaporator with the air temperature at 75 F or 24 C.

15-14 Solve Example 15-10 using the following conditions:
Condensing temperature, 95 F
Evaporating temperature, 45 F
Strong solution temperature leaving absorber, 105 F
Strong solution temperature entering generator, 180 F
Generator temperature, 200 F

15-15 Solar energy is to be used to operate a lithium bromide-water absorption system. The cooling capacity is 3 tons. Analize the cycle following the approach used in Example 15-10 and the following conditions:
Condensing temperature, 100 F
Evaporating temperature, 40 F
Strong solution temperature leaving absorber, 105 F
Strong solution temperature entering generator, 140 F
Generator temperature, 160 F

CHAPTER SIXTEEN

Solar Heating and Cooling

An introduction to solar radiation was given in Chapter 5 and should be referred to for the definition of terms and basic information needed to understand the material that follows.

The shortage of fossil fuels that threatens the world and the rapid increase in fuel prices have caused a tremendous interest in the utilization of solar energy for space heating and cooling. It has been estimated that about 20 percent of the total energy consumed in the United States is for space heating and cooling; therefore, very large energy savings can be accomplished if solar energy can replace a significant part of this requirement. The rate of growth of this new industry will depend upon the rate at which fuel costs increase and upon the development of efficient and reliable equipment that can be built and installed at low cost. At the present time it appears that solar energy is economically feasible for water and space heating in some parts of the country and will become more competitive with time. Solar cooling is less competitive economically but this may improve with new technological breakthroughs and with a very large increase in fuel costs.

16-1   THE FLAT PLATE COLLECTOR

A common way to gather solar energy is to use a flat plate collector to heat either a liquid or a gas. A simple flat plate collector for heating liquid is shown in Fig. 16-1. A blackened or specially treated metal surface called the absorber plate is the heart of the collector. The sun's rays striking the absorber plate are absorbed and raise the plate temperature. Fluid in intimate contact with the plate is heated as it passes through the collector. Heat losses to the surroundings are kept small by the use of insulation behind the absorber plate and by one or more transparent covers, usually glass or plastic. The covers permit the relatively short wavelength radiation from the sun to enter the collector, but absorb the relatively long thermal radiation emitted by the absorber plate. This is sometimes called the greenhouse effect. The transparent covers also reduce the heat loss by convection to the surrounding atmo-

580

sphere. The number of transparent covers is determined by the desired operating temperature of the collector with two covers used only for temperatures above about 150 degrees F and one cover for lesser temperatures. In some cases, such as with swimming pool heaters where collector temperatures are quite low, it may be desirable to omit the cover.

Flat plate solar collectors are usually tilted from the horizontal plane toward the equator in order to collect the maximum amount of solar radiation. The best angle of tilt α measured from the vertical places the collector at right angles to sun's rays at solar noon. This best angle depends upon the latitude and the time of the year. Since most flat plate collectors must be fixed at some angle, there is an optimum fixed tilt angle for a given latitude that depends upon whether one wishes to optimize for winter heating, summer cooling, or for year-around collection.

For maximum year around insolation it is suggested that the collector be tilted at an angle Σ ($\Sigma = 90$-α) from the horizontal such that $\Sigma = 0.9l$ where l is the local latitude. For maximum wintertime insolation the best angle of tilt is approximately given by $\Sigma = l + 10$ degrees and for summertime the best tilt is approximately $\Sigma = l - 10$ degrees.

A few degrees deviation in Σ from the optimum or in the wall solar azimuth γ from straight south does not seriously affect the performance of a collector array.

The relationship between θ, the collector angle of tilt α, the solar altitude β, and the wall solar azimuth γ was given in Eq. (5-5a).

A similar relationship between θ, the latitude l, the sun's declination d, and the hour angle h can be derived for a south-facing collector tilted at an angle Σ from the horizontal, located in the northern hemisphere.

$$\cos \theta = \cos (l - \Sigma) \cos d \cos h + \sin (l - \Sigma) \sin d \qquad (16\text{-}1)$$

By integration of Eq. (16-1) it can be shown that the hourly radiation rate on a collector tilted to the south is given by

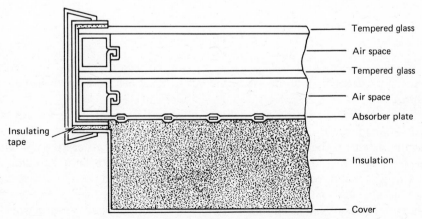

Figure 16-1. Cross section of a flat plate solar collector.

$$G_c = G_O \left(\frac{12}{\pi} \sin \frac{\pi}{24} \cos (l - \Sigma) \cos d \cos h + \sin (l - \Sigma) \sin d \right) \qquad (16\text{-}2)$$

where G_O = extraterrestrial radiation rate at that hour and h is the hour angle at the midpoint of the hour of interest. Equation (16-2) does not consider any absorption or scattering of the sun's rays by the atmosphere.

The hour angle at which the sun falls to the horizon (sunrise h_{SR} or sunset h_{SS}) can be calculated by assuming a horizontal surface, $\Sigma = 0$, in Eq. (16-1) and setting $\cos \theta$ equal to zero. The result is, since $h_{SS} = h_{SR}$,

$$\cos h_{SS} = \cos h_{SR} = -\tan l \tan d \qquad (16\text{-}3)$$

Similar quantities for a south-facing tilted surface are h'_{SS} and h'_{SR} where $h'_{SS} = h'_{SR}$:

$$\cos h'_{SS} = \cos h'_{SR} = -\tan (l - \Sigma) \tan d \qquad (16\text{-}4)$$

The daily radiation G_{OCD} incident upon a collector tilted to the south is given by

$$G_{OCD} = \frac{24}{\pi} G_O[\cos (l - \Sigma) \cos d \sin h_{SS} + h_{SS} \sin (l - \Sigma) \sin d]$$

$$\text{where} \qquad h_{SS} \leq h'_{SS} \qquad (16\text{-}5)$$

or

$$G_{OCD} = \frac{24}{\pi} G_O[\cos (l - \Sigma) \cos d \sin h_{SS} + h'_{SS} \sin (l - \Sigma) \sin d]$$

$$\text{when} \qquad h_{SS} \geq h'_{SS} \qquad (16\text{-}6)$$

The ratio of the daily radiation on a collector tilted at an angle Σ from horizontal and facing toward the south to the daily radiation on a horizontal surface is given by

$$\frac{G_{OCD}}{G_{OHT}} = \frac{\cos (l - \Sigma)(\sin h_{SS} - h_{SS} \cos h'_{SS})}{\cos l (\sin h_{SS} - h_{SS} \cos h_{SS})} \qquad h_{SS} \leq h'_{SS} \quad (16\text{-}7a)$$

$$= \frac{\cos (l - \Sigma)(\sin h'_{SS} - h'_{SS} \cos h'_{SS})}{\cos l (\sin h_{SS} - h_{SS} \cos h_{SS})} \qquad h_{SS} \geq h'_{SS} \quad (16\text{-}7b)$$

Values of solar declination for use in the above equations were previously given in Chapter 5.

The values of the hourly and daily extraterrestrial radiation on a south facing tilted surface given by Eq. (16-2), (16-5), and (16-6) are not used to design solar collector systems without proper correction for atmospheric absorption and cloudiness. They can be used with measured data with reasonable accuracy to determine the ratio of actual direct daily radiation on a tilted surface to the actual direct daily radiation on a horizontal surface. Allowance must be made for the portion of the incoming radiation that is diffuse rather than direct. A tilted surface will see a smaller portion of the sky and more ground objects than a horizontal surface.

Table 16-1 STILLWATER, OKLAHOMA AVERAGE
DAILY SOLAR IRRADIATION ON A
HORIZONTAL SURFACE

Month	cal/cm²	Btu/ft²	kW-hr/m²	MJ/m²
January	205	756	2.38	8.58
February	289	1065	3.36	12.1
March	390	1438	4.53	16.3
April	454	1674	5.28	19.0
May	504	1858	5.86	21.1
June	600	2212	6.98	25.1
July	596	2197	6.93	24.9
August	545	2009	6.34	22.8
September	455	1677	5.29	19.0
October	354	1305	4.12	14.8
November	269	992	3.13	11.3
December	209	770	2.43	8.75
Average	406	1496	4.72	17.0

The ratio of diffuse radiation to direct radiation for a horizontal surface depends on the cloudiness. It might, for example, be as low as 12 percent on a clear day and nearly 100 percent on a very cloudy day (7).

The approach to determining the ratio of diffuse to direct radiation, and how each type of radiation is treated was given in Chapter 5.

Table 16-1 gives the average daily solar irradiation on a horizontal surface at Stillwater, Oklahoma, for each month of the year. Weather data are typically collected and tabulated for a horizontal surface and the values desired for a tilted collector surface must often be calculated. Reference (6) is particularly useful for making such calculations.

Example 16-1. Using the data in Table 16-1 determine the average daily radiation on a south-facing solar collector surface tilted at 45 degrees located at Stillwater, Oklahoma, for the month of January. Neglect ground reflection.

Solution. For Stillwater, latitude 36 degrees in January the declination is -21 degrees 5 minutes $= -21.08$ degrees. The sunset hour angle h_{SS} is given by Eq. (16-3):

$$\cos h_{SS} = -\tan 36 \tan (-21.08)$$
$$\cos h_{SS} = -(0.726)(-0.385) = +0.280$$
$$h_{SS} = 73.74 \text{ degrees } (4.916 \text{ hrs} = 4{:}55 \text{ P.M.})$$
$$h_{SS} = 1.28 \text{ radians}$$

for the tilted surface

$$\cos h'_{ss} = -\tan (l - \Sigma) \tan d = -\tan (-9) \tan (-21.08)$$
$$\cos h'_{ss} = -(-0.1584)(-0.385) = -0.0610$$
$$h'_{ss} = 93.5 \text{ degrees} = 1.63 \text{ radians}$$

because $h'_{ss} > h_{ss}$ use Eq. (16-7a)

$$\frac{G_{OCD}}{G_{OHD}} = \frac{\cos (l - \Sigma)(\sin h_{ss} - h_{ss} \cos h'_{ss})}{\cos l (\sin h_{ss} - h_{ss} \cos h_{ss})}$$

$$\frac{G_{OCD}}{G_{OHD}} = \frac{\cos (-9)(\sin (74.74) - 1.28 \cos (93.5))}{\cos (36)\sin (73.74) - 1.28 \cos (73.74))}$$

$$= \frac{(0.988)(0.960 + (1.28)(0.0610))}{(0.809)(0.960 - (1.28)(0.280))}$$

$$= 2.10$$

From Table 16-1, $G_{OHD} = 756$ Btu/ft^2. The extraterrestrial radiation on a horizontal surface is required so that Fig. (5-10) can be used to determine the ratio of diffuse and direct radiation. Either Eq. (16-5) or (16-6) may be used to determine this extraterrestrial radiation because $h_{ss} = h'_{ss}$. G_O is determined from Table 5-3. For a horizontal surface $\Sigma = 0$ and

$$G_{OCD} = \frac{24}{\pi} G_o[\cos l \cos d \sin h_{ss} + h_{ss} \sin l \sin d]$$

$$= \frac{24}{\pi} (442.7)[\cos (36) \cos (-21.08) \sin [(1.28)(57.3)]$$

$$+ (1.28) \sin (36) \sin (-21.08)]$$

$$= (3382)[(0.809)(0.933)(0.958) + (1.28)(0.588)(-0.360)]$$

$$= 1528 \frac{\text{Btu}}{\text{hr ft}^2}$$

Then, to use Fig. 5-10

$$\frac{756}{1528} = 0.495$$

and from Fig. 5-10 the diffuse fraction is 0.38. The direct or beam component is then

$$(1 - 0.38)(756) = 469 \text{ Btu/day ft}^2$$

Assuming that the ratio of direct insolation on a tilted surface to that on a horizontal surface is the same on the earth's surface as it is outside the earth's atmosphere, the total daily direct irradiation on the tilted surface is

$$(468)(2.10) = 983 \text{ Btu/(day ft}^2)$$

The diffuse radiation on the tilted collector is found using a configuration factor from Eq. (5-14)

$$G_{CdD} = (287)(1 + \cos 45)/2 = 245 \text{ Btu/(day-ft}^2)$$

The total radiation on the collector is the sum of the direct and diffuse components

$$G_{CD} = 983 + 245 = 1228 \text{ Btu/(day-ft}^2)$$

The collector performance may be evaluated in terms of instantaneous collector efficiency, the ratio of the energy collected to that incident on the collector, usually expressed as a percent. At a fixed rate of solar insolation the collector efficiency of a given collector decreases with the temperature difference between the collector and the surrounding air. Thus, there is a tradeoff between temperature of collection and amount of energy collected. If high collection temperatures are desired, a larger amount of collector surface would be needed than is required to gather the same amount of energy at a lower collection temperature. Because the major cost of most solar energy systems is in the collectors, it is important to keep both the unit cost of the collectors and the total amount of collector surface as small as possible.

Typical efficiency curves for flat plate collectors are shown in Figs. 16-2 and 16-3. Figure 16-4 shows the effect of varying the amount of solar insolation on a collector and also the effect of variation in ambient temperature. It can be seen that for larger amounts of radiation striking a collector surface, higher efficiencies are obtained at a fixed temperature difference between absorber plate and ambient air. It can also be seen that for a fixed absorber plate temperature the collectors perform better on warmer days.

Example 16-2.   Compare the instantaneous collector efficiency of the *PPG* standard collector shown in Fig. 16-2 for ambient temperatures of 0 and 50 F. In both cases assume that the absorber plate temperature is 200 F and the solar insolation rate is 300 Btu/(hr-ft²).

Solution.   For an ambient temperature of 0 F and an absorber plate temperature of 200 F the temperature difference is 200 F. From Fig. 16-2*a* with G_{ND} = 300 Btu/(hr-ft²), $\eta = 0.2$. For an ambient temperature of 50 F the temperature difference is 150 F and from Fig. 16-2*b*, $\eta = 0.30$. Thus, the collector would gather 50 percent more energy with the higher ambient temperature.

Ordinarily a surface that is a good absorber is also a good emitter. Collector efficiency can be improved by the use of a selective surface, one with a high absorptance for sunlight but a low radiation emittance. Selective surfaces are usually prepared by special plating or deposition processes. At a fixed collector temperature, collector efficiency may be more than doubled over that of an ordinary surface when selective surfaces are utilized. Selective surfaces also permit the collection of energy at a higher than normal temperature without as large a decrease in collector efficiency as would occur for a nonselective absorber surface. The effect of varying the number of cover plates on the collector and the effect of changing the absorber surface from a nonselective to a selective coating is seen in Fig. 16-3 (2). Here the efficiency is obtained from a theoretical analysis and is plotted as a function of inlet fluid temperature.

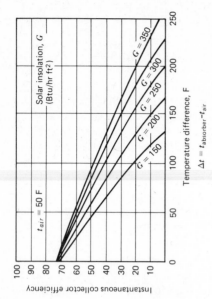

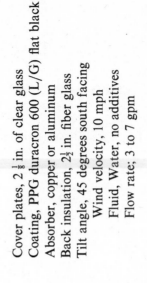

Cover plates, 2 $\frac{1}{8}$ in. of clear glass
Coating, PPG duracron 600 (L/G) flat black
Absorber, copper or aluminum
Back insulation, 2$\frac{1}{2}$ in. fiber glass
Tilt angle, 45 degrees south facing
Wind velocity, 10 mph
Fluid, Water, no additives
Flow rate; 3 to 7 gpm

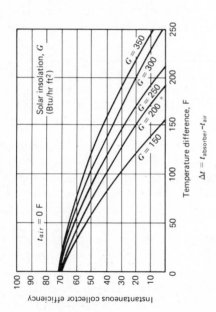

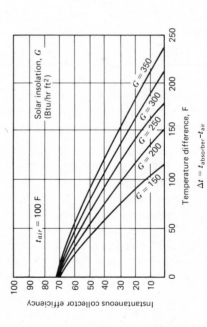

Figure 16-2. Instantaneous efficiency of PPG standard solar collector for various ambient temperatures and rates of insolation. (Courtesy of PPG Industries, Inc., Pittsburgh, Pa.)

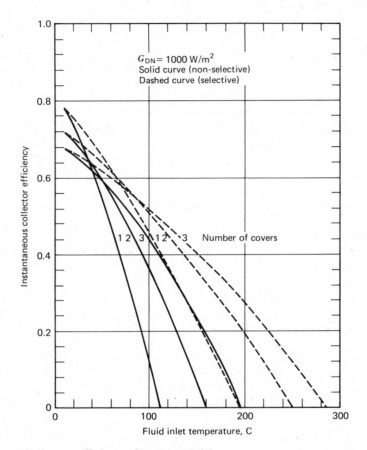

$G_{DN} = 1000 \ W/m^2$
Solid curve (non-selective)
Dashed curve (selective)

1 2 3 1 2 3   Number of covers

Instantaneous collector efficiency

Fluid inlet temperature, C

Collector efficiency factor = 0.95
Collector heat removal factor = 0.90
Tilt = 45 degrees; Wind velocity = 5 m/s
Ambient temperature = 10 C; Sky temperature = 10 C
Emittance of nonselective absorber = 0.95
Emittance of selective absorber = 0.10
Transmittance absorptance product
    1 cover 0.87, 2 covers 0.80, 3 covers 0.75
No gas absorptance

Figure 16-3.   Instantaneous collector efficiency at various fluid inlet temperatures for selective and nonselective absorber surface and varying number of cover plates. (Reprinted by permission from *Solar Energy Thermal Processes,* by John A. Duffie and W. A. Beckman. © 1974 by John Wiley and Sons, Inc., New York.)

A more common way of plotting collector efficiency is shown in Fig. 16-4, where the abscissa is $(T_i - T_a)/G_C$ and

$$T_i = \text{temperature of fluid at inlet to collector}$$
$$T_a = \text{ambient temperature}$$
$$G_C = \text{irradiation on the collector}$$

The units may be English or SI, in Fig. 16-4 English units are used.

This type of efficiency curve is useful because the parameters given above are generally known or specified. The data for most collectors plot as a straight line over the range of interest. The vertical intercept of the efficiency line can be shown to be equal to $(F_R\tau\alpha)$ and the slope of the line is $-F_R U_L$ where

τ = transmittance of glazing
α = absorptance of absorber plate
F_R = collector heat removal factor, ratio of the actual rate of heat transfer from
    collector to working fluid to that which would occur if all of the collector
    were at the inlet fluid temperature
U_L = overall loss coefficient of the collector, a measure of its rate of heat loss to
    the surroundings for a given temperature difference between the collector
    and surroundings

Notice that once the parameters F_R, U_L, τ, and α are specified for a collector, an efficiency curve similar to Fig. 16-4 can be drawn. Curves such as Fig. 16-4 are usually given for normal incidence of the sunlight. For non-normal incidence of the sun's rays the performance would be described by a line below and parallel to the given line.

An alternative type of solar collector to the liquid heater shown in Fig. 16-1 is the air heating collector. Because air does not freeze and because it can leak in small

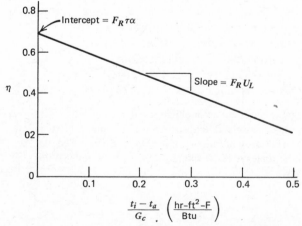

Figure 16-4. A typical flat plate solar collector efficiency curve.

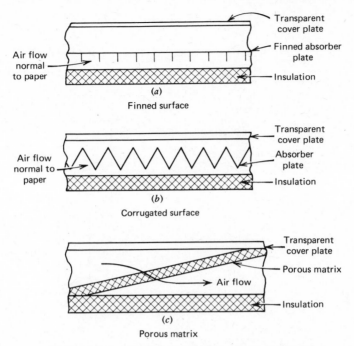

Figure 16-5.   Typical air heating collector designs.

amounts without doing damage, it is a popular type of collector fluid with many engineers and builders. Because air collectors require a relatively large surface area for convective heat transfer, their design is different from that of liquid collectors. Typical air collector designs are shown in Fig. 16-5.

16-2   SOLAR HEATING SYSTEMS

A simple solar energy space heating system using water as the collecting medium is shown in Fig. 16-6. A storage tank is needed to permit utilization of the collected energy at night and during brief periods of cloudiness. The size of the solar collector and the storage tank is usually determined by economics. In the usual situation the system is not capable of providing all of the energy required, and a full-size auxiliary back-up system is necessary. The warm water from the storage tank is passed through a heat exchanger placed in the air duct upstream of a conventional furnace. The fan in the furnace furnishes the air flow whenever heat is required, and the furnace is regulated to be turned on if the solar system is not meeting the demand.

The big drawback with the system shown is that water will freeze in the collector when there is no insolation and outside temperatures are low. The large volume of the storage tank (500 to 1000 gallons) precludes the use of antifreeze. In some systems this freezing problem is solved by dumping the collector liquid to the storage

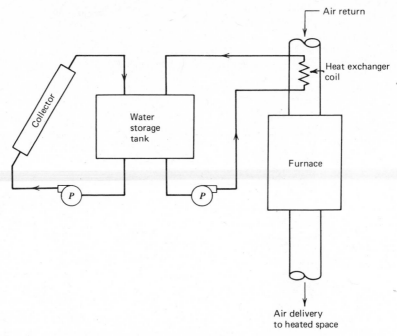

Figure 16-6. Solar heating system with water storage.

tank whenever the circulating pump is off. Figure 16-7 shows how the freezing problem may be easily overcome by the use of a heat exchanger coil in the storage tank, isolating the pure water in the tank from the ethylene glycol mixture passing through the collector.

Some engineers and builders favor the use of the air-type collector and storage systems such as can be seen in Fig. 16-8. The design shown utilizes only one blower, but is capable of carrying out all of the necessary modes of operation for direct heating, storing, and reclaiming heat from storage. Rock beds are the most common storage devices for air systems. It is very important that the air pass through the rock bed in one direction (down) during storage and the opposite direction during removal; otherwise during the removal cycle the air will simply transfer heat from one end of the rock bed to the other.

Research has shown that rocks of about $1-1\frac{1}{2}$ in. diameter are of optimum size, giving the best combination of surface area for heat transfer, internal thermal resistance, and fluid pressure drop through the bed.

A major problem with air systems is that the blower can consume a prohibitive amount of electricity as a result of the rock bed and the extensive duct system required.

The system shown in Fig. 16-9 utilizes a heat pump that draws its energy from a storage tank. The combination of heat pump and a solar system leads to the possibility of high collector efficiencies, high *COP* of the heat pump, and the possibility

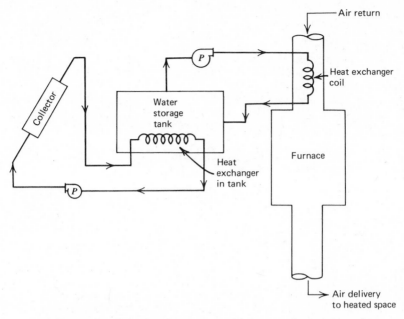

Figure 16-7. Solar heating system with heat exchanger in tank.

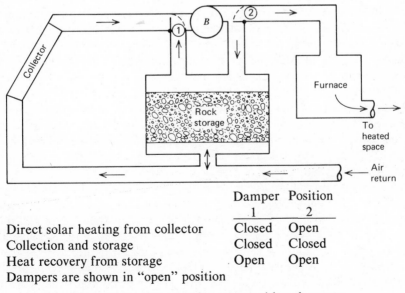

	Damper Position	
	1	2
Direct solar heating from collector	Closed	Open
Collection and storage	Closed	Closed
Heat recovery from storage	Open	Open

Dampers are shown in "open" position

Figure 16-8. Solar heating system with rock storage.

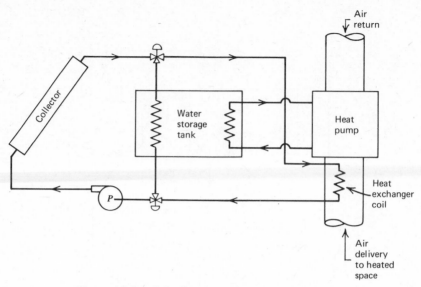

Figure 16-9.   Solar heating system with heat pump.

of storing energy to provide heating directly from the storage tank (without electricity) during hours when there is a heavy demand on the utility.

Some useful information on preliminary design and economic analysis of solar energy systems is given in references 9–13. In the actual design of a solar heating or cooling system, a computer analysis is usually necessary to determine the optimum size of collector and storage although there are some excellent computational procedures available that use "hand" calculations. One of the best known of these methods is the F-chart method (4). The single most important independent variable in the computation, the rate of solar irradiation, varies hourly, usually in very random manner. The calculation must therefore consider not only a typical day but a typical sequence of days from the standpoint of sunlight and ambient temperature.

The design involves an economic analysis and optimization with particular emphasis on the area of the solar collector system, since collectors are the most expensive part of a typical solar heating or cooling system. A system that is too large requires investment in surplus collector surface that is not effective in reducing fuel bills. On the other hand, a solar system with inadequate collector area will have too large an investment in pumps, controls, piping, and storage for the amount of energy collected. An optimum design results in a system with the least total operating cost, considering both amortization of the investment and the annual fuel and maintenance costs. The solar system for the optimum case will usually furnish only a fraction, such as 60 percent, of the total energy needs for a structure.

Generally in the design of a solar heating or cooling system it is presumed that there will be a full-sized back-up system using conventional fuel to allow for either long periods without sunlight or to permit operation when the solar system may fail to function. This full-sized back-up system may also furnish a portion of the energy not supplied by the solar system during routine operation under average conditions.

Any economic analysis and optimization study for a solar heating and cooling system must be carried out in combination with a thermal analysis of the building. Generally this combined computational scheme is carried out by use of a digital computer. When a building is poorly designed from a thermal standpoint or poorly insulated, often the most economical approach is to modify the building rather than to invest in the excessively large solar system that might be required. For example, the addition of insulation or storm windows on a building might be much less expensive than the additional solar collector area that would be required to take care of the higher thermal losses or gains.

Systems described in this chapter are often called "active" solar systems because they use pumps, blowers, and controls to store and utilize the solar energy collected by solar collectors. "Passive" solar design involves designing the structure itself so that it takes maximum advantage of the sunlight and other factors by "natural" means, that is, without pumps and blowers. Passive designs utilize the material of Chapter 5 extensively to help locate windows, overhangs, and shading devices in the optimum way. Reference 14 gives more detailed information on both the concept and the analysts.

REFERENCES

1. *ASHRAE Handbook and Product* Directory—Applications, American Society of Heating, Ventilating, and Air-Conditioning Engineers, New York, 1978.

2. John A. Duffie and William A. Beckman, *Solar Thermal Processes,* Wiley-Interscience, New York, 1974.

3. Frank Kreith and Jan R. Kreider, *Principles of Solar Engineering,* McGraw-Hill, New York, 1978.

4. William A Beckman, S. A. Klein, and J. A. Duffie, *Solar Heating Design by the F-Chart Method,* Wiley-Interscience, 1979.

5. "Solar Energy, A Bibliography," U.S. Energy Research and Development Administration, Office of Public Affairs, Technical Information Center, TID-3351-R1P1 and TID-3351-R1P2, March 1976.

6. *Applications of Solar Energy for Heating and Cooling of Buildings,* ASHRAE GRP 170, American Society of Heating, Refrigerating and Air-Conditioning Engineers, New York, 1977.

7. *ASHRAE Handbook of Fundamentals,* American Society of Heating, Refrigerating and Air-Conditioning Engineers, New York, 1977.

8. Peter J. Lunde, *Solar Thermal Engineering,* Wiley, New York, 1980.

9. Aden B. Meinel and Marjorie P. Meinel, *Applied Solar Energy—An Introduction,* Addison-Wesley, Reading, Massachusetts, 1976.

10. *Solar Heating Systems Design Manual,* International Telephone and Telegraph, Morton Grove, Illinois, 1976.

11. *ERDA Facilities Solar Design Handbook,* ERDA 77-65, 1977, Available from NTIS.

12. *Design of Solar Heating and Cooling Systems,* Tech. Report E-139, Construction Engineering Research Laboratory, Dept. of Army, Champaign, Illinois, 1978. Available from NTIS.

13. *Introduction to Solar Heating and Cooling Design and Sizing,* DOE/CS-0011, August 1978. Available from NTIS.

14. *Passive Solar Design Handbook, Volume I, Passive Solar Design Concepts DOE/ CS-0127/1, Volume II, Passive Solar Design Analysis DOE/CS-0127/2,* 1980. Available from NTIS.

PROBLEMS

16-1 One author suggests as a rule of thumb a collector coolant capacity rate per square foot of collector ($\dot{m}c/A_{col}$) of 20 Btu/(hr-F-ft^2). Assume the coolant to be water ($c = 1$ Btu/(lbm-F). (a) Determine the suggested flow rate for a solar panel containing 22 ft^2 of collector surface. (b) Compute the temperature rise of the water passing through the collector if the normal incident irradiation is 300 Btu/(hr-ft^2) and the instantaneous collector efficiency is 40 percent.

16-2 Assume a nominal building heating load of 12 Btu/(ft^2-F-day) based on the floor area. For a solar collector area to floor area ratio of 0.25 determine the fraction of the total heating load supplied by solar energy to a house in Stillwater, Oklahoma, in January. Assume an average collector efficiency of 30 percent and a tilt of 45 degrees. See Example 16-1 for the average daily total irradiation for January.

16-3 Determine the rate of direct irradiation on a south-facing solar collector tilted at latitude angle at solar noon on February 21 on a clear day.

16-4 Determine for each month of the year the average ratio of daily irradiation on a collector to the daily irradiation on a horizontal surface. Assume the latitude is 40 degrees and the collector is tilted from the horizontal at an angle Σ of (a) 40 degrees, (b) 30 degrees, and (c) 50 degrees.

16-5 The mean daily solar insolation on a horizontal surface at Indianapolis, Indiana, by month is given below in Langleys (1 Langley = 1 cal/cm^2 = 4.186 J/cm^2)

Jan	144	May	488	Sept	405
Feb	213	June	543	Oct	293
Mar	316	July	541	Nov	177
Apr	396	Aug	490	Dec	132

Determine the total amount of energy that can be collected annually by a collector tilted at 40 degrees to the horizontal and having an average collector efficiency of 36 percent.

16-6 Work Problem 16-5 for a collector tilted at 30 degrees to the horizontal.

16-7   Work Problem 16-5 for a collector tilted at 50 degrees to the horizontal.

16-8   Using the computed angle of incidence θ between the sun's rays and a normal to a collector surface, calculate the rate at which radiation should strike a south-facing collector tilted at 45 degrees and located at a latitude of 35 degrees at 12:00 solar noon on a clear day in December. Express your answer in Btu/(hr-ft^2).

16-9   Repeat Problem 16-8 for the hours of 9:00, 10:00, and 11:00 (solar time) and make a rough estimate of the total irradiation (direct and diffuse) striking the collector on a typical clear day.

16-10 If a typical solar collector of 600 ft^2 area were gathering 1400 Btu/(ft^2-day) and transferring the energy to a storage tank containing 1000 gallons of water, estimate the temperature increase in the tank during the day, assuming no losses and no energy removal from the tank.

16-11 Domestic water is available at 50 F. If the hot water requirement of a home is 100 gallons per day of water at 130 F, determine the amount of collector area required for a solar water heater. Assume an average collector efficiency of 40 percent, heat losses at 10 percent of the energy actually collected, and an average insolation on the collector of 2000 Btu/(day-ft^2).

16-12 Electricity is assumed to cost $0.03 per kilowatt hour and will increase in cost at an average rate of 8 percent per year. Money to build a solar collector system can be borrowed at 9 percent and the system is presumed to last 20 years. If the system can be presumed to collect an average over the entire year of 1000 Btu/(day-ft^2) of energy that can be utilized to replace electricity, estimate the amount that can be invested in the solar system (per square foot of collector) for a break-even cost. It will be necessary to put the electricity cost on a present worth basis to make the comparison. Assume no maintenance expenses.

16-13 For the collector shown in Fig. 16-4, assume that the transmittance of the glass cover is 0.9 and the absorptance of the collector plate is 0.90. Calculate (a) the value of the collector heat removal factor F_R, and (b) the overall heat loss coefficient.

16-14 For the collector described in Fig. 16-4, (a) determine the rate at which useful energy can be extracted (in Btu/hr-ft^2), assuming the insolation to be 220 Btu/(hr-ft^2), the inlet water temperature to be 100 F, and the ambient temperature to be 35 F. (b) Determine the extraction rate if the inlet temperature drops to 60 F. (c) Determine the extraction rate if the inlet temperature remains at 60 F, but the ambient temperature rises to 80 F.

16-15 Using a rule of thumb of 0.025 gpm of coolant flow per square foot of collector, determine the exit temperature of the coolant for each case in Problem 16-14.

16-16 For a temperature rise of 15 F in an air-type solar collector, calculate the air flow rate in cfm for a 20 ft^2 collector with a solar radiation input of 300 Btu/(hr-ft^2) and a collector efficiency of 30 percent.

16-17 Using Table 16-1, determine for the month of December the average daily direct, diffuse, and total radiation on a collector tilted at 30 degrees from the horizontal and facing due south at Stillwater, Oklahoma. Assume ground reflectance of 0.2.

16-18 Work Problem 16-17 for the month of (a) February, and (b) June.

16-19 Work Problem 16-17 for a tilt angle of 50 degrees from the horizontal.

16-20 At what angle will the direct sun's rays be striking a solar collector facing due south, tilted at 45 degrees, and located in Kansas City, Missouri, at 11 A.M., central standard time on January 21?

16-21 For the collector described in Problem 16-20, at what time (CDST) will the sun's direct rays start to irradiate the collector in the morning on June 21?

16-22 One rule of thumb for sizing rock bed storage for solar air heating systems is 0.75 ft^3 of rock per square foot of collector. If the heat capacity of rock is assumed to be about 36 Btu/(ft^3-F), calculate the required storage size, in ft^3, and the total useful energy stored in a rock storage system sized by the above rule, for a collector system of 400 ft^2. Assume that the rock bed is heated to a uniform temperature of 130 F and may be cooled down to a temperature of 100 F before the auxiliary heat is turned on.

16-23 If rock weighs about 110 lb/ft^3, determine the weight of the required rock bed storage system of Problem 16-22.

16-24 Using the information from Problem 16-17 and the collector performance of Fig. 16-4, estimate the amount of collector area required for furnishing 30 percent of the December heating energy to a house with a heat loss of 25,000 Btu/DD. Make whatever assumptions seem necessary and proper.

Thermodynamic Properties

Table A-1   WATER AT SATURATION—ENGLISH UNITS*

Fahrenheit Temperature	Pressure P psia	Specific Volume ft³/lbm		Enthalpy Btu/lbm			Entropy Btu/(lbm-R)		
		Saturated Solid v_i	Saturated Vapor $v_g \times 10^{-3}$	Saturated Solid i_i	Sublimation i_{ig}	Saturated Vapor i_g	Saturated Solid s_i	Sublimation s_{ig}	Saturated Vapor s_g
32.018	0.0887	0.01747	3.302	−143.34	1218.7	1075.4	−0.292	2.479	2.187
30	0.0808	0.01747	3.607	−144.35	1218.9	1074.5	−0.294	2.489	2.195
25	0.0641	0.01746	4.506	−146.84	1219.1	1072.3	−0.299	2.515	2.216
20	0.0505	0.01745	5.655	−149.31	1219.4	1070.1	−0.304	2.542	2.238
15	0.0396	0.01745	7.13	−151.75	1219.7	1067.9	−0.309	2.569	2.260
10	0.0309	0.01744	9.04	−154.17	1219.9	1065.7	−0.314	2.597	2.283
5	0.0240	0.01743	11.52	−156.56	1220.1	1063.5	−0.320	2.626	2.306
0	0.0185	0.01743	14.77	−158.93	1220.2	1061.2	−0.325	2.655	2.330

Fahrenheit Temperature	Pressure P psia	Specific Volume ft³/lbm		Enthalpy Btu/lbm			Entropy Btu/(lbm-R)		
		Saturated Liquid v_f	Saturated Vapor v_g	Saturated Liquid i_f	Evaporation i_{fg}	Saturated Vapor i_g	Saturated Liquid s_f	Evaporation s_{fg}	Saturated Vapor s_g
32.018	0.08866	0.016022	3302	0.01	1075.4	1075.4	0.00000	2.1869	2.1869
36	0.10397	0.016021	2839	4.00	1073.1	1077.1	0.00810	2.1648	2.1729
40	0.12166	0.016020	2445	8.02	1070.9	1078.9	0.01617	2.1430	2.1592
44	0.14196	0.016021	2112	12.03	1068.6	1080.7	0.02418	2.1215	2.1457
48	0.16520	0.016023	1829	16.05	1066.4	1082.4	0.03212	2.1003	2.1324
52	0.19173	0.016026	1588.6	20.07	1064.1	1084.2	0.04000	2.0795	2.1195
56	0.2219	0.016030	1382.9	24.08	1061.9	1085.9	0.04781	2.0590	2.1068
60	0.2563	0.016035	1206.9	28.08	1059.6	1087.7	0.05555	2.0388	2.0943
64	0.2952	0.016040	1055.8	32.09	1057.3	1089.4	0.06323	2.0189	2.0821
68	0.3391	0.016047	925.8	36.09	1055.1	1091.2	0.07084	1.9993	2.0701
72	0.3887	0.016055	813.7	40.09	1052.8	1092.9	0.07839	1.9800	2.0584
76	0.4446	0.016063	716.8	44.09	1050.6	1094.7	0.08589	1.9610	2.0469
80	0.5073	0.016073	632.8	48.09	1048.3	1096.4	0.09332	1.9423	2.0356

598

Temp.									
84	0.5776	0.016083	559.8	52.08	1046.0	1098.1	0.10069	1.9238	2.0245
88	0.6562	0.016094	496.3	56.07	1043.8	1099.9	0.10801	1.9056	2.0136
92	0.7439	0.016105	440.9	60.06	1041.5	1101.6	0.11527	1.8877	2.0030
96	0.8416	0.016117	392.4	64.06	1039.2	1103.3	0.12248	1.8700	1.9925
100	0.9503	0.016130	350.0	68.05	1037.0	1105.0	0.12963	1.8526	1.9822
104	1.0708	0.016144	312.8	72.04	1034.7	1106.7	0.13674	1.8354	1.9722
108	1.2044	0.016158	280.0	76.03	1032.4	1108.4	0.14379	1.8185	1.9623
112	1.3520	0.016173	251.1	80.02	1030.1	1110.2	0.15079	1.8017	1.9525
116	1.5150	0.016189	225.6	84.01	1027.8	1111.9	0.15775	1.7852	1.9430
120	1.6945	0.016205	203.0	88.00	1025.5	1113.5	0.16465	1.7690	1.9336
124	1.8921	0.016221	183.1	91.99	1023.2	1115.2	0.17152	1.7529	1.9244
128	2.1090	0.016239	165.30	95.98	1020.9	1116.9	0.17833	1.7370	1.9154
132	2.347	0.016256	149.51	99.98	1018.6	1118.6	0.18510	1.7214	1.9065
136	2.607	0.016275	135.44	103.97	1016.3	1120.3	0.19183	1.7059	1.8978
140	2.892	0.016293	122.88	107.96	1014.0	1121.9	0.19851	1.6907	1.8892
144	3.203	0.016313	111.66	111.96	1011.6	1123.6	0.20515	1.6756	1.8807
148	3.541	0.016333	101.61	115.96	1009.3	1125.2	0.21175	1.6607	1.8725
152	3.910	0.016353	92.60	119.96	1006.9	1126.9	0.21831	1.6460	1.8643
156	4.310	0.016374	84.51	123.96	1004.6	1128.5	0.22482	1.6315	1.8563
160	4.745	0.016395	77.23	127.96	1002.2	1130.1	0.23130	1.6171	1.8484
164	5.216	0.016417	70.67	131.96	999.8	1131.8	0.23774	1.6029	1.8407
168	5.726	0.016439	64.76	135.97	997.4	1133.4	0.24414	1.5889	1.8330
172	6.277	0.016462	59.42	139.97	995.0	1135.0	0.25050	1.5750	1.8255
176	6.873	0.016485	54.58	143.98	992.6	1136.6	0.25682	1.5613	1.8181
180	7.515	0.016509	50.20	147.99	990.2	1138.2	0.26311	1.5478	1.8109
184	8.206	0.016533	46.23	152.01	987.7	1139.7	0.26936	1.5344	1.8037
188	8.951	0.016558	42.62	156.02	985.3	1141.3	0.27557	1.5211	1.7967
192	9.750	0.016583	39.34	160.04	982.8	1142.9	0.28175	1.5080	1.7898
196	10.609	0.016608	36.36	164.05	980.4	1144.4	0.28789	1.4951	1.7830
200	11.529	0.016634	33.63	168.07	977.9	1145.9	0.29400	1.4822	1.7762
204	12.515	0.016661	31.15	172.10	975.4	1147.5	0.30008	1.4695	1.7696
208	13.570	0.016688	28.88	176.13	972.9	1149.0	0.30612	1.4570	1.7631
212	14.698	0.016716	26.80	180.16	970.3	1150.5	0.31213	1.4446	1.7567

Table A-1a WATER AT SATURATION—SI UNITS*

Temperature C	K	Pressure Pa $P \times 10^{-5}$	Specific Volume m³/kg Saturated Solid $v_i \times 10^3$	Saturated Vapor v_g	Enthalpy kJ/kg Saturated Solid i_i	Sublimation i_{ig}	Saturated Vapor i_g	Entropy kJ/(kg-K) Saturated Solid s_i	Sublimation s_{ig}	Saturated Vapor s_g
0.01	273.16	0.006113	1.0908	2.061	−333.40	2834.8	2501.4	−1.221	10.378	9.156
−2	271.15	0.005176	1.0904	2.417	−337.62	2835.3	2497.7	−1.237	10.456	9.219
−4	269.15	0.004375	1.0901	2.838	−341.78	2835.7	2494.0	−1.253	10.536	9.283
−6	267.15	0.003689	1.0898	3.342	−345.91	2836.2	2490.3	−1.268	10.616	9.348
−8	265.15	0.003102	1.0894	3.944	−350.02	2836.6	2486.6	−1.284	10.698	9.414
−10	263.15	0.002602	1.0891	4.667	−354.09	2837.0	2482.9	−1.299	10.781	9.481
−12	261.15	0.002176	1.0888	5.537	−358.14	2837.3	2479.2	−1.315	10.865	9.550
−14	259.15	0.001815	1.0884	6.588	−362.15	2837.6	2475.5	−1.331	10.950	9.619
−16	257.15	0.001510	1.0881	7.860	−366.14	2837.9	2471.8	−1.346	11.036	9.690

Temperature C	K	Pressure Pa $P \times 10^{-5}$	Specific Volume m³/kg Saturated Liquid $v_f \times 10^3$	Saturated Vapor $v_g \times 10^3$	Enthalpy kJ/kg Saturated Liquid i_f	Evaporation i_{fg}	Saturated Vapor i_g	Entropy kJ/(kg-K) Saturated Liquid s_f	Evaporation s_{fg}	Saturated Vapor s_g
0.01	273.16	0.006113	1.0002	206 136	0.01	2501.3	2501.4	0.0000	9.1562	9.1562
2	275.15	0.007056	1.0001	179 889	8.37	2496.7	2505.0	0.0305	9.0730	9.1035
4	277.15	0.008131	1.0001	157 232	16.78	2491.9	2508.7	0.0610	8.9904	9.0514
6	279.15	0.009349	1.0001	137 734	25.20	2487.2	2512.4	0.0912	8.9090	9.0003
8	281.15	0.010724	1.0002	120 917	33.60	2482.5	2516.1	0.1212	8.8289	8.9501
10	283.15	0.012276	1.0004	106 379	42.01	2477.7	2519.8	0.1510	8.7498	8.9008
12	285.15	0.014022	1.0005	93 784	50.41	2473.0	2523.4	0.1806	8.6718	8.8524
14	287.15	0.015983	1.0008	82 848	58.80	2468.3	2527.1	0.2099	8.5949	8.8048

16	289.15	0.018181	1.0011	73 333	67.19	2463.6	2530.8	0.2390	8.5191	8.7582
18	291.15	0.020640	1.0014	65 038	75.58	2458.8	2534.4	0.2679	8.4443	8.7123
20	293.15	0.02339	1.0018	57 791	83.96	2454.1	2538.1	0.2966	8.3706	8.6672
22	295.15	0.02645	1.0022	51 447	92.33	2449.4	2541.7	0.3251	8.2979	8.6229
24	297.15	0.02985	1.0027	45 883	100.70	2444.7	2545.4	0.3534	8.2261	8.5794
26	299.15	0.03363	1.0032	40 994	109.07	2439.9	2549.0	0.3814	8.1552	8.5367
28	301.15	0.03782	1.0037	36 690	117.43	2435.2	2552.6	0.4093	8.0854	8.4946
30	303.15	0.04246	1.0043	32 894	125.79	2430.5	2556.3	0.4369	8.0164	8.4533
32	305.15	0.04759	1.0050	29 540	134.15	2425.7	2559.9	0.4644	7.9483	8.4127
34	307.15	0.05324	1.0056	26 571	142.50	2421.0	2563.5	0.4917	7.8811	8.3728
36	309.15	0.05947	1.0063	23 940	150.86	2416.2	2567.1	0.5188	7.8147	8.3336
38	311.15	0.06632	1.0071	21 602	159.21	2411.5	2570.7	0.5458	7.7492	8.2950
40	313.15	0.07384	1.0078	19 523	167.57	2406.7	2574.3	0.5725	7.6845	8.2570
42	315.15	0.08208	1.0086	17 671	175.91	2401.9	2577.9	0.5991	7.6206	8.2197
44	317.15	0.09111	1.0095	16 018	184.27	2397.2	2581.4	0.6255	7.5574	8.1829
46	319.15	0.10098	1.0103	14 540	192.62	2392.4	2585.0	0.6518	7.4950	8.1468
48	321.15	0.11175	1.0112	13 218	200.97	2387.6	2588.5	0.6779	7.4334	8.1113
50	323.15	0.12349	1.0121	12 032	209.33	2382.7	2592.1	0.7038	7.3725	8.0763
52	325.15	0.13628	1.0131	10 968	217.69	2377.9	2595.6	0.7296	7.3123	8.0419
54	327.15	0.15019	1.0141	10 011	226.04	2373.1	2599.1	0.7552	7.2528	8.0080
56	329.15	0.16529	1.0151	9149	234.41	2368.2	2602.6	0.7807	7.1940	7.9747
58	331.15	0.18166	1.0161	8372	242.77	2363.4	2606.1	0.8060	7.1359	7.9419
60	333.15	0.19940	1.0172	7671	251.13	2358.5	2609.6	0.8312	7.0784	7.9096
62	335.15	0.21860	1.0182	7037	259.49	2353.6	2613.1	0.8562	7.0216	7.8778
64	337.15	0.23934	1.0194	6463	267.86	2348.7	2616.5	0.8811	6.9654	7.8465
66	339.15	0.2617	1.0205	5943	276.23	2343.7	2620.0	0.9058	6.9098	7.8156
68	341.15	0.2859	1.0217	5471	284.61	2338.8	2623.4	0.9304	6.8548	7.7852
70	343.15	0.3119	1.0228	5042	292.98	2333.8	2626.8	0.9549	6.8004	7.7553
72	345.15	0.3399	1.0240	4652	301.36	2328.9	2630.2	0.9792	6.7466	7.7258
74	347.15	0.3699	1.0253	4297	309.74	2323.9	2633.6	1.0034	6.6934	7.6968
76	349.15	0.4022	1.0265	3973	318.13	2318.9	2637.0	1.0275	6.6407	7.6682
78	351.15	0.4368	1.0278	3677	326.51	2313.8	2640.3	1.0515	6.5885	7.6400

Table A-1a. WATER AT SATURATION—SI UNITS* *(continued)*

Temperature C	K	Pressure Pa $P \times 10^{-5}$	Specific Volume m³/kg Saturated Liquid $v_f \times 10^3$	Saturated Vapor $v_g \times 10^3$	Enthalpy kJ/kg Saturated Liquid i_f	Evaporation i_{fg}	Saturated Vapor i_g	Entropy kJ/(kg-K) Saturated Liquid s_f	Evaporation s_{fg}	Saturated Vapor s_g
80	353.15	0.4739	1.0291	3407	334.91	2308.8	2643.7	1.0753	6.5369	7.6122
82	355.15	0.5136	1.0305	3160	343.30	2303.7	2647.0	1.0990	6.4858	7.5848
84	357.15	0.5560	1.0318	2934	351.70	2298.6	2650.3	1.1225	6.4353	7.5578
86	359.15	0.6014	1.0332	2726	360.10	2293.5	2653.6	1.1460	6.3852	7.5312
88	361.15	0.6498	1.0346	2536	368.51	2288.3	2656.6	1.1693	6.3356	7.5050
90	363.15	0.7014	1.0360	2361	376.92	2283.2	2660.1	1.1925	6.2866	7.4791
92	365.15	0.7564	1.0375	2200	385.33	2278.0	2663.3	1.2156	6.2379	7.4536
94	367.15	0.8149	1.0389	2052	393.75	2272.8	2666.5	1.2386	6.1898	7.4284
96	369.15	0.8771	1.0404	1915.0	402.17	2267.6	2669.7	1.2615	6.1421	7.4036
98	371.15	0.9433	1.0420	1789.1	410.61	2262.3	2672.9	1.2842	6.0948	7.3791
100	373.15	1.0135	1.0435	1672.9	419.04	2257.0	2676.1	1.3069	6.0480	7.3549

*Abridged from, "Steam Tables (International Edition—Metric Units)," by Joseph H. Keenan, F. G. Keyes, P. G. Hill, and J. G. Moore. Copyright © 1969 by John Wiley and Sons, Inc., New York. Reprinted by permission of John Wiley and Sons, Inc.

Table A-2 MOIST AIR (STANDARD ATMOSPHERIC PRESSURE, 29.921 in. Hg)—ENGLISH UNITS*

Fahrenheit Temperature	Humidity Ratio $W_s \times 10^3$	Volume ft³/lbm dry air			Enthalpy Btu/lbm dry air			Entropy Btu/(F·lbm dry air)			Fahrenheit Temperature
		v_a	v_{as}	v_s	i_a	i_{as}	i_s	s_a	s_{as}	s_s	
0	0.7872	11.578	0.015	11.593	0.000	0.835	0.835	0.00000	0.00192	0.00192	0
5	1.020	11.705	0.019	11.724	1.201	1.085	2.286	0.00260	0.00246	0.00506	5
10	1.315	11.831	0.025	11.856	2.402	1.401	3.803	0.00518	0.00314	0.00832	10
15	1.687	11.958	0.032	11.990	3.603	1.800	5.403	0.00772	0.00399	0.01171	15
20	2.152	12.081	0.042	12.126	4.804	2.302	7.106	0.01023	0.00504	0.01527	20
25	2.733	12.211	0.054	12.265	6.005	2.929	8.934	0.01273	0.00635	0.01908	25
30	3.454	12.338	0.068	12.406	7.206	3.709	10.915	0.01519	0.00796	0.02315	30
32	3.788	12.388	0.075	12.463	7.686	4.072	11.758	0.01617	0.00870	0.02487	32
36	4.450	12.489	0.089	12.578	8.647	4.791	13.438	0.01812	0.01016	0.02828	36
40	5.213	12.590	0.105	12.695	9.608	5.622	15.230	0.02005	0.01183	0.03188	40
44	6.091	12.691	0.124	12.815	10.569	6.580	17.149	0.02197	0.01373	0.03570	44
48	7.100	12.792	0.146	12.938	11.530	7.681	19.211	0.02387	0.01591	0.03978	48
52	8.256	12.891	0.170	13.064	12.191	8.945	21.436	0.02575	0.01839	0.04414	52
56	9.575	12.995	0.200	13.195	13.452	10.39	23.84	0.02762	0.02121	0.04883	56
60	11.08	13.096	0.233	13.329	14.413	12.05	26.46	0.02948	0.02441	0.05389	60
64	12.80	13.197	0.271	13.468	15.374	13.94	29.31	0.03132	0.02803	0.05935	64
68	14.75	13.298	0.315	13.613	16.355	16.09	32.42	0.03314	0.03213	0.06527	68
72	16.97	13.398	0.364	13.762	17.297	18.53	35.83	0.03495	0.03675	0.07170	72
76	19.48	13.499	0.422	13.921	18.259	21.31	39.57	0.03675	0.04197	0.07872	76
80	22.33	13.601	0.486	14.087	19.221	24.47	43.69	0.03854	0.04784	0.08638	80
84	25.55	13.702	0.560	14.262	20.183	28.04	48.22	0.04031	0.05446	0.09477	84
88	29.19	13.803	0.645	14.448	21.144	32.09	53.23	0.04207	0.06189	0.10396	88
92	33.30	13.904	0.741	14.645	22.106	36.67	58.78	0.04382	0.07025	0.11407	92
96	37.95	14.005	0.851	14.856	23.068	41.85	64.92	0.04556	0.07963	0.12519	96
100	43.19	14.106	0.975	15.081	24.029	47.70	71.73	0.04729	0.09016	0.13745	100
104	49.11	14.207	1.117	15.324	24.991	54.32	79.31	0.04900	0.1020	0.1510	104

Table A-2   MOIST AIR (STANDARD ATMOSPHERIC PRESSURE, 29.921 in. Hg)—ENGLISH UNITS* (continued)

Fahrenheit Temperature	Humidity Ratio $W_s \times 10^3$	Volume ft³/lbm dry air			Enthalpy Btu/lbm dry air			Entropy Btu/(F-lbm dry air)			Fahrenheit Temperature
		v_a	v_{as}	v_s	i_a	i_{as}	i_s	s_a	s_{as}	s_s	
108	55.78	14.308	1.278	15.586	25.953	61.80	87.76	0.05070	0.1153	0.1660	108
112	63.33	14.409	1.460	15.896	26.915	70.27	97.18	0.05239	0.1302	0.1826	112
116	71.85	15.510	1.668	16.178	27.878	79.85	107.73	0.05407	0.1470	0.2011	116
120	81.49	14.611	1.905	16.516	28.841	90.70	119.54	0.05573	0.1659	0.2216	120
124	92.42	14.712	2.174	16.886	29.804	103.0	132.8	0.05739	0.1872	0.2446	124
128	104.8	14.813	2.482	17.295	30.766	117.0	147.8	0.05903	0.2113	0.2703	128
132	118.9	14.915	2.834	17.749	31.729	133.0	164.7	0.06067	0.2386	0.2993	132
136	135.0	15.016	3.237	18.253	32.692	151.2	183.9	0.06229	0.2695	0.3318	136
140	153.4	15.117	3.702	18.819	33.655	172.0	205.7	0.06390	0.3047	0.3686	140
144	174.5	15.218	4.239	19.457	34.618	196.0	230.6	0.06549	0.3449	0.4104	144
148	198.9	15.319	4.862	20.181	35.581	223.7	259.3	0.06708	0.3912	0.4583	148
152	227.1	15.420	5.587	21.007	36.545	255.9	292.4	0.06866	0.4445	0.5132	152
156	260.2	15.521	6.439	21.960	37.508	293.5	331.0	0.07023	0.5066	0.5768	156
160	299.0	15.622	7.446	23.068	38.472	337.8	376.3	0.07179	0.5793	0.6511	160
164	345.2	15.723	8.648	24.371	39.436	390.5	429.9	0.07334	0.6652	0.7385	164
168	400.7	15.824	10.098	25.922	40.400	454.0	494.4	0.07488	0.7680	0.8429	168
172	468.2	15.925	11.870	27.795	41.364	531.3	572.7	0.07641	0.8927	0.9691	172
176	551.9	16.026	14.074	30.100	42.328	627.1	669.4	0.07794	1.047	1.125	176
180	657.8	16.127	16.870	32.997	43.292	748.5	791.8	0.07946	2.140	1.319	180
184	795.3	16.228	20.513	36.741	44.257	906.2	950.5	0.08096	1.490	1.571	184
188	980.2	16.329	25.427	41.756	45.222	1119	1164	0.08245	1.825	1.907	188
192	1241	16.430	32.375	48.805	46.187	1418	1464	0.08394	2.296	2.380	192
196	1635	16.531	42.885	59.416	47.153	1871	1918	0.08542	3.002	3.087	196
200	2295	16.632	60.510	77.142	48.119	2629	2677	0.08689	4.179	4.266	200

*Compiled by John A. Goff and S. Gratch. Abridged by permission from *ASHRAE Handbook of Fundamentals*, 1977.

Table A-2a   MOIST AIR (STANDARD ATMOSPHERIC PRESSURE, 101.325 kPa)—SI UNITS*

Celsius Temperature	Humidity Ratio	Volume m³/kg dry air			Enthalpy kJ/kg dry air			Entropy kJ/(K-kg dry air)			Celsius Temperature
	$W_s \times 10^3$	v_a	v_{as}	v_s	i_a	i_{as}	i_s	s_a	s_{as}	s_s	
−18	0.771	0.7222	0.0009	0.7231	−18.103	1.901	−16.203	−0.0686	0.0079	−0.0607	−18
−16	0.930	0.7279	0.0011	0.7290	−16.092	2.297	−13.795	−0.0607	0.0094	−0.0513	−16
−14	1.119	0.7336	0.0013	0.7349	−14.080	2.767	−11.314	−0.0529	0.0113	−0.0416	−14
−12	1.342	0.7393	0.0016	0.7409	−12.069	3.324	−8.745	−0.0452	0.0134	−0.0318	−12
−10	1.606	0.7450	0.0019	0.7469	−10.058	3.984	−6.073	−0.0375	0.0159	−0.0216	−10
−8	1.916	0.7506	0.0023	0.7529	−8.046	4.761	−3.285	−0.0299	0.0189	0.0110	−8
−6	2.280	0.7563	0.0028	0.7591	−6.035	5.674	−0.360	−0.0223	0.0223	0.0000	−6
−4	2.707	0.7620	0.0033	0.7653	−4.023	6.748	2.724	−0.0148	0.0264	0.0115	−4
−2	3.206	0.7677	0.0039	0.7716	−2.012	8.003	5.991	−0.0074	0.0310	0.0236	−2
0	3.788	0.7734	0.0047	0.7781	0.000	9.470	9.470	0.0000	0.0364	0.0364	0
2	4.38	0.7791	0.0055	0.7845	2.012	10.966	12.978	0.0073	0.0419	0.0492	2
4	5.05	0.7847	0.0064	0.7911	4.023	12.669	16.692	0.0146	0.0480	0.0626	4
6	5.82	0.7904	0.0074	0.7978	6.035	14.604	20.639	0.0219	0.0550	0.0768	6
8	6.68	0.7961	0.0085	0.8046	8.047	16.801	24.848	0.0290	0.0628	0.0919	8
10	7.66	0.8018	0.0098	0.8116	10.059	19.289	29.348	0.0362	0.0716	0.1078	10
12	8.76	0.8075	0.0113	0.8188	12.071	22.102	34.172	0.0432	0.0815	0.1248	12
14	10.01	0.8131	0.0130	0.8262	14.083	25.279	39.362	0.0503	0.0926	0.1429	14
16	11.41	0.8188	0.0150	0.8338	16.095	28.860	44.955	0.0573	0.1051	0.1624	16
18	12.99	0.8245	0.0172	0.8417	18.107	32.891	50.998	0.0642	0.1190	0.1832	18
20	14.75	0.8302	0.0196	0.8498	20.119	37.424	57.544	0.0711	1.1345	0.2056	20
22	16.74	0.8359	0.0224	0.8583	22.132	42.514	64.646	0.0779	0.1518	0.2298	22
24	18.96	0.8415	0.0256	0.8671	24.144	48.222	72.366	0.0847	0.1711	0.2559	24
26	21.44	0.8472	0.0291	0.8763	26.157	54.620	80.777	0.0915	0.1926	0.2841	26
28	24.22	0.8529	0.0331	0.8860	28.170	61.782	89.952	0.0982	0.2166	0.3147	28
30	27.32	0.8586	0.0376	0.8961	30.183	69.794	99.977	0.1048	0.2431	0.3480	30
32	30.78	0.8643	0.0426	0.9068	32.196	78.750	110.946	0.1115	0.2726	0.3841	32
34	34.64	0.8699	0.0483	0.9182	34.209	88.758	122.968	0.1180	0.3054	0.4235	34
36	38.95	0.8756	0.0546	0.9302	36.223	99.938	136.161	0.1246	0.3418	0.4664	36

Table A-2a MOIST AIR (STANDARD ATMOSPHERIC PRESSURE, 101.325 kPa)—SI UNITS* (continued)

Celsius Temperature	Humidity Ratio $W_s \times 10^3$	Volume m³/kg dry air			Enthalpy kJ/kg dry air			Entropy kJ/(K·kg dry air)			Celsius Temperature
		v_a	v_{as}	v_s	i_a	i_{as}	i_s	s_a	s_{as}	s_s	
38	43.76	0.8813	0.0617	0.9430	38.236	112.423	150.660	0.1311	0.3822	0.5133	38
40	49.11	0.8870	0.0697	0.9567	40.250	126.364	166.615	0.1375	0.4271	0.5646	40
42	55.09	0.8926	0.0787	0.9713	42.264	141.936	184.200	0.1439	0.4769	0.6208	42
44	61.76	0.8983	0.0887	0.9871	44.278	159.331	203.610	0.1503	0.5322	0.6825	44
46	69.20	0.9040	0.1001	1.0040	46.292	178.775	225.068	0.1566	0.5936	0.7503	46
48	77.51	0.9097	0.1127	1.0224	48.307	200.521	248.823	0.1629	0.6620	0.8249	48
50	86.80	0.9153	0.1270	1.0424	50.322	224.876	275.198	0.1692	0.7381	0.9072	50
52	97.20	0.9210	0.1431	1.0641	52.337	252.175	304.512	0.1754	0.8229	0.9983	52
54	108.9	0.9267	0.1612	1.0879	54.352	282.830	337.182	0.1816	0.9176	1.0992	54
56	122.0	0.9324	0.1817	1.1141	56.367	317.312	373.679	0.1877	1.0235	1.2112	56
58	136.7	0.9381	0.2049	1.1429	58.383	356.188	414.572	0.1938	1.1423	1.3361	58
60	153.4	0.9437	0.2312	1.1749	60.399	400.137	460.536	0.1999	1.2759	1.4757	60
62	172.3	0.9494	0.2611	1.2105	62.415	449.976	512.391	0.2059	1.4265	1.6324	62
64	193.8	0.9551	0.2953	1.2504	64.431	506.713	571.144	0.2119	1.5971	1.8090	64
66	218.3	0.9608	0.3345	1.2953	66.448	571.555	638.003	0.2179	1.7910	2.0089	66
68	246.4	0.9664	0.3797	1.3462	68.465	656.066	714.531	0.2238	2.0126	2.2364	68
70	278.8	0.9721	0.4322	1.4043	70.482	732.161	802.643	0.2297	2.2674	2.4972	70
72	316.6	0.9778	0.4933	1.4711	72.499	832.353	904.852	0.2356	2.5624	2.7980	72
74	360.8	0.9835	0.5653	1.5488	74.517	949.905	1024.422	0.2414	2.9068	3.1482	74
76	413.2	0.9891	0.6508	1.6400	76.535	1089.154	1165.689	0.2472	3.3127	3.5599	76
78	476.0	0.9948	0.7536	1.7484	78.554	1256.050	1334.604	0.2530	3.7967	4.0497	78
80	552.0	1.0005	0.8787	1.8792	80.572	1458.841	1539.414	0.2587	4.3820	4.6407	80
82	646.1	1.0062	1.0338	2.0399	82.591	1709.528	1792.119	0.2644	5.1019	5.3663	82
84	764.7	1.0118	1.2300	2.2418	84.611	2026.112	2110.723	0.2701	6.0067	6.2768	84
86	918.5	1.0175	1.4849	2.5025	86.630	2436.686	2523.317	0.2757	7.1746	7.4503	86
88	1125	1.0232	1.8281	2.8513	88.650	2988.382	3077.032	0.2813	8.7363	9.0176	88
90	1416	1.0289	2.3124	3.3412	90.671	3765.876	3856.547	0.2869	10.9270	11.2139	90

*Reproduced by permission of Carrier Corporation, Syracuse, NY

Table A-3 REFRIGERANT 12 SATURATION PROPERTIES—ENGLISH UNITS*

Fahrenheit Temperature	Pressure psia P	Volume ft³/lbm Liquid v_f	Vapor v_g	Enthalpy Btu/lbm Liquid i_f	Latent i_{fg}	Vapor i_g	Entropy Btu/(lbm-R) Liquid s_f	Vapor s_g	Fahrenheit Temperature
−40	9.3076	0.010564	3.8750	0	72.913	72.913	0	0.17373	−40
−38	9.8035	0.010586	3.6922	0.4215	72.712	73.134	0.001000	0.17343	−38
−36	10.320	0.010607	3.5198	0.8434	72.511	73.354	0.001995	0.17313	−36
−34	10.858	0.010629	3.3571	1.2659	72.309	73.575	0.002988	0.17285	−34
−32	11.417	0.010651	3.2035	1.6887	72.106	73.795	0.003976	0.17257	−32
−30	11.999	0.010674	3.0585	2.1120	71.903	74.015	0.004961	0.17229	−30
−28	12.604	0.010696	2.9214	2.5358	71.698	74.234	0.005942	0.17203	−28
−26	13.233	0.010719	2.7917	2.9601	71.494	74.454	0.006919	0.17177	−26
−24	13.886	0.010741	2.6691	3.3848	71.288	74.673	0.007894	0.17151	−24
−22	14.564	0.010764	2.5529	3.8100	71.081	74.891	0.008864	0.17126	−22
−20	15.267	0.010788	2.4429	4.2357	70.874	75.110	0.009831	0.17102	−20
−18	15.996	0.010811	2.3387	4.6618	70.666	75.328	0.010795	0.17078	−18
−16	16.753	0.010834	2.2399	5.0885	70.456	75.545	0.011755	0.17055	−16
−14	17.536	0.010858	2.1461	5.5157	70.246	75.762	0.012712	0.17032	−14
−12	18.348	0.010882	2.0572	5.9434	70.036	75.979	0.013666	0.17010	−12
−10	19.189	0.010906	1.9727	6.3716	69.824	76.196	0.014617	0.16989	−10
−8	20.059	0.010931	1.8924	6.8003	69.611	76.411	0.015564	0.16967	−8
−6	20.960	0.010955	1.8161	7.2296	69.397	76.627	0.016508	0.16947	−6
−4	21.891	0.010980	1.7436	7.6594	69.183	76.842	0.017449	0.16927	−4
−2	22.854	0.011005	1.6745	8.0898	68.967	77.057	0.018388	0.16907	−2
0	23.849	0.011030	1.6089	8.5207	68.750	77.271	0.019323	0.16888	0
2	24.878	0.011056	1.5463	8.9522	68.533	77.485	0.020255	0.16869	2

Table A-3 REFRIGERANT 12 SATURATION PROPERTIES—ENGLISH UNITS (continued)

Fahrenheit Temperature	Pressure psia P	Volume ft³/lbm		Enthalpy Btu/lbm			Entropy Btu/(lbm-R)		Fahrenheit Temperature
		Liquid v_f	Vapor v_g	Liquid i_f	Latent i_{fg}	Vapor i_g	Liquid s_f	Vapor s_g	
4	25.939	0.011082	1.4867	9.3843	68.314	77.698	0.021184	0.16851	4
6	27.036	0.011107	1.4299	9.8169	68.094	77.911	0.022110	0.16833	6
8	28.167	0.011134	1.3758	10.250	67.873	78.123	0.023033	0.16815	8
10	29.335	0.011160	1.3241	10.684	67.651	78.335	0.023954	0.16798	10
12	30.539	0.011187	1.2748	11.118	67.428	78.546	0.024871	0.16782	12
14	31.780	0.011214	1.2278	11.554	67.203	78.757	0.025786	0.16765	14
16	33.060	0.011241	1.128	11.989	66.977	78.966	0.026699	0.16750	16
18	34.378	0.011268	1.1399	12.426	66.750	79.176	0.027608	0.16734	18
20	35.736	0.011296	1.0988	12.863	66.522	79.385	0.028515	0.16719	20
22	37.135	0.011324	1.0596	13.300	66.293	79.593	0.029420	0.16704	22
24	38.574	0.011352	1.0220	13.739	66.061	79.800	0.030322	0.16690	24
26	40.056	0.011380	0.98612	14.178	65.829	80.007	0.031221	0.16676	26
28	41.580	0.011409	0.95173	14.618	65.596	80.214	0.032118	0.16662	28
30	43.148	0.011438	0.91880	15.058	65.361	80.419	0.033013	0.16648	30
32	44.760	0.011468	0.88725	15.500	65.124	80.624	0.033905	0.16635	32
34	46.417	0.011497	0.85702	15.942	64.886	80.828	0.034796	0.16622	34
36	48.120	0.011527	0.82803	16.384	64.647	81.031	0.035683	0.16610	36
38	49.870	0.011557	0.80023	16.828	64.406	81.234	0.036569	0.16598	38
40	51.667	0.011588	0.77357	17.273	64.163	81.436	0.037453	0.16586	40
42	53.513	0.011619	0.74798	17.718	63.919	81.637	0.038334	0.16574	42
44	55.407	0.011650	0.72341	18.164	63.673	81.837	0.039213	0.16562	44
46	57.352	0.011682	0.69982	18.611	63.426	82.037	0.040091	0.16551	46

48	59.347	0.011714	0.67715	19.059	63.177	82.236	0.040966	0.16540	48
50	61.394	0.011746	0.65537	19.507	62.926	82.433	0.041839	0.16530	50
52	63.494	0.011779	0.63444	19.957	62.673	82.630	0.042711	0.16519	52
54	65.646	0.011811	0.61431	20.408	62.418	82.826	0.043581	0.16509	54
56	67.853	0.011845	0.59495	20.859	62.162	83.021	C.044449	0.16499	56
58	70.115	0.011879	0.57632	21.312	61.903	83.215	0.045316	0.16489	58
60	72.433	0.011913	0.55839	21.766	61.643	83.409	0.046180	0.16479	60
62	74.807	0.011947	0.54112	22.221	61.380	83.601	0.047044	0.16470	62
64	77.239	0.011982	0.52450	22.676	61.116	83.792	0.047905	0.16460	64
66	79.729	0.012017	0.50848	23.133	60.849	83.982	0.048765	0.16451	66
68	82.279	0.012053	0.49305	23.591	60.580	84.171	0.049624	0.16442	68
70	84.888	0.012089	0.47818	24.050	60.309	84.359	0.050482	0.16434	70
72	87.559	0.012126	0.46383	24.511	60.035	84.546	0.051338	0.16425	72
74	90.292	0.012163	0.45000	24.973	59.759	84.732	0.052193	0.16417	74
76	93.087	0.012201	0.43666	25.435	59.481	84.916	0.053047	0.16408	76
78	95.946	0.012239	0.42378	25.899	59.201	85.100	0.053900	0.16400	78
80	98.870	0.012277	0.41135	26.365	58.917	85.282	0.054751	0.16392	80
82	101.86	0.012316	0.39935	26.832	58.631	85.463	0.055602	0.16384	82
84	104.92	0.012356	0.38776	27.300	58.343	85.643	0.056452	0.16376	84
86	108.04	0.012396	0.37657	27.769	58.052	85.821	0.057301	0.16368	86
88	111.23	0.012437	0.36575	28.241	57.757	85.998	0.058149	0.16360	88
90	114.49	0.012478	0.35529	28.713	57.461	86.174	0.058997	0.16353	90
92	117.82	0.012520	0.34518	29.187	57.161	86.348	0.059844	0.16345	92
94	121.22	0.012562	0.33540	29.663	56.858	86.521	0.060690	0.16338	94
96	124.70	0.012605	0.32594	30.140	56.551	86.691	0.061536	0.16330	96
98	128.24	0.012649	0.31679	30.619	56.242	86.861	0.062381	0.16323	98
100	131.86	0.012693	0.30794	31.100	55.929	87.029	0.063227	0.16315	100
102	135.56	0.012738	0.29937	31.583	55.613	87.196	0.064072	0.16308	102
104	139.33	0.012783	0.29106	32.067	55.293	87.360	0.064916	0.16301	104

Table A-3 REFRIGERANT 12 SATURATION PROPERTIES—ENGLISH UNITS (*continued*)

Fahrenheit Temperature	Pressure psia P	Volume ft³/lbm Liquid v_f	Vapor v_g	Enthalpy Btu/lbm Liquid i_f	Latent i_{fg}	Vapor i_g	Entropy Btu/(lbm-R) Liquid s_f	Vapor s_g	Fahrenheit Temperature
106	143.18	0.012829	0.28303	32.553	54.970	87.523	0.065761	0.16293	106
108	147.11	0.012876	0.27524	33.041	54.643	87.684	0.066606	0.16286	108
110	151.11	0.012924	0.26769	33.531	54.313	87.844	0.067451	0.16279	110
112	155.19	0.012972	0.26037	34.023	53.978	88.001	0.068296	0.16271	112
114	159.36	0.013022	0.25328	34.517	53.639	88.156	0.069141	0.16264	114
116	163.61	0.013072	0.24641	35.014	53.296	88.310	0.069987	0.16256	116
118	167.94	0.013123	0.23974	35.512	52.949	88.461	0.070833	0.16249	118
120	172.35	0.013174	0.23326	36.013	52.597	88.610	0.071680	0.16241	120
122	176.85	0.013227	0.22698	36.516	52.241	88.757	0.072528	0.16234	122
124	181.43	0.013280	0.22089	37.021	51.881	88.902	0.073376	0.16226	124
126	186.10	0.013335	0.21497	37.529	51.515	89.044	0.074225	0.16218	126
128	190.86	0.013390	0.20922	38.040	51.144	89.184	0.075075	0.16210	128
130	195.71	0.013447	0.20364	38.553	50.768	89.321	0.075927	0.16202	130
132	200.64	0.013504	0.19821	39.069	50.387	89.456	0.076779	0.16194	132

134	205.67	0.013563	0.19294	39.588	50.000	89.588	0.077633	0.16185	134
136	210.79	0.013623	0.18782	40.110	49.608	89.718	0.078489	0.16177	136
138	216.01	0.013684	0.18283	40.634	49.210	89.844	0.079346	0.16168	138
140	221.32	0.013746	0.17799	41.162	48.805	89.967	0.080205	0.16159	140
142	226.72	0.013810	0.17327	41.693	48.394	90.087	0.081065	0.16150	142
144	232.22	0.013874	0.16868	42.227	47.977	90.204	0.081928	0.16140	144
146	237.82	0.013941	0.16422	42.765	47.553	90.318	0.082794	0.16130	146
148	243.51	0.014008	0.15987	43.306	47.122	90.428	0.083661	0.16120	148
150	249.31	0.014078	0.15564	43.850	46.684	90.534	0.084531	0.16110	150
152	255.20	0.014148	0.15151	44.399	46.238	90.637	0.085404	0.16099	152
154	261.20	0.014221	0.14750	44.951	45.784	90.735	0.086280	0.16088	154
156	267.30	0.014295	0.14358	45.508	45.322	90.830	0.087159	0.16077	156
158	273.51	0.014371	0.13976	46.068	44.852	90.920	0.088041	0.16065	158
160	279.82	0.014449	0.13604	46.633	44.373	91.006	0.088927	0.16053	160
162	286.24	0.014529	0.13241	47.202	43.885	91.087	0.089817	0.16040	162
164	292.77	0.014611	0.12886	47.777	43.386	91.163	0.090710	0.16027	164
166	299.40	0.014695	0.12540	48.355	42.879	91.234	0.091608	0.16014	166
168	306.15	0.014782	0.12202	48.939	42.360	91.299	0.092511	0.16000	168
170	313.00	0.014871	0.11873	49.529	41.830	91.359	0.093418	0.15985	170
172	319.97	0.014963	0.11550	50.123	41.290	91.413	0.094330	0.15969	172
174	327.06	0.015058	0.11235	50.724	40.736	91.460	0.095248	0.15953	174

Table A-3a REFRIGERANT 12 SATURATION PROPERTIES—SI UNITS*

Celsius Temperature	Pressure Pa $P \times 10^{-5}$	Enthalpy kJ/kg i_f	i_{fg}	i_g	Entropy kJ/(kg-K) s_f	s_g	Specific Volume m³/kg $v_f \times 10^3$	$v_g \times 10^3$
−40	0.6412	163.949	169.601	333.551	0.85804	1.58548	0.65947	242.094
−35	0.8065	168.369	167.489	335.858	0.87675	1.58004	0.66561	195.536
−30	1.0034	172.811	165.342	338.153	0.89516	1.57516	0.67198	159.487
−28	1.0919	174.595	164.472	339.067	0.90243	1.57334	0.67459	147.373
−26	1.1864	176.381	163.596	339.978	0.90967	1.57160	0.67724	136.374
−24	1.2871	178.172	162.713	340.885	0.91686	1.56993	0.67993	126.364
−22	1.3943	179.966	161.824	341.790	0.92400	1.56833	0.68267	117.243
−20	1.5083	181.765	160.927	342.692	0.93110	1.56680	0.68544	108.915
−18	1.6294	183.568	160.022	343.590	0.93816	1.56533	0.68827	101.303
−16	1.7577	185.375	159.110	344.485	0.94518	1.56393	0.69113	94.3375
−14	1.8938	187.186	158.189	345.375	0.95216	1.56258	0.69404	87.9461
−12	2.0377	189.003	157.259	346.262	0.95911	1.56129	0.69701	82.0803
−10	2.1898	190.823	156.322	347.144	0.96601	1.56005	0.70002	76.6898
−9	2.2691	191.735	155.849	347.584	0.96945	1.55945	0.70154	74.1563
−8	2.3505	192.648	155.374	348.022	0.97288	1.55886	0.70308	71.7262
−7	2.4341	193.562	154.898	348.460	0.97630	1.55829	0.70463	69.3948
−6	2.5199	194.478	154.418	348.896	0.97971	1.55773	0.70619	67.1525
−5	2.6080	195.395	153.936	349.331	0.98311	1.55718	0.70777	65.0005
−4	2.6986	196.313	153.452	349.765	0.98651	1.55664	0.70936	62.9298
−3	2.7913	197.233	152.965	350.198	0.98989	1.55611	0.71097	60.9418
−2	2.8866	198.154	152.475	350.629	0.99327	1.55560	0.71259	59.0285
−1	2.9841	199.077	151.983	351.059	0.99664	1.55509	0.71422	57.1910
0	3.0842	200.000	151.488	351.488	1.00000	1.55460	0.71587	55.4217

1	3.1869	200.926	150.990	351.916	1.00336	1.55411	0.71754	53.7176
2	3.2922	201.853	150.489	352.342	1.00670	1.55364	0.71922	52.0763
3	3.4000	202.781	149.985	352.766	1.01004	1.55317	0.72091	50.4973
4	3.5104	203.711	149.478	353.189	1.01338	1.55272	0.72262	48.9758
5	3.6236	204.642	148.969	353.611	1.01670	1.55227	0.72435	47.5095
6	3.7394	205.574	148.457	354.031	1.02002	1.55184	0.72609	46.0982
7	3.8580	206.509	147.941	354.450	1.02333	1.55140	0.72786	44.7358
8	3.9794	207.445	147.422	354.867	1.02663	1.55098	0.72963	43.4225
9	4.1035	208.383	146.900	355.283	1.02993	1.55058	0.73143	42.1578
10	4.2306	209.322	146.375	355.697	1.03322	1.55017	0.73324	40.9363
11	4.3609	210.264	145.845	356.109	1.03651	1.54977	0.73507	39.7551
12	4.4939	211.207	145.313	356.520	1.03978	1.54939	0.73692	38.6171
13	4.6298	212.152	144.777	356.929	1.04306	1.54901	0.73879	37.5192
14	4.7689	213.099	144.237	357.336	1.04632	1.54863	0.74068	36.4570
15	4.9111	214.047	143.694	357.741	1.04959	1.54826	0.74259	35.4317
16	5.0564	214.998	143.147	358.145	1.05284	1.54790	0.74451	34.4409
17	5.2051	215.951	142.596	358.546	1.05609	1.54755	0.74646	33.4816
18	5.3567	216.905	142.041	358.947	1.05934	1.54720	0.74843	32.5569
19	5.5119	217.863	141.482	359.344	1.06258	1.54685	0.75042	31.6603
20	5.6703	218.821	140.919	359.740	1.06581	1.54652	0.75243	30.7944
21	5.8319	219.782	140.352	360.134	1.06904	1.54619	0.75446	29.9570
22	5.9969	220.745	139.781	360.526	1.07227	1.54586	0.75652	29.1470
23	6.1653	221.710	139.206	360.916	1.07549	1.54554	0.75859	38.3634
24	6.3374	222.678	138.625	361.303	1.07871	1.54523	0.76069	27.6030
25	6.5131	223.649	138.040	361.688	1.08193	1.54491	0.76282	26.8672
26	6.6922	224.621	137.450	362.071	1.08514	1.54461	0.76497	26.1552
27	6.8750	225.595	136.856	362.451	1.08834	1.54430	0.76715	25.4649
28	7.0615	226.574	136.256	362.829	1.09155	1.54400	0.76935	24.7958
29	7.2516	227.554	135.652	363.206	1.09475	1.54370	0.77157	24.1481

Table A-3a   REFRIGERANT 12 SATURATION PROPERTIES—SI UNITS* (*continued*)

Celsius Temperature	Pressure Pa $P \times 10^{-5}$	Enthalpy kJ/kg			Entropy kJ/(kg-K)		Specific Volume m³/kg	
		i_f	i_{fg}	i_g	s_f	s_g	$v_f \times 10^3$	$v_g \times 10^3$
30	7.4455	228.537	135.043	363.579	1.09795	1.54341	0.77383	23.5200
31	7.6430	229.521	134.429	363.950	1.10114	1.54312	0.77611	22.9119
32	7.8451	230.511	133.806	364.317	1.10433	1.54283	0.77842	22.3192
33	8.0507	231.502	133.181	364.683	1.10752	1.54254	0.78075	21.7461
34	8.2598	232.495	132.551	365.046	1.11071	1.54226	0.78312	21.1912
35	8.4732	233.492	131.914	365.406	1.11390	1.54198	0.78552	20.6518
36	8.6909	234.493	131.270	365.763	1.11708	1.54170	0.78794	20.1276
37	8.9123	235.496	130.622	366.117	1.12027	1.54142	0.79040	19.6197
38	9.1384	236.503	129.965	366.468	1.12345	1.54114	0.79290	19.1251
39	9.3683	237.512	129.304	366.816	1.12663	1.54087	0.79542	18.6459
40	9.6026	238.526	128.635	367.161	1.12981	1.54059	0.79798	18.1799
41	9.8410	239.541	127.961	367.502	1.13299	1.54032	0.80058	17.7276
42	10.084	240.561	127.280	367.841	1.13617	1.54004	0.80321	17.2884
43	10.331	241.585	126.592	368.177	1.13935	1.53977	0.80588	16.8612

44	10.583	242.612	125.895	368.507	1.14253	1.53949	0.80858	16.4449
45	10.839	243.643	125.193	368.835	1.14571	1.53921	0.81133	16.0413
46	11.099	244.678	124.481	369.159	1.14889	1.53893	0.81411	15.6479
47	11.364	245.715	123.765	369.480	1.15207	1.53865	0.81694	15.2665
48	11.634	246.758	123.039	369.797	1.15525	1.53837	0.81981	14.8945
49	11.909	247.805	122.303	370.108	1.15844	1.53808	0.82272	14.5319
50	12.188	248.855	121.562	370.417	1.16162	1.53780	0.82568	14.1803
52	12.761	250.969	120.053	371.022	1.16799	1.53722	0.83173	13.5040
54	13.353	253.101	118.507	371.608	1.17438	1.53662	0.83799	12.8623
56	13.966	255.252	116.924	372.176	1.18078	1.53601	0.84446	12.2532
58	14.600	257.422	115.301	372.723	1.18719	1.53537	0.85115	11.6747
60	15.253	259.611	113.639	373.250	1.19361	1.53472	0.85808	11.1253
62	15.928	261.819	111.937	373.756	1.20005	1.53404	0.86528	10.6033
64	16.625	264.051	110.187	374.239	1.20650	1.53332	0.87275	10.1061
66	17.344	266.302	108.394	374.696	1.21297	1.53258	0.88053	9.63298
68	18.086	268.577	106.552	375.129	1.21947	1.53180	0.88863	9.18216
70	18.851	270.873	104.662	373.534	1.22598	1.53098	0.89709	8.75265
75	20.867	276.701	99.726	376.428	1.24231	1.52876	0.92001	7.76538
80	23.038	282.647	94.493	377.140	1.25869	1.52626	0.94603	6.88968

*Abstracted by permission from "Thermodynamic Table for Refrigerant 12 in SI-Units" International Institute of Refrigeration, Paris.

Table A-4   REFRIGERANT 22 SATURATION PROPERTIES—ENGLISH UNITS*

Fahrenheit Temperature	Pressure psia P	Volume ft³/lbm Liquid v_f	Volume ft³/lbm Vapor v_g	Enthalpy Btu/lbm Liquid i_f	Enthalpy Btu/lbm Latent i_{fg}	Enthalpy Btu/lbm Vapor i_g	Entropy Btu/(lbm-R) Liquid s_f	Entropy Btu/(lbm-R) Vapor s_g	Fahrenheit Temperature
−40	15.222	0.011363	3.2957	0.000	100.257	100.257	0.00000	0.23888	−40
−38	16.024	0.011389	3.1412	0.506	99.971	100.477	0.00120	0.23827	−38
−36	16.859	0.011415	2.9954	1.014	99.682	100.696	0.00240	0.23767	−36
−34	17.728	0.011442	2.8578	1.524	99.391	100.914	0.00359	0.23707	−34
−32	18.633	0.011469	2.7278	2.035	99.097	101.132	0.00479	0.23649	−32
−30	19.573	0.011495	2.6049	2.547	98.801	101.348	0.00598	0.23591	−30
−28	20.549	0.011523	2.4887	3.061	98.503	101.564	0.00716	0.23534	−28
−26	21.564	0.011550	2.3787	3.576	98.202	101.778	0.00835	0.23478	−26
−24	22.617	0.011578	2.2746	4.093	97.899	101.992	0.00953	0.23423	−24
−22	23.711	0.011606	2.1760	4.611	97.593	102.204	0.01072	0.23369	−22
−20	24.845	0.011634	2.0826	5.131	97.285	102.415	0.01189	0.23315	−20
−18	26.020	0.011662	1.9940	5.652	96.974	102.626	0.01307	0.23262	−18
−16	27.239	0.011691	1.9099	6.175	96.660	102.835	0.01425	0.23210	−16
−14	28.501	0.011720	1.8302	6.699	96.344	103.043	0.01542	0.23159	−14
−12	29.809	0.011749	1.7544	7.224	96.025	103.250	0.01659	0.23108	−12
−10	31.162	0.011778	1.6825	7.751	95.704	103.455	0.01776	0.23058	−10
−8	32.563	0.011808	1.6141	8.280	95.380	103.660	0.01892	0.23008	−8
−6	34.011	0.011838	1.5491	8.810	95.053	103.863	0.02009	0.22960	−6
−4	35.509	0.011868	1.4872	9.341	94.724	104.065	0.02125	0.22912	−4
−2	37.057	0.011899	1.4283	9.874	94.391	104.266	0.02241	0.22864	−2
0	38.657	0.011930	1.3723	10.409	94.056	104.465	0.02357	0.22817	0
2	40.309	0.011961	1.3189	10.945	93.718	104.663	0.02472	0.22771	2
4	42.014	0.011992	1.2680	11.483	93.378	104.860	0.02587	0.22725	4
6	43.775	0.012024	1.2195	12.022	93.034	105.056	0.02703	0.22680	6

616

8	45.591	0.012056	1.1732	12.562	92.688	105.250	0.02818	0.22636	8
10	47.464	0.012088	1.1290	13.104	92.338	105.442	0.02932	0.22592	10
12	49.396	0.012121	1.0869	13.648	91.986	105.633	0.03047	0.22548	12
14	51.387	0.012154	1.0466	14.193	91.630	105.823	0.03161	0.22505	14
16	53.438	0.012188	1.0082	14.739	91.272	106.011	0.03275	0.22463	16
18	55.551	0.012221	0.97144	15.288	90.910	106.198	0.03389	0.22421	18
20	57.727	0.012255	0.93631	15.837	90.545	106.383	0.03503	0.22379	20
22	59.967	0.012290	0.90270	16.389	90.178	106.566	0.03617	0.22338	22
24	62.272	0.012325	0.87055	16.942	89.807	106.748	0.03730	0.22297	24
26	64.644	0.012360	0.83978	17.496	89.433	106.928	0.03844	0.22257	26
28	67.083	0.012395	0.81031	18.052	89.055	107.107	0.03958	0.22217	28
30	69.591	0.012431	0.78208	18.609	88.674	107.284	0.04070	0.22178	30
32	72.169	0.012468	0.75503	19.169	88.290	107.459	0.04182	0.22139	32
34	74.818	0.012505	0.72911	19.729	87.903	107.632	0.04295	0.22100	34
36	77.540	0.012542	0.70425	20.292	87.512	107.804	0.04407	0.22062	36
38	80.336	0.012579	0.68041	20.856	87.118	107.974	0.04520	0.22024	38
40	83.206	0.012618	0.65753	21.422	86.720	108.142	0.04632	0.21986	40
42	86.153	0.012656	0.63557	21.989	86.319	108.308	0.04744	0.21949	42
44	89.177	0.012695	0.61448	22.558	85.914	108.472	0.04855	0.21912	44
46	92.280	0.012735	0.59422	23.129	85.506	108.634	0.04967	0.21876	46
48	95.463	0.012775	0.57476	23.701	85.094	108.795	0.05079	0.21839	48
50	98.727	0.012815	0.55606	24.275	84.678	108.953	0.05190	0.21803	50
52	102.07	0.012856	0.53808	24.851	84.258	109.109	0.05301	0.21768	52
54	105.50	0.012898	0.52078	25.429	83.834	109.263	0.05412	0.21732	54
56	109.02	0.012940	0.50414	26.008	83.407	109.415	0.05523	0.21697	56
58	112.62	0.012982	0.48813	26.589	82.975	109.564	0.05634	0.21662	58
60	116.31	0.013025	0.47272	27.172	82.540	109.712	0.05745	0.21627	60
62	120.09	0.013069	0.45788	27.757	82.100	109.857	0.05855	0.21592	62
64	123.96	0.013114	0.44358	28.344	81.656	110.000	0.05966	0.21558	64
66	127.92	0.013159	0.42981	28.932	81.208	110.140	0.06076	0.21524	66
68	131.97	0.013204	0.41653	29.523	80.755	110.278	0.06186	0.21490	68

Table A-4   REFRIGERANT 22 SATURATION PROPERTIES—ENGLISH UNITS* *(continued)*

Fahrenheit Temperature	Pressure psia P	Volume ft³/lbm Liquid v_f	Volume ft³/lbm Vapor v_g	Enthalpy Btu/lbm Liquid i_f	Enthalpy Btu/lbm Latent i_{fg}	Enthalpy Btu/lbm Vapor i_g	Entropy Btu/(lbm-R) Liquid s_f	Entropy Btu/(lbm-R) Vapor s_g	Fahrenheit Temperature
70	136.12	0.013251	0.40373	30.116	80.298	110.414	0.06296	0.21456	70
72	140.37	0.013297	0.39139	30.170	79.836	110.547	0.06406	0.21422	72
74	144.71	0.013345	0.37949	31.307	79.370	110.677	0.06516	0.21388	74
76	149.15	0.013393	0.36800	31.906	78.899	110.805	0.06626	0.21355	76
78	153.69	0.013442	0.35691	32.506	78.423	110.930	0.06736	0.21321	78
80	158.33	0.013492	0.34621	33.109	77.943	111.052	0.06846	0.21288	80
82	163.07	0.013543	0.33587	33.714	77.457	111.171	0.06956	0.21255	82
84	167.92	0.013594	0.32588	34.322	76.966	111.288	0.07065	0.21222	84
86	172.87	0.013647	0.31623	34.931	76.470	111.401	0.07175	0.21188	86
88	177.93	0.013700	0.30690	35.543	75.968	111.512	0.07285	0.21155	88
90	183.09	0.013754	0.29789	36.158	75.461	111.619	0.07394	0.21122	90
92	188.37	0.013809	0.28917	36.774	75.949	111.723	0.07504	0.21089	92
94	193.76	0.013864	0.28073	37.394	74.430	111.824	0.07613	0.21056	94
96	199.26	0.013921	0.27257	38.016	73.905	111.921	0.07723	0.21023	96
98	204.87	0.013979	0.26467	38.640	74.375	112.015	0.07832	0.20989	98
100	210.60	0.014038	0.25702	39.267	72.838	112.105	0.07942	0.20956	100
102	216.45	0.014098	0.24962	39.897	72.294	112.192	0.08052	0.20923	102
104	222.42	0.014159	0.24244	40.530	71.744	112.274	0.08161	0.20889	104
106	228.50	0.014221	0.23549	41.166	71.187	112.353	0.08271	0.20855	106
108	234.71	0.014285	0.22875	41.804	70.623	112.427	0.08381	0.20821	108
110	241.04	0.014350	0.22222	42.446	70.052	112.498	0.08491	0.20787	110
112	247.50	0.014416	0.21589	43.091	69.473	112.564	0.08601	0.20753	112
114	254.08	0.014483	0.20974	43.739	68.886	112.626	0.08711	0.20718	114
116	260.79	0.014552	0.20378	44.391	68.291	112.682	0.08821	0.20684	116

118	0.20649	0.08932	112.735	67.688	45.046	118
120	0.20613	0.09042	112.782	67.077	45.705	120
122	0.20578	0.09153	112.824	66.456	46.368	122
124	0.20542	0.09264	112.860	65.826	47.034	124
126	0.20505	0.09375	112.891	65.186	47.705	126
128	0.20468	0.09487	112.917	64.537	48.380	128
130	0.20431	0.09598	112.936	63.877	49.059	130
132	0.20393	0.09711	112.949	63.206	49.743	132
134	0.20354	0.09823	112.955	62.523	50.432	134
136	0.20315	0.09936	112.954	61.829	51.125	136
138	0.20275	0.10049	112.947	61.123	51.824	138
140	0.20235	0.10163	112.931	60.403	52.528	140
142	0.20194	0.10277	112.908	59.670	53.238	142
144	0.20152	0.10391	112.877	59.922	53.955	144
146	0.20109	0.10507	112.836	58.159	54.677	146
148	0.20065	0.10622	112.787	57.380	55.406	148
150	0.20020	0.10739	112.728	56.585	56.143	150
152	0.19974	0.10856	112.658	55.771	56.887	152
154	0.19926	0.10974	112.577	54.939	57.638	154
156	0.19878	0.11093	112.485	54.087	58.399	156
158	0.19828	0.11213	112.381	53.213	59.168	158
160	0.19776	0.11334	112.263	52.316	59.948	160
162	0.19723	0.11456	112.131	51.394	60.737	162
164	0.19668	0.11580	111.984	50.446	61.538	164
166	0.19611	0.11705	111.820	49.469	62.351	166
168	0.19552	0.11831	111.639	48.461	63.178	168
170	0.19490	0.11959	111.438	47.419	64.019	170
172	0.19425	0.12089	111.216	46.340	64.875	172
174	0.19358	0.12222	110.970	45.221	65.750	174

118	267.63	0.014622	0.19800	45.046	67.688
120	274.60	0.014694	0.19238	45.705	67.077
122	281.71	0.014768	0.18692	46.368	66.456
124	288.95	0.014843	0.18163	47.034	65.826
126	296.33	0.014920	0.17648	47.705	65.186
128	303.84	0.014999	0.17147	48.380	64.537
130	311.50	0.015080	0.16661	49.059	63.877
132	319.29	0.015163	0.16187	49.743	63.206
134	327.23	0.015248	0.15727	50.432	62.523
136	335.32	0.015336	0.15279	51.125	61.829
138	343.56	0.015426	0.14843	51.824	61.123
140	351.94	0.015518	0.14418	52.528	60.403
142	360.48	0.015613	0.14004	53.238	59.670
144	369.17	0.015712	0.13600	53.955	59.922
146	378.02	0.015813	0.13207	54.677	58.159
148	387.03	0.015917	0.12823	55.406	57.380
150	396.19	0.016025	0.12448	56.143	56.585
152	405.52	0.016137	0.12083	56.887	55.771
154	415.02	0.016252	0.11726	57.638	54.939
156	424.68	0.016372	0.11376	58.399	54.087
158	434.52	0.016497	0.11035	59.168	53.213
160	444.53	0.016627	0.10710	59.948	52.316
162	454.71	0.016762	0.10374	60.737	51.394
164	465.07	0.016902	0.10054	61.538	50.446
166	475.61	0.017050	0.97393	62.351	49.469
168	486.34	0.017204	0.094309	63.178	48.461
170	497.26	0.017367	0.091279	64.019	47.419
172	508.37	0.017538	0.088299	64.875	46.340
174	519.67	0.017719	0.085365	65.750	45.221

Table A-4a REFRIGERANT 22 SATURATION PROPERTIES—SI UNITS*

Celsius Temperature	Pressure Pa $P \times 10^{-5}$	Enthalpy kJ/kg			Entropy kJ/(kg-K)		Specific Volume m³/kg	
		i_f	i_{fg}	i_g	s_f	s_g	$v_f \times 10^3$	$v_g \times 10^3$
−40	1.0490	155.624	233.204	388.828	0.82489	1.82512	0.70935	205.841
−35	1.3162	160.923	230.162	391.085	0.84742	1.81388	0.71679	166.470
−30	1.6340	166.291	227.008	393.299	0.86975	1.80337	0.72451	135.907
−28	1.7768	168.458	225.712	394.170	0.87863	1.79934	0.72768	125.613
−26	1.9291	170.636	224.398	395.034	0.88748	1.79542	0.73091	116.262
−24	2.0912	172.826	223.064	395.890	0.89630	1.79160	0.73418	107.747
−22	2.2639	175.027	221.709	396.736	0.90509	1.78786	0.73751	99.9741
−20	2.4472	177.239	220.335	397.574	0.91385	1.78423	0.74090	92.8825
−18	2.6418	179.463	218.939	398.402	0.92259	1.78066	0.74435	86.3881
−16	2.8481	181.698	217.522	399.220	0.93129	1.77719	0.74785	80.4427
−14	3.0666	183.945	216.083	400.028	0.93997	1.77378	0.75142	74.9848
−12	3.2976	186.204	214.622	400.826	0.94862	1.77046	0.75505	69.9736
−10	3.5416	188.473	213.140	401.613	0.95725	1.76720	0.75875	65.3646
−9	3.6686	189.612	212.390	402.002	0.96155	1.76560	0.76062	63.1994
−8	3.7992	190.755	211.634	402.388	0.96585	1.76401	0.76251	61.1184
−7	3.9331	191.900	210.873	402.772	0.97014	1.76245	0.76442	59.1227
−6	4.0708	193.049	210.104	403.153	0.97442	1.76089	0.76635	57.2017
−5	4.2118	194.199	209.332	403.531	0.97870	1.75935	0.76830	55.3612
−4	4.3568	195.354	208.551	403.905	0.98297	1.75782	0.77026	53.5871
−3	4.5054	196.511	207.765	404.277	0.98724	1.75631	0.77225	51.8828
−2	4.6577	197.671	206.974	404.645	0.99150	1.75482	0.77425	50.2452
−1	4.8138	198.834	206.177	405.011	0.99575	1.75334	0.77628	48.6700
0	4.9740	200.000	205.372	405.373	1.00000	1.75187	0.77832	47.1523
1	5.1382	201.170	204.562	405.731	1.00424	1.75041	0.78039	45.6920
2	5.3065	202.343	203.744	406.087	1.00848	1.74896	0.78248	44.2848
3	5.4786	203.517	202.922	406.439	1.01271	1.74753	0.78459	42.9319
4	5.6549	204.696	202.092	406.788	1.01694	1.74612	0.78672	41.6278

5	5.8360	205.878	201.254	407.132	1.02115	1.74470	0.78887	40.3677
6	6.0207	207.062	200.412	407.474	1.02537	1.74331	0.79105	39.1576
7	6.2100	208.250	199.562	407.812	1.02958	1.74192	0.79325	37.9890
8	6.4036	209.441	198.706	408.147	1.03378	1.74055	0.79548	36.8624
9	6.6020	210.636	197.841	408.477	1.03798	1.73918	0.79773	35.7742
10	6.8045	211.834	196.971	408.804	1.04218	1.73782	0.80001	34.7261
11	7.0123	213.036	196.090	409.126	1.04637	1.73647	0.80231	33.7109
12	7.2239	214.240	195.207	409.447	1.05056	1.73514	0.80464	32.7357
13	7.4408	215.448	194.314	409.761	1.05474	1.73380	0.80699	31.7909
14	7.6626	216.660	193.412	410.072	1.05892	1.73248	0.80937	30.8778
15	7.8890	217.874	192.504	410.378	1.06309	1.73116	0.81179	29.9967
16	8.1203	219.092	191.589	410.681	1.06726	1.72985	0.81423	29.1453
17	8.3566	220.314	190.666	410.979	1.07142	1.72855	0.81670	28.3224
18	8.5980	221.539	189.735	411.274	1.07559	1.72726	0.81920	27.5268
19	8.8446	222.768	188.795	411.563	1.07974	1.72597	0.82173	26.7567
20	9.0967	224.002	187.846	411.847	1.08390	1.72468	0.82430	26.0109
21	9.3535	225.237	186.891	412.128	1.08805	1.72341	0.82689	25.2910
22	9.6158	226.477	185.926	412.403	1.09220	1.72214	0.82953	24.5937
23	9.8840	227.722	184.951	412.673	1.09634	1.72086	0.83219	23.9174
24	10.157	228.970	183.970	412.939	1.10049	1.71960	0.83489	23.2643
25	10.436	230.222	182.978	413.200	1.10463	1.71834	0.83763	22.6307
26	10.720	231.477	181.978	413.455	1.10876	1.71708	0.84041	22.0177
27	11.011	232.738	180.967	413.705	1.11290	1.71582	0.84322	21.4229
28	11.306	234.001	179.949	413.950	1.11703	1.71457	0.84608	20.8476
29	11.608	235.270	178.920	414.190	1.12117	1.71332	0.84897	20.2890
30	11.915	236.543	177.881	414.424	1.12530	1.71207	0.85191	19.7478
31	12.229	237.820	176.831	414.651	1.12943	1.71082	0.85489	19.2222
32	12.549	239.102	175.771	414.873	1.13356	1.70957	0.85791	18.7129
33	12.874	240.389	174.700	415.089	1.13769	1.70832	0.86099	18.2183
34	13.206	241.678	173.621	415.300	1.14181	1.70708	0.86410	17.7397
35	13.544	242.976	172.527	415.502	1.14594	1.70582	0.86727	17.2732
36	13.889	244.276	171.424	415.700	1.15007	1.70457	0.87049	16.8217
37	14.239	245.580	170.310	415.891	1.15420	1.70332	0.87376	16.3832

Table A-4a   REFRIGERANT 22 SATURATION PROPERTIES—SI UNITS* (continued)

Celsius Temperature	Pressure Pa $P \times 10^{-5}$	Enthalpy kJ/kg			Entropy kJ/(kg-K)		Specific Volume m³/kg	
		i_f	i_{fg}	i_g	s_f	s_g	$v_f \times 10^3$	$v_g \times 10^3$
38	14.597	246.892	169.182	416.074	1.15833	1.70206	0.87708	15.9564
39	14.960	248.207	168.044	416.251	1.16246	1.70080	0.88046	15.5425
40	15.331	249.530	166.890	416.419	1.16659	1.69953	0.88390	15.1396
41	15.708	250.857	165.725	416.581	1.17073	1.69826	0.88739	14.7482
42	16.092	252.190	164.546	416.735	1.17487	1.69699	0.89095	14.3677
43	16.483	253.528	163.355	416.882	1.17901	1.69571	0.89457	13.9980
44	16.881	254.873	162.148	417.020	1.18315	1.69442	0.89825	13.6380
45	17.286	256.223	160.927	417.150	1.18730	1.69312	0.90201	13.2880
46	17.698	257.581	159.690	417.271	1.19145	1.69181	0.90583	12.9471
47	18.117	258.944	158.440	417.384	1.19561	1.69050	0.90973	12.6157
48	18.543	260.314	157.173	417.487	1.19977	1.68918	0.91371	12.2929
49	18.977	261.692	155.889	417.581	1.20394	1.68784	0.91777	11.9784
50	19.418	263.073	154.594	417.667	1.20811	1.68650	0.92190	11.6731
52	20.323	265.866	151.938	417.804	1.21648	1.68377	0.93044	11.0839
54	21.258	268.686	149.214	417.900	1.22488	1.68099	0.93936	10.5251
56	22.227	271.546	146.400	417.946	1.23334	1.67812	0.94869	9.99235
58	23.227	274.441	143.502	417.943	1.24183	1.67518	0.95847	9.48558
60	24.260	277.373	140.514	417.887	1.25038	1.67215	0.96874	9.00332
62	25.326	280.348	137.423	417.771	1.25899	1.66903	0.97957	8.54301
64	26.428	283.373	134.216	417.589	1.26768	1.66577	0.99101	8.10286
66	27.566	286.451	130.883	417.334	1.27647	1.66238	1.00314	7.68155
68	28.740	289.585	127.417	417.002	1.28535	1.65884	1.01604	7.27841
70	29.952	292.786	123.795	416.581	1.29436	1.65512	1.02983	6.89127
75	33.153	301.133	113.925	415.058	1.31757	1.64480	1.06911	5.98528
80	36.614	310.139	102.511	412.650	1.34222	1.63249	1.11803	5.15095

*Abstracted by permission from "Thermodynamic Table for Refrigerant R22 in SI-Units," International Institute of Refrigeration, Paris.

Thermophysical Properties

Table B-1   REFRIGERANT 12—ENGLISH UNITS*

Fahrenheit Temperature	Viscosity, μ lbm/(ft-hr)			Thermal Conductivity, k Btu/(hr-ft-F)			Specific Heat, c_p Btu/(lbm-F)			Fahrenheit Temperature
	Saturated Liquid	Saturated Vapor	Gas $P = 1$ atm $\mu \times 10^2$	Saturated Liquid	Saturated Vapor	Gas $P = 1$ atm $k \times 10^3$	Saturated Liquid	Saturated Vapor	Gas $P = 0$	
−140	2.44			0.0655			0.199		0.1085	−140
−120	1.95			0.0631			0.202		0.1233	−120
−100	1.593			0.0608			0.204		0.1160	−100
−80	1.331			0.0585			0.207	0.126	0.1196	−80
−60	1.132			0.0561			0.209	0.133	0.1230	−60
−40	0.978			0.0538			0.212	0.139	0.1264	−40
−20	0.856	0.0246	2.46	0.0514	0.0040	4.00	0.214	0.145	0.1296	−20
0	0.758	0.0262	2.58	0.0490	0.0043	4.31	0.217	0.150	0.1327	0
20	0.679	0.0276	2.69	0.0467	0.0046	4.63	0.220	0.157	0.1356	20
40	0.613	0.0287	2.80	0.0443	0.0050	4.95	0.224	0.164	0.1385	40
60	0.557	0.0297	2.90	0.0420	0.0053	5.28	0.229	0.164	0.1413	60
80	0.511	0.0308	3.01	0.0397	0.0056	5.61	0.234	0.174	0.1439	80

100	0.471	0.0320	3.12	0.0373	0.0060	5.94	0.240	0.185	0.1465	100
120	0.436	0.0335	3.22	0.0350	0.0064	6.27	0.251	0.199	0.1490	120
140	0.403	0.0354	3.32	0.0326	0.0068	6.60	0.266	0.216	0.1513	140
160	0.366	0.0380	3.42	0.0302	0.0072	6.94	0.288	0.235	0.1536	160
180	0.325	0.0412	3.52	0.0276	0.0076	7.28	0.317	0.290	0.1558	180
200	0.270	0.0453	3.62	0.0246	0.0083	7.63	0.356	0.362	0.1579	200
220	0.198	0.051	3.72	0.0204	0.0093	7.98	0.406		0.1599	220
230	0.148	0.060	3.77	0.0161	0.0107	8.16			0.1609	230
234	0.075	0.075	3.79	0.0130	0.130	8.23			0.1612	234
240			3.82			8.34			0.1618	240
260			3.92			8.71			0.1637	260
280			4.01			9.08			0.1654	280
300			4.10			9.45			0.1671	300
320			4.20			9.82			0.1687	320
340			4.29			10.1			0.1703	340
360			4.38			10.5			0.1718	360
380			4.47			10.8			0.1732	380
400			4.56			11.2			0.1746	400

*Reprinted by permission from *ASHRAE Handbook of Fundamentals*, 1972.

Table B-1a REFRIGERANT 12—SI UNITS*

Celsius Temperature	Viscosity, μ (N-s)/m²			Thermal Conductivity, k W/(m-K)			Specific Heat, c_p kJ/(kg-K)			Kelvin Temperature
	Saturated Liquid $\mu \times 10^3$	Saturated Vapor $\mu \times 10^3$	Gas $P = 101.33$ kPa $\mu \times 10^3$	Saturated Liquid	Saturated Vapor	Gas $P = 101.33$ kPa $k \times 10$	Saturated Liquid	Saturated Vapor	Gas $P = 0$	
−95	1.01			0.113			0.833		0.4543	178
−84	0.806			0.109			0.846		0.4702	189
−73	0.659			0.105			0.854		0.4857	200
−62	0.550			0.101			0.867		0.5007	211
−51	0.468			0.097			0.875	0.528	0.5150	222
−40	0.404			0.093			0.888	0.557	0.5292	233
−29	0.354	0.0102	0.0102	0.089	0.0069	0.069	0.896	0.582	0.5426	244
−18	0.313	0.0108	0.0107	0.085	0.0074	0.075	0.909	0.607	0.5556	255
−7	0.281	0.0114	0.0111	0.081	0.0080	0.080	0.921	0.628	0.5677	266
4	0.253	0.0119	0.0116	0.077	0.0087	0.086	0.938	0.657	0.5799	277
16	0.230	0.0123	0.0120	0.073	0.0920	0.091	0.959	0.687	0.5916	289
27	0.211	0.0127	0.0124	0.069	0.0097	0.097	0.980	0.729	0.6025	300

38	0.195	0.0132	0.0129	0.065	0.0104	0.103	1.00	0.775	0.6113	311
49	0.180	0.0138	0.0133	0.061	0.0111	0.109	1.05	0.833	0.6238	322
60	0.167	0.0146	0.0137	0.056	0.0118	0.114	1.11	0.904	0.6335	333
71	0.151	0.0157	0.0141	0.052	0.0125	0.120	1.21	0.984	0.6431	344
82	0.134	0.0170	0.0146	0.048	0.0132	0.126	1.33	1.21	0.6523	355
93	0.112	0.0187	0.0150	0.043	0.0173	0.132	1.49	1.52	0.6611	366
104	0.082	0.021	0.0154	0.035	0.0161	0.138	1.70		0.6695	377
110	0.061	0.025	0.0156	0.028	0.0185	0.141			0.6737	383
112	0.031	0.031	0.0157	0.022	0.0225	0.142			0.6749	385
116			0.0158			0.144			0.6774	389
127			0.0162			0.151			0.6854	400
138			0.0166			0.157			0.6925	411
149			0.0169			0.164			0.6996	422
160			0.0174			0.170			0.7063	433
171			0.0177			0.175			0.7130	444
183			0.0181			0.182			0.7193	456
194			0.0185			0.187			0.7252	467
205			0.0189			0.194			0.7310	478

*Adapted by permission from *ASHRAE Handbook of Fundamentals*, 1972.

Table B-2 REFRIGERANT 22—ENGLISH UNITS*

Fahrenheit Temperature	Viscosity, μ lbm/(ft-hr)			Thermal Conductivity, k Btu/(hr-ft-F)			Specific Heat, c_p Btu/(lbm-F)			Fahrenheit Temperature
	Saturated Liquid	Saturated Vapor	Gas $P = 1$ atm $\mu \times 10^2$	Saturated Liquid	Saturated Vapor	Gas $P = 1$ atm $k \times 10^3$	Saturated Liquid	Saturated Vapor	Gas $P = 0$	
−100	1.153			0.0789			0.255		0.1260	−100
−80	1.002			0.0757			0.256		0.1292	−80
−60	0.883			0.0725			0.259	0.139	0.1324	−60
−40	0.788	0.0242	2.42	0.0693	0.0040	4.04	0.262	0.146	0.1356	−40
−20	0.710	0.0254	2.54	0.0661	0.0044	4.43	0.266	0.152	0.1388	−20
0	0.646	0.0266	2.65	0.0630	0.0048	4.81	0.271	0.158	0.1420	0
20	0.592	0.0279	2.76	0.0598	0.0052	5.20	0.276	0.165	0.1452	20
40	0.546	0.0292	2.88	0.0566	0.0056	5.58	0.283	0.175	0.1484	40
60	0.507	0.0306	2.99	0.0534	0.0060	5.97	0.291	0.187	0.1515	60
80	0.474	0.0321	3.10	0.0502	0.0064	6.35	0.300	0.204	0.1546	80
100	0.444	0.0339	3.21	0.0471	0.0068	6.74	0.313	0.226	0.1577	100
120	0.422	0.0360	3.33	0.0439	0.0072	7.12	0.332	0.253	0.1608	120

140	0.387	0.0383	3.44	0.0407	0.0077	7.51	0.357	0.288	0.1638	140
160	0.340	0.0411	3.54	0.0371	0.0084	7.90	0.390	0.332	0.1668	160
180	0.285	0.045	3.65	0.0318	0.0105	8.28	0.433		0.1697	180
190	0.244	0.049	3.71	0.0288	0.0119	8.48			0.1712	190
200	0.182	0.058	3.76	0.0238	0.0140	8.67			0.1726	200
205	0.074	0.074	3.79	0.0177	0.0177	8.76			0.1733	205
220			3.87			9.05			0.1754	220
240			3.98			9.44			0.1782	240
260			4.08			9.82			0.1810	260
280			4.19			10.21			0.1836	280
300			4.29			10.59			0.1863	300
320			4.39			11.0			0.1888	320
340			4.50			11.4			0.1913	340
360			4.60			11.8			0.1937	360
380			4.70			12.1			0.1960	380
400			4.80			12.5			0.1983	400
420			4.90			12.9			0.2005	420
440			5.00			13.3			0.2026	440

*Reprinted by permission from *ASHRAE Handbook of Fundamentals,* 1972.

Table B-2a   REFRIGERANT 22—SI UNITS*

Celsius Temperature	Viscosity, μ (N-s)/m^2			Thermal Conductivity, k W/(m-K)			Specific Heat, c_p kJ/(kg-K)			Kelvin Temperature
	Saturated Liquid $\mu \times 10^3$	Saturated Vapor $\mu \times 10^3$	Gas $P = 101.33$ kPa $\mu \times 10^5$	Saturated Liquid	Saturated Vapor	Gas $P = 101.33$ kPa $k \times 10$	Saturated Liquid	Saturated Vapor	Gas $P = 0$	
−73	0.4766			0.137			1.07		0.5275	200
−62	0.4142			0.131			1.07		0.5409	211
−51	0.365			0.125			1.08	0.582	0.5543	222
−40	0.326	0.0100	1.00	0.120	0.0069	0.0699	1.10	0.611	0.5677	233
−29	0.294	0.0105	1.05	0.114	0.0076	0.0767	1.11	0.636	0.5811	244
−18	0.267	0.0110	1.10	0.109	0.0083	0.0832	1.13	0.662	0.5945	255
−7	0.245	0.0115	1.14	0.103	0.0090	0.0900	1.16	0.691	0.6079	266
4	0.226	0.0121	1.19	0.0980	0.0097	0.0966	1.18	0.733	0.6213	277
16	0.210	0.0126	1.24	0.0924	0.010	0.103	1.22	0.783	0.6343	289
27	0.196	0.0133	1.28	0.0869	0.011	0.110	1.26	0.854	0.6473	300
38	0.184	0.0140	1.33	0.0815	0.012	0.117	1.31	0.946	0.6603	311
49	0.174	0.0149	1.38	0.0760	0.012	0.123	1.39	1.06	0.6732	322
60	0.160	0.0158	1.42	0.0704	0.013	0.130	1.49	1.21	0.6858	333

71	0.141	0.0170	1.46	0.0642	0.015	0.137	1.63	1.39	0.6984	344
82	0.118	0.019	1.51	0.0550	0.018	0.143	1.81		0.7105	355
88	0.101	0.020	1.53	0.0498	0.021	0.147			0.7168	361
94	0.075	0.024	1.55	0.0412	0.024	0.150			0.7226	367
96	0.031	0.031	1.57	0.0306	0.031	0.152			0.7256	369
105			1.60			0.157			0.7344	378
116			1.65			0.163			0.7461	389
127			1.69			0.170			0.7578	400
138			1.73			0.1767			0.7687	411
149			1.77			0.1833			0.7800	422
160			1.81			0.190			0.7905	433
171			1.86			0.197			0.8009	444
183			1.90			0.204			0.8110	456
194			1.94			0.209			0.8206	467
205			1.98			0.216			0.8302	478
216			2.03			0.223			0.8395	489
227			2.07			0.230			0.8482	500

*Adapted by permission from *ASHRAE Handbook of Fundamentals*, 1972.

Table B-3 WATER—ENGLISH UNITS*

Fahrenheit Temperature	Viscosity, μ lbm/(ft-hr)			Thermal Conductivity, k Btu/(hr-ft-F)			Specific Heat, c_p Btu/(lbm-F)				Fahrenheit Temperature
	Saturated Liquid	Saturated Vapor	Gas, $P = 1$ atm $\mu \times 10^2$	Saturated Liquid	Saturated Vapor	Gas, $P = 1$ atm $k \times 10^3$	Saturated Liquid	Saturated Vapor	Gas $P = 0$	Gas $P = 1$ atm	
32	4.23	0.0192		0.329	0.0100		1.007		0.4438		32
40	3.65	0.0197		0.334	0.0103		1.005		0.4442		40
50	3.09	0.0202		0.339	0.0105		1.003		0.4445		50
60	2.64	0.0207		0.345	0.0107		1.001	0.4446	0.4448		60
70	2.29	0.0213		0.350	0.0109		1.000	0.4454	0.4452		70
80	2.01	0.0218		0.354	0.0112		0.999	0.4464	0.4456		80
90	1.77	0.0224		0.359	0.0114		0.998	0.4475	0.4459		90
100	1.58	0.0229		0.363	0.0116		0.998	0.4488	0.4463		100
120	1.29	0.0240		0.371	0.0121		0.998	0.4522	0.4472		120
140	1.080	0.0251		0.378	0.0125		1.000	0.4567	0.4480		140
160	0.927	0.0261		0.383	0.0130		1.001	0.4626	0.4490		160
180	0.804	0.0272		0.388	0.0135		1.003	0.4699	0.4500		180

200	0.708	0.0283		0.392	0.0140		1.006	0.4791	0.4510		200
212	0.660	0.0289	2.89	0.394	0.0143	14.3	1.008	0.4859	0.4516	0.4892	212
220	0.631	0.0292	2.94	0.395	0.0145	14.4	1.009	0.4902	0.4521	0.4867	220
240	0.567	0.0302	3.05	0.397	0.0151	14.7	1.013	0.5019	0.4532	0.4819	240
260	0.515	0.0312	3.15	0.398	0.0158	15.3	1.017	0.5157	0.4544	0.4780	260
280	0.471	0.0322	3.26	0.398	0.0165	15.8	1.022	0.5319	0.4556	0.4750	280
300	0.433	0.0331	3.37	0.397	0.0172	16.4	1.029	0.5508	0.4569	0.4729	300
320	0.401	0.0341	3.48	0.396	0.0181	17.0	1.040	0.573	0.4582	0.4714	320
340	0.374	0.0350	3.59	0.394	0.0190	17.5	1.053	0.598	0.4595	0.4706	340
360	0.350	0.0359	3.69	0.391	0.0199	18.1	1.066	0.626	0.4609	0.4703	360
400	0.311	0.0377	3.91	0.383	0.0222	19.4	1.085	0.695	0.4638	0.4709	400
500	0.248	0.0427	4.45	0.349	0.0306	22.6	1.180	0.968	0.4715	0.4745	500
600	0.197	0.0504	4.99	0.296	0.0486	26.2	1.528	1.71	0.4798	0.4817	600
700			5.53	0.188	0.1203	29.8			0.4885	0.4895	700
706	0.101	0.101	5.56	0.139	0.139	30.0			0.4890	0.4900	706
750			5.80			31.7			0.4929	0.4935	750
800			6.07			33.7			0.4973	0.4977	800
1000			7.15			41.8			0.5150	0.5152	1000

*Reprinted by permission from *ASHRAE Handbook of Fundamentals*, 1972.

Table B-3a WATER—SI UNITS*

Celsius Temperature	Viscosity, μ (N-s)/m²			Thermal Conductivity, k W/(m-K)			Specific Heat, c_p kJ/(kg-K)				Kelvin Temperature
	Saturated Liquid $\mu \times 10^3$	Saturated Vapor $\mu \times 10^3$	Gas $P = 101.33$ kPa $\mu \times 10^5$	Saturated Liquid	Saturated Vapor	Gas $P = 101.33$ kPa $k \times 10^3$	Saturated Liquid	Saturated Vapor	Gas $P = 0$ kPa	Gas $P = 101.33$ kPa	
0	1.75	0.00794		0.569	0.0173		4.216		1.858		273
4	1.51	0.00814		0.578	0.0178		4.208		1.860		277
10	1.28	0.00835		0.587	0.0182		4.199		1.861		283
16	1.09	0.00856		0.597	0.0185		4.191	1.861	1.862		289
21	0.947	0.00880		0.606	0.0189		4.187	1.865	1.864		294
27	0.831	0.00901		0.613	0.0194		4.183	1.869	1.866		300
32	0.732	0.00926		0.621	0.0197		4.178	1.874	1.867		305
38	0.653	0.00947		0.628	0.0201		4.178	1.879	1.869		311
49	0.533	0.00992		0.642	0.0209		4.178	1.893	1.872		322
60	0.446	0.0104		0.654	0.0216		4.187	1.912	1.876		333
71	0.383	0.0108		0.663	0.0225		4.191	1.937	1.880		344

82	0.332	0.0112		0.672	0.0234		4.199	1.967	1.884		355
93	0.293	0.0117		0.678	0.0242		4.212	2.006	1.888		366
100	0.273	0.0119	1.19	0.682	0.0247	24.7	4.220	2.034	1.891	2.048	373
104	0.261	0.121	1.22	0.684	0.0251	24.9	4.224	2.052	1.893	2.038	377
116	0.234	0.0125	1.26	0.687	0.0261	25.4	4.241	2.101	1.897	2.018	389
127	0.213	0.0129	1.30	0.689	0.0273	26.5	4.258	2.159	1.902	2.001	400
138	0.195	0.0133	1.35	0.689	0.0286	27.3	4.280	2.227	1.908	1.989	411
149	0.179	0.0137	1.39	0.687	0.0298	28.4	4.308	2.306	1.913	1.980	422
160	0.166	0.0141	1.44	0.685	0.0313	29.4	4.354	2.399	1.918	1.974	433
171	0.155	0.0145	1.48	0.682	0.0329	30.3	4.409	2.504	1.924	1.970	444
183	0.145	0.0148	1.53	0.677	0.0344	31.3	4.463	2.621	1.930	1.969	456
205	0.129	0.0156	1.62	0.663	0.0384	33.6	4.543	2.910	1.942	1.972	478
260	0.103	0.0177	1.84	0.604	0.0530	39.1	4.940	4.053	1.974	1.987	533
316	0.0814	0.208	2.06	0.512	0.0841	45.3	6.397	7.159	2.009	2.017	589
371			2.29	0.325	0.208	51.6			2.045	2.049	644
375	0.0418	0.0418	2.30	0.241	0.241	51.9			2.047	2.052	648
399			2.40			54.9			2.064	2.066	672
427			2.51			58.3			2.082	2.084	700
538			2.96			72.3			2.156	2.157	811

*Adapted by permission from *ASHRAE Handbook of Fundamentals*, 1972.

Table B-4 AIR—ENGLISH UNITS

Fahrenheit Temperature	Viscosity, μ lbm/(ft-hr)			Thermal Conductivity, k Btu/(hr-ft-F)			Specific Heat, c_p Btu/(lbm-F)				Fahrenheit Temperature
	Saturated Liquid	Saturated Vapor	Gas $P = 1$ atm $\mu \times 10^2$	Saturated Liquid	Saturated Vapor	Gas $P = 1$ atm $k \times 10^3$	Saturated Liquid	Saturated Vapor	Gas $P = 0$	Gas $P = 1$ atm	
−352	0.7865			0.1040	0.00312						−352
−334	0.5348			0.0942	0.00370						−334
−316	0.3993	0.0133		0.0838	0.00433		0.4690		0.2394		−316
−298	0.3194	0.0157	1.54	0.0740	0.00497	4.80	0.4962	0.2676	0.2394		−298
−280	0.2664	0.0182	1.71	0.0636	0.00584	5.32	0.5268	0.2891	0.2394	0.2456	−280
−262	0.2297	0.0208	1.88	0.0537	0.00705	5.89	0.5784	0.3440	0.2394	0.2442	−262
−244	0.1815	0.0247	2.04	0.0439	0.00890	6.41	0.6689	0.4778	0.2394	0.2430	−244
−266	0.1016	0.0346	2.20	0.0312	0.01213	6.93			0.2394	0.2422	−226
−220.6	0.05009	0.05009	2.25	0.01964	0.01964	7.11			0.2394	0.2420	−220.6
−208			2.36			7.45			0.2394	0.2418	−208
−190			2.51			7.97			0.2394	0.2415	−190
−172			2.66			8.44			0.2394	0.2411	−172

−136	2.95	9.48	0.2394	0.2406	−136
−100	3.24	10.5	0.2394	0.2403	−100
−64	3.51	11.4	0.2396	0.2403	−64
−28	3.75	12.4	0.2396	0.2401	−28
8	4.02	13.4	0.2396	0.2401	8
44	4.26	14.2	0.2399	0.2403	44
80	4.48	15.1	0.2401	0.2403	80
116	4.72	16.0	0.2403	0.2406	116
152	4.94	16.8	0.2406	0.2408	152
188	5.16	17.6	0.2411	0.2413	188
224	5.35	18.3	0.2415	0.2418	224
260	5.54	19.1	0.2420	0.2422	260
296	5.76	19.9	0.2427	0.2430	296
332	5.93	20.6	0.2434	0.2437	332
368	6.12	21.4	0.2442	0.2444	368
404	6.32	22.1	0.2449	0.2451	404
440	6.49	22.8	0.2458	0.2461	440
620	7.33	26.4	0.2511	0.2513	620

*Adapted by permission from *ASHRAE Handbook of Fundamentals*, 1972.

Table B-4a AIR—SI UNITS*

Celsius Temperature	Viscosity, μ (N-s)/m²			Thermal Conductivity, k W/(m-K)			Specific Heat, c_p kJ/(kg-K)				Kelvin Temperature
	Saturated Liquid $\mu \times 10^3$	Saturated Vapor $\mu \times 10^3$	Gas $P = 101.33$ kPa $\mu \times 10^6$	Saturated Liquid	Saturated Vapor	Gas $P = 101.33$ kPa	Saturated Liquid	Saturated Vapor	Gas $P = 0$ kPa	Gas $P = 101.33$ kPa	
−213	0.325			0.180	0.0054						60
−203	0.221			0.163	0.0064						70
−193	0.165	0.0055		0.145	0.0075		1.963		1.002		80
−183	0.132	0.0065	6.35	0.128	0.0086	0.0083	2.077	1.12	1.002		90
−173	0.1101	0.0075	7.06	0.110	0.0101	0.0092	2.205	1.21	1.002	1.028	100
−163	0.0949	0.0086	7.75	0.093	0.0122	0.0102	2.421	1.44	1.002	1.022	110
−153	0.0750	0.0102	8.43	0.076	0.0154	0.0111	2.80	2.00	1.002	1.017	120
−143	0.0420	0.0143	9.09	0.054	0.021	0.0120			1.002	1.014	130
−140	0.0207	0.0207	9.29	0.034	0.034	0.0123			1.002	1.013	133
−133			9.74			0.0129			1.002	1.012	140
−123			10.38			0.0138			1.002	1.011	150

−113	11.0	0.0146	1.002	1.009	160
−93	12.2	0.0164	1.002	1.007	180
−73	13.4	0.0181	1.002	1.006	200
−53	14.5	0.0198	1.003	1.006	220
−33	15.5	0.0215	1.003	1.005	240
−13	16.6	0.0231	1.003	1.005	260
7	17.6	0.0246	1.004	1.006	280
27	18.5	0.0261	1.005	1.006	300
47	19.5	0.0276	1.006	1.007	320
67	20.4	0.0290	1.007	1.008	340
87	21.3	0.0304	1.009	1.010	360
107	22.1	0.0317	1.011	1.012	380
127	22.9	0.0331	1.013	1.014	400
147	23.8	0.0344	1.016	1.017	420
167	24.5	0.0357	1.019	1.020	440
187	25.3	0.0370	1.022	1.023	460
207	26.1	0.0383	1.025	1.026	480
227	26.8	0.0395	1.029	1.030	500
327	30.3	0.0456	1.051	1.052	600

*Reprinted by permission from *ASHRAE Handbook of Fundamentals*, 1972.

Weather Data

Table C-1 CLIMATIC CONDITIONS FOR THE UNITED STATES AND CANADA*

State and Station	Latitude, Degree and Minutes	Longitude, Degree and Minutes	Elevation ft	Elevation m	Winter 99 Percent F	Winter 99 Percent C	Winter 97½ Percent F	Winter 97½ Percent C	Summer Dry Bulb 2½ Percent F	Summer Dry Bulb 2½ Percent C	Summer Wet Bulb 2½ Percent F	Summer Wet Bulb 2½ Percent C	Summer Daily Range F	Summer Daily Range C
Alabama, Birmingham AP	33 3	86 5	610	186	17	−8	21	−6	94	34	75	24	21	12
Alaska, Anchorage AP	61 1	150 0	90	27	−23	−31	−18	−28	68	20	58	14	15	8
Arizona, Tucson AP	32 1	111 0	2584	788	28	−2	32	0	102	39	66	19	26	14
Arkansas, Little Rock AP	34 4	92 1	257	78.3	15	−9	20	−7	96	36	77	25	22	12
California, San Francisco AP	37 4	122 2	8	2.4	35	2	38	3	77	25	63	17	20	11
Colorado, Denver AP	39 5	104 5	5283	1610	−5	−21	1	−17	91	33	59	15	28	16
Connecticut, Bridgeport AP	41 1	73 1	7	2.1	−6	−21	9	−13	84	29	71	22	18	10
Delaware, Wilmington AP	39 4	75 3	78	24	10	−12	14	−10	89	32	74	23	20	11
District of Columbia, Washington National	38 5	77 0	14	4.3	14	−10	17	−8	91	33	74	23	18	10
Florida, Tallahassee AP	30 2	84 2	58	18	27	−3	30	−1	92	33	76	24	19	11
Georgia, Atlanta AP	33 4	84 3	1005	306	17	−8	22	−6	92	33	74	23	19	11
Hawaii, Honolulu AP	21 2	158 0	7	2.1	62	17	63	17	86	30	73	23	12	7
Idaho, Boise AP	43 3	116 1	2842	866	3	−16	10	−12	94	34	64	18	31	17
Illinois, Chicago O'Hare AP	42 0	87 5	658	201	−8	−22	−4	−20	89	32	74	23	20	11
Indiana, Indianapolis AP	39 4	86 2	793	242	−2	−19	2	−17	90	32	74	23	22	12
Iowa, Sioux City AP	42 2	96 2	1095	334	−11	−24	−7	−22	92	33	74	23	24	13
Kansas, Wichita AP	37 4	97 3	1321	403	3	−16	7	−14	98	37	73	23	23	13
Kentucky, Louisville AP	38 1	85 4	474	144	5	−15	10	−12	93	34	75	24	23	13
Louisiana, Shreveport AP	32 3	93 5	252	76.8	20	−7	25	−4	96	36	76	24	20	11
Maine, Caribou AP	46 5	68 0	624	190	−18	−28	−13	−25	81	27	67	19	21	12
Maryland, Baltimore AP	39 1	76 4	146	44.5	10	−12	13	−11	91	33	75	24	21	12
Massachusetts, Boston AP	42 2	71 0	15	4.6	−6	−14	9	−13	88	31	71	22	16	9
Michigan, Lansing AP	42 5	84 4	852	260	−3	−19	1	−17	87	31	72	22	24	13
Minnesota, Minneapolis/St. Paul	44 5	93 1	822	251	−16	−27	−12	−24	89	32	73	23	22	12
Mississippi, Jackson AP	32 2	90 1	330	101	21	−6	25	−4	95	35	76	24	21	12

Location	Lat.	Long.	Elev.										
Missouri, Kansas City AP	39 1	94 4	742	226	2	−17	6	−14	96	36	74	23	20	11
Montana, Billings AP	45 4	108 3	3567	1087	−15	−26	−10	−23	91	33	64	18	31	17
Nebraska, Lincoln CO	40 5	96 5	1150	351	−5	−21	−2	−19	95	35	74	23	24	13
Nevada, Las Vegas AP	36 1	115 1	2162	659	25	−4	28	−12	106	41	65	18	30	17
New Hampshire, Concord	43 1	71 3	339	103	−8	−22	−3	−19	87	31	70	21	26	14
New Jersey, Atlantic City CO	39 3	74 3	11	3.4	10	−12	13	−11	89	32	74	23	18	10
New Mexico, Albuquerque AP	35 0	106 4	5310	1618	12	−11	16	−9	94	34	61	16	30	17
New York, Syracuse AP	43 1	76 1	424	129	−3	−19	2	−17	87	31	71	22	20	11
North Carolina, Charlotte AP	35 1	81 0	735	224	18	−8	22	−6	93	34	74	23	20	11
North Dakota, Bismarck AP	46 5	100 5	1647	502	−23	−31	−19	−28	91	33	68	20	27	15
Ohio, Cleveland AP	41 2	81 5	777	237	1	−17	5	−15	88	31	72	22	22	12
Oklahoma, Stillwater	36 1	97 1	884	269	8	−13	13	−11	96	36	74	23	24	13
Oregon, Pendleton AP	45 4	118 5	1492	455	−2	−19	5	−15	93	34	64	18	29	16
Pennsylvania, Pittsburgh AP	40 3	80 0	1137	347	1	−17	5	−15	86	30	71	22	22	12
Rhode Island, Providence AP	41 4	71 3	55	17	5	−15	9	−13	86	30	72	22	19	11
South Carolina, Charleston AFB	32 5	80 0	41	12	24	−4	27	−3	91	33	78	26	18	10
South Dakota, Rapid City	44 0	103 0	3165	965	−11	−24	−7	−22	92	33	65	18	28	16
Tennessee, Memphis AP	35 0	90 0	263	80.2	13	−11	18	−8	95	35	76	24	21	12
Texas, Dallas AP	32 5	96 5	481	147	18	−8	22	−6	100	38	75	24	20	11
Utah, Salt Lake City	40 5	112 0	4220	1286	3	−16	8	−13	95	35	62	17	32	18
Vermont, Burlington AP	44 3	73 1	331	101	−12	−24	−7	−22	85	29	70	21	23	13
Virginia, Norfolk AP	36 5	76 1	26	7.9	20	−7	22	−6	91	33	76	24	18	10
Washington, Spokane AP	47 4	117 3	2357	718	−6	−21	2	−17	90	32	63	17	28	16
West Virginia, Charleston AP	38 2	81 4	939	286	7	−14	11	−12	90	32	73	23	20	11
Wisconsin, Milwaukee AP	43 0	87 5	672	205	−8	−22	−4	−20	87	31	73	23	21	12
Wyoming, Casper AP	42 5	106 3	5319	1621	−11	−24	−5	−21	90	32	57	14	31	17
Alberta, Calgary AP	51 1	114 1	3540	1079	−27	−33	−23	−31	81	27	61	16	25	14
British Columbia, Vancouver AP	49 1	123 1	16	4.9	15	−9	19	−7	77	25	66	19	17	9
Manitoba, Winnipeg AP	49 5	97 1	786	240	−30	−34	−27	−33	86	30	71	22	22	12
New Brunswick, Fredericton	45 5	66 3	74	23	−16	−27	−11	−24	85	29	69	21	23	13
Nova Scotia, Halifax AP	44 4	63 3	136	41.5	1	−17	5	−15	76	24	71	18	16	9
Ontario, Ottawa AP	45 2	76 0	339	103	−17	−27	−13	−25	87	31	71	22	21	12
Quebec, Montreal AP	45 3	74 0	98	30	−16	−27	−10	−23	85	29	72	22	18	10
Saskatchewan, Regina AP	50 3	105 0	1884	574	−33	−36	−29	−34	88	31	68	20	27	15

*Abridged by permission from *ASHRAE Handbook of Fundamentals*, 1977.

Table C-2   AVERAGE DEGREE DAYS[a] FOR CITIES IN THE UNITED STATES AND CANADA*

State and Station	Average Winter Temperature F	C	July	August	September	October	November	December	January	February	March	April	May	June	Yearly Total
Alabama, Birmingham	54.2	12.7	0	0	6	93	363	555	592	462	363	108	9	0	2551
Alaska, Anchorage	23.0	5.0	245	291	516	930	1284	1572	1631	1316	1293	879	592	315	10864
Arizona, Tucson	58.1	14.8	0	0	0	25	231	406	471	344	242	75	6	0	1800
Arkansas, Little Rock	50.5	10.6	0	0	9	127	465	716	756	577	434	126	9	0	3219
California, San Francisco	53.4	12.2	81	78	60	143	306	462	508	395	363	279	214	126	3015
Colorado, Denver	37.6	3.44	6	9	117	428	819	1035	1132	938	887	558	288	66	6283
Connecticut, Bridgeport	39.9	4.72	0	0	66	307	615	986	1079	966	853	510	208	27	5617
Delaware, Wilmington	42.5	6.17	0	0	51	270	588	927	980	874	735	387	112	6	4930
District of Columbia, Washington	45.7	7.94	0	0	33	217	519	834	871	762	626	288	74	0	4224
Florida, Tallahassee	60.1	15.9	0	0	0	28	198	360	375	286	202	86	0	0	1485
Georgia, Atlanta	51.7	11.28	0	0	18	124	417	648	636	518	428	147	25	0	2961
Hawaii, Honolulu	74.2	23.8	0	0	0	0	0	0	0	0	0	0	0	0	0
Idaho, Boise	39.7	4.61	0	0	132	415	792	1017	1113	854	722	438	245	81	5809
Illinois, Chicago	35.8	2.44	0	12	117	381	807	1166	1265	1086	939	534	260	72	6639
Indiana, Indianapolis	39.6	4.56	0	0	90	316	723	1051	1113	949	809	432	177	39	5699
Iowa, Sioux City	43.0	1.10	0	9	108	369	867	1240	1435	1198	989	483	214	39	6951
Kansas, Wichita	44.2	7.11	0	0	33	229	618	905	1023	804	645	270	87	6	4620
Kentucky, Louisville	44.0	6.70	0	0	54	248	609	890	930	818	682	315	105	9	4660
Louisiana, Shreveport	56.2	13.8	0	0	0	47	297	477	552	426	304	81	0	0	2184
Maine, Caribou	24.4	-3.89	78	115	336	682	1044	1535	1690	1470	1308	858	468	183	9767
Maryland, Baltimore	43.7	6.83	0	0	48	264	585	905	936	820	679	327	90	0	4654
Massachusetts, Boston	40.0	4.40	0	9	60	316	603	983	1088	972	846	513	208	36	5634
Michigan, Lansing	34.8	1.89	6	22	138	431	813	1163	1262	1142	1011	579	273	69	6909
Minnesota, Minneapolis	28.3	-1.72	22	31	189	505	1014	1454	1631	1380	1166	621	288	81	8382
Mississippi, Jackson	55.7	13.5	0	0	0	65	315	502	546	414	310	87	0	0	2239
Missouri, Kansas City	43.9	6.94	0	0	39	220	612	905	1032	818	682	294	109	0	4711
Montana, Billings	34.5	1.72	6	15	186	487	897	1135	1296	1100	970	570	285	102	7049

Location	°C	°F													Yearly Total
Nebraska, Lincoln	38.8	4.11	0	6	75	301	726	1066	1237	1016	834	402	171	30	5864
Nevada, Las Vegas	53.5	12.28	0	0	0	78	387	617	688	487	335	111	6	0	2709
New Hampshire, Concord	33.0	0.60	6	50	177	505	822	1240	1358	1184	1032	636	298	75	7383
New Jersey, Atlantic City	43.2	6.56	0	0	39	251	549	880	936	848	741	420	133	15	4812
New Mexico, Albuquerque	45.0	7.20	0	0	12	229	642	868	930	703	595	288	81	0	4348
New York, Syracuse	35.2	2.11	6	28	132	415	744	1153	1271	1140	1004	570	248	45	6756
North Carolina, Charlotte	50.4	10.56	0	0	6	124	438	691	691	582	481	156	22	0	3191
North Dakota, Bismarck	26.6	−2.67	34	28	222	577	1083	1463	1708	1442	1203	645	329	117	8851
Ohio, Cleveland	37.2	3.22	9	25	105	384	738	1088	1159	1047	918	552	260	66	6351
Oklahoma, Stillwater	48.3	9.39	0	0	15	164	498	766	868	664	527	189	34	0	3725
Oregon, Pendleton	42.6	6.22	0	0	111	350	711	884	1017	773	617	396	205	63	5127
Pennsylvania, Pittsburgh	38.4	3.89	0	9	105	375	726	1063	1119	1002	874	480	195	39	5987
Rhode Island, Providence	38.8	4.11	0	16	96	372	660	1023	1110	988	868	534	236	51	5954
South Carolina, Charleston	56.4	13.9	0	0	0	59	282	471	487	389	291	54	0	0	2033
South Dakota, Rapid City	33.4	1.11	22	12	165	481	897	1172	1333	1145	1051	615	326	126	7345
Tennessee, Memphis	50.5	10.6	0	0	18	130	447	698	729	585	456	147	22	0	3232
Texas, Dallas	55.3	13.3	0	0	0	62	321	524	601	440	319	90	6	0	2363
Utah, Salt Lake City	38.4	3.89	0	0	81	419	849	1082	1172	910	763	459	233	84	6052
Vermont, Burlington	29.4	−1.11	28	65	207	539	891	1349	1513	1333	1187	714	353	90	8269
Virginia, Norfolk	49.2	9.89	0	0	0	136	408	698	738	655	533	216	37	0	3421
Washington, Spokane	36.5	2.83	9	25	168	493	879	1082	1231	980	834	531	288	135	6655
West Virginia, Charleston	44.8	7.44	0	0	63	254	591	865	880	770	648	300	96	9	4476
Wisconsin, Milwaukee	32.6	0.667	43	47	174	471	876	1252	1376	1193	1054	642	372	135	7635
Wyoming, Casper	33.4	1.11	6	16	192	524	942	1169	1290	1084	1020	657	381	129	7410
Alberta, Calgary	—	—	109	186	402	719	1110	1389	1575	1379	1268	798	477	291	9703
British Columbia, Vancouver	—	—	81	87	219	456	657	787	862	723	676	501	310	156	5515
Manitoba, Winnipeg	—	—	38	71	322	683	1251	1757	2008	1719	1465	813	405	147	10679
New Brunswick, Fredericton	—	—	78	68	234	592	915	1392	1541	1379	1172	753	406	141	8671
Nova Scotia, Halifax	—	—	58	51	180	457	710	1074	1213	1122	1030	742	487	237	7361
Ontario, Ottawa	—	—	25	81	222	567	936	1469	1624	1441	1231	708	341	90	8735
Quebec, Montreal	—	—	9	43	165	521	882	1392	1566	1381	1175	684	316	69	8203
Saskatchewan, Regina	—	—	78	93	360	741	1284	1711	1965	1687	1473	804	409	201	10806

*Abridged by permission from *ASHRAE Handbook and Product Directory—Systems*, 1980.

ªBased on degrees F, quantities may be converted to degree days based on degrees C by multiplying by $\tfrac{5}{9}$. This assumes 18 C corresponds to 65 F.

Pipe and Tube Data

Table D-1 STEEL PIPE DIMENSIONS*—ENGLISH AND SI UNITS

Nominal Pipe Size in.	Schedule Number	Diameter				Wall Thickness		Inside Cross-Sectional Area	
		OD		ID					
		in.	mm × 10⁻³	in.	mm × 10⁻³	in.	mm	ft²	m² × 10³
$\frac{1}{4}$	40	0.540	0.0137	0.364	0.00925	0.088	2.23	0.00072	0.067
	80			0.302	0.00767	0.119	3.02	0.00050	0.046
$\frac{3}{8}$	40	0.675	0.0171	0.493	0.0125	0.091	2.31	0.00133	0.124
	80			0.423	0.0107	0.126	3.20	0.00098	0.091
$\frac{1}{2}$	40	0.840	0.0213	0.622	0.0158	0.109	2.77	0.00211	0.196
	80			0.546	0.0139	0.147	3.73	0.00163	0.151
$\frac{3}{4}$	40	1.050	0.0267	0.824	0.0209	0.113	2.87	0.00371	0.345
	80			0.742	0.0188	0.154	3.91	0.00300	0.279
1	40	1.315	0.0334	1.049	0.0266	0.133	3.38	0.00600	0.557
	80			0.957	0.0243	0.179	4.55	0.00499	0.464
$1\frac{1}{2}$	40	1.900	0.0483	1.610	0.0409	0.145	3.68	0.01414	1.314

	Sched.	O.D.		I.D.					
	80			1.500	0.0381	0.200	5.08	0.01225	1.138
2	40	2.375	0.0603	2.067	0.0525	0.154	3.91	0.02330	2.165
	80			1.939	0.0493	0.218	5.54	0.02050	1.905
$2\frac{1}{2}$	40	2.875	0.0730	2.469	0.0627	0.203	5.16	0.03322	3.086
	80			2.323	0.0590	0.276	7.01	0.02942	2.733
3	40	3.500	0.0889	3.068	0.0779	0.216	5.49	0.05130	4.766
	80			2.900	0.0737	0.300	7.62	0.04587	4.262
4	40	4.500	0.1143	4.026	0.1023	0.237	6.02	0.08840	8.213
	80			3.826	0.0972	0.337	8.56	0.07986	7.419
5	40	5.563	0.1413	5.047	0.1281	0.258	6.55	0.1390	12.91
	80			4.813	0.1223	0.375	9.53	0.1263	11.73
6	40	6.625	0.1683	6.065	0.1541	0.280	7.11	0.2006	18.64
	80			5.761	0.1463	0.432	11.0	0.1810	16.82
8	40	8.625	0.2191	7.981	0.2027	0.322	8.18	0.3474	32.28
	80			7.625	0.1937	0.500	12.7	0.3171	29.46
10	40	10.75	0.2731	10.020	0.2545	0.365	9.27	0.5475	50.86
	80			9.750	0.2477	0.500	12.7	0.5185	48.17

*Adapted from A.S.A. Standards B36.10.

Table D-2   TYPE L COPPER TUBE DIMENSIONS*—ENGLISH AND SI UNITS

| Nominal Pipe Size in. | Diameter | | | | Wall Thickness | | Inside Cross-Sectional Area | |
| | OD | | ID | | | | | |
	in.	mm	in.	mm	in.	mm	in.^{2}	m$^2 \times 10^3$
$\frac{1}{4}$	0.375	9.53	0.315	8.00	0.030	0.762	0.0779	0.0503
$\frac{3}{8}$	0.500	12.7	0.43	11	0.035	0.889	0.145	0.094
$\frac{1}{2}$	0.625	15.9	0.545	13.8	0.040	1.02	0.233	0.150
$\frac{5}{8}$	0.750	19.1	0.666	16.9	0.042	1.07	0.348	0.225
$\frac{3}{4}$	0.875	22.2	0.785	19.9	0.045	1.14	0.484	0.312
1	1.125	28.58	1.025	26.04	0.050	1.27	0.825	0.532
$1\frac{1}{4}$	1.375	34.93	1.265	32.13	0.055	1.40	1.26	0.813
$1\frac{1}{2}$	1.625	41.28	1.505	38.23	0.060	1.52	1.78	1.15
2	2.125	53.98	1.985	50.42	0.070	1.78	3.10	2.00
$2\frac{1}{2}$	2.625	66.68	2.465	62.61	0.080	2.03	4.77	3.08
3	3.125	79.38	2.945	74.80	0.090	2.29	6.81	4.39
$3\frac{1}{2}$	3.625	92.08	3.425	87.00	0.100	2.54	9.21	5.94
4	4.125	104.8	3.905	99.19	0.110	2.79	12.0	7.74

*Based on ASTM A-88.

Useful Data

Table E-1 CONVERSION FACTORS

Length
1 ft = 30.48 cm
1 in. = 2.54 cm
1 m = 39.37 in.
1 micron = 10^{-6} m = 3.281×10^{-6} ft
1 mile = 5280 ft

Area
1 m^2 = 1550.1472 $in.^2$
1 m^2 = 10.76392 ft^2

Volume
1 ft^3 = 7.48 U.S. gallons = 1728 $in.^3$
1 m^3 = 6.1×10^4 $in.^3$
1 m^3 = 35.3147 ft^3
1 m^3 = 264.154 U.S. gallons

Mass
1 kg = 2.20462 lbm
1 lbm = 7000 grains = 453.5924 g

Force
1 N = 0.224809 lbf
1 lbf = 4.44822 N

Table E-1   CONVERSION FACTORS (*continued*)

Energy
1 Btu = 778.28 ft-lbf
1 kilocalorie = 10^3 calories = 3.968 Btu
1 J = 9.48×10^{-4} Btu = 0.73756 ft-lbf
1 kW-hr = 3412 Btu = 2.6562×10^6 ft-lbf

Power
1 hp = 33,000 (ft-lbf)/min
1 hp = 745.7 W
1 W = 3.412 Btu/hr
1 W = 0.001341 hp
1 W = 0.0002843 tons of refrigeration

Pressure
1 atm = 14.6959 psia = 2116 lbf/ft^2 = 101325 N/m^2
1 in. of water = 249.08 Pa
1 in. of mercury = 3376.85 Pa
1 lbf/in.^{2} = 6894.76 Pa
1 Pa = 1 N/m^2 = 1.4504×10^{-4} lbf/in.^{2}

Temperature
1 degree R difference = 1 degree F difference = $\frac{5}{9}$ degree C difference = $\frac{5}{9}$ degree K difference
degree F = $\frac{9}{5}$ (degree C) + 32
degree C = $\frac{5}{9}$ (degree F − 32)

Thermal Conductivity

$$1\,\frac{\text{Btu}}{\text{hr-ft-F}} = 0.004134\,\frac{\text{calorie}}{\text{s-cm-C}} = 1.7307\,\frac{\text{W}}{\text{m-C}}$$

$$1\,\frac{\text{W}}{\text{m-C}} = 0.5778\,\frac{\text{Btu}}{\text{hr-ft-F}}$$

Table E-1   CONVERSION FACTORS(*continued*)

$$1 \frac{\text{Btu-in.}}{\text{hr-ft}^2\text{-F}} = 0.1442 \frac{\text{W}}{\text{m-C}}$$

$$1 \frac{\text{W}}{\text{m-C}} = 6.933 \frac{\text{Btu-in.}}{\text{hr-ft}^2\text{-F}}$$

Heat Transfer Coefficient

$$1 \frac{\text{Btu}}{\text{hr-ft}^2\text{-F}} = 5.678 \frac{\text{W}}{\text{m}^2\text{-C}}$$

$$1 \frac{\text{W}}{\text{m}^2\text{-C}} = 0.1761 \frac{\text{Btu}}{\text{hr-ft}^2\text{-F}}$$

Viscosity, absolute
1 poise = 100 centipoises

$$1 \frac{\text{lbm}}{\text{sec-ft}} = 1490 \text{ centipoises} = 1.49 \frac{\text{N-s}}{\text{m}^2}$$

$$1 \frac{\text{lbf-sec}}{\text{ft}^2} = 47,800 \text{ centipoises}$$

$$1 \text{ centipoise} = 0.001 \frac{\text{N-s}}{\text{m}^2}$$

Viscosity, kinematic
1 ft²/sec = 0.0929 m²/s
1 m²/s = 10.764 ft²/sec

Specific heat

$$1 \frac{\text{calorie}}{\text{gm-C}} = 1 \frac{\text{Btu}}{\text{lbm-F}}$$

$$1 \frac{\text{Btu}}{\text{lbm-F}} = 4186.8 \frac{\text{J}}{\text{kg-C}}$$

$$1 \frac{\text{kJ}}{\text{kg-C}} = 0.2388 \frac{\text{Btu}}{\text{lbm-F}}$$

Index

657